TRANSLATIONAL IMMUNOTHERAPY OF BRAIN TUMORS

TRANSLATIONAL IMMUNOTHERAPY OF BRAIN TUMORS

Edited by

JOHN H. SAMPSON

Robert H. and Gloria Wilkins Distinguished Professor of Neurosurgery
Chair, Department of Neurosurgery
Duke University Medical Center
Durham, NC, United States

ELSEVIER

ACADEMIC PRESS

An imprint of Elsevier
elsevier.com

Academic Press is an imprint of Elsevier
125 London Wall, London EC2Y 5AS, United Kingdom
525 B Street, Suite 1800, San Diego, CA 92101-4495, United States
50 Hampshire Street, 5th Floor, Cambridge, MA 02139, United States
The Boulevard, Langford Lane, Kidlington, Oxford OX5 1GB, United Kingdom

Library of Congress Cataloging-in-Publication Data
A catalog record for this book is available from the Library of Congress

British Library Cataloguing-in-Publication Data
A catalogue record for this book is available from the British Library

ISBN: 978-0-12-802420-1

Publisher: Mara Conner
Acquisition Editor: Melanie Tucker
Editorial Project Manager: Kristi Anderson
Production Project Manager: Lucía Pérez
Designer: Christian Bilbow

Typeset by TNQ Books and Journals
Transferred to Digital Printing in 2017

Contents

II

STUDYING BRAIN TUMOR IMMUNOTHERAPY

III

EXPERIMENTAL BRAIN TUMOR IMMUNOTHERAPIES

List of Contributors

J.P. Antonios David Geffen School of Medicine at UCLA, Los Angeles, CA, United States

O.J. Becher Duke University Medical Center, Durham, NC, United States

J. Berry-Candelario Duke University Medical Center, Durham, NC, United States

D.D. Bigner Duke University Medical Center, Durham, NC, United States

A.B. Carpenter Duke University Medical Center, Durham, NC, United States

V. Chandramohan Duke University Medical Center, Durham, NC, United States

A.L. Chang The University of Chicago, Chicago, IL, United States

B.D. Choi Massachusetts General Hospital and Harvard Medical School, Boston, MA, United States

N.A. Colwell Florida International University Herbert Wertheim College of Medicine, Miami, FL, United States

K.L. Congdon Duke University Medical Center, Durham, NC, United States

M. Dey The University of Chicago, Chicago, IL, United States

R.G. Everson David Geffen School of Medicine at UCLA, Los Angeles, CA, United States

S.H. Farber Duke University Medical Center, Durham, NC, United States

P.E. Fecci Duke University Medical Center, Durham, NC, United States

C. Flores University of Florida, Gainesville, FL, United States

K. Gabrusiewicz The University of Texas MD Anderson Cancer Center, Houston, TX, United States

P. Gedeon Duke University Medical Center, Durham, NC, United States

A.B. Heimberger The University of Texas MD Anderson Cancer Center, Houston, TX, United States

T. Hennika Duke University Medical Center, Durham, NC, United States

Y. He Duke University Medical Center, Durham, NC, United States

R.Y. Huang Brigham and Women's Hospital, Boston, MA, United States

L.A. Johnson University of Pennsylvania, Philadelphia, PA, United States

S. Khagi Duke University Medical Center, Durham, NC, United States

J.M. Komisarow Duke University Medical Center, Durham, NC, United States

M.S. Lesniak The University of Chicago, Chicago, IL, United States

L.M. Liau David Geffen School of Medicine at UCLA, Los Angeles, CA, United States

J. Miska The University of Chicago, Chicago, IL, United States

D. Mitchell University of Florida, Gainesville, FL, United States

I.H. Pastan National Institutes of Health, Bethesda, MD, United States

C.J. Pirozzi Duke University Medical Center, Durham, NC, United States

J. Qiao The University of Chicago, Chicago, IL, United States

D.A. Reardon Dana-Farber Cancer Institute, Boston, MA, United States

K.A. Riccione Duke University Medical Center, Durham, NC, United States

P. Roth University Hospital Zurich and University of Zurich, Zurich, Switzerland

J.H. Sampson Duke University Medical Center, Durham, NC, United States

L. Sanchez-Perez Duke University Medical Center, Durham, NC, United States

T.H. Schaller Duke University Medical Center, Durham, NC, United States

C.M. Suryadevara Duke University Medical Center, Durham, NC, United States

A.M. Swartz Duke University Medical Center, Durham, NC, United States

A.M. Tucker David Geffen School of Medicine at UCLA, Los Angeles, CA, United States

G. Vlahovic Duke University Medical Center, Durham, NC, United States

M. Weller University Hospital Zurich and University of Zurich, Zurich, Switzerland

P.A. Wells University of Florida, Gainesville, FL, United States

H. Yan Duke University Medical Center, Durham, NC, United States

Introduction to Translational Immunotherapy for Brain Tumors

A.M. Swartz, T.H. Schaller, J.H. Sampson
Duke University Medical Center, Durham, NC, United States

TRANSLATIONAL IMMUNOTHERAPY FOR BRAIN TUMORS SUMMARY

Malignant gliomas are an especially aggressive group of diseases that are largely recalcitrant to conventional therapies. Even with modern advancements in treatment, the median overall survival for patients afflicted with these diseases is roughly sixteen months.[1] The inability of standard therapies to safely and specifically eliminate all cancerous cells leaves patients vulnerable to tumor relapse and underscores the demand for alternative treatment modalities. In response, researchers across the globe are now attempting to exploit the immune system's natural cytotoxic capacity to mount an aggressive immunological siege upon cancer—a field known as immunotherapy. From this therapeutic paradigm, a variety of novel treatment strategies were born, several of which are showing promise against multiple cancers. In *Translational Immunotherapy for Brain Tumors*, we provide an overview of immunotherapy, with a particular emphasis on its application to malignant gliomas. General topics of discussion will include the immunological landscape of malignant gliomas, the approaches by which researchers and clinicians evaluate immunotherapies, and an in-depth look at the various experimental immunotherapies that are being pursued for malignant gliomas. Although great progress is being made in this area of research, the use of immunotherapy for malignant gliomas is still a relatively new concept. Furthermore, these central nervous system (CNS)-resident neoplasms are proving to be particularly resilient in the face of potent immunological intervention. Despite these obstacles, studies have elucidated the undeniable connection between the immune system and CNS-resident tumors. With each immunotherapeutic attempt that is made, the evasive mechanisms of these tumors becomes more clear, providing hope that one day immunotherapy will be at least a component of the successful eradication of malignant gliomas.

PRIMER ON MALIGNANT GLIOMAS

In *Translational Immunotherapy for Brain Tumors*, we pay special focus to immunotherapy for brain tumors. More specifically, most of the discussion is directed toward malignant gliomas, which are among the most difficult tumors to treat. Gliomas arise from glial cells. The three types of glial cells that are known to produce tumors are the astrocytes, oligodendrocytes, and ependymal cells. Astrocytes, which make up 20–40% of all glial cells, are regarded as the supportive cells of the CNS and have a wide range of functions including, but not limited to, metabolic and biochemical regulation, structural support, and maintenance of the blood–brain barrier (BBB). Oligodendrocytes also play a supportive role, albeit not as extensive as astrocytes, by providing electrical insulation by encasing neuronal axons within the CNS with a lipid-rich myelin sheath. Finally, ependymal cells line the ventricular system and central canal and are involved in the production of cerebrospinal fluid. The exact etiological mechanisms that cause these cells to transform into cancerous cells with uncontrolled growth have not been fully clarified, though several intrinsic and environmental factors have been proposed.

Gliomas represent the second most common primary intracranial tumor (~27%), behind only meningiomas (~36%). However, given that most meningiomas are benign, gliomas hold the title of the most common primary malignant CNS-resident tumor by a wide margin (~80%). Although they are not staged, gliomas are classified by a World Health Organization grading system that reflects their organization and structure relative to normal tissue (i.e., differentiation), growth potential, and aggressiveness. Grade I and II gliomas generally have a more differentiated phenotype, are slower-growing, and are less aggressive. These low-grade gliomas make up the majority of brain tumors that afflict children and adolescents. Alternatively, grade III and IV gliomas are poorly differentiated, grow rapidly, and are highly aggressive. These gliomas are designated as malignant or high-grade gliomas, occur more frequently in adults, and make up >90% of all diagnosed gliomas.[2]

GBM is the most common and aggressive primary malignant brain tumor, representing ~55% of all gliomas (Fig. 1), and has an incidence of ~2–3 in 100,000 people in the United States and Europe.[2] Multidimensional genomic analyses have shown that GBM represents a very diverse class of tumors and can be divided into several subtypes, each with their own unique signature. The classical subtype is frequently characterized by aberrations in the EGFR gene. The mesenchymal subtype typically exhibits deletions in the NF1 gene. The proneural subtype is typified by alterations in the PDGFRA gene, as well as point mutations in isocitrate dehydrogenase 1 (IDH1); this group has been shown to make up a large fraction of secondary GBMs. Finally, the neural subtype is distinguished by the expression of a number of neuronal markers.[3] However, it is important to note that these subtypes are determined by a tumor's bulk expression profile, and studies have shown that an individual GBM tumor can contain cells of varying subtypes.[4]

The standard of care therapy for GBM is surgical resection, if tumor is operable, followed by external beam radiation therapy (60 Gy delivered in 30 fractions) and concomitant temozolomide (TMZ) chemotherapy (75 mg/m^2/day) administered over a period of six weeks. This is followed by adjuvant TMZ chemotherapy (150–200 mg/m^2/day) on days one to five of every 28-day cycle for up to six to twelve cycles.[5,6] Despite this aggressive treatment regimen, patient prognosis is poor. Treated GBM patients almost invariably succumb to tumor relapse, surviving approximately 15.6 months (95% CI: 13.3-19.1 months).[1,5] Although there is no established standard of care for alternative (non-GBM) malignant gliomas, treatment is generally similar (i.e., surgical resection, radiation therapy, chemotherapy). As with GBM, therapy only slightly prolongs survival in most cases. Not only are the conventional therapies used to treat malignant gliomas incapacitating and damaging to healthy tissues,

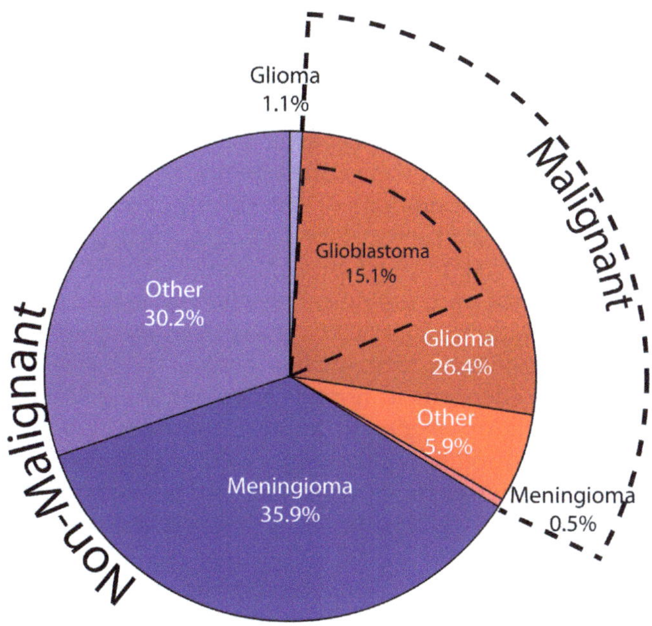

FIGURE 1 Distribution of malignant and nonmalignant brain and CNS-resident tumors.[2]

they ultimately suffer from a lack of tumor selectivity. A hallmark feature of malignant gliomas is their ability to infiltrate and diffuse into normal, healthy CNS tissues. Consequently, any residual tumor cells with stem cell-like attributes that remain following therapy have the potential to repopulate a new tumor. To overcome these limitations, a novel modality that can selectively target malignant glioma cells is likely required.

THE POTENTIAL OF CANCER IMMUNOTHERAPY

The notion that the immune system can detect and eliminate cancerous cells has been around since 1909, when Paul Ehrlich made this proposition.[7] In the 1950s, Sir Frank MacFarlane Burnet and Lewis Thomas elaborated upon this theory by claiming that malignant cells arise within a host quite regularly but are quickly eliminated by the immune system—a concept known as immunosurveillance.[8] This theory was dealt a significant blow in the 1970s when it was realized that athymic nude mice, which at that time were considered to be devoid of a functioning immune system, did not develop cancer at an increased rate.[9] However, it was later determined that these mice were not so immunodeficient after all,[10,11] and studies with mice lacking immunological effector molecules (e.g., IFN-γ) did exhibit a higher incidence of cancer.[12,13] More recently, in 2004, Gavin Dunn, Lloyd Old, and Robert Schreiber put forth a theory known as immunoediting,[14] positing that tumors are sculpted by the immune system, selecting for tumor cells with a low-immunogenic phenotype. Thus, it would appear that cancer has managed to exploit an evolutionary weak point in the immune system, leaving many to question whether the immune system truly has the capacity to eradicate all malignant cells within a host. Fortunately, however, there seems to be more to the story.

One of the major considerations of cancer immunotherapy regards the sufficiency of the host immune system to target malignant cells; in other words, the immune system must possess immune cells that can detect tumor cells in order for immunological rejection to occur. From an immunotherapeutic standpoint, cancer truly is a wolf in sheep's clothing, insomuch as its components are more or less normal cellular material. Most tumor antigens are identical to normal antigens to such an extent that tumor cells easily evade detection by innate immune cells, which scout for pathogen-associated motifs. Fortunately, evolution has provided us with an adaptive immune system that can be educated de novo to target specific molecules. But again, given the similarity between most tumor and normal antigens, most adaptive immune cells that recognize these homologous antigens are thought to be destroyed or rendered nonresponsive (i.e., anergic) to avoid the consequence of autoimmunity. Nevertheless, several decades worth of preclinical studies have repeatedly demonstrated that adaptive immune cells do indeed possess the capacity to target tumor cells, proofed by the abrogation of antitumor effects in the context of adaptive immune cell depletion.

Early attempts to target tumor antigens took a brute force approach by using vaccines consisting of autologous, typically irradiated, cancer cells. Despite an exhaustive variety of approaches, these therapies induced only moderate and short-lived antitumor benefits, at best. Then, in the early 1990s, in the wake of cytokine-gene cloning, researchers began evaluating the antitumor effects of tumor cells transfected with genes encoding various immunological signaling molecules. Though results often varied among studies depending on the cancer type and site of tumor challenge, striking antitumor effects were frequently seen in preclinical studies using the immunomodulator granulocyte-macrophage colony stimulating factor.[15,16] As an alternative to the brute force approach, investigators at that time also began identifying specific tumor antigens that could be recognized by adaptive immune cells.[17–19] Notably, several melanocyte-specific proteins (e.g., MART-1) were found to stimulate tumor infiltrating lymphocytes (TILs) isolated from melanoma patients,[18] and vaccines consisting of these melanocytic proteins were able to produce significant antitumor responses in preclinical models of melanoma. This was a rather striking observation considering that clonal deletion of self-reactive lymphocytes was demonstrated a few years earlier.[20,21] We now know these processes are incomplete, and self-reactive immune cells do indeed exist in the periphery.[22] However, activation of these cells does pose the risk of autoimmunity, which is a critical concern in the field of immunotherapy.[23] The readministration of ex vivo-expanded TILs into melanoma patients, while effective in several cases, frequently induced autoimmune effects (e.g., vitiligo)[24,25]—a not too surprising outcome considering the nature of the recognized antigens. Though the treatment of one disease with another is not unprecedented, autoimmunity can have just as debilitating effects as malignant cancer. Consequently, researchers are beginning to focus their attention on safer tumor-specific antigens, which only became possible in recent years.

Upon the advent of next generation sequencing (NGS) in the early 21st century, large-scale tumor sequencing efforts quickly ensued. Using this technology, it became possible to peer inside tumor cells at the molecular level to determine their exact genetic and transcriptomic compositions. Although it was largely known that cancer was a disease caused by genetic mutations, the comprehensive landscape of these alterations was unknown prior to the availability of NGS. Much of the findings of these studies were not all too surprising: (1) genetic mutations were found in coding and noncoding DNA, (2) oncogenic and tumor-suppressor genes were frequently perturbed, and (3) tumors thought to develop from environmental stressors (e.g., melanoma and lung cancer) had a far higher number of mutations. However, it became evident that it may be possible to target the expressed mutations, or neoantigens, within tumors. The benefit of using neoantigens as immunotherapeutic targets is two-fold: they are inherently tumor-specific and neoantigen-cognate adaptive immune cells are less likely to be subject to deletion or quiescence. Consequently, efforts are now being made to target the repertoire of tumor-specific neoantigens within tumors.

Numerous studies have demonstrated the immunogenicity of tumor-specific and tumor-associated antigens. How, then, are endogenous immune responses to these antigens prevented? The answer lies, in part, in the developmental process of a tumor. From incipience, cancers evolve an immunosuppressive phenotype that develops gradually, balancing on the cusp of immunological rejection with constant signals relayed to the immune system that it is simply a normal, healthy cluster of cells. Paracrine-acting suppressive molecules secreted by the tumor subdue local effector functions, whereas endocrine signaling molecules diminish activation of new antitumor immune cells in the secondary lymphoid organs, as well as promote systemic immunosuppression. In those unlucky few, the balance between immunological rejection and immunosuppression skews in the favor of tumor development. By the time of diagnosis, the arsenal of immunosuppressive mechanisms exhibited by the tumor is frequently evidenced by the infiltration of immunosuppressive immune cells (e.g., regulatory T-cells [Tregs]), the reeducation of the stroma into an immunosuppressive phenotype, and the elaboration of immunosuppressive signaling molecules from the tumor cells themselves.

Overturning the effects of tumor-mediated immunosuppression is now a primary focus of immunotherapy. The history of this strategy dates back to the late 1800s, when a bone surgeon by the name of William B. Coley noticed that a patient suffering from a recurrent neck sarcoma experienced tumor regression following a case of erysipelas, or *Streptococcus pyogenes*, infection.[26] In response to this observation, Dr. Coley crafted his own treatment cocktail consisting of killed *S. pyogenes* and *Serretia marcescens*—eponymously termed Coley's toxin. Although the antitumor effects of Coley's toxin are mixed, several cases of tumor regression following treatment have been documented. Several explanations of the antitumor effects of these bacterial components have been proposed, including the activation of innate toll-like receptors; however, the exact mechanisms are unknown. Suffice it to say that the stimulation of the host's immune system by Coley's toxin seems to be sufficient to reverse immunological suppression, thereby promoting antitumor responses, in some instances. As the field of immunology progressed, the suppressive pathways of the immune system began to develop understanding. In the late 1980s, a molecule known as cytotoxic T lymphocyte associated antigen 4 (CTLA-4) was discovered,[27] and shortly thereafter, the receptor programmed death 1 (PD-1)[28] and its ligands (PD-L1[29] and PD-L2[30]) were identified. These molecules are collectively known as checkpoints due to their role in keeping the immune system in check by promoting immunosuppression. Checkpoint inhibition via antibody-mediated blockade is gaining considerable attention in the field of immunotherapy in light of the potent antitumor responses it engenders against several tumors (e.g., melanoma and lung cancer).[31,32] These effects highlight the role of immunological suppression in subduing endogenous antitumor responses. Unfortunately, these therapies do not seem to be equally efficacious in all cases or against all tumors. Further optimization of clinical protocols using checkpoint inhibitors is required before their true effectiveness can be confirmed. It is unlikely, however, that checkpoint blockade alone will be the "magic bullet" envisioned by Dr. Ehrlich in the 1900s. The list of immunosuppressive mechanisms utilized by cancer is an extensive one, several of which will be discussed throughout this book, and therapeutic remediation of these pathways is showing great promise against certain cancers.

The immune system truly is a remarkably complex, and at times seemingly inextricable but equally astonishing, network of cells and cellular processes that staggers the imagination when its ontogeny is contemplated. From a therapeutic perspective, the immune system boasts potent cytotoxic potential with the ability to resolve structures at the nanometer level, giving it a distinct advantage over conventional cancer treatment strategies. Leveraging these capabilities to produce safe, selective, and durable antitumor responses is the foremost goal of immunotherapy, and, while this field has been mired with hurdles, great progress has been made and intriguing discoveries unearthed along the way. Studies are continuously showing that the body does indeed have a defense network in place that can eradicate malignant cells, but cancer—masquerading as a normal cell—is a formidable expert at attrition warfare and has decisively co-opted the immune system for its own benefit. However, there have been several strong indications that cancer's immunological evasive mechanisms are reversible, as will be illustrated throughout this book. We truly are in the eve of the "Golden Age" of cancer immunotherapy, and these next few years will be critical in establishing immunotherapy as a respected modality in the armamentarium of cancer therapeutics.

CHALLENGES OF BRAIN TUMOR IMMUNOTHERAPY

From the serendipitous discovery of bacteria-mediated tumor suppression in the 19th century to the striking response rates promoted by modern molecular-guided immunotherapies, immunotherapy is proving to be a powerful modality for the treatment of cancer. Nevertheless, hopefulness is met with reasonable skepticism, particularly with regard to brain tumor immunotherapy. Although some types of cancer do seem to be amenable to immunotherapeutic intervention, there are several aspects of brain tumors that make the use of immunotherapy a controversial proposition.

CNS tissues represent a very unique immunological environment compared to peripheral tissues (Fig. 2). Seminal studies in the early 20th century demonstrated that grafts implanted within the CNS are rejected much slower than grafts placed in the periphery,[33] leading to the notion that the CNS was, in essence, an immunologically privileged site. This concept was further supported by the apparent lack of draining lymphatics[34] in the brain and CNS tissues and the presence of a highly restrictive blood–brain barrier BBB.[35] From an evolutionary standpoint, it is reasonable to assume that the CNS has developed means to safeguard itself from the destructive effects of inflammation, given the indispensability of these tissues; however, we now know that the immune privilege of these tissues is not absolute. Studies have shown that immune cells do indeed have the potential to access these compartments, as evidenced by disease

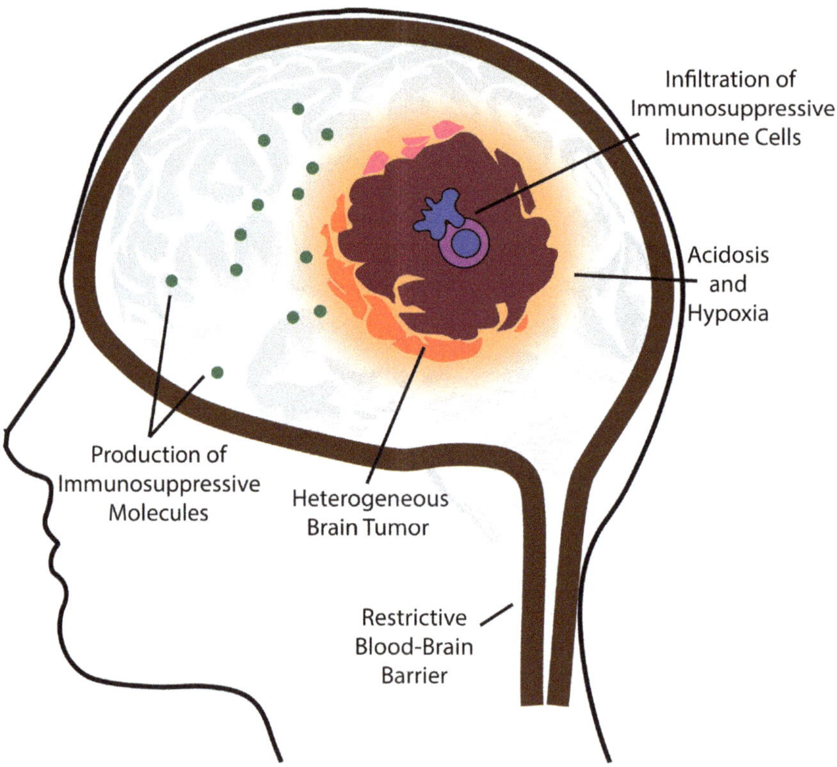

FIGURE 2 The immunological barriers to CNS-resident tumors.

states such as multiple sclerosis and cancer. The ability of immune cells to access the CNS-resident tumors can further be aided by the disruption of the BBB by invading glioma cells.[36] Additionally, a lymphatic system was recently elucidated in the dural sinuses of mice,[37] which carry immune cells and antigen to the deep cervical lymph nodes.[38] Thus, it is currently well-established that there exists an appreciable degree of immunological surveillance of CNS tissues and CNS-derived antigen, albeit not as extensive as that which occurs in peripheral tissues. Nevertheless, it is also clear that these mechanisms are not sufficient to promote complete endogenous rejection of CNS-resident tumors.

A second aspect that can limit the effectiveness of brain tumor immunotherapy has less to do with the anatomy of the CNS, but rather the architecture and behavior of gliomas (Fig. 2). Immunological evasion is one of the hallmarks of cancer,[39] and malignant gliomas are true experts. Much like viruses that frequently mutate to avoid immunological detection, malignant gliomas exhibit profound heterogeneity[40–42] that enables them to escape monovalent immunotherapeutic intervention.[43] Even upon the activation of high numbers of diverse tumor-reactive immune cells, malignant gliomas may still be protected by additional lines of defense. For example, the malignant glioma milieu is often associated with acidosis and hypoxia. This environment is extremely detrimental to effector immune cell function but is able to sustain immunosuppressive immune cells (e.g., Tregs[44]), as well as stimulate angiogenesis.[45] Additionally, malignant gliomas are known to produce a host of immunosuppressive molecules, such as transforming growth factor (TGF)-β, interleukin (IL)-10, and indoleamine-2,3-dioxygenase (IDO), that dramatically inhibit the immune response. These factors are also thought to contribute to the profound immunosuppression identified in malignant glioma patients, characterized by lymphopenia and T-cell dysfunction.

The aforementioned characteristics of brain tumors arguably represent the worst case scenario for immunotherapy. Few other cancers are situated behind such a vital organ that also maintains partial immunological seclusion. Nevertheless, the need for safer and more selective therapies for brain tumors is undeniable, and the evaluation of brain tumor immunotherapy has, thus far, not been a fruitless endeavor. The studies carried out in this pursuit have provided great knowledge into the immunobiology of the CNS and CNS-resident tumors. It is now clear that immunotherapy has the potential to stimulate immunological changes within patients afflicted with brain tumors and, in several cases, impart molecular changes within the tumor itself.[43] The next hurdles will be to overcome the issues of tumor-mediated immunosuppression and tumor heterogeneity, and there are several strategies in the therapeutic pipeline that are being tested to fulfill this demand.

CHAPTER OVERVIEW

Translational Immunotherapy for Brain Tumors is divided into several sections to help guide the reader through the rapidly expanding field of brain tumor immunotherapy. In Section I, we look at the immunological features of brain tumors, including many of the immunosuppressive aspects associated with these cancers. Section II details the methods that researchers and clinicians use to evaluate brain tumor immunotherapies. Finally, in Section III, we review several experimental immunotherapies that are currently being evaluated for the treatment of brain tumors.

Section I–Immunological Features of Brain Tumors focuses on the salient immunological aspects associated with malignant gliomas. We begin with a basic immunology overview in Chapter 1, including a brief discussion of the innate and adaptive arms of the immune system. This chapter also includes a detailed look at the physiological components that give the CNS partial seclusion from the immune system, including such elements as a restrictive BBB, limited antigen presentation, and the once thought absent CNS-resident lymphatic system that was only recently discovered in the murine brain.

Chapters 3, 4, and 5 focus on an issue that is proving to be a formidable obstacle to brain tumor immunotherapy: profound immunosuppression. Chapter 2 begins with a brief history of the seminal studies that first observed immunological dysfunction in patients with malignant glioma. These early studies were at times contradictory but most demonstrated compromised T-cell functionality and lymphopenia. Although the finer details of these deficiencies remain to be elucidated, it appears as though malignant gliomas have managed to co-opt the immunological signaling networks and immunosuppressive pathways to protect themselves from immunological rejection. Two well-known players in malignant glioma immunosuppression—regulatory T-cells (Tregs) and IDO—are further described in Chapter 3. Tregs, demarcated by the expression of the transcription factor Foxp3, are an immunosuppressive subset of CD4 T-cells that are selectively recruited to malignant gliomas through the production of chemo-attractants by tumor-resident cells. Tregs, which can persist within the hypoxic tumor microenvironment, have been shown to diminish effector T-cell function by several mechanisms, including contact-mediated cytolysis, IL-2 growth factor consumption, elaboration of immunosuppressive cytokines (e.g., TGF-β and IL-10), as well as several alternative mechanisms. Another mediator of immunosuppression that is frequently upregulated in the glioma microenvironment and known to inhibit effector cell function is IDO, the rate-limiting enzyme involved in the catabolism of the essential amino acid tryptophan to kynurenine. Together, Tregs and IDO represent two promising targets for immunotherapeutic intervention.

In Chapter 4 we discuss the role of myeloid-derived suppressor cells (MDSCs), a heterogeneous population of immature myeloid cells with impaired ability to differentiate into macrophages, granulocytes, and dendritic cells, in immunosuppression. MDSCs are found in GBM where they suppress innate and adaptive antitumor immunity by subduing T-cell functions and other mechanisms that have not yet been fully understood. This chapter describes the origins, activation, expansion, and known mechanisms of MDSCs.

Finally, this section concludes in Chapter 5, with an overview of the conserved tumor-specific mutations, or neoantigens, that are found in gliomas. Large-scale sequencing efforts have demonstrated that most genetic mutations are unique to an individual tumor; however, several mutations occur in gliomas at disproportionate frequencies, including the R132H isocitrate dehydrogenase mutation in low-grade and secondary glioblastomas, EGFRvIII in malignant gliomas, and histone mutations found in pediatric gliomas. The incidence of these mutations suggests that they may play a role in oncogenesis, and, given their conserved nature and tumor specificity, they may serve as ideal immunotherapeutic targets.

In Section II–Studying Brain Tumor Immunotherapy, the approaches that make it possible to study brain tumor immunotherapy are reviewed. Chapter 6 takes a look at the various preclinical models that are available to researchers for the study of brain tumor immunotherapy. These include a wide range of mostly murine spontaneous, chemically induced, virally induced, or genetically altered glioma models. Additionally, severely immunocompromised mice are available that enable the engraftment of human cells, generating reliable xenograft and humanized models. Although murine glioma models have been extremely useful for preliminary immunotherapeutic studies, limitations arise from their dissimilarities with human gliomas (e.g., lack of heterogeneity and spontaneity). To bridge the gap between murine and human gliomas, several researchers have recently turned their preclinical focus toward spontaneous canine gliomas, which occur at a relatively high frequency and afford the benefits of heterogeneity and spontaneity. Preclinical studies in these nonhuman models are essential for determining the safety and therapeutic potential of various immunotherapies. However, these factors are often determined using highly invasive methods that are not typically translatable to human studies.

Given the vital nature of the brain, noninvasive methods of determining an immunotherapeutic response at the early stages of treatment are needed to limit unnecessary outcomes. Although this requirement can be fulfilled by advanced imaging techniques, assessment of CNS-resident responses is frequently complicated by pseudoprogressive inflammation, which is a sign of an active immune response and may not be associated with clinical decline. Conventional assessment criteria (e.g., Response Evaluation

Criteria in Solid Tumors - RECIST) do not consider pseudoprogression in their analyses; therefore, effective immunotherapies could be inadvisably discontinued. In response to this issue, the Immunotherapy Response Assessment in Neuro-Oncology criteria, which will be discussed in detail in Chapter 7, was developed to help discern tumor growth from pseudoprogression, among other limitations that conflict prior assessment criteria.

We finish this section in Chapter 8 with an overview of clinical trial design, including such topics as overall study design, candidate selection, safety assessment, and response evaluation. This chapter also includes a look at the various completed and ongoing immunotherapy-focused clinical trials that relate to neuro-oncology. To date, no immunotherapeutic modality has been approved for the treatment of glial neoplasms; however, great progress has been made and valuable lessons learned from the clinical evaluation of immunotherapies. Future clinical studies will likely rely on patient stratification based on novel biomarkers and glioma subtypes, as well as more informed response criteria, which may reveal positive antitumor effects that have been obscured among this highly heterogeneous class of tumors.

This book concludes, in Section III–Experimental Brain Tumor Immunotherapies with an extensive look at the various experimental immunotherapies that are being evaluated for the treatment of brain tumors. The list of immunotherapies outlined in this section is by no means comprehensive. The topics were chosen based on the breadth of information available for a particular immunotherapy; however, we acknowledge that there are several other promising immunotherapeutic strategies that are emerging for the treatment of malignant glioma (e.g., virus-based therapy). In the first three chapters of this section, we focus on acellular therapies and follow up with cell-based therapies. We begin with an examination of peptide-specific vaccines in Chapter 9. Peptide vaccines are generally synthesized in vitro to encompass a tumor antigen and administered with a potent adjuvant. Given their ease to generate, peptide vaccines represent ideal "off-the-shelf" therapies for targeting conserved tumor-specific antigens and overexpressed tumor-associated antigens, which are described throughout this chapter. Multipeptide vaccine regimens are also being rigorously explored in an attempt to combat tumor recurrence resulting from the immunological escape of antigen-loss tumor cell variants. In light of their inherent simplicity, peptide vaccines will likely remain the cornerstone of numerous immunotherapeutic clinical trials for years to come; however, a major concern with active immunotherapies such as vaccines is their suboptimal performance in immunosuppressed glioma patients. An alternative approach is to use passive immunotherapies with preformed effector molecules (e.g., antibodies). A clever utilization of this approach involves the use of immunotoxins, which will be discussed in Chapter 10.

Immunotoxins are generally composed of an antibody or ligand conjugated to a plant or bacterial toxin, which has the effect of delivering a cytotoxic payload to cells expressing the targeted antigen or receptor. Traditional immunotoxins were generated by the chemical coupling of these two components; however, modern immunotoxins exhibit a more rational design, incorporating only the essential elements of each constituent. These third-generation immunotoxins are typically smaller in size and have a decreased susceptibility to immunological rejection, thereby permitting increased access to the tumor and multiple rounds of therapy, respectively. Several cell-surface antigens have been discovered on malignant gliomas that are either tumor-specific (e.g., EGFRvIII) or overexpressed on tumor cells, making these the leading targets for immunotoxin-based therapies. Tumor-focused targeting can also be achieved by delivery of immunotoxins directly into the tumor, which also affords the benefit of bypassing the BBB. Researchers are now testing combination therapies with immunotoxins in an effort to overcome the limitations caused by intratumoral heterogeneity of tumor antigen expression. An alternative antibody-based therapy that has gained significant momentum in recent years is antibody-mediated modulation of immune checkpoints, which includes, but is not limited to, the molecules CTLA-4 and PD-1. Immune checkpoints mediate immunosuppressive effects on the immune response through various mechanisms, and their inhibition via antibody blockade is frequently able to alleviate immunosuppression to a significant degree. Chapter 11 focuses on the progress and challenges of immune checkpoint modulators in GBM. Immune checkpoint inhibitors have shown astounding outcomes in melanoma and non-small-cell lung cancer by curbing tumor-mediated immunosuppression and promoting endogenous antitumor T-cell response. These therapies are now being explored in a wide variety of cancers. This chapter provides an overview of antibody-based immunomodulation feasibility in CNS malignancies and summarizes the ongoing progress of checkpoint blockade for GBM treatment.

We turn our focus to cell-based immunotherapies for the remainder of Section III. We start, in Chapter 12, with an introduction to the use of dendritic cell (DC) vaccines for the treatment of gliomas, which have been studied extensively for the past 20 years. DCs, also referred to as "nature's adjuvant," are specialized immune cells considered to have the unique capacity to prime naïve T-cells. DC vaccines have demonstrated the ability to induce immune responses with an excellent safety profile in patients with malignant gliomas. One of the major appeals of using DCs as a vaccine platform is their ability to be loaded with numerous tumor antigens using various methods (tumor lysate pulsation, tumor RNA transfection, tumor fusion, etc.), making them a promising tool for addressing the issue of intratumoral antigenic heterogeneity.

However, as with most active immunotherapies, the induction of potent endogenous immune responses within immunosuppressed glioma patients remains a significant challenge. Alternatively, repletion of a host's immune cell compartment via adoptive cell transfer of ex vivo-expanded effector cells has the benefit of furnishing high numbers of tumor-reactive immune cells—a strategy that is less reliant on all stages of the endogenous immune response.

In Chapter 13, we introduce four common adoptive therapy approaches: tumor infiltrating leukocytes, ex vivo activated tumor-specific T-cells, designer T-cells, and chimeric antigen receptor (CAR) T-cells. Furthermore, we contrast the success of adoptive immunotherapy in metastatic melanoma and renal cancer with the difficulty of applying these methods to brain tumors, given the lack of ubiquitously expressed single conserved tumor-specific and tumor-associated antigens in brain tumors.

We conclude this section with an in-depth discussion of CAR T-cell therapy for brain tumors in Chapter 14. Though their design has evolved significantly since their introduction in 1989, CARs are typically composed of the variable region of an antibody, targeting a specific molecule, fused to the constant region of a T-cell receptor, thereby permitting T-cell responses in an MHC-independent manner. Although this modality has mediated striking antitumor benefits against several hematogenous cancers in clinical trials, their effectiveness against human solid tumors remains largely unknown. Several Phase I clinical trials using CARs targeting various glioma antigens have recently been initiated to assess the safety of these treatments, so we will likely soon get an indication of their true immunotherapeutic potential.

Together, these sections provide an extensive overview of the field of brain tumor immunotherapy. Although we attempted to make this book as thorough as possible, this field is expanding at a meteoric rate, and an evaluation of every relevant topic would reach an encyclopedic level. Even during the time it took to draft this book, new discoveries and insights were made that bolster, or even contradict, some of the claims made throughout. Despite the alarming rate of growth of this subject, at the current time, the possibilities of immunotherapy seem endless, and there is no downturn in progress in sight. These next few years are met with excitement, as several new immunotherapeutic modalities go into clinical testing, and our understanding of immunotherapy will undoubtedly grow from these studies. Much is left to learn about the interplay between the immune system and CNS-resident cancers—a dynamic that is, at this point, irrefutable. Whether these mechanisms can be leveraged to safely promote the complete rejection of a glioma is uncertain; however, there have yet to be any decisive indications that they cannot.

References

1. Stupp R, Idbaih A, Steinberg DM, et al. LTBK-01: prospective, multi-center phase III trial of tumor treating fields together with temozolomide compared to temozolomide alone in patients with newly diagnosed glioblasoma. *Neuro-oncology*. 2016;18(suppl 6):i1.
2. Ostrom QT, Gittleman H, Fulop J, et al. CBTRUS statistical report: primary brain and central nervous system tumors diagnosed in the United States in 2008-2012. *Neuro Oncol*. October 2015;17(suppl 4):iv1–iv62.
3. Verhaak RG, Hoadley KA, Purdom E, et al. Integrated genomic analysis identifies clinically relevant subtypes of glioblastoma characterized by abnormalities in PDGFRA, IDH1, EGFR, and NF1. *Cancer Cell*. January 19, 2010;17(1):98–110.
4. Patel AP, Tirosh I, Trombetta JJ, et al. Single-cell RNA-seq highlights intratumoral heterogeneity in primary glioblastoma. *Science*. June 20, 2014;34(6190):1396–1401.
5. Stupp R, Taillibert S, Kanner AA, et al. Maintenance therapy with tumor-treating fields plus temozolomide vs temozolomide alone for glioblastoma a randomized clinical trial. *JAMA*. December 15, 2015;314(23):2535–2543.
6. Stupp R, Mason WP, van den Bent MJ, et al. Radiotherapy plus concomitant and adjuvant temozolomide for glioblastoma. *N Engl J Med*. March 10, 2005;352(10):987–996.
7. Strebhardt K, Ullrich A. Paul Ehrlich's magic bullet concept: 100 years of progress. Nature reviews. *Cancer*. June 2008;8(6):473–480.
8. Burnet M. Cancer: a biological approach. III. Viruses associated with neoplastic conditions. IV. Practical applications. *Br Med J*. April 13, 1957;1(5023):841–847.
9. Stutman O. Tumor development after 3-methylcholanthrene in immunologically deficient athymic-nude mice. *Science*. February 8, 1974;183(4124):534–536.
10. Maleckar JR, Sherman LA. The composition of the T cell receptor repertoire in nude mice. *J Immunol*. June 1, 1987;138(11):3873–3876.
11. Ikehara S, Pahwa RN, Fernandes G, et al. Functional T cells in athymic nude mice. *Proc Natl Acad Sci U S A*. February 1984;81(3):886–888.
12. Street SE, Trapani JA, MacGregor D, et al. Suppression of lymphoma and epithelial malignancies effected by interferon gamma. *J Exp Med*. July 1, 2002;196(1):129–134.
13. Street SE, Cretney E, Smyth MJ. Perforin and interferon-gamma activities independently control tumor initiation, growth, and metastasis. *Blood*. January 1, 2001;97(1):192–197.
14. Dunn GP, Old LJ, Schreiber RD. The three Es of cancer immunoediting. *Annu Rev Immunol*. 2004;22:329–360.
15. Dranoff G, Jaffee E, Lazenby A, et al. Vaccination with irradiated tumor cells engineered to secrete murine granulocyte-macrophage colony-stimulating factor stimulates potent, specific, and long-lasting anti-tumor immunity. *Proc Natl Acad Sci USA*. April 15, 1993;90(8):3539–3543.
16. Sampson JH, Archer GE, Ashley DM, et al. Subcutaneous vaccination with irradiated, cytokine-producing tumor cells stimulates $CD8^+$ cell-mediated immunity against tumors located in the "immunologically privileged" central nervous system. *Proc Natl Acad Sci U S A*. September 17, 1996;93(19):10399–10404.
17. Gaugler B, Van den Eynde B, van der Bruggen P, et al. Human gene MAGE-3 codes for an antigen recognized on a melanoma by autologous cytolytic T lymphocytes. *J Exp Med*. March 1, 1994;179(3):921–930.
18. Anichini A, Maccalli C, Mortarini R, et al. Melanoma cells and normal melanocytes share antigens recognized by HLA-A2-restricted cytotoxic T cell clones from melanoma patients. *J Exp Med*. April 1, 1993;177(4):989–998.
19. Brichard V, Van Pel A, Wolfel T, et al. The tyrosinase gene codes for an antigen recognized by autologous cytolytic T lymphocytes on HLA-A2 melanomas. *J Exp Med*. August 1, 1993;178(2):489–495.
20. Nemazee DA, Burki K. Clonal deletion of B lymphocytes in a transgenic mouse bearing anti-MHC class I antibody genes. *Nature*. February 9, 1989;337(6207):562–566.

21. Kappler JW, Roehm N, Marrack P. T cell tolerance by clonal elimination in the thymus. *Cell.* April 24, 1987;49(2):273–280.
22. Yu W, Jiang N, Ebert PJ, et al. Clonal deletion prunes but does not eliminate self-specific alphabeta CD8(+) T Lymphocytes. *Immunity.* May 19, 2015;42(5):929–941.
23. Bigner DD, Pitts OM, Wikstrand CJ. Induction of lethal experimental allergic encephalomyelitis in nonhuman primates and guinea pigs with human glioblastoma multiforme tissue. *J Neurosurg.* July 1981;55(1):32–42.
24. Dudley ME, Wunderlich JR, Robbins PF, et al. Cancer regression and autoimmunity in patients after clonal repopulation with antitumor lymphocytes. *Science.* October 25, 2002;298(5594):850–854.
25. Yeh S, Karne NK, Kerkar SP, et al. Ocular and systemic autoimmunity after successful tumor-infiltrating lymphocyte immunotherapy for recurrent, metastatic melanoma. *Ophthalmology.* May 2009;116(5):981–989. e981.
26. Coley WB. The treatment of inoperable sarcoma by bacterial toxins (the mixed toxins of the streptococcus erysipelas and the Bacillus prodigiosus). *Proc R Soc Med.* 1910;3(Surg Sect):1–48.
27. Brunet JF, Denizot F, Luciani MF, et al. A new member of the immunoglobulin superfamily–CTLA-4. *Nature.* July 16-22, 1987;328(6127):267–270.
28. Ishida Y, Agata Y, Shibahara K, et al. Induced expression of PD-1, a novel member of the immunoglobulin gene superfamily, upon programmed cell death. *EMBO J.* November 1992;11(11):3887–3895.
29. Freeman GJ, Long AJ, Iwai Y, et al. Engagement of the PD-1 immunoinhibitory receptor by a novel B7 family member leads to negative regulation of lymphocyte activation. *J Exp Med.* October 2, 2000;192(7):1027–1034.
30. Latchman Y, Wood CR, Chernova T, et al. PD-L2 is a second ligand for PD-1 and inhibits T cell activation. *Nat Immunol.* March 2001;2(3):261–268.
31. Hodi FS, O'Day SJ, McDermott DF, et al. Improved survival with ipilimumab in patients with metastatic melanoma. *N Engl J Med.* August 19, 2010;363(8):711–723.
32. Brahmer JR, Drake CG, Wollner I, et al. Phase I study of single-agent anti-programmed death-1 (MDX-1106) in refractory solid tumors: safety, clinical activity, pharmacodynamics, and immunologic correlates. *J Clin Oncol.* July 1, 2010;28(19):3167–3175.
33. Medawar PB. Immunity to homologous grafted skin; the fate of skin homografts transplanted to the brain, to subcutaneous tissue, and to the anterior chamber of the eye. *Br J Exp Pathol.* February 1948;29(1):58–69.
34. Louveau A, Harris TH, Kipnis J. Revisiting the mechanisms of CNS immune privilege. *Trends Immunol.* October 2015;36(10):569–577.
35. Saunders NR, Dreifuss JJ, Dziegielewska KM, et al. The rights and wrongs of blood–brain barrier permeability studies: a walk through 100 years of history. *Front Neurosci.* 2014;8:404.
36. Watkins S, Robel S, Kimbrough IF, Robert SM, Ellis-Davies G, Sontheimer H. Disruption of astrocyte-vascular coupling and the blood–brain barrier by invading glioma cells. *Nat Commun.* 2014;5:4196.
37. Louveau A, Smirnov I, Keyes TJ, et al. Structural and functional features of central nervous system lymphatic vessels. *Nature.* July 16, 2015;523(7560):337–341.
38. Harris MG, Hulseberg P, Ling C, et al. Immune privilege of the CNS is not the consequence of limited antigen sampling. *Sci Rep.* 2014;4:4422.
39. Hanahan D, Weinberg RA. Hallmarks of cancer: the next generation. *Cell.* March 4, 2011;144(5):646–674.
40. Kumar A, Boyle EA, Tokita M, et al. Deep sequencing of multiple regions of glial tumors reveals spatial heterogeneity for mutations in clinically relevant genes. *Genome Biol.* 2014;15(12):530.
41. Soeda A, Hara A, Kunisada T, Yoshimura S, Iwama T, Park DM. The evidence of glioblastoma heterogeneity. *Sci Rep.* 2015;5:7979.

42. Sottoriva A, Spiteri I, Piccirillo SG, et al. Intratumor heterogeneity in human glioblastoma reflects cancer evolutionary dynamics. *Proc Natl Acad Sci USA*. March 5, 2013;110(10):4009–4014.
43. Sampson JH, Heimberger AB, Archer GE, et al. Immunologic escape after prolonged progression-free survival with epidermal growth factor receptor variant III peptide vaccination in patients with newly diagnosed glioblastoma. *J Clin Oncol*. November 1, 2010;28(31):4722–4729.
44. Fecci PE, Sweeney AE, Grossi PM, et al. Systemic anti-CD25 monoclonal antibody administration safely enhances immunity in murine glioma without eliminating regulatory T cells. *Clin Cancer Res*. July 15, 2006;12(14 Pt 1):4294–4305.
45. Fukumura D, Xu L, Chen Y, Gohongi T, Seed B, Jain RK. Hypoxia and acidosis independently up-regulate vascular endothelial growth factor transcription in brain tumors in vivo. *Cancer Res*. August 15, 2001;61(16):6020–6024.

SECTION I

IMMUNOLOGICAL FEATURES OF BRAIN TUMORS

CHAPTER

1

An Introduction to Immunotherapy in the Treatment of Brain Tumors

J.M. Komisarow, J.H. Sampson

Duke University Medical Center, Durham, NC, United States

OUTLINE

INTRODUCTION

Despite aggressive treatment, the prognosis for malignant brain tumors remains relatively poor.[1] Glioblastoma (GBM) is regarded as the most aggressive of the malignant brain tumors. Unfortunately, it is also the most common primary brain tumor to afflict the adult population.[2,3] Standard treatment consists of the maximum surgical resection safely possible, followed by radiation and chemotherapy (specifically the alkylating agent temozolomide).[1]

http://dx.doi.org/10.1016/B978-0-12-802420-1.00001-6

It should be noted that this standard of care continues to evolve. In addition to the exciting therapies outlined below, a promising clinical trial is currently underway coupling the established standard of care with the NovoTTF device. Preliminary data suggests a survival benefit when compared to the use of temozolomide alone. This treatment is relatively nonselective and results in significant morbidity related to damage of nonmalignant cells. These off-target effects result in both short-term (meaning while therapy is taking place) and long-term toxicity.[4] Therapies selectively targeting cancer cells are desperately needed to augment or replace the current standard of care. Immunotherapy has been shown to be capable of selectively targeting malignant cells without damaging surrounding normal brain tissue, thus sparing patients the side effects that limit quality of life.[5–7] Effectively designing these therapies demands an understanding of the normal immune response, how that response is altered in the setting of malignancy, and how to appropriately select targets on cancer cells.[8] Heterogeneity of the tumor's antigen complement may also have profound effects on these efforts.

THE IMMUNE RESPONSE

The immune system is a collection of specialized cells and signaling networks designed to defend against, identify, and eliminate disease. Elaborate mechanisms have evolved to defend against a nearly constant onslaught of unwanted foreign or damaged host cells. The immune system is most often thought of in the context of infectious disease, but the immune system plays a pervasive role in almost all manners of illness, including cancer.[9–11]

The rapid identification, and eventual safe elimination, of nonhost cells is key to the survival of an organism. The human immune system consists of multiple layers with distinct roles in protecting the body. These range from physical barriers that separate the internal environment from the external, to complex and rapidly adaptive systems that can identify and target certain molecular patterns for safe destruction.

The human immune system has been historically divided into innate and adaptive responses. The components of the innate immune system include physical barriers from the environment (epithelial barriers) and cells such as phagocytes that recognize and mark for elimination preprogrammable molecular patterns that are known to be deleterious. The adaptive aspects of the immune system provide incredibly specific and often long-lasting protection against virtually any class of molecule (termed antigen). The innate and adaptive divisions of the immune system do not exist in isolation. The signaling mechanisms activated in response to antigen recognition by cells in either branch of the immune system reinforce and help to regulate each other. Because of its ability to select, target, and eliminate unwanted cells, the adaptive immune system has been a logical target in the design of immunotheraputics.[12]

Adaptive immunity is most directly mediated by lymphocytes and can be further divided into humoral and cellular immunity. Humoral immunity is mediated by a subset of lymphocytes called B cells, so named for the Bursa of Fabricius, the site of their maturation within the bird. B cells secrete specialized proteins called antibodies. Antibodies can be thought of as bipolar proteins, with one end recognizing a specific antigen, and the other end designed to activate a certain class of immune response.

Cell-mediated immunity is carried out by a subset of lymphocytes called T cells (analogously named for their site of maturation within the thymus). Although B cells are capable of secreting into circulation their antigen recognizing protein, T cells possess a membrane bound antigen receptor appropriately named the T cell receptor. Several lines of research have established the T cell as an important mediator of immune protection against cancer development.[13,14]

IMMUNE TOLERANCE

One of the fundamental problems the immune system deals with is differentiating self from nonself antigens. Many mechanisms have evolved to prevent immune activation by self-antigen (broadly termed autoimmunity). The deletion of maturing autoreactive T cells in the thymus is a key step in preventing destruction of normal self cells.[15] The activation of appropriately selected T cells can result in damage to surrounding tissue if the immune response is allowed to progress unchecked (autoimmunity). For this reason, mechanisms exist to suppress even appropriate antigen recognition, thereby regulating both the specificity and scope of the immune response. These mechanisms include both secreted signaling molecules and specialized cells. T regulatory cells are a subset of T lymphocytes whose function appears to suppress immune activation, serving as an important guard against autoimmunity.[16] Other cell subsets, notably myeloid-derived suppressor cells, may also contribute. Tolerance is an important concept within immunotherapy as it can be co-opted in disease states in an effort to evade the immune response.[17]

UNIQUE ASPECTS OF THE CENTRAL NERVOUS SYSTEM IMMUNE RESPONSE

The central nervous system (CNS) exists in a tightly regulated environment relatively shielded from systemic inflammation.[18] This relative sequestration of the CNS from the rest of the body, and thus a generalized immune response, creates a relative (although not absolute) immune privileged zone.[19] Some of the earliest evidence for this came from work by Sir Peter Medawar, 1960 Nobel Prize winner in Physiology or Medicine, who demonstrated that xenografts transplanted in the CNS or the eye persisted for longer periods of time than those transplanted into other parts of the body.[20] Several different factors contribute to the relative immune deficiency of the CNS. The main factors thought to play a role in immune isolation include the blood–brain barrier (BBB), lack of a classic lymphatic system, reduced major histocompatibility complex expression, and limited presence of resident antigen presenting cells.[21]

The Blood–Brain Barrier

One of the structural contributors to this physiologic distinction is a unique vascular system known as the BBB.[22] A relative separation of the CNS from the periphery has been noted since the late 1800s when German physiologist Paul Ehrlich classically demonstrated that the injection of albumin tagged dyes distributed and stained nearly all body tissues except for the brain parenchyma. Shortly thereafter, Goldmann conducted a similar experiment in which Trypan blue dye injected intravenously failed to stain the CNS, but dye injected directly into the cerebrospinal fluid (CSF) spaces was rapidly taken up and stained the brain parenchyma.[23] In the 1960s, the exclusion of molecules from crossing the BBB was visualized with the use of electron microscopy, when Karnovksy injected high density dyes and showed their collection adjacent to the cells of the BBB.[24]

The BBB is characterized by vessels with specialized endothelial cell (EC) tight junctions that limit the intercellular flow of cells, proteins, and electrolytes. These ECs are invested by astrocytes and pericytes that contribute to the vascular integrity of the BBB.[22] Although the BBB serves to protect the CNS from the harmful effects of the inflammatory process, it also creates challenges when designing immunotherapeutics targeted at CNS pathology.

The integrity of the BBB is not absolute, but rather the selective exclusion of molecules is dynamically regulated. Brain tumors induce increased vascular permeability.[25,26] This has been visualized grossly via computed tomography scan and magnetic resonance imaging as well as on a microscopic level. The mechanism by which this occurs is not completely understood but likely stems in large part by changes in Aquaporin expression.[27] The aquaporins are a class of integral membrane proteins that help regulate the intracellular and extracellular fluid balance by allowing for the selective passage of water molecules.

Specific proteins that contribute to the integrity and specialized structure of the BBB include the occludins and the claudins. Because of their specificity in regulating the flow of molecules across the BBB, these proteins (in particular the claudins) have become potential therapeutic targets.[28,29]

The CNS Lymphatic System

The lymphatic system plays an important role in immune surveillance by allowing for the circulation and coalescing of immune cells and antigen. The CNS has been thought of as lacking a classical lymphatic system thus limiting these opportunities for antigen–immune interaction. Paradoxically, despite the lack of a conventional lymphatic system within the CNS, molecules injected into the CNS concentrate within the cervical lymph nodes. This was previously thought to occur via the passage of molecules through the cribriform plate into the lymph channels residing in the nasal mucosa.[30,31] A specific population of macrophages is known to reside within the CNS and sample the antigen space in similar fashion to immune cells within the circulating lymph.[32–35]

There has been recent challenge to these long-held beliefs. New data argue for the existence of a lymphatic system within the CNS.[36] Functional lymphatic vessels, expressing many of the classical molecular markers of lymphatic ECs, have been found lining the murine dural venous sinuses. These channels are structurally similar to initial afferent lymphatic vessels and are capable of draining solute within CSF, presumable through the arachnoid granulations that line the cerebral ventricles. Ultimately, this drainage leads to the deep cervical lymph nodes. This data provides mechanistic insight to support prior observations regarding lymphatic drainage of the CNS. Specifically, the data presented by the Kipnis group suggest that lymphatic drainage does not occur via passage through the cribiform plate as previously thought, but rather through the meningeal lymphatic vessels they elegantly elucidated. Although the data come from murine models, it provides strong impetus for further investigation and may provide alternate explanations for previous observations. If validated in humans, it is an important step forward in our understanding of immune cell trafficking into and out of the CNS.

Antigen Presentation in the CNS

The classical model of lymphocyte activation involves the appropriate presentation of a nonself antigen by a cell. An integral part of this presentation process is the proper intracellular processing of the antigen and its ultimate display on the cellular surface in conjunction with a specific group of molecules. The major histocompatibility complex (MHC) comprises a group of cell surface molecules responsible for antigen presentation. The MHC, also termed the human leukocyte antigen (HLA) in humans, is broadly grouped into class I and class II. This distinction is based on the ultimate composition of the MHC. Class I and class II MHC are expressed on different cell types and are ultimately involved in activating different types of immune responses. The MHC class I is nearly ubiquitously expressed on all cell types, thus allowing for the signaling of an infected cell or a malignant cell. MHC class I interacts with $CD8^+$ T cells to activate a cellular immune response. In this manner "sick" cells—cells that are either infected with a pathogen or malignantly transformed may be targeted for cytotoxic destruction. The expression of MHC class II is generally restricted to what are generally thought of as "professional" antigen presenting cells.[37] These include B cells, dendritic cells, and macrophages. Presentation of antigen in context with an MHC class II invokes a humoral immune response via the activation of $CD4^+$ T cells. There is mixed evidence concerning the expression of MHC components in CNS cells, but the available data suggest that expression is either markedly reduced or absent on CNS cells. This is true of both MHC class I and MHC class II. MHC class II expression seems to be most significant on antigen presenting cells that migrate from the periphery to the CNS as opposed to resident CNS cells. The reduced ability to present antigen contributes to the relative immunodeficiency within the CNS.

The immune response in the CNS is limited by the low levels of MHC expression per cell, as well as by the limited presence of professional antigen presenting cells that express MHC.[38] The CNS appears to lack resident dendritic cells or macrophages. Microglia appear to be the predominant antigen presenting immunocompetent cells resident to the CNS. Microglia are monocyte-derived cells that have been shown to scan the CNS for potential antigens of interest. Upon activation they transform morphologically into an activated form capable of eliciting an immune response. Microglia appear to be centered in the leptomeninges and perivascular spaces.[35]

The various means by which the CNS is isolated from a systemic immune response provides important protection from immune mediated damage. Certainly multiple pathological processes are rooted in an inappropriately robust or nonselective inflammatory response in the CNS. One of the best-studied models of this is experimental or allergic encephamyelitis. Although protective to the normal cells of the CNS, it also serves to shield malignantly transformed cells from immune detection and clearance.

CANCER IMMUNOTHERAPIES

The last decade has seen marked success in the use of targeted immunotherapies in the treatment of malignancy. The checkpoint inhibitors serve as an additional example of successful immunotheraputics.[6,39] The term immune checkpoint broadly refers to the signaling mechanisms involved in maintaining self-tolerance or the prevention of autoimmunity.[40] These pathways are key in generation of an immune response that is of appropriate scope. Because immune checkpoints are capable of down regulating or "putting the brakes" on an immune response, they are an attractive evolutionary target in the oncogenic process and have been shown to be misregulated as part of tumor progression. For similar reasons, they are attractive drug targets. Several immune checkpoint modulators are in various stages of development and clinical use. Ipilimumab (antibody targeted at the blockage of CTLA-4) is one of the best known of these.[41] The development of ipilimumab follows the observation that blockade of CTLA-4 enhances both immediate antitumor immunity as well as protection against subsequent exposure to tumor. After a series of successful clinical trials, ipilimumab was ultimately approved by the Food and Drug Administration for the treatment of melanoma. A second monoclonal antibody, pembrolizumab (also known as lambrolizumab), targets PD-1 and is also approved in the treatment of melanoma.[42] More recently, lambrolizumab has won approval for the treatment of non-small-cell lung cancer subtypes that express PD-L1.[43] A third agent, nivolumab, which also targets PD-1 was recently found to be superior in the treatment of squamous cell carcinoma of the lung when compared to the long used docetaxel.[44,45] Although all these studies bring with them caveats and cautions, they provide a great deal of excitement regarding the harnessing of the immune system toward the treatment of cancer.

The specialized immune response associated with the CNS provides important protective measures. This protection creates challenges when designing immunotherapy targeted at CNS malignancy that are not present when targeting non-CNS cancers. Despite the challenges, there have been successful attempts at targeting CNS malignancy with immunotherapy. Arguably the most successful of these is the creation of a vaccine against the epidermal growth factor receptor variant type III (EGFRvIII). EGFRvIII is a tumor-specific mutation of the EGFR that is expressed in a large percentage of primary GBMs and some other common malignancies.[46] A peptide-based vaccine against this tumor-specific mutation has been created and verified to be immunogenic in multiple animal and human studies. This vaccine, by Celldex Therapeutics under the name rindopepimut, has been shown to be safe and potentially efficacious in three phase II trials. A randomized phase II trial in patients with recurrent GBM did show a survival advantage. An international multicenter phase III trial, however, failed to show efficacy in newly-diagnosed patients. These studies were conducted alongside standard of care treatment for GBM that includes the potent alkylating agent temozolomide. Temozolomide administration results in profound lymphocyte loss.[47] *A priori*, it would be reasonable to expect this loss of lymphocytes to diminish response to the vaccine, as lymphocytes are needed for immunological memory. Interestingly, this is not case and can likely be explained by a basis in the antigen preference obtained during immune reconstitution.

The successful use of the checkpoint inhibitors (outlined earlier) has not been limited to non-CNS tumors. A significant body of research within animal models has demonstrated that both CTLA-4 and PD-1 play a key role in glioma pathogenesis. Furthermore, data suggest that blockage of these pathways enhances immune-mediated tumor control in mice. Clinical trials exploring the use of nivolumab and pembrolizumab (both PD-1 antibodies) are currently underway.

References

1. Stupp R, Mason WP, van den Bent MJ, et al. Radiotherapy plus concomitant and adjuvant temozolomide for glioblastoma. *N Engl J Med*. 2005;352(10):987–996. http://dx.doi.org/10.1056/NEJMoa043330.
2. Fisher JL, Schwartzbaum JA, Wrensch M, Wiemels JL. Epidemiology of brain tumors. *Neurol Clin*. 2007;25(4):867–890. http://dx.doi.org/10.1016/j.ncl.2007.07.002. vii.
3. Ostrom QT, Gittleman H, Fulop J, et al. CBTRUS statistical report: primary brain and central nervous system tumors diagnosed in the United States in 2008–2012. *Neuro-oncology*. 2015;17(suppl 4):iv1–iv62. http://dx.doi.org/10.1093/neuonc/nov189.
4. Lawrence YR, Wang M, Dicker AP, et al. Early toxicity predicts long-term survival in high-grade glioma. *Br J Cancer*. 2011;104(9):1365–1371. http://dx.doi.org/10.1038/bjc.2011.123.
5. Okada H, Weller M, Huang R, et al. Immunotherapy response assessment in neuro-oncology: a report of the RANO working group. *Lancet Oncol*. 2015;16(15):e534–e542. http://dx.doi.org/10.1016/S1470-2045(15)00088-1.
6. Preusser M, Lim M, Hafler DA, Reardon DA, Sampson JH. Prospects of immune checkpoint modulators in the treatment of glioblastoma. *Nat Rev Neurol*. 2015;11(9):504–514. http://dx.doi.org/10.1038/nrneurol.2015.139.
7. Suryadevara CM, Verla T, Sanchez-Perez L, et al. Immunotherapy for malignant glioma. *Surg Neurol Int*. 2015;6(suppl 1):S68–S77. http://dx.doi.org/10.4103/2152-7806.151341.
8. Rolle CE, Sengupta S, Lesniak MS. Challenges in clinical design of immunotherapy trials for malignant glioma. *Neurosurg Clin N Am*. 2010;21(1):201–214. http://dx.doi.org/10.1016/j.nec.2009.08.002.
9. Koebel CM, Vermi W, Swann JB, et al. Adaptive immunity maintains occult cancer in an equilibrium state. *Nature*. 2007;450(7171):903–907. http://dx.doi.org/10.1038/nature06309.
10. Dunn GP, Bruce AT, Ikeda H, Old LJ, Schreiber RD. Cancer immunoediting: from immunosurveillance to tumor escape. *Nat Immunol*. 2002;3(11):991–998. http://dx.doi.org/10.1038/ni1102-991.
11. de Visser KE, Eichten A, Coussens LM. Paradoxical roles of the immune system during cancer development. *Nat Rev Cancer*. 2006;6(1):24–37. http://dx.doi.org/10.1038/nrc1782.
12. Vesely MD, Kershaw MH, Schreiber RD, Smyth MJ. Natural innate and adaptive immunity to cancer. *Annu Rev Immunol*. 2011;29:235–271. http://dx.doi.org/10.1146/annurev-immunol-031210-101324.
13. Smyth MJ, Thia KY, Street SE, et al. Differential tumor surveillance by natural killer (NK) and NKT cells. *J Exp Med*. 2000;191(4):661–668.
14. Smyth MJ, Thia KY, Street SE, MacGregor D, Godfrey DI, Trapani JA. Perforin-mediated cytotoxicity is critical for surveillance of spontaneous lymphoma. *J Exp Med*. 2000;192(5):755–760.
15. Kappler JW, Roehm N, Marrack P. T cell tolerance by clonal elimination in the thymus. *Cell*. 1987;49(2):273–280.
16. Josefowicz SZ, Lu L-F, Rudensky AY. Regulatory T cells: mechanisms of differentiation and function. *Annu Rev Immunol*. 2012;30:531–564. http://dx.doi.org/10.1146/annurev.immunol.25.022106.141623.
17. Mapara MY, Sykes M. Tolerance and cancer: mechanisms of tumor evasion and strategies for breaking tolerance. *J Clin Oncol*. 2004;22(6):1136–1151. http://dx.doi.org/10.1200/JCO.2004.10.041.

18. Pachter JS, de Vries HE, Fabry Z. The blood–brain barrier and its role in immune privilege in the central nervous system. *J Neuropathol Exp Neurol*. 2003;62(6):593–604.
19. Ransohoff RM, Engelhardt B. The anatomical and cellular basis of immune surveillance in the central nervous system. *Nat Rev Immunol*. 2012;12(9):623–635. http://dx.doi.org/10.1038/nri3265.
20. Medawar PB. Immunity to homologous grafted skin; the fate of skin homografts transplanted to the brain, to subcutaneous tissue, and to the anterior chamber of the eye. *Br J Exp Pathol*. 1948;29(1):58–69.
21. Wilson EH, Weninger W, Hunter CA. Trafficking of immune cells in the central nervous system. *J Clin Invest*. 2010;120(5):1368–1379. http://dx.doi.org/10.1172/JCI41911.
22. Abbott NJ, Rönnbäck L, Hansson E. Astrocyte-endothelial interactions at the blood–brain barrier. *Nat Rev Neurosci*. 2006;7(1):41–53. http://dx.doi.org/10.1038/nrn1824.
23. Bentivoglio M, Kristensson K. Tryps and trips: cell trafficking across the 100-year-old blood–brain barrier. *Trends Neurosci*. 2014;37(6):325–333. http://dx.doi.org/10.1016/j.tins.2014.03.007.
24. Reese TS, Karnovsky MJ. Fine structural localization of a blood–brain barrier to exogenous peroxidase. *J Cell Biol*. 1967;34(1):207–217.
25. Long DM. Capillary ultrastructure and the blood–brain barrier in human malignant brain tumors. *J Neurosurg*. 1970;32(2):127–144. http://dx.doi.org/10.3171/jns.1970.32.2.0127.
26. Dubois LG, Campanati L, Righy C, et al. Gliomas and the vascular fragility of the blood–brain barrier. *Front Cell Neurosci*. 2014;8:418. http://dx.doi.org/10.3389/fncel.2014.00418.
27. Papadopoulos MC, Saadoun S, Binder DK, Manley GT, Krishna S, Verkman AS. Molecular mechanisms of brain tumor edema. *Neuroscience*. 2004;129(4):1011–1020. http://dx.doi.org/10.1016/j.neuroscience.2004.05.044.
28. Morin PJ. Claudin proteins in human cancer: promising new targets for diagnosis and therapy. *Cancer Res*. 2005;65(21):9603–9606. http://dx.doi.org/10.1158/0008-5472.CAN-05-2782.
29. Singh AB, Sharma A, Dhawan P. Claudin family of proteins and cancer: an overview. *J Oncol*. 2010;2010:541957. http://dx.doi.org/10.1155/2010/541957.
30. Cserr HF, Harling-Berg CJ, Knopf PM. Drainage of brain extracellular fluid into blood and deep cervical lymph and its immunological significance. *Brain Pathol*. 1992;2(4):269–276.
31. Goldmann J, Kwidzinski E, Brandt C, Mahlo J, Richter D, Bechmann I. T cells traffic from brain to cervical lymph nodes via the cribroid plate and the nasal mucosa. *J Leukoc Biol*. 2006;80(4):797–801. http://dx.doi.org/10.1189/jlb.0306176.
32. Kreutzberg GW. Microglia: a sensor for pathological events in the CNS. *Trends Neurosci*. 1996;19(8):312–318.
33. Graeber MB, Scheithauer BW, Kreutzberg GW. Microglia in brain tumors. *Glia*. 2002;40(2):252–259. http://dx.doi.org/10.1002/glia.10147.
34. Nimmerjahn A, Kirchhoff F, Helmchen F. Resting microglial cells are highly dynamic surveillants of brain parenchyma in vivo. *Science*. 2005;308(5726):1314–1318. http://dx.doi.org/10.1126/science.1110647.
35. Soulet D, Rivest S. Microglia. *Curr Biol*. 2008;18(12):R506–R508. http://dx.doi.org/10.1016/j.cub.2008.04.047.
36. Louveau A, Smirnov I, Keyes TJ, et al. Structural and functional features of central nervous system lymphatic vessels. *Nature*. 2015;523(7560):337–341. http://dx.doi.org/10.1038/nature14432.
37. Zinkernagel RM, Doherty PC. The discovery of MHC restriction. *Immunol Today*. 1997;18(1):14–17.
38. Massa PT. Specific suppression of major histocompatibility complex class I and class II genes in astrocytes by brain-enriched gangliosides. *J Exp Med*. 1993;178(4):1357–1363.
39. Pardoll DM. The blockade of immune checkpoints in cancer immunotherapy. *Nat Rev Cancer*. 2012;12(4):252–264. http://dx.doi.org/10.1038/nrc3239.
40. Zou W, Chen L. Inhibitory B7-family molecules in the tumour microenvironment. *Nat Rev Immunol*. 2008;8(6):467–477. http://dx.doi.org/10.1038/nri2326.

41. Hodi FS, O'Day SJ, McDermott DF, et al. Improved survival with ipilimumab in patients with metastatic melanoma. *N Engl J Med*. 2010;363(8):711–723. http://dx.doi.org/10.1056/NEJMoa1003466.
42. Watson I, Dominguez PP, Donegan E, Charles Z, Robertson J, Adam EJ. NICE guidance on pembrolizumab for advanced melanoma. *Lancet Oncol*. 2016;17(1):21–22. http://dx.doi.org/10.1016/S1470-2045(15)00547-1.
43. Herbst RS, Baas P, Kim D-W, et al. Pembrolizumab versus docetaxel for previously treated, PD-L1-positive, advanced non-small-cell lung cancer (KEYNOTE-010): a randomised controlled trial. *Lancet*. December 2015. http://dx.doi.org/10.1016/S0140-6736(15)01281-7.
44. Borghaei H, Paz-Ares L, Horn L, et al. Nivolumab versus docetaxel in advanced nonsquamous non-small-cell lung cancer. *N Engl J Med*. 2015;373(17):1627–1639. http://dx.doi.org/10.1056/NEJMoa1507643.
45. Brahmer J, Reckamp KL, Baas P, et al. Nivolumab versus docetaxel in advanced squamous-cell non-small-cell lung cancer. *N Engl J Med*. 2015;373(2):123–135. http://dx.doi.org/10.1056/NEJMoa1504627.
46. Wong AJ, Ruppert JM, Bigner SH, et al. Structural alterations of the epidermal growth factor receptor gene in human gliomas. *Proc Natl Acad Sci USA*. 1992;89(7):2965–2969.
47. Sengupta S, Marrinan J, Frishman C, Sampath P. Impact of temozolomide on immune response during malignant glioma chemotherapy. *Clin Dev Immunol*. 2012;2012(3):1–7. http://dx.doi.org/10.1155/2012/831090.

CHAPTER

2

Immune Constitution of Patients With Brain Tumors

J. Berry-Candelario, S.H. Farber, P.E. Fecci

Duke University Medical Center, Durham, NC, United States

OUTLINE

Translational Immunotherapy of Brain Tumors
http://dx.doi.org/10.1016/B978-0-12-802420-1.00002-8

BACKGROUND AND DISCOVERY

Immune deficits in patients with glioblastoma (GBM) have been documented for many years. Much of the pioneering work to document cell-mediated immunity (CMI) impairment in patients with GBM was performed in the 1970s and 1980s and is attributable to William Brooks, Thomas Roszman, and Steve Mahaley. The earliest publication, one in 1972 by Brooks et al.[1] analyzed host immune responsiveness among 23 patients with benign and malignant intracranial tumors. The patients enrolled in this study were skin tested for delayed hypersensitivity and, when compared to 20 healthy volunteers, demonstrated reduced sensitization to dinitrochlorobenzene, as well as significantly diminished responsiveness to three of four common skin-test antigens [purified protein derivative [PPD], streptokinase–streptodornase (SK-SD), and trichophyton (TRICH), but not *Candida* (CAND)]. Patient lymphocyte proliferative responses were also assessed, and these were found to parallel skin-test results, with significantly fewer patients responding to PPD and SK-SD. No significant differences were observed for CAND, TRICH, or phytohemagglutinin (PHA), although 5/22 patients' lymphocytes did not respond to PHA. In this earliest study, no anemia, granulocytopenia, or lymphopenia was observed. Patient lymphocytes performed well in mixed leukocyte reactions, or when cultured with autologous tumor cells, but these responses were significantly hindered when patient plasma was substituted for normal plasma in the reactions. Patient plasma also demonstrated the ability to hinder normal lymphocyte responses, prompting suspicion that an immunosuppressive factor was present in patient sera. The authors were able to localize the suppressor factor to the IgG fraction of patients' sera, and they concluded that the suppressor factor might be an antibody that could bind receptors on lymphocyte membranes, restricting their function.

A pursuant paper in 1974[2] repeated the previously mentioned findings while examining only the patients with glioma (one GBM, seven astrocytoma). Attempts to correlate the amount of suppression observed with the degree of malignancy were not statistically significant (although a paper by Young et al. in 1976 found that immunocompetence is altered in patients with Grade III and IV but not Grade I and II tumors[3]). One finding that did emerge in this study, however, was that the "humoral" suppression of CMI was lost subsequent to tumor removal, raising the possibility that immunosuppression might become a clinically useful diagnostic tool for recurrence.

The initial installation of a major foray into the immunobiology of primary intracranial tumors was published in 1977 by Mahaley et al.[4] and this introductory portion examined certain parameters of both the cellular and humoral immune competence of brain tumor patients. The study enrolled 42 patients with GBM, 17 with anaplastic gliomas, and 17 with meningioma. Cellular immune competence was reflected in preoperative lymphocyte counts and delayed hypersensitivity reactions (DHRs) to SK-SD, mumps antigen, TRICH, and PPD. Humoral immunity was assayed through measurements of serum IgG, IgM, and IgA levels, as well as tetanus and influenza antibody titers. The group of patients with GBM possessed the smallest number of DHR responders to two or more antigens (31% vs. 95% for controls), and was the only group demonstrating significant lymphopenia at preoperative baseline and serial bimonthly postoperative time points. Patients with GBM had mean IgG, IgM, and IgA levels that were consistently within the normal range, although IgM levels experienced a decline over time after an initial elevation. Serum tetanus antibody responded to boost, but serum influenza antibody eventually declined progressively, despite boosting. This paper was the first to examine humoral immunity, and it was one of the first studies to report lymphopenia in brain tumor patients, whereas some of the earlier studies had discussed the presence of normal lymphocyte counts.[1,5]

Studies of patients with GBM continued to document depressed cutaneous reactivity to common antigens, as well as the ability of patient plasma to inhibit lymphocyte responsiveness in mixed leukocyte reaction or to mitogens such as PHA. In 1976, Young et al. added the observation that lymphocytes from approximately 50% of patients with GBM were significantly inhibited in their response to either PHA or concanavalin A (ConA), even when cultured in the presence of normal plasma.[3] This observation, combined with a study by Brooks et al. that noted a decreased propensity for patient T cells to form rosettes with sheep erythrocytes (E-RFC), introduced the possibility that additional defects existed that were intrinsic to the lymphocyte compartment. A subsequent report was published in 1977,[6] and this provided a more specific analysis of lymphocyte subpopulations in patients with primary brain tumors. The study confirmed the presence of lymphopenia in patients with glial tumors, the lymphopenia whose severity correlated with tumor grade such that patients with GBM presented with a 37% reduction in the number of circulating lymphocytes. Furthermore, the lymphopenia was characterized by a selective depletion of T cells, whereas B cell levels remained normal (when corrected for the lymphopenia).

In 1980, it became clearer that circulating humoral factors could not account for much of the impaired CMI observed in patients with primary brain tumors. Although the activation of lymphocytes obtained from patients could be somewhat improved when cultured with normal sera, the response remained significantly less than expected, implying a defect intrinsic to lymphocytes. Roszman and Brooks reported that at all concentrations of PHA, the response of peripheral blood lymphocytes (PBL) from normal individuals was 3–3.7 times greater than that of PBL obtained from patients, despite culture in normal human sera.[7] Importantly, the existence of T cell lymphopenia could not explain these observations, as purified T cell populations revealed similar results. Neither increasing the number of lymphocytes in culture nor altering the duration of the culture period corrected the observed proliferative defects. Thus, the first bit of conclusive evidence was offered for alterations in the functional qualities of lymphocytes obtained from patients with glioma, and much subsequent work was aimed at explaining the now center-stage T cell dysfunction.

CURRENT UNDERSTANDING: ANALYZING SYSTEMIC IMMUNE DYSFUNCTION

Lymphopenia

Pretreatment of lymphopenia in GBM patients, specifically, was first suggested in the articles discussed earlier, beginning with the work of Mahaley et al. in 1977.[4] Others have also documented this phenomenon more recently and demonstrated that T cells, and particularly cluster of differentiation 4+($CD4^+$) helper T cells, are seemingly the populations most affected.[8] In contrast, some groups have advanced that lymphopenia only arises posttreatment with radiation and/or chemotherapy,[9,10] creating some debate as to the timing and etiology of lymphocyte disappearance in these patients. Certainly the etiology remains unelucidated. In 1999, Morford et al. offered one of the potential mechanisms.[11] They used a fluorescent staining technique to show that peripheral T cells appeared to be undergoing apoptosis in patients with glioma. Furthermore, T cells exposed in vitro to glioma cell culture supernatant exhibited apoptosis with T cell receptor (TCR) stimulation.[11] More recent observations, however, suggest that T cells may indeed remain available, albeit sequestered in alternative immunologic compartments, fostering T cell ignorance.[12]

Lymphocyte Dysfunction

PBLs from patients with glioma exhibit varying degrees of unresponsiveness to T cell mitogens, including ConA, PHA, and anti-CD3 monoclonal antibody (mAb).[13,14] With evidence that purified T cell populations respond poorly to stimulation, T cells became the focus of studies highlighting lymphocyte dysfunction in patients with brain tumors. The dysfunction was progressively narrowed to within the $CD4^+$ helper T cell subset, as T cells from patients proved unable to provide adequate helper activity in the pokeweed mitogen (PWM) response. Accordingly, it was found that $CD4^+$ T cells were significantly reduced among patient PBL (40% vs. 55% for controls), whereas the proportion of $CD8^+$ T cells remained normal.[15,16] Furthermore, in a later study, patient-derived $CD4^+$ T cells demonstrated a diminished capacity for proliferation in response to mitogen stimulation than $CD8^+$ T cells obtained from the same patient or $CD4^+$ T cells obtained from healthy controls (Table 2.1).

TABLE 2.1 Systemic and Local Immune Defects in Patients With GBM

SYSTEMIC IMMUNE DEFECTS
Diminished DTH responses
Lymphopenia
Decreased lymphocyte proliferative responses
TH2-skewed cytokine production
Altered humoral immunity
Increased percentages of regulatory T cells amidst CD4 T cell compartment
Increased myeloid-derived suppressor cells
Soluble circulating NKG2D ligands
Intrinsic T Cell Deficits
Decreased functional responsiveness, anergy
Apoptosis
Sequestration in alternative immune compartments
Increased expression of inhibitory receptors/immune checkpoints
Defective interleukin (IL)-2 production and IL-2R signaling
Disrupted T cell receptor signaling
TH2-skewed cytokine production
LOCAL IMMUNE DEFECTS
Tumor Microenvironment
Increased percentages of regulatory T cells amidst TILS
Increased local immunosuppressive cytokines: TGF-β, IL-10, and PGE2
Decreased TIL function, increased TIL exhaustion
Inhibitory tumor-associated macrophages and microglia
Increased myeloid-derived suppressor cells
Tumor Cell Intrinsic
PD-L1 expression on tumor cell surface
STAT3 expression
IDO1 expression
Disrupted antigen presentation
Decreased costimulatory molecules
Release of soluble NKG2D ligands
HLA-E and HLA-G expression

DTH, delayed-type hypersensitivity; *HLA*, human leukocyte antigen; *IDO*, indoleamine 2,3 dioxygenase; *NKG2D*, natural killer group 2D; *PGE2*, prostaglandin E2; *PD-L1*, programmed death receptor ligand 1; *STAT3*, signal transducer and activator of transcription-3; *TGF-β*, transforming growth factor beta; *TH*,T helper; *TILs*, tumor-infiltrating lymphocytes.

Although allogeneic PWM cultures were used to localize defects to within the $CD4^+$ subset of T cells, cytokinetic analysis of PHA-induced [^{3}H]-thymidine incorporation indicated defects in the ability of PBL from patients with brain tumor patients to undergo clonal expansion. Measurements of interleukin (IL)-2 activity in culture supernatants demonstrated significantly lower levels of the cytokine in cultures containing patient cells (levels were not measured directly but as a function of the ability of supernatants to support the growth of an IL-2–dependent, cloned mouse cytotoxic cell line). Addition of exogenous IL-2 was unable to reverse the existing proliferative defects.[17,18] The diminished IL-2 production by patient lymphocytes has proved to be a crucial observation in the work to gain understanding of the mechanisms behind T cell failures (although one study saw diminished IL-2 production in response to *Candida* mannoprotein (MP), but could not reproduce the defects seen with PHA stimulation[18]). Furthermore, the demonstration that exogenous IL-2 does not improve the mitogen responsiveness of patient lymphocytes suggested that IL-2 receptor (IL-2R) signaling or expression might also be defective. One of the earliest works in this regard examined the percentage of IL-$2R^+$ lymphocytes at various time points after PHA stimulation and found that this number was reduced when compared to normal values.[19] This study used an indirect immunofluorescence assay and began reporting levels after 24h of stimulation; no data on the levels of IL-2R on resting lymphocytes were available. Therefore, the conclusion was advanced that brain tumor patients demonstrate a quantitative deficiency in the number of inducible IL-2R–bearing PBLs. A later study by the same authors advocated a role for an unidentified "glioma-derived suppressor factor" in eliciting the diminished IL-2 production and reduced induction of IL-2R expression on activated patient T cells.[20]

Further investigation into the nature of T cell anergy observed among patients with malignant glioma (MG) has yielded that TCR-mediated signaling is defective in patient T cells, as well. For example, in a study by Morford et al. in 1997,[20,21] both PHA and anti-CD3 mAb-stimulated PBL and T cells obtained from patients exhibited marked defects in early transmembrane signaling events, measured in the tyrosine phosphorylation of a variety of proteins. For instance, western blot analysis of T cell lysates revealed decreased phosphorylation of pp100 and phospholipase Cg1 (PLCg1). Additionally, PLCg1 and $p56^{lck}$ protein levels were reduced in patient T cells, and the cells mobilized less calcium than normal T cells after PHA stimulation or ionomycin treatment. Proliferative capacity was not restored following stimulation with phorbol 12-myristate 13-acetate (PMA) and ionomycin, indicating that observed anergy in patient T cells was attributable to defects in early transmembrane signaling events associated with TCR/CD3 stimulation.

Regulatory T Cells

Brooks and Roszman long noted that purified T cells from glioma patients were unresponsive and that the impairment of CMI was in part due to defects within the T cell compartment.[7] They eventually concluded the existence of a nonspecific suppressor cell population, inducible in patients with primary brain tumors.[22] As early as 1971, such a suppressor cell population had been hypothesized by Gershon.[23] Further clues were provided by the pioneering work of Sakaguchi[24] and Shevach[25], which brought a paradigm shift to modern-day immunology by revealing immunoregulatory cell activity focused in the $CD25^+$ (IL-2a $receptor^+$) fraction of $CD4^+$ T cells. These regulatory T cells (Tregs) represent the $CD25^+FOXP3^+$ subset of CD4 T cells. These cells account for 5–10% of $CD4^+$ T cells in the blood of normal individuals but represent an increased component of the CD4 compartment in patients with a variety of cancers, including GBM.[8,26–29] Tregs are generated in the thymus via their avidity for self-peptides and prove responsible for the peripheral inhibition of self- and even foreign-antigen reactive $CD4^+$ and $CD8^+$ T cells. Tregs limit T cell activation and proliferative responses and cause downregulation of the production of IL-2 and interferon-gamma (IFN-γ) in favor of the Th2 cytokines, IL-10, and transforming growth factor-beta (TGF-β). Hence, Fecci et al. have shown a reversal to systemic T cell dysfunction in patients with GBM when Tregs are depleted, and in vivo experimentation in mice with Tregs depletion strongly reverses T cell deficits and even improves survival.[8,30,31] Additional work by El Andaloussi analyzed tumor-infiltrating lymphocytes (TILs) of 10 GBM patients and found that over 50% of $CD4^+$ TILs exhibited FOXP3 expression. These data provide strong evidence for the presence of Tregs among the glioma TILs, which further supports the critical nexus of the glioma microenvironment.

CURRENT UNDERSTANDING: ANALYZING THE TUMOR

Cytokine Dysregulation

GBM appears to shift cytokine production from an antitumor pathway to one that is dysregulated and immunosuppressive. Notably, GBM cells produce TGF-β, IL-10, and prostaglandin E2 (PGE2). TGF-β1 and β2, in particular, have far reaching effects, including suppression of IL-2–dependent generation of cytotoxic T-lymphocytes (CTL) from PBLs and TILs, inhibition of IL-2 receptor expression on T cells, reduction of IL-1–and IL-2–dependent proliferation of T and B cells, reduction of IFN-γ production, and elimination of CMI.[32] In addition, peripheral lymphocytes and TILs from glioma patients develop a highly tumor-favorable profile to their own cytokine production. Peripheral lymphocytes produce higher levels of IL-10 while minimizing IL-12.[27] TILs isolated from glioma specimens express IL-4 and granulocyte macrophage colony-stimulating factor but do not express IFN-γ, IL-2, or TNF-β above normal PBL.[33] Overall there is a shift from cytotoxicity-inducing Th1 cytokines to Th2-type cytokine production. Most recently, evidence has also surfaced that GBM may directly negatively regulate T cell cytokine production and proliferation via cell surface expression of programmed death receptor ligand 1 (PD-L1, also called B7-H1), which would interact with PD-1 on tumor-specific T cells and potentially permit tumor escape from the host immune system.[34]

Signal Transducer and Activator of Transcription-3

Signal transducer and activator of transcription-3 (STAT3) has been shown to play a major role in the tumorigenicity of various cancers.[35–37] Mechanistically, STAT3 is found to be constitutively activated in glioma models and appears to promote oncogenic properties such as angiogenesis, cell proliferation, and antiapoptosis, as well as assisting the tumor-promoting effects of epidermal growth factor receptor variant III (EGFRvIII) expression.[37–39] Finally, STAT3 may play an upstream role in the proliferation of Tregs and dendritic cell attenuation via induction of tumor-based cytokines such as vascular endothelial growth factor (VEGF), IL-10, and IL-6, which leads to STAT3-mediated impairment of cytotoxicity in the innate and adaptive immune compartments.[40,41] Despite these data, the molecular story of STAT3 is complicated. Mounting evidence suggests that its role factors in on both sides of the tumor/antitumor coin. On one side, Fujita et al. revealed that systemic inhibition of STAT3 enhanced immune-mediated survival in a murine glioma model.[42] Conversely, STAT3 was found to behave like a tumor suppressor in the absence of phosphatase and tensin homolog tumor suppressor.[43,44] Together, these data point to critical context-dependent immunological and nonimmunological functions of STAT3 in gliomagenesis and warrant its careful study in the clinical setting.

Indoleamine 2,3 Dioxygenase

Indoleamine 2,3 dioxygenase 1 (IDO1) is an inducible and rate-limiting enzyme of tryptophan catabolism, whose role in GBM pathogenesis is being diligently delineated. An identified role for tryptophan catabolism in cancer dates back to the 1950s when urinary catabolites of tryptophan were found in breast and prostate cancer.[45] Friberg et al. introduced the idea that tumor IDO1 expression might foster immune evasion in a sentinel paper revealing a greater allogeneic T cell response against Lewis lung carcinoma cells cultured with an IDO1 inhibitor.[46] Furthermore, the authors observed delayed tumor growth with systemic administration of the same inhibitor.

Nakamura et al. looked at the correlation between IDO1 and Treg abundance in the development and progression of uterine cervical cancer.[47] The cervical intraepithelial neoplasia cell cultures revealed high levels of IDO1 expression, and Tregs proved to be in close proximity. Wainwright et al. used a syngeneic GL261 intracranial model of MG to demonstrate that tumor-derived dendritic cells with IDO1 expression were required for Treg recruitment and suppression of the antibrain tumor immune response. When IDO1 was genetically silenced in GL261 brain tumors, there was a definitive decrease in Treg recruitment and increased long-term survival.[48] Regarding translatable strategies, IDO1 inhibition alone showed no evidence of tumor rejection, but triple therapy of radiotherapy (RT), temozolomide (TMZ), and IDO1 inhibition led to a survival benefit over RT and TMZ alone.[48,49] Clinically, at the mRNA level, IDO1 upregulation has been associated with decreased overall survival in GBM.[50] IDO1 has been found with high frequency in gliomas with stronger expression present in the higher-grade tumors.[50,51] This is also true when secondary GBM is compared to their low-grade counterparts. Ultimately, the clinical significance of upregulated IDO1 and its impact on GBM continues to be elucidated. These data, however, certainly suggest an immunosuppressive role.

CLINICAL OPPORTUNITIES

A great deal of preclinical work in GBM has focused on developing therapies to direct an activated immune response against tumor and tumor antigens through various methods. Likewise, resultant clinical trials have now used therapies such as peptide and dendritic cell vaccines, as well as adoptive lymphocyte transfers (ALTs) in patients with GBM. Often, however, these therapies do not incorporate means for countering the immunosuppressive capacities of GBM. Therefore they remain susceptible to, and are thus limited by, the GBM-induced mechanisms of immunosuppression. In a vehicle, the driver must remove his/her foot from the brake before stepping on the gas. As we continue to gain an understanding of the immune deficits linked to GBM, we must leverage this knowledge to develop therapies that allow us to both activate the immune system (step on the gas) as well as inhibit the various mediators of immune dysfunction (remove the brakes) (Table 2.2).

TABLE 2.2 Ongoing Clinical Trials

Number	Title	Intervention	Status	Phase	Site	Disease
NCT02529072	Nivolumab with DC vaccines for recurrent brain tumors (AVERT)	DC vaccine + nivolumab	Currently recruiting	I	Duke University Medical Center, NC, USA	Recurrent MG
NCT02526017	Study of FPA008 in combination with nivolumab in patients with selected advance cancers (FPA008-003)	Nivolumab + FPA008	Currently recruiting	I	Multiple sites	Selected advanced cancers (including GBM)
NCT02311582	MK-3475 in combination with MRI-guided laser ablation in recurrent malignant gliomas	MK-3475 + laser ablation	Currently recruiting	I/II	Multiple sites	MG
NCT02311920	Ipilimumab and/or nivolumab in combination with temozolomide in treating patients with newly diagnosed GBM or gliosarcoma	Ipilimumab, nivolumab, TMZ	Currently recruiting	I	Multiple sites	GBM, gliosarcoma
NCT02550249	Neoadjuvant nivolumab in GBM (neo-nivo)	Nivolumab	Currently recruiting	II	Clinica Universidad de Navarra, Spain	GBM
NCT02017717	A study of the efficacy and safety of nivolumab vs. bevacizumab and of nivolumab with or without ipilimumab in GBM patients at different stages of treatment (CheckMate 143)	Ipilimumab, nivolumab, bevacizumab	Currently recruiting	III	Multiple sites	Recurrent GBM
NCT02617589	Study of nivolumab versus temozolomide, given with radiation therapy, for newly diagnosed patients with GBM (malignant brain cancer) (CheckMate 498)	Nivolumab, TMZ, radiotherapy	Currently recruiting	III	Multiple sites	GBM

NCT01582269	A study in recurrent GBM	LY2157299, lomustine	Ongoing, not recruiting	II	Multiple sites	GBM
NCT02502708	Study of the IDO pathway inhibitor, indoximod, and temozolomide for pediatric patients with primary malignant brain tumors	Indoximod, TMZ, radiation	Currently recruiting	I	Georgia Regents University, GA, USA	Malignant brain tumors
NCT02327078	A study of the safety, tolerability, and efficacy of epacadostat administered in combination with nivolumab in select advanced cancers	Nivolumab + epacadostat	Currently recruiting	I/II	Multiple sites	Advanced solid tumors and lymphomas (including GBM)
NCT02648633	Stereotactic radiosurgery with nivolumab and valproate in patients with recurrent glioblastoma	SRS + nivolumab and valproate	Currently recruiting	I	University of Virginia, Charottesville, VA	GBM
NCT02667587	Study of temozolomide plus radiation therapy with nivolumab or placebo, for newly diagnosed patients with glioblastoma (GBM, a malignant brain cancer). (CheckMate548)	Nivolumab, temozolomide, radiotherapy	Currently recruiting	II	Multiple sites	Brain neoplasms
NCT02658981	Anti-LAG-3 or urelumab alone and in combination with nivolumab in treating patients with recurrent glioblastoma	Anti-LAG3, anti-PD1	Currently recruiting	I	Multiple sites	Recurrent GBM
NCT02335918	A dose escalation and cohort expansion study of anti-CD27 (varlilumab) and anti-PD-1 (nivolumab) in advanced refractory solid tumors	Varlilumab and nivolumab	Currently recruiting	II	Multiple sites	Solid tumors (including GBM)

Trial information current as of December 2016.
GBM, glioblastoma; *IDO*, indoleamine 2,3 dioxygenase; *MG*, malignant glioma; *MRI*, magnetic resonance imaging; *TMZ*, temozolomide.

Countering T Cell Dysfunction: Immune Checkpoint Blockade

To this end, one of the emerging fields in immunotherapy for GBM is that of immune checkpoint blockade. Immune checkpoints are upregulated on T cells subsequent to their activation and act to mitigate immune responses. Examples include programmed cell death-1 (PD-1) and cytotoxic T-lymphocyte–associated antigen 4 (CTLA-4). Binding of PD-1 to its ligands (PD-L1 and PD-L2), expressed on tumors such as GBM,[52] instigates T cell shutdown and apoptosis, whereas CTLA-4 is responsible for downregulating T cell proliferation and abrogating activated T cell responses.[53] Nivolumab and pembrolizumab are monoclonal antibodies directed against PD-1, and ipilimumab is a monoclonal anti–CTLA-4 antibody. All three have been approved by the Food and Drug Administration for treatment of metastatic melanoma,[54,55] thus prompting the study of their use in other cancers including GBM. Clinical trials are currently underway that are testing these therapies in combination with various other treatment modalities. A phase I trial is currently assessing the dose-limiting toxicity of ipilimumab and nivolumab in combination with TMZ, in patients with newly diagnosed GBM (NCT02311920). Another phase I trial is currently recruiting and will assess the safety of nivolumab in combination with dendritic cell vaccination for patients with recurrent MG (NCT02529072). Nivolumab is also being tested in combination with FPA008—a monoclonal antibody that inhibits immunosuppressive tumor-associated macrophages. This phase I trial will assess the safety, recommended dose, and objective response rate of this regimen (NCT02526017). Neoadjuvant nivolumab is currently being tested in patients with both primary and recurrent GBM prior to surgical resection of the tumor. This phase II study will measure the changes in level of expression of PD-L1 by tumor cells and lymphocytes at baseline and following treatment. The safety and efficacy of this treatment will also be assessed using the common toxicity criteria and the response assessment in neurooncology criteria, respectively (NCT02550249). A combined phase I/II study will assess the synergistic combination of magnetic resonance imaging–guided laser ablation (MLA) with MK-3475 (pembrolizumab). This protocol is hypothesized to increase access to the tumor via disruption of the blood–brain barrier (BBB) using MLA while maximizing tumor-specific reactions by effectors using pembrolizumab. This study is currently recruiting participants (NCT02311582).

The efficacy of these therapies is also being tested in two randomized phase III trials. The first study, CheckMate 143, compares nivolumab to bevacizumab (anti-VEGF) in patients with recurrent GBM, as well as nivolumab given alone or in combination with ipilimumab. This study will reach completion in January 2018 (NCT02017717). The second study, CheckMate 498, compares the overall survival at three years and progression-free survival at 2 years for patients treated with nivolumab and RT compared to standard therapy with TMZ and RT in patients with newly diagnosed GBM. This study is currently recruiting participants (NCT02617589).

Countering Regulatory T Cells and Lymphopenia

Given the impact of Tregs in disrupting normal T cell function in GBM patients, depletion of Tregs has been a not infrequent therapeutic goal. The first attempts targeted IL-2R-alpha(α)/CD25 due to its constitutive expression on Tregs.[56–59] Research from our lab and others has shown in mouse models that Tregs can be deactivated or depleted using an anti–IL-2Rα monoclonal antibody.[60,61] Clinical evidence exists as well. Jacobs et al. used humanized anti–IL-2Rα daclizumab to deplete Tregs in metastatic melanoma patients receiving antitumor vaccination. Unfortunately, despite this depletion of Tregs, the patients were unable to mount any effector T cell response following vaccination.[62]

Future studies in GBM patients may be able to avail of intentionally myeloablative doses of TMZ. One of the largest lessons learned in the ALT arena has been the antitumor benefit of myeloconditioning prior to adoptive transfer.[63] Lymphodepletion using total body irradiation or myeloablative chemotherapy in mice and humans has been applied to optimize immunotherapy by leveraging the induction of homeostatic cytokines and the consequent T cell proliferation.[64] In fact, this treatment has been shown to induce massive T cell proliferation and significant amplification of tumor-specific immune responses.[64] Lymphocytes specific for antigens that predominate during this recovery period, like those provided in the form of a vaccine, have a competitive advantage and become disproportionately overrepresented in the recovering lymphocyte population.[65,66] In the case of GBM, standard of care TMZ, a potent methylating agent, also induces lymphodepletion, which theoretically could eradicate the production of an immune response, especially in the setting of serial cycles. However, published preclinical data demonstrate that when the vaccine is timed during homeostatic T cell recovery after TMZ-induced nadirs, TMZ administration synergizes with vaccination and adoptive T cell therapies by increasing levels of circulating homeostatic cytokines.[63] These effects are dose dependent, with enhanced efficacy observed at the highest doses of TMZ, presumably due to a greater degree of host lymphodepletion.[67]

Regarding the reversal of immunosuppressive mechanisms, lymphodepletion may also permit specific eradication of Tregs. Tregs are known to be uniquely dependent on the high affinity IL-2Rα (CD25) for their function and survival, and defects in IL-2Rα in mice and humans lead to multiorgan inflammatory infiltrates. mAbs that block IL-2Rα abrogate Treg function in animal models but can also inhibit effective antitumor immune responses in mice and humans with autoimmune disorders and organ transplants.[31] This negative impact on activated immune responses is likely due to a requirement for IL-2Rα signaling by effector T cells or a recently described indirect elimination of effector T cells by regulatory NK cells activated by IL-2Rα–specific MAbs. Preliminary data demonstrate that anti–IL-2Rα mAbs have very different effects on activated immune responses in mice undergoing homeostatic proliferation in response to transient lymphodepletion, as might be seen after therapeutic cycles of cytotoxic chemotherapies such as TMZ. In the context of lymphodepletion-induced homeostatic γ common cytokines, IL-2 may be dispensable for effector T cells, whereas Tregs remain absolutely dependent on it. Under these conditions, anti-CD25 in the setting of homeostatic expansion may instigate the depletion of Tregs while sparing effector T cells.[68] Indeed, a single dose of the anti–IL-2Rα mAb daclizumab administered concurrently with EGFRvIII targeted vaccination led to depletion of $CD4^+FOXP3^+$ Tregs in TMZ-treated patients with GBM without impairing vaccine-induced immune responses.[69]

Countering Cytokine Dysregulation

Another possible mediator of glioma-associated immunosuppression that can be targeted is TGF-β. There has been consistent interest in developing strategies to neutralize TGF-β or its signal transduction pathways in the setting of GBM.[70] Although early results were poor,[71] this first clinical trial aimed to suppress TGF-β transcription in GBM tissue by convection-enhanced delivery of TGF-β antisense oligonucleotides into the tumor. There were, however, limitations to this study. In particular, trabedersen (AP 12009), the synthetic antisense phosphorothioate oligodeoxynucleotide complementary to the messenger RNA of the human TGF-β2 gene, suffered insufficient penetration into the target tissue. Drug development has slowed due to the systemic side effect profile and evidence that aneurysms associated with Marfan's syndrome appear to benefit from suppression of TGF-β signal transduction. Most recently, a phase Ib/IIa trial (NCT01582269) using LY2157299, a TGF-β signaling blocker, is enlisting an alternating dosing schedule of 150 mg taken twice daily for 14 days followed by a pause for 14 days in GBM patients. The drug is added to standard chemoradiotherapy with TMZ in patients with newly diagnosed GBM without aortic aneurysms. In another trial, the same dosing schedule is being used in a three-arm phase 2–randomized placebo-controlled trial (NCT01582269). LY2157299 or placebo is administered as monotherapy or in combination with lomustine for recurrent GBM without preexisting aortic aneurysms. Both trials have completed accrual. LY2157299 appears to be safe and biologically active in the periphery.[72]

Countering Tumor STAT3 and IDO Expression

As discussed earlier, there is growing evidence of the connection between IDO and GBM. In fact, a preclinical model has revealed that adding IDO-blocking drugs to TMZ plus radiation significantly prolonged survival. Both NCT02502708 and NCT02052648 are phase I pediatric and adult studies examining indoximod, an inhibitor of IDO. In the pediatric study, indoximod is being used in combination with TMZ-based therapy in all primary pediatric brain tumors. In the adult study, indoximod plus TMZ is evaluated in patients with GBM. Both trials are open and accruing patients. Another nonrandomized study, NCT02327078, is evaluating the safety, tolerability, and efficacy of epacadostat, an oral hydroxyamidine and IDO1 inhibitor. In this dose escalation trial, epacadostat is coadministered with nivolumab in patients with a variety of solid tumors (including GBM) and lymphoma. The trial is currently open and accruing patients.

Currently, there are no open clinical trials targeting the STAT3 pathway in GBM. However, on the horizon, there is a planned clinical trial (NCT01904123) which has been approved as a dose escalation phase I trial of WP1066, a small molecule inhibitor designed to target the STAT3 pathway in cancer cells. The patient population to be enrolled includes patients with central nervous system (CNS) melanoma and recurrent GBM.

CONCLUSION

We have herein explored the evolution of our understanding of immune-restricting mechanisms in GBM. The disruptions to the immune system in GBM patients are severe and system-wide, resulting in lymphopenia and/or innate and cellular immune dysfunction. Regarding lymphopenia, the debate over its timing and etiology remains unresolved, but its presence is well accepted. Further research is necessary, and better characterization of the various immune compartments and lymphoid organs may enlighten our understanding. The evidence for innate and cellular immune dysfunction, however, is quite robust and quantifiably linked to the presence of GBM and its cadre of tumor-derived cytokine signals. Studies aimed at identifying the factors orchestrating a tumor-induced path of decline in immune function have provided numerous markers such as STAT3 and IDO for further research. Likewise, evidence continues to emphasize the significance of Tregs as a critical population fostering T cell inhibition and tolerance to tumor.

Given the severity of defects to GBM patient immune function, a crucial component to the success of immunotherapy moving forward must be a gained understanding of the mechanisms and means for reversal of these derangements. Several clinical trials highlighted earlier are indeed designed to develop and assess such strategies, including immune checkpoint blockade, eradication of Tregs and/or inhibition of their function, blockade of counterproductive tumor-intrinsic signaling pathways fostering immune escape, and neutralization of TGF-β and other cytokine signaling cascades. The success of these strategies will depend heavily on improving our understanding of the connections between the CNS and the immune system.

References

1. Brooks WH, Netsky MG, Normansell DE, Horwitz DA. Depressed cell-mediated immunity in patients with primary intracranial tumors. Characterization of a humoral immunosuppressive factor. *J Exp Med*. December 1, 1972;136(6):1631–1647.
2. Brooks WH, Caldwell HD, Mortara RH. Immune responses in patients with gliomas. *Surg Neurol*. November 1974;2(6):419–423.
3. Young HF, Sakalas R, Kaplan AM. Inhibition of cell-mediated immunity in patients with brain tumors. *Surg Neurol*. January 1976;5(1):19–23.
4. Mahaley Jr MS, Brooks WH, Roszman TL, Bigner DD, Dudka L, Richardson S. Immunobiology of primary intracranial tumors. Part 1: studies of the cellular and humoral general immune competence of brain-tumor patients. *J Neurosurg*. April 1977;46(4):467–476.
5. Brooks WH, Roszman TL, Rogers AS. Impairment of rosette-forming T lymphocytes in patients with primary intracranial tumors. *Cancer*. 1976;37(4):1869–1873.
6. Brooks WH, Roszman TL, Mahaley MS, Woosley RE. Immunobiology of primary intracranial tumours. II. Analysis of lymphocyte subpopulations in patients with primary brain tumours. *Clin Exp Immunol*. July 1977;29(1):61–66.
7. Roszman TL, Brooks WH. Immunobiology of primary intracranial tumours. III. Demonstration of a qualitative lymphocyte abnormality in patients with primary brain tumours. *Clin Exp Immunol*. February 1980;39(2):395–402.
8. Fecci PE, Mitchell DA, Whitesides JF, et al. Increased regulatory T-cell fraction amidst a diminished CD4 compartment explains cellular immune defects in patients with malignant glioma. *Cancer Res*. March 15, 2006;66(6):3294–3302.
9. Yovino S, Grossman SA. Severity, etiology and possible consequences of treatment-related lymphopenia in patients with newly diagnosed high-grade gliomas. *CNS Oncol*. November 2012;1(2):149–154.
10. Yovino S, Kleinberg L, Grossman SA, Narayanan M, Ford E. The etiology of treatment-related lymphopenia in patients with malignant gliomas: modeling radiation dose to circulating lymphocytes explains clinical observations and suggests methods of modifying the impact of radiation on immune cells. *Cancer Invest*. February 2013;31(2):140–144.
11. Morford LA, Dix AR, Brooks WH, Roszman TL. Apoptotic elimination of peripheral T lymphocytes in patients with primary intracranial tumors. *J Neurosurg*. December 1999;91(6):935–946.
12. Schietinger A, Greenberg PD. Tolerance and exhaustion: defining mechanisms of T cell dysfunction. *Trends Immunol*. February 2014;35(2):51–60.
13. Grulich AE, van Leeuwen MT, Falster MO, Vajdic CM. Incidence of cancers in people with HIV/AIDS compared with immunosuppressed transplant recipients: a meta-analysis. *Lancet*. July 7, 2007;370(9581):59–67.
14. Dix AR, Brooks WH, Roszman TL, Morford LA. Immune defects observed in patients with primary malignant brain tumors. *J Neuroimmunol*. December 1999;100(1–2):216–232.

15. Roszman TL, Brooks WH, Steele C, Elliott LH. Pokeweed mitogen-induced immunoglobulin secretion by peripheral blood lymphocytes from patients with primary intracranial tumors. Characterization of T helper and B cell function. *J Immunol*. March, 1985;134(3):1545–1550.
16. Elliott LH, Brooks WH, Roszman TL. Activation of immunoregulatory lymphocytes obtained from patients with malignant gliomas. *J Neurosurg*. August 1987;67(2):231–236.
17. Elliott LH, Brooks WH, Roszman TL. Cytokinetic basis for the impaired activation of lymphocytes from patients with primary intracranial tumors. *J Immunol*. March 1984;132(3):1208–1215.
18. Ausiello CM, Palma C, Maleci A, et al. Cell mediated cytotoxicity and cytokine production in peripheral blood mononuclear cells of glioma patients. *Eur J Cancer*. 1991;27(5):646–650.
19. Elliott L, Brooks W, Roszman T. Role of interleukin-2 (IL-2) and IL-2 receptor expression in the proliferative defect observed in mitogen-stimulated lymphocytes from patients with gliomas. *J Natl Cancer Inst*. May 1987;78(5):919–922.
20. Elliott LH, Brooks WH, Roszman TL. Suppression of high affinity IL-2 receptors on mitogen activated lymphocytes by glioma-derived suppressor factor. *J Neurooncol*. September 1992;14(1):1–7.
21. Morford LA, Elliott LH, Carlson SL, Brooks WH, Roszman TL. T cell receptor-mediated signaling is defective in T cells obtained from patients with primary intracranial tumors. *J Immunol*. November 1, 1997;159(9):4415–4425.
22. Roszman TL, Brooks WH, Elliott LH. Immunobiology of primary intracranial tumors. VI. Suppressor cell function and lectin-binding lymphocyte subpopulations in patients with cerebral tumors. *Cancer*. October 1, 1982;50(7):1273–1279.
23. Gershon RK, Kondo K. Infectious immunological tolerance. *Immunology*. December 1971;21(6):903–914.
24. Sakaguchi S, Sakaguchi N, Asano M, Itoh M, Toda M. Immunologic self-tolerance maintained by activated T cells expressing IL-2 receptor alpha-chains (CD25). Breakdown of a single mechanism of self-tolerance causes various autoimmune diseases. *J Immunol*. August 1, 1995;155(3):1151–1164.
25. Suri-Payer E, Amar AZ, Thornton AM, Shevach EM. $CD4^+CD25^+$ T cells inhibit both the induction and effector function of autoreactive T cells and represent a unique lineage of immunoregulatory cells. *J Immunol*. February 1, 1998;160(3):1212–1218.
26. El Andaloussi A, Lesniak MS. An increase in $CD4^+CD25^+FOXP3^+$ regulatory T cells in tumor-infiltrating lymphocytes of human glioblastoma multiforme. *Neuro Oncol*. July 2006;8(3):234–243.
27. Zou JP, Morford LA, Chougnet C, et al. Human glioma-induced immunosuppression involves soluble factor(s) that alters monocyte cytokine profile and surface markers. *J Immunol*. April 15, 1999;162(8):4882–4892.
28. Bodmer S, Strommer K, Frei K, et al. Immunosuppression and transforming growth factor-beta in glioblastoma. Preferential production of transforming growth factor-beta 2. *J Immunol*. November 15, 1989;143(10):3222–3229.
29. Wrann M, Bodmer S, de Martin R, et al. T cell suppressor factor from human glioblastoma cells is a 12.5-kd protein closely related to transforming growth factor-beta. *EMBO J*. June 1987;6(6):1633–1636.
30. El Andaloussi A, Han Y, Lesniak MS. Prolongation of survival following depletion of $CD4^+CD25^+$ regulatory T cells in mice with experimental brain tumors. *J Neurosurg*. September 2006;105(3):430–437.
31. Learn CA, Fecci PE, Schmittling RJ, et al. Profiling of CD4, CD8, and CD4 CD25 CD45RO FoxP3 T cells in patients with malignant glioma reveals differential expression of the immunologic transcriptome compared with T cells from healthy volunteers. *Clin Cancer Res*. 2006;12(24):7306–7315.

32. Torre-Amione G, Beauchamp RD, Koeppen H, et al. A highly immunogenic tumor transfected with a murine transforming growth factor type beta 1 cDNA escapes immune surveillance. *Proc Natl Acad Sci USA*. February 1990;87(4):1486–1490.
33. Roussel E, Gingras MC, Grimm EA, Bruner JM, Moser RP. Predominance of a type 2 intratumoural immune response in fresh tumour-infiltrating lymphocytes from human gliomas. *Clin Exp Immunol*. 1996;105(2):344–352. 1996/8.
34. Wintterle S, Schreiner B, Mitsdoerffer M, et al. Expression of the B7-related molecule B7-H1 by glioma cells: a potential mechanism of immune paralysis. *Cancer Res*. November 1, 2003;63(21):7462–7467.
35. Haura EB, Turkson J, Jove R. Mechanisms of disease: Insights into the emerging role of signal transducers and activators of transcription in cancer. *Nat Clin Pract Oncol*. June 2005;2(6):315–324.
36. Brantley EC, Benveniste EN. Signal transducer and activator of transcription-3: a molecular hub for signaling pathways in gliomas. *Mol Cancer Res*. May 2008;6(5):675–684.
37. Carro MS, Lim WK, Alvarez MJ, et al. The transcriptional network for mesenchymal transformation of brain tumours. *Nature*. January 21, 2010;463(7279):318–325.
38. Buettner R, Mora LB, Jove R. Activated STAT signaling in human tumors provides novel molecular targets for therapeutic intervention. *Clin Cancer Res*. April 2002;8(4):945–954.
39. Inda MM, Bonavia R, Mukasa A, et al. Tumor heterogeneity is an active process maintained by a mutant EGFR-induced cytokine circuit in glioblastoma. *Genes Dev*. August 15, 2010;24(16):1731–1745.
40. Yu H, Kortylewski M, Pardoll D. Crosstalk between cancer and immune cells: role of STAT3 in the tumour microenvironment. *Nat Rev Immunol*. January 2007;7(1):41–51.
41. Wang T, Niu G, Kortylewski M, et al. Regulation of the innate and adaptive immune responses by Stat-3 signaling in tumor cells. *Nat Med*. January 2004;10(1):48–54.
42. Fujita M, Zhu X, Sasaki K, et al. Inhibition of STAT3 promotes the efficacy of adoptive transfer therapy using type-1 CTLs by modulation of the immunological microenvironment in a murine intracranial glioma. *J Immunol*. February 15, 2008;180(4):2089–2098.
43. de la Iglesia N, Konopka G, Lim K-L, et al. Deregulation of a STAT3-interleukin 8 signaling pathway promotes human glioblastoma cell proliferation and invasiveness. *J Neurosci*. June 4, 2008;28(23):5870–5878.
44. de la Iglesia N, Konopka G, Puram SV, et al. Identification of a PTEN-regulated STAT3 brain tumor suppressor pathway. *Genes Dev*. February 15, 2008;22(4):449–462.
45. Boyland E, Williams DC. The estimation of tryptophan metabolites in the urine of patients with cancer of the bladder. *Biochem J*. March 19, 1955;60 (Annual General Meeting):v.
46. Friberg M, Maria F, Ronald J, et al. Indoleamine 2,3-dioxygenase contributes to tumor cell evasion of T cell-mediated rejection. *Int J Cancer*. 2002;101(2):151–155.
47. Nakamura T, Takafumi N, Tomoko S, et al. Expression of indoleamine 2, 3-dioxygenase and the recruitment of Foxp3-expressing regulatory T cells in the development and progression of uterine cervical cancer. *Cancer Sci*. 2007;98(6):874–881.
48. Wainwright DA, Chang AL, Dey M, et al. Durable therapeutic efficacy utilizing combinatorial blockade against IDO, CTLA-4, and PD-L1 in mice with brain tumors. *Clin Cancer Res*. 2014;20(20):5290–5301.
49. Li M, Bolduc AR, Hoda MN, et al. The indoleamine 2,3-dioxygenase pathway controls complement-dependent enhancement of chemo-radiation therapy against murine glioblastoma. *J Immunother Cancer*. July 7, 2014;2:21.
50. Wainwright DA, Balyasnikova IV, Chang AL, et al. IDO expression in brain tumors increases the recruitment of regulatory T cells and negatively impacts survival. *Clin Cancer Res*. 2012;18(22):6110–6121.
51. Mitsuka K, Kentaro M, Tomoyuki K, et al. Expression of indoleamine 2,3-dioxygenase and correlation with pathological malignancy in gliomas. *Neurosurgery*. 2013;72(6):1031–1039.

52. Berghoff AS, Kiesel B, Widhalm G, et al. Programmed death ligand 1 expression and tumor-infiltrating lymphocytes in glioblastoma. *Neuro Oncol*. August 2015;17(8):1064–1075.
53. Sharma P, Allison JP. The future of immune checkpoint therapy. *Science*. April 3, 2015;348(6230):56–61.
54. Hodi FS, O'Day SJ, McDermott DF, et al. Improved survival with ipilimumab in patients with metastatic melanoma. *N Engl J Med*. August 19, 2010;363(8):711–723.
55. Sunshine J, Taube JM. PD-1/PD-L1 inhibitors. *Curr Opin Pharmacol*. August 2015;23:32–38.
56. Morse MA, Hobeika AC, Osada T, et al. Depletion of human regulatory T cells specifically enhances antigen-specific immune responses to cancer vaccines. *Blood*. August 1, 2008;112(3):610–618.
57. Dannull J, Su Z, Rizzieri D, et al. Enhancement of vaccine-mediated antitumor immunity in cancer patients after depletion of regulatory T cells. *J Clin Invest*. December 2005;115(12):3623–3633.
58. Attia P, Maker AV, Haworth LR, Rogers-Freezer L, Rosenberg SA. Inability of a fusion protein of IL-2 and diphtheria toxin (Denileukin Diftitox, DAB389IL-2, ONTAK) to eliminate regulatory T lymphocytes in patients with melanoma. *J Immunother*. November–December 2005;28(6):582–592.
59. Powell Jr DJ, Felipe-Silva A, Merino MJ, et al. Administration of a CD25-directed immunotoxin, LMB-2, to patients with metastatic melanoma induces a selective partial reduction in regulatory T cells in vivo. *J Immunol*. October 1, 2007;179(7):4919–4928.
60. Kohm AP, McMahon JS, Podojil JR, et al. Cutting Edge: Anti-CD25 monoclonal antibody injection results in the functional inactivation, not depletion, of $CD4^+CD25^+$ T regulatory cells. *J Immunol*. March 15, 2006;176(6):3301–3305.
61. Fecci PE, Sweeney AE, Grossi PM, et al. Systemic anti-CD25 monoclonal antibody administration safely enhances immunity in murine glioma without eliminating regulatory T cells. *Clin Cancer Res*. July 15, 2006;12(14 Pt. 1):4294–4305.
62. Jacobs JF, Punt CJ, Lesterhuis WJ, et al. Dendritic cell vaccination in combination with anti-CD25 monoclonal antibody treatment: a phase I/II study in metastatic melanoma patients. *Clin Cancer Res*. October 15, 2010;16(20):5067–5078.
63. Sanchez-Perez LA, Choi BD, Archer GE, et al. Myeloablative temozolomide enhances $CD8^+$T-cell responses to vaccine and is required for efficacy against brain tumors in mice. *PLoS One*. March 18, 2013;8(3):e59082.
64. Dudley ME, Wunderlich JR, Robbins PF, et al. Cancer regression and autoimmunity in patients after clonal repopulation with antitumor lymphocytes. *Science*. October 25, 2002;298(5594):850–854.
65. Dummer W, Niethammer AG, Baccala R, et al. T cell homeostatic proliferation elicits effective antitumor autoimmunity. *J Clin Invest*. July 2002;110(2):185–192.
66. Asavaroengchai W, Kotera Y, Koike N, Pilon-Thomas S, Mulé JJ. Augmentation of antitumor immune responses after adoptive transfer of bone marrow derived from donors immunized with tumor lysate-pulsed dendritic cells. *Biol Blood Marrow Transplant*. August 2004;10(8):524–533.
67. Sampson JH, Aldape KD, Archer GE, et al. Greater chemotherapy-induced lymphopenia enhances tumor-specific immune responses that eliminate EGFRvIII-expressing tumor cells in patients with glioblastoma. *Neuro Oncol*. March 2011;13(3):324–333.
68. Mitchell DA, Cui X, Schmittling RJ, et al. Monoclonal antibody blockade of IL-2 receptor alpha during lymphopenia selectively depletes regulatory T cells in mice and humans. *Blood*. September 15, 2011;118(11):3003–3012.
69. Sampson JH, Schmittling RJ, Archer GE, et al. A pilot study of IL-2Ralpha blockade during lymphopenia depletes regulatory T-cells and correlates with enhanced immunity in patients with glioblastoma. *PLoS One*. 2012;7(2):e31046.
70. Platten M, Wick W, Weller M. Malignant glioma biology: role for TGF-beta in growth, motility, angiogenesis, and immune escape. *Microsc Res Tech*. February 15, 2001;52(4):401–410.

71. Bogdahn U, Hau P, Stockhammer G, et al. Targeted therapy for high-grade glioma with the TGF-β2 inhibitor trabedersen: results of a randomized and controlled phase IIb study. *Neuro Oncol*. January 2011;13(1):132–142.
72. Rodon J, Carducci MA, Sepulveda-Sanchez JM, et al. First-in-human dose study of the novel transforming growth factor-β receptor I kinase inhibitor LY2157299 monohydrate in patients with advanced cancer and glioma. *Clin Cancer Res*. 2014;21(3):553–560. February 1, 2015.

CHAPTER

3

The Role of Regulatory T Cells and Indoleamine-2, 3-dioxygenase in Brain Tumor Immunosuppression

M. Dey, A.L. Chang, J. Miska, J. Qiao, M.S. Lesniak

The University of Chicago, Chicago, IL, United States

OUTLINE

Translational Immunotherapy of Brain Tumors
http://dx.doi.org/10.1016/B978-0-12-802420-1.00003-X

BACKGROUND AND DISCOVERY

One of the biggest hurdles in the way of developing effective anti-glioma therapeutics is the profound immunosuppression associated with malignant gliomas (MGs). All the therapeutic agents currently in use for the treatment of MG, one of the most common primary adult brain tumors, have 100% failure rate. Even though there can be considerable immune cell infiltrate in MG, the environment is largely immunosuppressive, thus stymieing potential antitumor responses. Several factors contribute to this profound immune suppression, including (1) immunosuppressive cells such as regulatory T cells (Tregs),[1–3] tumor-associated myeloid-derived suppressor cells (MDSCs),[4,5] plasmacytoid dendritic cells (pDCs),[6,7] etc.; (2) cytokines such as TGF-β,[8] IL-10,[9] etc.; and (3) enzymes such as heme-oxygenase-1 (HO1),[1,10] indoleamine-2,3-dioxygenase (IDO),[11,12] etc. Perhaps the most studied of these mechanisms are Tregs and factors that regulate their function.

ORIGIN OF TREGS

Tregs were discovered as a subset of $CD4^+$ T cells with the potent ability to suppress T cell proliferation.[13] Although this finding was considered dubious for a number of years, the discovery of a master regulatory factor for Treg function, Foxp3, cemented the existence of this population of cells.[14] After these seminal discoveries, widespread effort was put in to determine: the origin of these suppressive cells, their functional mechanisms of immunosuppression, and how they could be dysfunctioning in disease.

Based on their origin, Tregs are divided into two subsets: natural Tregs (nTregs) and induced Tregs (iTregs). nTregs undergo the same "education" processes that all T cells go through when developing in the thymus. What is thought to determine their fate during selection is based on the "affinity window" hypothesis. This refers to the range of affinities between a developing thymocyte's T cell receptor (TCR) and a peptide–major histocompatibility complex (MHC) on thymic antigen presenting cells (APCs).[15] If the affinity of the TCR-MHC interaction is too high (or too low), then the thymocyte gets removed; however, there is an optimum range of affinities that are permissive to T cell development. Where the TCR affinity lies within this "window" determines whether or not the thymocyte will develop into a nTreg.[16] This theory is not absolute; however, as yet other undefined mechanisms are thought to be at play.[17] These thymic-derived cells are considered to be the majority of Tregs.

In addition, under several circumstances, T cells in the periphery can also differentiate into subsets of Tregs called iTregs. The notion that peripheral $CD4^+$ T cells could be pushed to a Treg lineage was first demonstrated in vitro.[18] Chen et al. demonstrated that polarizing conditions of transforming growth factor beta (TGF-β) treatment during TCR stimulation could "convert" a naïve T cell into a $Foxp3^+$ suppressor T cell. It was later determined that interleukin (IL)-2 and retinoic acid are important accessory signals for the generation of these iTregs.[19] This was eventually validated using TGF-β knockout mice,[20] and anti–TGF-β antibodies,[21] justifying the relevance of this phenomenon. In the next section, we will describe the wide variety of suppressive modalities that Tregs use to suppress immune responses.

FUNCTION OF TREGS

Determination of Foxp3 as a lineage-specific marker provided a platform to understand how mechanistically Tregs suppress their targets. Tregs boast an impressive array of mechanisms to suppress effector responses, all of which have some utility depending on the setting. Since their in vitro functionality was first described as being contact dependent, perforin and granzyme expression (contacted mediated cytolysis of target cells) by Tregs was one of the initial findings.[22] This was demonstrated to have a role in tumor microenvironment as well, highlighting the in vivo relevance of this pathway.[23] Cytotoxic T lymphocyte antigen-4 (CTLA-4) is a pioneering marker of Treg function[24] and is constitutively expressed by Tregs. Their ablation on the regulatory compartment alone causes lethal autoimmunity in mice.[25] This is thought to influence T cell:DC interactions as well as Treg:Teff (effector T cells) interactions, which is outside the scope of this chapter.

Potently immunosuppressive cytokines such as TGF-β[26] and IL-10[27] are also produced by Tregs, enhancing their suppressive capabilities. Other potential mechanisms include physical restraint of effector cells,[28] action as an IL-2 sink,[29] the production of extracellular AMP through Treg CD39/73 (a pair of ectonucleotidase enzymes) expression,[30] and many others. These probably serve as redundant mechanisms and perhaps suggest different roles based on location (i.e., the draining lymph node vs. the target tissue). As one might expect, in the context of malignancy, hijacking these cells to suppress antitumor responses has become a central tenant in tumor-induced immunosuppression.

TREGS IN THE CANCER IMMUNITY CYCLE

Historically, the immune system was thought to be ignorant of tumors. However, in the past two decades, it has become abundantly clear that the immune system is constantly surveilling for malignancies, and is entirely capable of eliminating them.[31] Theoretically, the release of danger signals by dying tumor cells, the activation of antigen presentation, and the recruitment of immune cells to the malignancy should result in an immunity "cycle" that can effectively remove tumors from our bodies.[32] Although this is working in many instances, in many other instances tumors do grow out, and a number of mechanisms have been elucidated for how the tumor prevents its own destruction. Of these multitudinous mechanisms (for those interested, the *Hallmarks of Cancer* article[33] is suggested for further reading), we will focus here on the role of Tregs and their role in suppressing antitumor immunity.

The first clue that Tregs might be important in cancer was their preferential abundance within many different malignancies. They were found to be enriched within head and neck squamous cell carcinoma (HNSCC),[34] colorectal cancer,[35] liver cancers,[36] gastric cancers,[37] melanoma,[38] glioma (which we will discuss further) (Fig. 3.1),[39] just to name a few. This widespread utilization of Tregs highlights their central role in the suppression of antitumor immune responses.

Mechanistically, there have been some explanations as to why Tregs are found enriched within the tumor microenvironment. Differential chemokine receptor expression by Tregs, such as C-C chemokine receptor type 2 (CCR2), 4, and 8 causes their preferential recruitment to the tumor microenvironment.[40] One explanation of this phenomenon is that tumor-derived chemokines selectively recruit Tregs to the tumor site.[37] Tumors are also well-known to recruit and co-opt immunosuppressive myeloid cells, such as MDSCs, to produce chemokines that will increase the recruitment of Tregs.[41]

Other than recruitment, Treg intrinsic properties allow for their selective abundance within tumor tissues. TGF-β, which is produced by both tumor-associated macrophages and Tregs themselves, serves not only to enhance Treg numbers within the tumors, but also to suppress the effector responses of T cells that do make it to the tissue.[42] Tregs also have different metabolic needs that might influence their survival within tumor tissue,[43] which is a burgeoning field of current research. Further, Tregs have differential expression of genes that allow for their survival in a hypoxic environment, such as HO1,[10] or hypoxic inducible factor.[44,45] These are just some of the mechanisms that can explain the abundance of Tregs within tumor tissues and can prime us for a discussion of how these factors play in glioma.

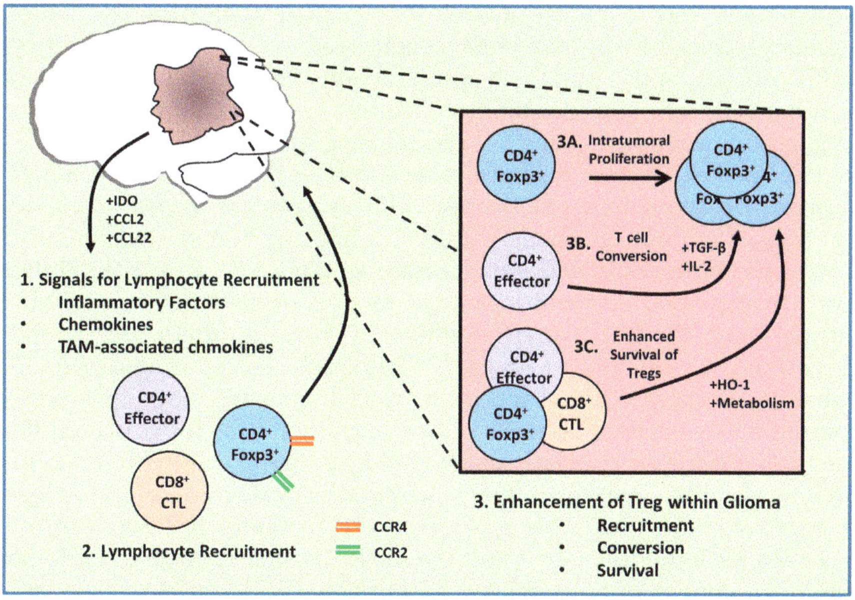

FIGURE 3.1 Factors involved in regulatory T cell (Treg) recruitment in glioma. A number of mechanisms participate in the accumulation of Tregs within the glioblastoma microenvironment. (1) In the innate immune response to malignancy a number of cytokines and chemokines that attract lymphocytes to the tumor microenvironment are released. (2) There are studies suggesting that Tregs can be preferentially recruited to the tumor site through differential chemokine receptor expression and other soluble factors. (3) Within the glioma microenvironment a number of potential mechanisms have been posited that can also explain the preponderance of Tregs such as the intratumoral proliferation of Tregs, the conversion of Naïve $CD4^+$ T cells into Tregs, and the intrinsic ability of Tregs to survive the harsh glioma microenvironment. All of these mechanisms are likely important in the accumulation of Tregs seen within the tumors of many glioma patients.

TREGS AND GLIOMA

Up until very recently the notion of lymphatic drainage by the brain was a tenuous notion;[46] therefore, T cell activation and recruitment in the context of glioma was a difficult phenomenon to explain. However, before this breakthrough was made, it was discovered that Tregs accumulate within glioma tissue in both humans and mouse models.[39,47,48] In murine models of glioma, Treg accumulation was robust within these tumors, and ablation of them caused a significant increase in animal survival.[49] The reason for this accumulation shares many aspects of other established tumor models. Differential chemokine production, such as C-C motif chemokine ligand 2 (CCL2) and CCL22 by the glioma (and associated macrophages/microglia), selectively recruits Tregs in both in vitro human experiments and in mouse models.[49–51] Furthermore, different soluble factors produced by glioma enhance survival and proliferation of Tregs while concomitantly causing apoptosis in effector cells.[52] The conversion of naïve cells into iTregs at the tumor tissue is still a controversial topic, with groups both supporting[52,53] and refuting this notion[2] in glioma.

Treg intrinsic factors may also play a role in their survival in the glioma microenvironment. For example, HO1 is a protein enzyme expressed by Tregs that protects them within the hypoxic tumor microenvironment. It is both correlated with disease progression[1] and its inhibition leads to increased survival in animal models of glioblastoma (GBM).[10] Furthermore, the hypoxic environment within the tumor can actually drive the development of Tregs.[44,54] Although animal models provide valuable information about specific aspects of glioma biology, only with a comprehensive genetic analysis of human tumors was it discovered that four unique tumor subtypes exist: classical, neural, mesenchymal, and proneural.[55] Each subtype differs in its hypothetical cell of origin, its level of aggressiveness, and what type of milieu surrounds it. Interestingly, it has been recently discovered that the mesenchymal subtype of GBM has the most amount of immune cell recruitment, corresponding to a large influx of Tregs in this tumor tissue.[56] These data suggest that these cells play a differential role in gliomagenesis depending on the glioma subtype, and thus targeting them in certain groups of patients will be more beneficial compared to others.

TRYPTOPHAN CATABOLISM IN IMMUNITY AND IN CANCER

One of the rapidly emerging fields of cancer biology is focused on the metabolic needs and regulation of tumor cells. The catabolism of tryptophan (Trp) to *N*-formylkynurenine (Kyn) (Fig. 3.2) is a major mode of regulation for both innate and adaptive immune responses. This metabolic immune regulation proceeds through the oxidative catabolism of Trp along the Kyn pathway. Three major enzymes can catalyze the degradation of Trp to Kyn: IDO1, IDO2, and tryptophan-2,3-dioxygenase (TDO). The IDO pathway comprises the three enzymes that degrade Trp, the immediate metabolite of Kyn, and metabolic products that exist downstream of Kyn (e.g., quinolinic acid and picolinic acid). Trp depletion triggers amino acid sensing pathways to limit immune responses within effector cells.[57] Additionally, Kyn is an endogenous ligand for the aryl hydrocarbon receptor (AhR), which mediates downstream immunomodulatory functions.[58] The AhR was classically identified as a receptor for environmental toxins, but its role in immunomodulation has been

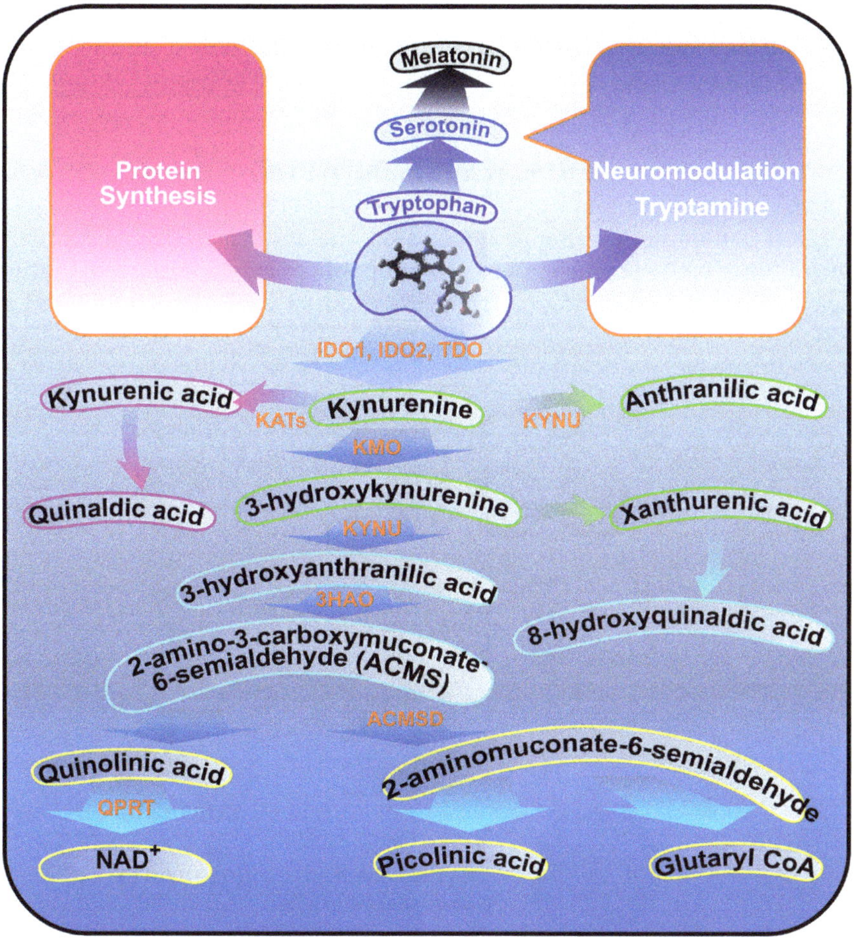

FIGURE 3.2 The kynurenine pathway. The essential amino acid L-tryptophan can be used in protein synthesis, converted into tryptamine, or shunted into serotonin synthesis pathway. The kynurenine pathway is initiated by the conversion of tryptophan to N-formylkynurenine by either indoleamine-2,3-dioxygenase 1 (IDO1), IDO2, or tryptophan-2,3-dioxygenase. Three downstream pathways can then act on kynurenine. The irreversible transamination to the product kynurenic acid occurs through a family of kynurenine aminotransferases. Kynurenase (KYNU) further degrades kynurenine to anthranilic acid, which can reenter the kynurenine pathway further downstream after conversion to 3-hydroxyanthranilic acid through non-specific hydroxylation. The enzyme kynurenine-3-monooxygenase converts kynurenine to 3-hydroxykynurenine, which is then degraded to 3-hydroxylanthranilic acid by KYNU activity. 3-Hydroxylanthranilic acid is a substrate for 3-hydroxyanthranilate oxidase, which results in the production of 2-amino-3-carboxymuconate-6-semialdehyde (ACMS). ACMS undergoes rearrangement to quinolinic acid, which can be used for NAD^+ synthesis through quinolate phosphoribosyltransferase. Alternatively, 2-amino-3-carboxymuconate-6-semialdehyde-decarboxylase can continue the catabolic cascade of ACMS to picolinic acid or to glutaryl-coenzyme A (CoA).

recently appreciated. The capacity of the IDO pathway for attenuating ongoing immune responses is dysregulated in the context of cancer and is used as a powerful resistance mechanism against antitumor immunity.

IDO1, IDO2, AND TDO

IDO1 is the rate-limiting enzyme in the metabolism of Trp and among one of three well-established enzymes capable of catalyzing the oxidative cleavage of the indole ring of Trp to the product Kyn[59] (Fig. 3.3). IDO1 is

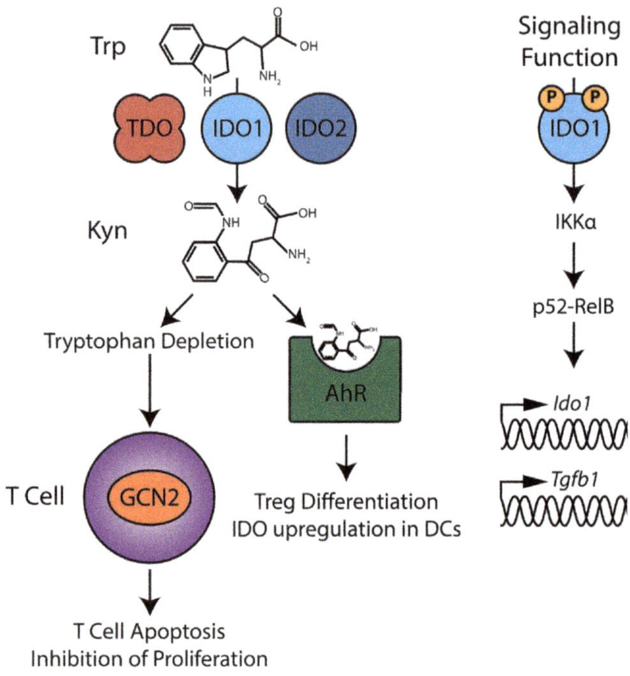

FIGURE 3.3 Tryptophan (Trp) is catabolized into N-formylkynurenine (Kyn) by three major enzymes, indoleamine-2,3-dioxygenase 1 (IDO1), IDO2, and tryptophan-2,3-dioxygenase (TDO). TDO does not share sequence or structural homology with the two closely related IDO genes and functions as a tetramer rather than a monomer. The effects of tryptophan catabolism are mediated through both tryptophan depletion and binding of kynurenine to the aryl hydrocarbon receptor (AhR). Tryptophan depletion is sensed in T cells by the general control nonderepressible 2 (GCN2) kinase, which results in the inhibition of T cell proliferation and the induction of apoptosis. Upon binding of kynurenine, the AhR translocates to the nucleus and acts as a transcription factor. Downstream effects of AhR activity include supporting Treg differentiation and the induction of IDO expression in dendritic cells. IDO1 also has signaling activity independent of its metabolic activity. IDO1 can be phosphorylated at two immunoreceptor tyrosine-based inhibitory motifs, which signals downstream through the IKKα pathway, increasing the transcription of IDO and TGF-β.

expressed in a wide range of cell types, including APCs such as DCs or macrophages, epithelial cells, endothelium, and some tumor cells.[60] In the late 1970s, early work identified IDO1 as the first interferon-activated gene to be described.[61] The expression of IDO1 is strongly induced by interferons in monocytes and some stromal cell types.[62–64] Interferon-γ (IFN-γ)-dependent induction of IDO1 expression in monocytes requires the activity of signal transducer and activator of transcription 1 (STAT1) and interferon regulatory factor 4 (IRF4). A separate study demonstrated that the upregulation of IDO1 in splenic $CD19^+$ DCs could be induced by the binding of the costimulatory molecules CD80/86 by soluble CTLA4-Ig.[65] In addition, stimulation of Toll-like receptor 9 (TLR9) by CpG oligonucleotides also resulted in the upregulation of IDO1 in the same $CD19^+$ splenic DC population.[66] Both the TLR9 and CD80/86 modes of IDO1 regulation were dependent on the production of IFN-α. One of the earliest demonstrations of the role for IDO1 in immunity was in the context of preventing allogeneic fetal rejection in mice. Treatment of mice pregnant with allogeneic conception with the IDO inhibitor 1-methyltryptophan (1-MT) resulted in rejection, suggesting that IDO activity was critical for prevention of maternal lymphocyte–mediated fetal rejection.[67] IDO1 can also act as a signaling molecule independent of its enzymatic activity through the recruitment of SHP1 and SHP2. This complex is required for the sustained tolerogenic effect of TGF-β on the pDC phenotype.[68]

Trp catabolism can also be accomplished by IDO2, a gene closely related to IDO1 that is located near the *IDO1* gene on chromosome 8 in both humans and mice.[69–73] The expression patterns of IDO1 and IDO2 exhibit little overlap between tissues. The expression of IDO1 is predominantly localized to the colon and epididymis, whereas IDO2 expression is primarily found in the cerebral cortex, liver, and kidney. Furthermore, the two enzymes may be regulated independently, as the expression of full-length IDO2 is not induced by IFN-γ in mouse mesenchymal stem cells.[64] IDO2 can be regulated by certain inflammatory stimuli such as pathogen challenge; both IDO2 and IDO1 were upregulated in the brain during *Toxoplasma gondii* infection.[74] The K_m for IDO2 is almost 100-fold higher than the physiological concentration of Trp, which suggests that IDO2 does not play a primary role for Trp degradation. Additional support for the reduced role of IDO2 can be found in two polymorphisms in the IDO2 that occur in humans, R248W and Y359X. Despite the fact that the R248W polymorphism encodes IDO2 with drastically reduced enzyme activity and the Y359X polymorphism encodes a truncated nonfunctional protein, these polymorphisms are present in up to 50% of the Caucasian population.[70] However, IDO2 does play a role in certain immune contexts, including Treg generation and as a mediator of autoimmune arthritis in a murine model.[65,75–78]

In the liver, the conversion of Trp to Kyn can occur due to the activity of TDO, an enzyme structurally unrelated to IDO in terms of primary and quaternary protein structure. In contrast to the monomeric structures of IDO1 and IDO2, TDO functions as a tetramer.[72] TDO functions as the main enzyme responsible for dietary Trp metabolism and is positively regulated by the presence of Trp in the blood. There is also emerging evidence that TDO is expressed in the brain, where it may function as a regulator of neuroactive kynurenine pathway metabolites.[79] In parallel with the role of TDO in the liver, expression of TDO can be found at high levels in hepatocellular carcinomas. Constitutive expression of TDO has also been described in several cancer cell lines spanning GBM, colorectal carcinoma, head-and-neck carcinoma, and gallbladder carcinoma origins.[80,81]

THE AHR

The AhR is a ligand-activated transcription factor that was originally described as a receptor for 2,3,7,8-tetrachlorodibenzo-*p*-dioxin (TCDD) and additional polycyclic aromatic hydrocarbons.[82] Upon ligand binding, cytoplasmic AhR translocate to the nucleus and promotes the transcription of cytochrome P450 genes, many of which encode enzymes that primarily metabolize toxins. The first link of AhR activity to immunity was established through observations that TCDD exposure resulted in thymic involution and immunosuppression.[83] When an immunoregulatory role was described for the kynurenine products from tryptophan metabolism, the AhR was investigated as a potential receptor for kynurenines since indoles and other Trp derivatives were known as potential ligands for the AhR. Indeed, kynurenine was found to be an endogenous ligand for the AhR in T cells, an interaction that is critical for Treg generation in vitro.[84,85] AhR ligands also regulate the expression of IDO in DCs and play an immunoregulatory role in this cellular population as well.[85,86]

MECHANISMS OF IMMUNE REGULATION BY THE IDO PATHWAY

Regulation of the immune system by the IDO pathway is thought to occur through two main mechanisms: tryptophan depletion and the downstream activity of kynurenine pathway metabolites. T cells sense the depletion of Trp through uncharged tRNAs, which activate the general control nonderepressible 2 (GCN2) kinase, an effector kinase that leads to cell cycle arrest and the induction of anergy.[87] GCN2 phosphorylates the initiation factor eIF-2α, effectively blunting protein translation for most messenger RNA (mRNA) transcripts. However, the translation of certain transcripts becomes favored during this process, such as the immunoregulatory transcription factor liver inhibitory protein (LIP), which induces the production of IL-6 and

CCL2.[88] The effects of Trp depletion on T cells also involve the abrogation of T cell proliferation through activation of the mechanistic target of rapamycin complex (mTOR) pathway.[89] Here, the catabolism of Trp by IDO inhibits mTOR complex 1 (mTORC1), ultimately triggering autophagy in the Trp-starved T cells. The role of L-Kyn as an endogenous AhR ligand was further examined in detail in a study that found a requirement for AhR, functional IDO1, and TGF-β in a model of tolerance to ligand lipopolysaccharide (LPS) rechallenge.[90] In this study, the authors found that IDO pathway–dependent tolerance to LPS rechallenges was critical for immunoprotection against *Salmonella typhimurium* infection and group B *Streptococcus*–induced septic arthritis. The downstream activity of the AhR is highly context dependent: certain Ahr ligands are capable of enhancing the generation of Tregs or supporting T helper 17 cell (Th17) differentiation.[91,92]

THE IDO PATHWAY IN CANCER AND IN GLIOMA

Elevated tryptophan catabolism has been linked to cancer since early observations in cancer patients during the 1950s.[93] The overexpression of IDO1 has been found in a wide spectrum of cancer types and is associated with poor prognosis.[94–96] IDO1 overexpression in cancer is potentially caused by the loss of the tumor suppressor Bin1, a tumor suppressor that is commonly attenuated during cancer progression.[97] Further studies in mouse models of lung cancer and metastatic breast cancer found that IDO drives cancer progression through affecting the inflammatory microenvironment.[98] In the context of cancer, the immunosuppressive properties of the IDO pathway can be due to its activation in the cancer cells or in cells within the tumor microenvironment. Three patterns for IDO1 immunohistochemical staining were found in a large study covering 866 tumors of 15 cancer types: tumor cells (20% of samples), interstitial cells located in lymphocyte-infiltrated sections of the tumor stroma (47% of samples), and endothelial cells (14% of samples) observed both individually as well as in combination.[95] IDO1 expression in both tumor and host cells was found to be a resistance mechanism against checkpoint blockade immunotherapy.[99] In gliomas, a role for TDO, Kyn, and the AhR was identified in promoting tumor cell motility and survival in addition to suppressing antitumor T cell responses.[80] In this study, overexpression of the AhR was also found to correlate with poor prognosis across several different cancers. Our own work has shown that IDO contributes to the accumulation of Tregs in brain tumors as part of the highly immunosuppressive glioma microenvironment.[11] Given the IFN-dependent regulation of IDO1, cancers may express IDO1 in response to IFN-γ produced by activated lymphocytes within the tumor microenvironment. However, tumors can also acquire IDO1 expression independent of inflammation, potentially during oncogenic transformation as a defense mechanism against the immune system and as a mechanism that promotes tumorigenesis. Particularly in the brain, it appears that the Trp metabolism may exist as part of central nervous system (CNS) "immune privilege", a way of preventing ectopic immune responses from destroying CNS tissue.[80,100,101] Trp metabolism in the IDO pathway could be co-opted by the developing brain tumor and amplified as a potent mechanism for immune evasion.

TARGETING TREGS IN PRECLINICAL MODEL

Since Tregs have been shown to infiltrate and induce immunosuppression, even in early neoplastic development[102,103] different strategies have been explored to therapeutically deplete and functionally inactivate Tregs in vivo to enhance antitumor immunity and immunotherapy.[104–107] In tumor-bearing mice, CD25 monoclonal antibody (mAb) can be used to deplete CD25-positive Tregs to promote immune surveillance and enhance host tumor immunity and efficacy of immunotherapies.[104–106] CD25 mAb not only decreases the frequency of tumor-infiltrating $CD4^{+}CD25^{+}$ T cells, but also eliminates their suppressor activity and inhibition of Treg function resulting in enhanced lymphocyte proliferation and IFN-γ production with as much as 80% lysis of glioma cells in vitro.[108] Our group along with several others have shown that in GL261 cell–based murine model of glioma, mice pretreated with CD25 mAb lived significantly longer than those bearing tumor and receiving control IgG antibody.[109] Furthermore, compared to the only systemic administration of CD25 mAb, combining intraperitoneal and intracranial administration of CD25 mAb further enhanced long-term survival, led to complete tumor rejection.[108,110] Treg depletion with CD25 mAb has also been shown to strongly enhance the efficacy of DC vaccination.[111] However, one of the major limitations of targeting CD25 is that it is also expressed by activated effector T cells, or even transiently expressed by myeloid DCs upon activation.[112,113] Therefore, a CD25 antibody–based approach may result in the depletion of other immune cells that carry out antitumor responses.[114]

Since Foxp3 is the only gene product known to be almost exclusively expressed by Tregs in mice, it logically emerged as a prime target for Treg depletion. CD25 mAb mediated versus Foxp3 targeted Treg depletion has slightly different effects in tumor-bearing mice. CD25 mAb depletes Treg systemically; whereas Foxp3 vaccination leads to Treg depletion intratumorally with sparing of Treg in the periphery.[115] Foxp3 in Tregs can be targeted by vaccination of mice with Foxp3 mRNA-transfected DCs, which elicits a robust Foxp3-specific T cell response and vaccine-associated protective immunity. Vaccine-programmed effector T cells can differentially target aberrantly programmed, phenotypically distinct intratumoral Tregs.[11,116–118] Another strategy for targeting Foxp3 in Tregs utilizes a synthetic peptide, P60, which directly binds to Foxp3 and inhibits Foxp3 nuclear translocation, decreasing its ability to suppress the transcription of nuclear factor kappa B and nuclear factor of activated T cells (NFAT). P60 is effective in inducing immune-mediated protection against tumor implantation in preclinical murine model.[119] Although this has been a promising strategy in the preclinical models, it has yet to be tested in the clinical setting.

One of the most common and well-described side effects of chemotherapeutic agents is lymphopenia. Interestingly, several chemotherapeutic agents have been shown to cause preferential depletion of Tregs in a time-dependent or dose-dependent manner. A widely used DNA-alkylating agent cyclophosphamide (CTX) eliminates both regulatory and effector T cells in the vaccination model; however, ultimately enhances tumor-specific vaccination due to an earlier reconstitution of effector T cells than that of Tregs.[107] Furthermore, a low dose of CTX (also known as metronomic CTX) has been shown to induce Treg apoptosis without interfering with effector $CD4^+$ T cell viability,[120–122] expand antigen-specific tumor-reactive T cells, and restore DC homeostasis.[123] CTX-induced Treg suppression is found to be a result of imbalance among various DC compartments, preferential depletion of $CD8^+$ lymphoid–resident DC and increased migratory, cytokine production, and antigen presentation capacity of DC.[124] Combination of IL-12 and CTX has been shown to eliminate not only intratumoral Tregs but also MDSCs, while simultaneously inducing proinflammatory myeloid cells within the tumor microenvironment, leading to colon cancer rejection in the mouse model.[125] Similar to the effect of CTX, decrease in the total number and frequency of Treg in the peripheral blood, and simultaneous increase in serum IFN-α concentrations, was also reported in a canine cancer model.[126,127] Depletion of Tregs can restore natural killer (NK) cell function in the setting of malignancy, as selective Treg depletion using low-dose CTX has been shown to restore NK tumoricidal activity.[122] Intratumoral administration of mAb against OX40, a costimulatory receptor from the tumor necrosis factor (TNF) family, increases NK activity through inactivation of Treg cells, resulting in rejection of tumors in mice.[128] Another widely used chemotherapeutic agent, temozolomide, the first-line agent used for MG, is also known for its ability to decrease Tregs and is discussed in detail under the clinical trial section.

With the advancement in the understanding of the molecular and cellular basis for the development, activation, and function of Treg cells, the strategies that can manipulate Treg function have also been developed. TGF-β has been shown to promote the proliferation of nTreg cells and also play a key role in the induction of naïve $CD4^+$ T helper cells into induced (iTreg) cells.[129] Enhanced antitumor immunity can be achieved by use of TGF-β antibody in the treatment of tumors.

Another potential way to modulate Tregs is through the suppressive molecule CTLA-4.[25] Engagement of CTLA-4 on T cells with its ligand on

DCs induces the activation of Tregs, prevents effector T cell activation, hinders DC function as well as the generation of Tregs.[130] CTLA-4 is not only associated with glioma progression and prognosis, but CTLA-4 polymorphisms are also potential clinically relevant biomarkers for identifying risk of developing glioma.[131] In the setting of DC vaccines, combining Treg depletion with CTLA-4 neutralization results in partial tumor rejection by glioma-specific cytotoxic T cells and Tconv effector T cell responses, while also increasing antiglioma antibody titers.[132] CTLA-4 neutralizing mAb, 9H10, has been used successfully in preclinical glioma model and has been shown to confer long-term survival benefit with re-establishment of normal CD4 counts concomitant with a decreased Treg fraction.[133] In tumors, PD-L1 expressing DCs have been also reported to induce Tregs and to promote the proliferation of Tregs.[134] Furthermore, in other malignancies, combining targeting of CTLA4 and/or PD-L1 to impair Treg induction and expansion, results in even more effective antitumor immune responses.[135] In addition, combining antibodies against CTLA-4 and PD-1[136,137] with chemotherapeutic agents such as lenalidomide and pomalidomide[138] promotes natural killer cell activation through inhibition of proliferation and function of Treg cells to enhance antitumor effects.

So far, a lot of research effort is currently dedicated to the development of strategies to specifically deplete and modulate the function of Treg cells in cancer. However, the removal of Treg cells may also generate pathologic changes such as autoimmune sequelae. Further understanding of the mechanisms of interactions of Tregs with other immune cells will provide valuable information to develop the most effective Tregs depletion strategy, which may preserve the positive effects of both effector (NK or T) cells and Treg cells to achieve antitumor efficacy without induction of significant side effects.

TARGETING IDO IN PRECLINICAL MODELS

Several studies recently demonstrated that there is a strong correlation between IDO expression and Treg abundance in the development and progression of tumors,[77,139] in particular, IDO-expressing DCs regulate the proliferation and activation of Tregs.[77,140]

The systemic administration of the IDO inhibitors has been shown to delay tumor growth in mice.[141] More importantly, in mice bearing p815B tumors that are transfected to express IDO1 and resistant to immune-mediated tumor rejection, administration of IDO1 inhibitor reversed this immune resistance, leading to an efficient tumor rejection.[96] In syngeneic murine GL261 glioma model, we found that tumor-derived IDO1 increases Treg recruitment into tumors, whereas the genetic silencing of IDO1 significantly decreases Tregs infiltrating into tumors, as well as leads to a long-term mice survival.[11,12] Although the use of IDO1 inhibitor (L-1-MT) alone did not elicit tumor rejection responses in our study,[12] the combination of another inhibitor DL-1-MT with radiotherapy (RT) and TMZ led to a significant survival benefit as compared to the combination of RT and TMZ.[142]

CLINICAL TRIAL DATA

Determining the optimal monotherapy and/or combinatorial therapy that will lead to the greatest impact on decreasing Treg numbers and/or function has been a very active area of research. Since most cytotoxic chemotherapeutic agents are known to cause myelosuppression and associated lymphopenia, it was assumed that combining chemotherapy with immunotherapy will ablate the beneficial effect of immunotherapy. However, emerging literature that further dissects general lymphopenia into specific subsets of T cell depletion outlines a more encouraging story.

Temozolomide, a second-generation DNA-alkylating agent that methylates the O6 position of guanine causing double-stranded DNA cross-linking and DNA damage, is the first-line of chemotherapy treatment for MG. Although it is well tolerated and prolongs patient survival, it has also been well described to induce immunosuppression in the various forms of lymphopenia.[143,144] More interestingly the lymphopenia associated with TMZ is a preferential depletion of $CD4^+CD25^+$ Treg.[145] Heimberger et al. cleverly utilized this property of TMZ to optimally time a peptide vaccine targeting the epidermal growth factor receptor variant III (EGFRvIII) when TMZ-induced Treg depletion was at its maximum. Using this strategy, they demonstrated that chemotherapy and immunotherapy can be delivered concurrently without negating the effects of immunotherapy.[146] This initial observation was further tested in a larger-Phase II multicenter trial of EGFRvIII vaccine in patients with newly-diagnosed GBM (ACT III trial) and showed that proper timing of Treg depletion in combination with immune-boosting vaccination provides added survival benefit to current standard of care.[147] A phase III, double-blind, randomized trial ("ACT IV") is currently under way to further validate this strategy. Since TMZ is the current standard of care for glioma patients, manipulating the beneficial TMZ-induced side effect(s) is particularly appealing in this patient cohort.

Another alkylating agent, CTX, has been recently shown to induce antitumor immune responses by stimulating the effector arm of the immune response, while simultaneously inhibiting the suppressive arm in a preclinical model.[148,149] In a clinical study of 12 patients with treatment-refractory metastatic breast cancer, single-agent CTX resulted in transient initial reduction in circulating Treg by more than 40%. However, ultimately Treg levels returned to pre-CTX treatment levels due to increased proliferative activity.[150] Nonetheless, this overall transient Treg depletion led to an increase in breast tumor–reactive T cells that remained at high levels throughout the treatment duration and correlated with disease stabilization and overall survival. Clinical Treg depletion using metronomic CTX has been attempted in several human cancers including MGs;[151–155] however, the optimal timing of this strategy still requires better definition with regard to pathology and location of the tumor.

Another strategy of depleting Tregs that has been shown effective in preclinical studies is based on targeting CD25, a constitutively expressed cell surface marker, present on Tregs. In the clinical setting, a phase I study by Kreitman et al. using LMB-2, a CD25 mAb conjugated to truncated *Pseudomonas* exotoxin, was found to elicit a promising clinical response in patients with hematological malignancy.[156] However, the same antibody in the context of metastatic melanoma, failed to augment the immune response to cancer vaccination despite inducing a transient decrease in Treg.[157] Similar mixed results were reported regarding recombinant IL-2-diphtheria toxin conjugate, DAB (389) IL-2 (also known as denileukin diftitox and ONTAK), which was also designed as a Treg-depleting agent.[158–160] On the one hand, in a phase III trial in the context of cutaneous T cell lymphoma, it had significant and durable effect on overall response rate and progression-free survival with a favorable safety profile.[161] On the other hand, in the setting of non–Hodgkin's lymphoma, combination of denileukin diftitox with rituximab significantly increased the toxicity of the combination without an improvement in response rate or time to progression.[162] In the context of MG, a randomized placebo-controlled pilot study combining administration of humanized CD25 mAb, Daclizumab with peptide vaccination against the EGFRvIII and TMZ safely and selectively depleted Treg[163] with no symptoms of autoimmune toxicity.

Based on several promising preclinical studies, humanized CTLA-4 mAb has been successfully tested in clinical trials for the treatment of metastatic melanoma.[164,165] However, these early clinical studies also showed several instances of immune-related adverse effects associated with anti–CTLA-4 antibody treatment such as hypophysitis, meningitis, and Guillain–Barré syndrome.[166] Anti–CTLA-4 antibody has been tried for the treatment of MG and the results from these clinical trials have yet to be reported.[167–169]

Glucocorticoid-induced tumour necrosis factor receptor (GITR) is constitutively expressed at a high level on Tregs.[116] GITR–GITRL interaction renders responder T cells more susceptible to suppression by Treg and also directly inhibits the ability of Tregs to mediate suppression of responder T cells.[170] In preclinical studies, T cell stimulation by agonistic GITR mAb attenuated Treg-mediated suppression and enhanced tumor killing by cytotoxic and conventional T cells via increased secretion of IFN-γ. Currently, a humanized GITR mAb (TRX518) is in a phase I safety and tolerability dose–escalation clinical trial for late stage (III and IV) melanoma patients with resectable tumors. OX40 is another member of the TNF receptor family expressed on Tregs that can be clinically targeted. Although it has yet to be tested for MGs, OX40 mAb–mediated therapy is currently being tested in a phase I/II trial for patients with metastatic melanoma (NCT01689870), as well as a phase I trial for patients with advanced forms of cancer (NCT01644968).

Activity of IDO1, IDO2, and TDO can be inhibited by several chemical inhibitors. L and D-stereoisomers of 1-MT are two popular compounds that are being tested in the clinical setting. Both of these inhibitors bind with IDO1, IDO2, and TDO with various affinity and exhibit a complex relationship between binding and inhibition of enzymatic activity. IDO1 has a 10-fold higher-binding affinity for the L-1-MT than D-1-MT,[171] whereas IDO2 is more effectively inhibited by L-1-MT,[172–174] and the treatment effect of D-1-MT was found to be IDO1 dependent.[175] Based on these findings D-1-MT has been tested in Phase I clinical trial for the treatment of metastatic and/or refractory solid tumors (NCT00567931) and results are pending. Given the significant role played by IDO in immunosuppression, the search for specific IDO1, IDO2, and TDO inhibitor is a highly active area of research identifying some promising options such as INCB024360 and Amg-1, that selectively inhibit IDO1,[176,177] tenatoprazole, a specific IDO2 inhibitor,[178] and LM10, a selective TDO inhibitor.[81] There are several ongoing clinical trials that are assessing safety and efficacy of IDO inhibitors in various cancers (Table 3.1).

TABLE 3.1 Various Indoleamine-2,3-dioxygenase (IDO) Inhibitors Currently Being Tested in Clinical Trials

Trial Identification	IDO Inhibitor	Type of Cancer	Trial Description	Status
NCT02042430	INCB024360	Ovarian, fallopian tube, or primary peritoneal cancer	INCB024360 before surgery in treating patients with newly diagnosed stage III–IV epithelial ovarian, fallopian tube, or primary peritoneal cancer	Recruiting
NCT01982487	INCB024360	Ovarian, fallopian tube, or primary peritoneal cancer	Vaccine therapy and IDO1 inhibitor INCB024360 in treating patients with epithelial ovarian, fallopian tube, or primary peritoneal cancer who are in remission	Withdrawn
NCT01219348	IDO peptide vaccination	Non–small cell lung cancer	IDO peptide vaccination for stage III–IV non-small-cell lung cancer patients (IDO vaccine)	Completed

Continued

TABLE 3.1 Various Indoleamine-2,3-dioxygenase (IDO) Inhibitors Currently Being Tested in Clinical Trials—cont'd

Trial Identification	IDO Inhibitor	Type of Cancer	Trial Description	Status
NCT02166905	INCB024360	Ovarian, fallopian tube, or primary peritoneal cancer	DEC-205/NY-ESO-1 fusion protein CDX-1401, poly ICLC, and IDO1 inhibitor INCB024360 in treating patients with ovarian, fallopian tube, or primary peritoneal cancer in remission	Recruiting
NCT01961115	INCB024360	Melanoma	INCB024360 and vaccine therapy in treating patients with stage III–IV melanoma	Recruiting
NCT01685255	INCB024360	Ovarian, fallopian tube, or primary peritoneal cancer	Phase II study of the IDO inhibitor INCB024360 versus tamoxifen for subjects with biochemical recurrent-only EOC, PPC, or FTC following complete remission with first-line chemotherapy	Completed
NCT02118285	INCB024360	Ovarian, fallopian tube, or primary peritoneal cancer	Intraperitoneal natural killer cells and INCB024360 for recurrent ovarian, fallopian tube, and primary peritoneal cancer	Recruiting
NCT01792050	D-1-MT (D-1-methyltryptophan)	Breast cancer	Study of chemotherapy in combination with IDO inhibitor in metastatic breast cancer	Recruiting
NCT02460367	D-1-MT	Non–small cell lung cancer	Immunotherapy combination study in advanced previously treated non-small-cell lung cancer	Not yet recruiting
NCT00739609	D-1-MT	Solid cancer	IDO inhibitor study for relapsed or refractory solid tumors (D-1-MT)	Terminated
NCT02502708	D-1-MT	Glioma	Study of the IDO pathway inhibitor, indoximod, and temozolomide for pediatric patients with primary malignant brain tumors	Not yet recruiting

TABLE 3.1 Various Indoleamine-2,3-dioxygenase (IDO) Inhibitors Currently Being Tested in Clinical Trials—cont'd

Trial Identification	IDO Inhibitor	Type of Cancer	Trial Description	Status
NCT02052648	D-1-MT	Glioma	Study of IDO inhibitor and temozolomide for adult patients with primary malignant brain tumors	Recruiting
NCT02048709	GDC-0919	Solid cancer	IDO inhibitor in advanced solid tumors	Recruiting
NCT02077881	D-1-MT	Pancreatic cancer	Study of IDO inhibitor in combination with gemcitabine and nab-paclitaxel in patients with metastatic pancreatic cancer	Recruiting
NCT00567931	D-1-MT	Solid cancer	1-methyl-D-tryptophan in treating patients with metastatic or refractory solid tumors that cannot be removed by surgery	Completed
NCT02559492	INCB024360	Solid cancer	INCB039110 combined with INCB024360 and/or INCB039110 combined with INCB050465 in advanced solid tumors	Not yet recruiting
NCT01543464	IDO/Survivin peptide vaccine	Melanoma	Peptide vaccine and temozolomide for metastatic melanoma patients	Terminated
NCT02073123	D-1-MT	Melanoma	Study of IDO inhibitor in combination with ipilimumab for adult patients with metastatic melanoma	Recruiting
NCT01604889	INCB024360	Melanoma	Phase I/II randomized, blinded, placebo controlled study of ipilimumab in combination with INCB024360 or placebo in subjects with unresectable or metastatic melanoma	Active, not recruiting

CONCLUSION

Based on recent preclinical and clinical studies, it is evident that for the success of immunotherapy the issue of immunosuppression has to be addressed. In the case of MG, Tregs and IDO are two of the well-studied players that are intricately involved in various processes that result in a profoundly immunosuppressive tumor microenvironment. The exact mechanism of their involvement and interaction is not yet fully understood, and neither is the optimal way to target and influence them. Since many of these clinical trials are already underway, we will get important answers in the near future regarding both the efficacy of these treatments and the steps we must take to continue targeting tumor-induced immunosuppression.

References

1. El Andaloussi A, Lesniak MS. CD4(+)CD25(+)FoxP3(+) T-cell infiltration and heme oxygenase-1 expression correlate with tumor grade in human gliomas. *J Neuro Oncol*. 2007;83:145–152.
2. Wainwright DA, Sengupta S, Han Y, Lesniak MS. Thymus-derived rather than tumor-induced regulatory T cells predominate in brain tumors. *Neuro Oncol*. 2011;13:1308–1323.
3. Wainwright DA, Dey M, Chang A, Lesniak MS. Targeting Tregs in malignant brain cancer: overcoming IDO. *Front Immunol*. 2013;4:116.
4. Fujita M, Kohanbash G, Fellows-Mayle W, et al. COX-2 blockade suppresses gliomagenesis by inhibiting myeloid-derived suppressor cells. *Cancer Res*. 2011;71:2664–2674.
5. Kohanbash G, Okada H. Myeloid-derived suppressor cells (MDSCs) in gliomas and glioma-development. *Immunol Invest*. 2012;41:658–679.
6. Wang R, Zhang JL, Wei B, et al. Upregulation of plasmacytoid dendritic cells in glioma. *Tumour Biol*. 2014;35:9661–9666.
7. Dey M, Chang AL, Miska J, et al. Dendritic cell-based vaccines that utilize myeloid rather than plasmacytoid cells offer a superior survival advantage in malignant glioma. *J Immunol*. 2015;195:367–376.
8. Han J, Alvarez-Breckenridge CA, Wang QE, Yu J. TGF-beta signaling and its targeting for glioma treatment. *Am J Cancer Res*. 2015;5:945–955.
9. Qiu B, Zhang D, Wang C, et al. IL-10 and TGF-beta2 are overexpressed in tumor spheres cultured from human gliomas. *Mol Biol Rep*. 2011;38:3585–3591.
10. Dey M, Chang AL, Wainwright DA, et al. Heme oxygenase-1 protects regulatory T cells from hypoxia-induced cellular stress in an experimental mouse brain tumor model. *J Neuroimmunol*. 2014;266:33–42.
11. Wainwright DA, Balyasnikova IV, Chang AL, et al. IDO expression in brain tumors increases the recruitment of regulatory T cells and negatively impacts survival. *Clin Cancer Res*. 2012;18:6110–6121.
12. Wainwright DA, Chang AL, Dey M, et al. Durable therapeutic efficacy utilizing combinatorial blockade against IDO, CTLA-4, and PD-L1 in mice with brain tumors. *Clin Cancer Res*. 2014;20:5290–5301.
13. Sakaguchi S, Sakaguchi N, Asano M, Itoh M, Toda M. Immunologic self-tolerance maintained by activated T cells expressing IL-2 receptor alpha-chains (CD25). Breakdown of a single mechanism of self-tolerance causes various autoimmune diseases. *J Immunol*. 1995;155:1151–1164.

14. Hori S, Nomura T, Sakaguchi S. Control of regulatory T cell development by the transcription factor Foxp3. *Science*. 2003;299:1057–1061.
15. Alam SM, Travers PJ, Wung JL, et al. T-cell-receptor affinity and thymocyte positive selection. *Nature*. 1996;381:616–620.
16. Moran AE, Holzapfel KL, Xing Y, et al. T cell receptor signal strength in Treg and iNKT cell development demonstrated by a novel fluorescent reporter mouse. *J Exp Med*. 2011;208:1279–1289.
17. Benoist C, Mathis D. Treg cells, life history, and diversity. *Cold Spring Harb Perspect Biol*. 2012;4:a007021.
18. Chen W, Jin W, Hardegen N, et al. Conversion of peripheral $CD4^+CD25^-$ naive T cells to $CD4^+CD25^+$ regulatory T cells by TGF-beta induction of transcription factor *Foxp3*. *J Exp Med*. 2003;198:1875–1886.
19. Fu S, Zhang N, Yopp AC, et al. TGF-beta induces $Foxp3^+$ T-regulatory cells from $CD4^+$ $CD25^-$ precursors. *Am J Transplant*. 2004;4:1614–1627.
20. Marie JC, Letterio JJ, Gavin M, Rudensky AY. TGF-beta1 maintains suppressor function and Foxp3 expression in $CD4^+CD25^+$ regulatory T cells. *J Exp Med*. 2005;201:1061–1067.
21. Moo-Young TA, Larson JW, Belt BA, et al. Tumor-derived TGF-beta mediates conversion of $CD4^+Foxp3^+$ regulatory T cells in a murine model of pancreas cancer. *J Immunother*. 2009;32:12–21.
22. Grossman WJ, Verbsky JW, Barchet W, Colonna M, Atkinson JP, Ley TJ. Human T regulatory cells can use the perforin pathway to cause autologous target cell death. *Immunity*. 2004;21:589–601.
23. Cao XF, Cai SF, Fehniger TA, et al. Granzyme B and perforin are important for regulatory T cell-mediated suppression of tumor clearance. *Immunity*. 2007;27:635–646.
24. Wu YQ, Borde M, Heissmeyer V, et al. Foxp3 controls regulatory T cell function through cooperation with NFAT. *Cell*. 2006;126:375–387.
25. Wing K, Onishi Y, Prieto-Martin P, et al. CTLA-4 control over Foxp3(+) regulatory T cell function. *Science*. 2008;322:271–275.
26. Stockis J, Colau D, Coulie PG, Lucas S. Membrane protein GARP is a receptor for latent TGF-beta on the surface of activated human Treg. *Eur J Immunol*. 2009;39:3315–3322.
27. O'Garra A, Vieira PL, Vieira P, Goldfeld AE. IL-10-producing and naturally occurring $CD4^+$ Tregs: limiting collateral damage. *J Clin Invest*. 2004;114:1372–1378.
28. Miska J, Abdulreda MH, Devarajan P, et al. Real-time immune cell interactions in target tissue during autoimmune-induced damage and graft tolerance. *J Exp Med*. 2014;211:441–456.
29. Busse D, de la Rosa M, Hobiger K, et al. Competing feedback loops shape IL-2 signaling between helper and regulatory T lymphocytes in cellular microenvironments. *Proc Natl Acad Sci USA*. 2010;107:3058–3063.
30. Borsellino G, Kleinewietfeld M, Di Mitri D, et al. Expression of ectonucleotidase CD39 by $Foxp3^+$ Treg cells: hydrolysis of extracellular ATP and immune suppression. *Blood*. 2007;110:1225–1232.
31. Swann JB, Smyth MJ. Immune surveillance of tumors. *J Clin Invest*. 2007;117:1137–1146.
32. Chen DS, Mellman I. Oncology meets immunology: the cancer-immunity cycle. *Immunity*. 2013;39:1–10.
33. Hanahan D, Weinberg RA. Hallmarks of cancer: the next generation. *Cell*. 2011;144:646–674.
34. Duray A, Demoulin S, Hubert P, Delvenne P, Saussez S. Immune suppression in head and neck cancers: a review. *Clin Dev Immunol*. 2010;2010:701657.
35. Lin YC, Mahalingam J, Chiang JM, et al. Activated but not resting regulatory T cells accumulated in tumor microenvironment and correlated with tumor progression in patients with colorectal cancer. *Int J Cancer*. 2013;132:1341–1350.

36. Pedroza-Gonzalez A, Verhoef C, Ijzermans JN, et al. Activated tumor-infiltrating $CD4^+$ regulatory T cells restrain antitumor immunity in patients with primary or metastatic liver cancer. *Hepatology*. 2013;57:183–194.
37. Mizukami Y, Kono K, Kawaguchi Y, et al. CCL17 and CCL22 chemokines within tumor microenvironment are related to accumulation of Foxp3(+) regulatory T cells in gastric cancer. *Int J Cancer*. 2008;122:2286–2293.
38. Jandus C, Bioley G, Speiser DE, Romero P. Selective accumulation of differentiated Foxp3(+) CD4(+) T cells in metastatic tumor lesions from melanoma patients compared to peripheral blood. *Cancer Immunol Immunother*. 2008;57:1795–1805.
39. Grauer OM, Nierkens S, Bennink E, et al. $CD4^+FoxP3^+$ regulatory T cells gradually accumulate in gliomas during tumor growth and efficiently suppress antiglioma immune responses in vivo. *Int J Cancer*. 2007;121:95–105.
40. Franciszkiewicz K, Boissonnas A, Boutet M, Combadiere C, Mami-Chouaib F. Role of chemokines and chemokine receptors in shaping the effector phase of the antitumor immune response. *Cancer Res*. 2012;72:6325–6332.
41. Schlecker E, Stojanovic A, Eisen C, et al. Tumor-infiltrating monocytic myeloid-derived suppressor cells mediate CCR5-dependent recruitment of regulatory T cells favoring tumor growth. *J Immunol*. 2012;189:5602–5611.
42. Chen ZB, Benoist C, Mathis D. How defects in central tolerance impinge on a deficiency in regulatory T cells. *Proc Natl Acad Sci USA*. 2005;102:14735–14740.
43. Michalek RD, Gerriets VA, Jacobs SR, et al. Cutting edge: distinct glycolytic and lipid oxidative metabolic programs are essential for effector and regulatory CD4(+) T cell subsets. *J Immunol*. 2011;186:3299–3303.
44. Clambey ET, McNamee EN, Westrich JA, et al. Hypoxia-inducible factor-1 alpha-dependent induction of FoxP3 drives regulatory T-cell abundance and function during inflammatory hypoxia of the mucosa. *Proc Natl Acad Sci USA*. 2012;109:E2784–E2793.
45. Ben-Shoshan J, Maysel-Auslender S, Mor A, Keren G, George J. Hypoxia controls CD4(+)CD25(+) regulatory T-cell homeostasis via hypoxia-inducible factor-1 alpha. *Eur J Immunol*. 2008;38:2412–2418.
46. Louveau A, Smirnov I, Keyes TJ, et al. Structural and functional features of central nervous system lymphatic vessels. *Nature*. 2015;523.
47. Jacobs JF, Idema AJ, Bol KF, et al. Prognostic significance and mechanism of Treg infiltration in human brain tumors. *J Neuroimmunol*. 2010;225:195–199.
48. Fecci PE, Mitchell DA, Whitesides JF, et al. Increased regulatory T-cell fraction amidst a diminished CD4 compartment explains cellular immune defects in patients with malignant glioma. *Cancer Res*. 2006;66:3294–3302.
49. Sonabend AM, Rolle CE, Lesniak MS. The role of regulatory T cells in malignant glioma. *Anticancer Res*. 2008;28:1143–1150.
50. Jordan JT, Sun W, Hussain SF, DeAngulo G, Prabhu SS, Heimberger AB. Preferential migration of regulatory T cells mediated by glioma-secreted chemokines can be blocked with chemotherapy. *Cancer Immunol Immunother*. 2008;57:123–131.
51. Vasco C, Canazza A, Rizzo A, et al. Circulating T regulatory cells migration and phenotype in glioblastoma patients: an in vitro study. *J Neuro Oncol*. 2013;115:353–363.
52. Crane CA, Ahn BJ, Han SJ, Parsa AT. Soluble factors secreted by glioblastoma cell lines facilitate recruitment, survival, and expansion of regulatory T cells: implications for immunotherapy. *Neuro Oncol*. 2012;14:584–595.
53. Xu S, Shao QQ, Sun JT, et al. Synergy between the ectoenzymes CD39 and CD73 contributes to adenosinergic immunosuppression in human malignant gliomas. *Neuro Oncol*. 2013;15:1160–1172.
54. Wei J, Wu A, Kong LY, et al. Hypoxia potentiates glioma-mediated immunosuppression. *PLoS One*. 2011;6.
55. Verhaak RG, Hoadley KA, Purdom E, et al. Integrated genomic analysis identifies clinically relevant subtypes of glioblastoma characterized by abnormalities in PDGFRA, IDH1, EGFR, and NF1. *Cancer Cell*. 2010;17:98–110.

56. Doucette T, Rao G, Rao A, et al. Immune heterogeneity of glioblastoma subtypes: extrapolation from the cancer genome atlas. *Cancer Immunol Res*. 2013;1:112–122.
57. McGaha TL, Huang L, Lemos H, et al. Amino acid catabolism: a pivotal regulator of innate and adaptive immunity. *Immunol Rev*. 2012;249:135–157.
58. Platten M, von Knebel Doeberitz N, Oezen I, Wick W, Ochs K. Cancer immunotherapy by targeting IDO1/TDO and their downstream effectors. *Front Immunol*. 2014;5:673.
59. Ball HJ, Jusof FF, Bakmiwewa SM, Hunt NH, Yuasa HJ. Tryptophan-catabolizing enzymes – party of three. *Front Immunol*. 2014;5:485.
60. Yeung AW, Terentis AC, King NJ, Thomas SR. Role of indoleamine 2,3-dioxygenase in health and disease. *Clin Sci*. 2015;129:601–672.
61. Yoshida R, Imanishi J, Oku T, Kishida T, Hayaishi O. Induction of pulmonary indoleamine 2,3-dioxygenase by interferon. *Proc Natl Acad Sci USA*. 1981;78:129–132.
62. Chon SY, Hassanain HH, Gupta SL. Cooperative role of interferon regulatory factor 1 and p91 (STAT1) response elements in interferon-gamma-inducible expression of human indoleamine 2,3-dioxygenase gene. *J Biol Chem*. 1996;271:17247–17252.
63. Konan KV, Taylor MW. Importance of the two interferon-stimulated response element (ISRE) sequences in the regulation of the human indoleamine 2,3-dioxygenase gene. *J Biol Chem*. 1996;271:19140–19145.
64. Croitoru-Lamoury J, Lamoury FM, Caristo M, et al. Interferon-gamma regulates the proliferation and differentiation of mesenchymal stem cells via activation of indoleamine 2,3 dioxygenase (IDO). *PLoS One*. 2011;6:e14698.
65. Baban B, Hansen AM, Chandler PR, et al. A minor population of splenic dendritic cells expressing CD19 mediates IDO-dependent T cell suppression via type I IFN signaling following B7 ligation. *Int Immunol*. 2005;17:909–919.
66. Mellor AL, Baban B, Chandler PR, Manlapat A, Kahler DJ, Munn DH. Cutting edge: CpG oligonucleotides induce splenic $CD19^+$ dendritic cells to acquire potent indoleamine 2,3-dioxygenase-dependent T cell regulatory functions via IFN Type 1 signaling. *J Immunol*. 2005;175:5601–5605.
67. Munn DH, Zhou M, Attwood JT, et al. Prevention of allogeneic fetal rejection by tryptophan catabolism. *Science*. 1998;281:1191–1193.
68. Pallotta MT, Orabona C, Volpi C, et al. Indoleamine 2,3-dioxygenase is a signaling protein in long-term tolerance by dendritic cells. *Nat Immunol*. 2011;12:870–878.
69. Ball HJ, Sanchez-Perez A, Weiser S, et al. Characterization of an indoleamine 2,3-dioxygenase-like protein found in humans and mice. *Gene*. 2007;396:203–213.
70. Metz R, Duhadaway JB, Kamasani U, Laury-Kleintop L, Muller AJ, Prendergast GC. Novel tryptophan catabolic enzyme IDO2 is the preferred biochemical target of the antitumor indoleamine 2,3-dioxygenase inhibitory compound D-1-methyl-tryptophan. *Cancer Res*. 2007;67:7082–7087.
71. Prendergast GC, Metz R, Muller AJ, Merlo LM, Mandik-Nayak L. IDO2 in immunomodulation and autoimmune disease. *Front Immunol*. 2014;5:585.
72. van Baren N, Van den Eynde BJ. Tryptophan-degrading enzymes in tumoral immune resistance. *Front Immunol*. 2015;6:34.
73. Yuasa HJ, Takubo M, Takahashi A, Hasegawa T, Noma H, Suzuki T. Evolution of vertebrate indoleamine 2,3-dioxygenases. *J Mol Evol*. 2007;65:705–714.
74. Divanovic S, Sawtell NM, Trompette A, et al. Opposing biological functions of tryptophan catabolizing enzymes during intracellular infection. *J Infect Dis*. 2012;205:152–161.
75. Merlo LM, Pigott E, DuHadaway JB, et al. IDO2 is a critical mediator of autoantibody production and inflammatory pathogenesis in a mouse model of autoimmune arthritis. *J Immunol*. 2014;192:2082–2090.
76. Metz R, Smith C, DuHadaway JB, et al. IDO2 is critical for IDO1-mediated T-cell regulation and exerts a non-redundant function in inflammation. *Int Immunol*. 2014;26:357–367.
77. Sharma MD, Baban B, Chandler P, et al. Plasmacytoid dendritic cells from mouse tumor-draining lymph nodes directly activate mature Tregs via indoleamine 2,3-dioxygenase. *J Clin Invest*. 2007;117:2570–2582.

78. Trabanelli S, Ocadlikova D, Ciciarello M, et al. The SOCS3-independent expression of IDO2 supports the homeostatic generation of T regulatory cells by human dendritic cells. *J Immunol*. 2014;192:1231–1240.
79. Kanai M, Funakoshi H, Takahashi H, et al. Tryptophan 2,3-dioxygenase is a key modulator of physiological neurogenesis and anxiety-related behavior in mice. *Mol Brain*. 2009;2:8.
80. Opitz CA, Litzenburger UM, Sahm F, et al. An endogenous tumour-promoting ligand of the human aryl hydrocarbon receptor. *Nature*. 2011;478:197–203.
81. Pilotte L, Larrieu P, Stroobant V, et al. Reversal of tumoral immune resistance by inhibition of tryptophan 2,3-dioxygenase. *Proc Natl Acad Sci USA*. 2012;109:2497–2502.
82. Julliard W, Fechner JH, Mezrich JD. The aryl hydrocarbon receptor meets immunology: friend or foe? A little of both. *Front Immunol*. 2014;5:458.
83. Nohara K, Pan X, Tsukumo S, et al. Constitutively active aryl hydrocarbon receptor expressed specifically in T-lineage cells causes thymus involution and suppresses the immunization-induced increase in splenocytes. *J Immunol*. 2005;174:2770–2777.
84. Funatake CJ, Marshall NB, Steppan LB, Mourich DV, Kerkvliet NI. Cutting edge: activation of the aryl hydrocarbon receptor by 2,3,7,8-tetrachlorodibenzo-p-dioxin generates a population of $CD4^+$ $CD25^+$ cells with characteristics of regulatory T cells. *J Immunol*. 2005;175:4184–4188.
85. Mezrich JD, Fechner JH, Zhang X, Johnson BP, Burlingham WJ, Bradfield CA. An interaction between kynurenine and the aryl hydrocarbon receptor can generate regulatory T cells. *J Immunol*. 2010;185:3190–3198.
86. Vorderstrasse BA, Kerkvliet NI. 2,3,7,8-Tetrachlorodibenzo-p-dioxin affects the number and function of murine splenic dendritic cells and their expression of accessory molecules. *Toxicol Appl Pharmacol*. 2001;171:117–125.
87. Munn DH, Sharma MD, Baban B, et al. GCN2 kinase in T cells mediates proliferative arrest and anergy induction in response to indoleamine 2,3-dioxygenase. *Immunity*. 2005;22:633–642.
88. Hu HM, Tian Q, Baer M, et al. The C/EBP bZIP domain can mediate lipopolysaccharide induction of the proinflammatory cytokines interleukin-6 and monocyte chemoattractant protein-1. *J Biol Chem*. 2000;275:16373–16381.
89. Metz R, Rust S, Duhadaway JB, et al. IDO inhibits a tryptophan sufficiency signal that stimulates mTOR: a novel IDO effector pathway targeted by D-1-methyl-tryptophan. *Oncoimmunology*. 2012;1:1460–1468.
90. Bessede A, Gargaro M, Pallotta MT, et al. Aryl hydrocarbon receptor control of a disease tolerance defence pathway. *Nature*. 2014;511:184–190.
91. Quintana FJ, Basso AS, Iglesias AH, et al. Control of T(reg) and T(H)17 cell differentiation by the aryl hydrocarbon receptor. *Nature*. 2008;453:65–71.
92. Veldhoen M, Hirota K, Westendorf AM, et al. The aryl hydrocarbon receptor links TH17-cell-mediated autoimmunity to environmental toxins. *Nature*. 2008;453:106–109.
93. Boyland E, Williams DC. The metabolism of tryptophan. 2. The metabolism of tryptophan in patients suffering from cancer of the bladder. *Biochem J*. 1956;64:578–582.
94. Godin-Ethier J, Hanafi LA, Piccirillo CA, Lapointe R. Indoleamine 2,3-dioxygenase expression in human cancers: clinical and immunologic perspectives. *Clin Cancer Res*. 2011;17:6985–6991.
95. Theate I, van Baren N, Pilotte L, et al. Extensive profiling of the expression of the indoleamine 2,3-dioxygenase 1 protein in normal and tumoral human tissues. *Cancer Immunol Res*. 2015;3:161–172.
96. Uyttenhove C, Pilotte L, Theate I, et al. Evidence for a tumoral immune resistance mechanism based on tryptophan degradation by indoleamine 2,3-dioxygenase. *Nat Med*. 2003;9:1269–1274.
97. Muller AJ, DuHadaway JB, Donover PS, Sutanto-Ward E, Prendergast GC. Inhibition of indoleamine 2,3-dioxygenase, an immunoregulatory target of the cancer suppression gene Bin1, potentiates cancer chemotherapy. *Nat Med*. 2005;11:312–319.

98. Smith C, Chang MY, Parker KH, et al. IDO is a nodal pathogenic driver of lung cancer and metastasis development. *Cancer Discov*. 2012;2:722–735.
99. Holmgaard RB, Zamarin D, Munn DH, Wolchok JD, Allison JP. Indoleamine 2,3-dioxygenase is a critical resistance mechanism in antitumor T cell immunotherapy targeting CTLA-4. *J Exp Med*. 2013;210:1389–1402.
100. Galea I, Bechmann I, Perry VH. What is immune privilege (not)? *Trends Immunol*. 2007;28:12–18.
101. Kwidzinski E, Bunse J, Aktas O, et al. Indolamine 2,3-dioxygenase is expressed in the CNS and down-regulates autoimmune inflammation. *FASEB J*. 2005;19:1347–1349.
102. Woo EY, Chu CS, Goletz TJ, et al. Regulatory CD4(+)CD25(+) T cells in tumors from patients with early-stage non-small cell lung cancer and late-stage ovarian cancer. *Cancer Res*. 2001;61:4766–4772.
103. Sakaguchi S, Yamaguchi T, Nomura T, Ono M. Regulatory T cells and immune tolerance. *Cell*. 2008;133:775–787.
104. Shimizu J, Yamazaki S, Sakaguchi S. Induction of tumor immunity by removing $CD25^+CD4^+$ T cells: a common basis between tumor immunity and autoimmunity. *J Immunol*. 1999;163:5211–5218.
105. Sutmuller RP, van Duivenvoorde LM, van Elsas A, et al. Synergism of cytotoxic T lymphocyte-associated antigen 4 blockade and depletion of CD25(+) regulatory T cells in antitumor therapy reveals alternative pathways for suppression of autoreactive cytotoxic T lymphocyte responses. *J Exp Med*. 2001;194:823–832.
106. Onizuka S, Tawara I, Shimizu J, Sakaguchi S, Fujita T, Nakayama E. Tumor rejection by in vivo administration of anti-CD25 (interleukin-2 receptor alpha) monoclonal antibody. *Cancer Res*. 1999;59:3128–3133.
107. Ghiringhelli F, Larmonier N, Schmitt E, et al. $CD4^+CD25^+$ regulatory T cells suppress tumor immunity but are sensitive to cyclophosphamide which allows immunotherapy of established tumors to be curative. *Eur J Immunol*. 2004;34:336–344.
108. Fecci PE, Sweeney AE, Grossi PM, et al. Systemic anti-CD25 monoclonal antibody administration safely enhances immunity in murine glioma without eliminating regulatory T cells. *Clin Cancer Res*. 2006;12:4294–4305.
109. El Andaloussi A, Han Y, Lesniak MS. Prolongation of survival following depletion of $CD4^+CD25^+$ regulatory T cells in mice with experimental brain tumors. *J Neurosurg*. 2006;105:430–437.
110. Poirier MD, Haban H, El Andaloussi A. A combination of systemic and intracranial anti-CD25 immunotherapy elicits a long-time survival in murine model of glioma. *J Oncol*. 2009;2009:963037.
111. Maes W, Rosas GG, Verbinnen B, et al. DC vaccination with anti-CD25 treatment leads to long-term immunity against experimental glioma. *Neuro Oncol*. 2009;11:529–542.
112. Jacobs JF, Punt CJ, Lesterhuis WJ, et al. Dendritic cell vaccination in combination with anti-CD25 monoclonal antibody treatment: a phase I/II study in metastatic melanoma patients. *Clin Cancer Res*. 2010;16:5067–5078.
113. Wuest SC, Edwan JH, Martin JF, et al. A role for interleukin-2 trans-presentation in dendritic cell-mediated T cell activation in humans, as revealed by daclizumab therapy. *Nat Med*. 2011;17:604–609.
114. Colombo MP, Piconese S. Regulatory-T-cell inhibition versus depletion: the right choice in cancer immunotherapy. *Nat Rev Cancer*. 2007;7:880–887.
115. Nair S, Boczkowski D, Fassnacht M, Pisetsky D, Gilboa E. Vaccination against the forkhead family transcription factor Foxp3 enhances tumor immunity. *Cancer Res*. 2007;67:371–380.
116. Cohen AD, Schaer DA, Liu C, et al. Agonist anti-GITR monoclonal antibody induces melanoma tumor immunity in mice by altering regulatory T cell stability and intratumor accumulation. *PLoS One*. 2010;5:e10436.

117. Wainwright DA, Sengupta S, Han Y, Ulasov IV, Lesniak MS. The presence of IL-17A and T helper 17 cells in experimental mouse brain tumors and human glioma. *PLoS One*. 2010;5:e15390.
118. Blatner NR, Mulcahy MF, Dennis KL, et al. Expression of RORgammat marks a pathogenic regulatory T cell subset in human colon cancer. *Sci Transl Med*. 2012;4:164ra159.
119. Casares N, Rudilla F, Arribillaga L, et al. A peptide inhibitor of Foxp3 impairs regulatory T cell activity and improves vaccine efficacy in mice. *J Immunol*. 2010;185:5150–5159.
120. Lutsiak ME, Semnani RT, De Pascalis R, Kashmiri SV, Schlom J, Sabzevari H. Inhibition of $CD4^+25^+$ T regulatory cell function implicated in enhanced immune response by low-dose cyclophosphamide. *Blood*. 2005;105:2862–2868.
121. Motoyoshi Y, Kaminoda K, Saitoh O, et al. Different mechanisms for anti-tumor effects of low- and high-dose cyclophosphamide. *Oncol Rep*. 2006;16:141–146.
122. Ghiringhelli F, Menard C, Puig PE, et al. Metronomic cyclophosphamide regimen selectively depletes $CD4^+CD25^+$ regulatory T cells and restores T and NK effector functions in end stage cancer patients. *Cancer Immunol Immunother*. 2007;56:641–648.
123. Radojcic V, Bezak KB, Skarica M, et al. Cyclophosphamide resets dendritic cell homeostasis and enhances antitumor immunity through effects that extend beyond regulatory T cell elimination. *Cancer Immunol Immunother*. 2010;59:137–148.
124. Nakahara T, Uchi H, Lesokhin AM, et al. Cyclophosphamide enhances immunity by modulating the balance of dendritic cell subsets in lymphoid organs. *Blood*. 2010;115:4384–4392.
125. Medina-Echeverz J, Fioravanti J, Zabala M, Ardaiz N, Prieto J, Berraondo P. Successful colon cancer eradication after chemoimmunotherapy is associated with profound phenotypic change of intratumoral myeloid cells. *J Immunol*. 2011;186:807–815.
126. Burton JH, Mitchell L, Thamm DH, Dow SW, Biller BJ. Low-dose cyclophosphamide selectively decreases regulatory T cells and inhibits angiogenesis in dogs with soft tissue sarcoma. *J Vet Intern Med*. 2011;25:920–926.
127. Mitchell L, Thamm DH, Biller BJ. Clinical and immunomodulatory effects of toceranib combined with low-dose cyclophosphamide in dogs with cancer. *J Vet Intern Med*. 2012;26:355–362.
128. Piconese S, Valzasina B, Colombo MP. OX40 triggering blocks suppression by regulatory T cells and facilitates tumor rejection. *J Exp Med*. 2008;205:825–839.
129. Valzasina B, Piconese S, Guiducci C, Colombo MP. Tumor-induced expansion of regulatory T cells by conversion of $CD4^+CD25^-$ lymphocytes is thymus and proliferation independent. *Cancer Res*. 2006;66:4488–4495.
130. Vignali DA, Collison LW, Workman CJ. How regulatory T cells work. *Nat Rev Immunol*. 2008;8:523–532.
131. Wu Q, Zhan X, Dou T, et al. CTLA4 A49G polymorphism shows significant association with glioma risk in a Chinese population. *Biochem Genet*. 2011;49:190–201.
132. Grauer O, Poschl P, Lohmeier A, Adema GJ, Bogdahn U. Toll-like receptor triggered dendritic cell maturation and IL-12 secretion are necessary to overcome T-cell inhibition by glioma-associated TGF-beta2. *J Neuro Oncol*. 2007;82:151–161.
133. Fecci PE, Ochiai H, Mitchell DA, et al. Systemic CTLA-4 blockade ameliorates glioma-induced changes to the $CD4^+$ T cell compartment without affecting regulatory T-cell function. *Clin Cancer Res*. 2007;13:2158–2167.
134. Selenko-Gebauer N, Majdic O, Szekeres A, et al. B7-H1 (programmed death-1 ligand) on dendritic cells is involved in the induction and maintenance of T cell anergy. *J Immunol*. 2003;170:3637–3644.
135. Wolchok JD, Kluger H, Callahan MK, et al. Nivolumab plus ipilimumab in advanced melanoma. *N Engl J Med*. 2013;369:122–133.
136. Peggs KS, Quezada SA, Chambers CA, Korman AJ, Allison JP. Blockade of CTLA-4 on both effector and regulatory T cell compartments contributes to the antitumor activity of anti-CTLA-4 antibodies. *J Exp Med*. 2009;206:1717–1725.

137. Curran MA, Montalvo W, Yagita H, Allison JP. PD-1 and CTLA-4 combination blockade expands infiltrating T cells and reduces regulatory T and myeloid cells within B16 melanoma tumors. *Proc Natl Acad Sci USA*. 2010;107:4275–4280.
138. Galustian C, Meyer B, Labarthe MC, et al. The anti-cancer agents lenalidomide and pomalidomide inhibit the proliferation and function of T regulatory cells. *Cancer Immunol Immunother*. 2009;58:1033–1045.
139. Nakamura T, Shima T, Saeki A, et al. Expression of indoleamine 2,3-dioxygenase and the recruitment of Foxp3-expressing regulatory T cells in the development and progression of uterine cervical cancer. *Cancer Sci*. 2007;98:874–881.
140. Chung DJ, Rossi M, Romano E, et al. Indoleamine 2,3-dioxygenase-expressing mature human monocyte-derived dendritic cells expand potent autologous regulatory T cells. *Blood*. 2009;114:555–563.
141. Friberg M, Jennings R, Alsarraj M, et al. Indoleamine 2,3-dioxygenase contributes to tumor cell evasion of T cell-mediated rejection. *Int J Cancer*. 2002;101:151–155.
142. Li M, Bolduc AR, Hoda MN, et al. The indoleamine 2,3-dioxygenase pathway controls complement-dependent enhancement of chemo-radiation therapy against murine glioblastoma. *J Immunother Cancer*. 2014;2:21.
143. Lanzetta G, Campanella C, Rozzi A, et al. Temozolomide in radio-chemotherapy combined treatment for newly-diagnosed glioblastoma multiforme: phase II clinical trial. *Anticancer Res*. 2003;23:5159–5164.
144. Kocher M, Kunze S, Eich HT, Semrau R, Muller RP. Efficacy and toxicity of postoperative temozolomide radiochemotherapy in malignant glioma. *Strahlenther Onkol*. 2005;181:157–163.
145. Su YB, Sohn S, Krown SE, et al. Selective $CD4^+$ lymphopenia in melanoma patients treated with temozolomide: a toxicity with therapeutic implications. *J Clin Oncol*. 2004;22:610–616.
146. Heimberger AB, Sun W, Hussain SF, et al. Immunological responses in a patient with glioblastoma multiforme treated with sequential courses of temozolomide and immunotherapy: case study. *Neuro Oncol*. 2008;10:98–103.
147. Schuster J, Lai RK, Recht LD, et al. A phase II, multicenter trial of rindopepimut (CDX-110) in newly diagnosed glioblastoma: the ACT III study. *Neuro Oncol*. 2015;17:854–861.
148. Langroudi L, Hassan ZM, Ebtekar M, Mahdavi M, Pakravan N, Noori S. A comparison of low-dose cyclophosphamide treatment with artemisinin treatment in reducing the number of regulatory T cells in murine breast cancer model. *Int Immunopharmacol*. 2010;10:1055–1061.
149. Sharabi A, Laronne-Bar-On A, Meshorer A, Haran-Ghera N. Chemoimmunotherapy reduces the progression of multiple myeloma in a mouse model. *Cancer Prev Res*. 2010;3:1265–1276.
150. Ge Y, Domschke C, Stoiber N, et al. Metronomic cyclophosphamide treatment in metastasized breast cancer patients: immunological effects and clinical outcome. *Cancer Immunol Immunother*. 2012;61:353–362.
151. Vermeij R, Leffers N, Hoogeboom BN, et al. Potentiation of a p53-SLP vaccine by cyclophosphamide in ovarian cancer: a single-arm phase II study. *Int J Cancer*. 2012;131:E670–E680.
152. Peng S, Lyford-Pike S, Akpeng B, et al. Low-dose cyclophosphamide administered as daily or single dose enhances the antitumor effects of a therapeutic HPV vaccine. *Cancer Immunol Immunother*. 2013;62:171–182.
153. Huijts CM, Santegoets SJ, van den Eertwegh AJ, et al. Phase I–II study of everolimus and low-dose oral cyclophosphamide in patients with metastatic renal cell cancer. *BMC Cancer*. 2011;11:505.
154. Berd D, Maguire Jr HC, McCue P, Mastrangelo MJ. Treatment of metastatic melanoma with an autologous tumor-cell vaccine: clinical and immunologic results in 64 patients. *J Clin Oncol*. 1990;8:1858–1867.

155. Plautz GE, Miller DW, Barnett GH, et al. T cell adoptive immunotherapy of newly diagnosed gliomas. *Clin Cancer Res*. 2000;6:2209–2218.
156. Kreitman RJ, Wilson WH, White JD, et al. Phase I trial of recombinant immunotoxin anti-Tac(Fv)-PE38 (LMB-2) in patients with hematologic malignancies. *J Clin Oncol*. 2000;18:1622–1636.
157. Powell Jr DJ, Felipe-Silva A, Merino MJ, et al. Administration of a CD25-directed immunotoxin, LMB-2, to patients with metastatic melanoma induces a selective partial reduction in regulatory T cells in vivo. *J Immunol*. 2007;179:4919–4928.
158. Attia P, Maker AV, Haworth LR, Rogers-Freezer L, Rosenberg SA. Inability of a fusion protein of IL-2 and diphtheria toxin (Denileukin Diftitox, DAB389IL-2, ONTAK) to eliminate regulatory T lymphocytes in patients with melanoma. *J Immunother*. 2005;28:582–592.
159. Dannull J, Su Z, Rizzieri D, et al. Enhancement of vaccine-mediated antitumor immunity in cancer patients after depletion of regulatory T cells. *J Clin Invest*. 2005;115:3623–3633.
160. Mahnke K, Schonfeld K, Fondel S, et al. Depletion of $CD4^+CD25^+$ human regulatory T cells in vivo: kinetics of Treg depletion and alterations in immune functions in vivo and in vitro. *Int J Cancer*. 2007;120:2723–2733.
161. Prince HM, Duvic M, Martin A, et al. Phase III placebo-controlled trial of denileukin diftitox for patients with cutaneous T-cell lymphoma. *J Clin Oncol*. 2010;28:1870–1877.
162. Ansell SM, Tang H, Kurtin PJ, et al. Denileukin diftitox in combination with rituximab for previously untreated follicular B-cell non-Hodgkin's lymphoma. *Leukemia*. 2012;26:1046–1052.
163. Sampson JH, Schmittling RJ, Archer GE, et al. A pilot study of IL-2Ralpha blockade during lymphopenia depletes regulatory T-cells and correlates with enhanced immunity in patients with glioblastoma. *PLoS One*. 2012;7:e31046.
164. Mathew M, Tam M, Ott PA, et al. Ipilimumab in melanoma with limited brain metastases treated with stereotactic radiosurgery. *Melanoma Res*. 2013;23:191–195.
165. Wilgenhof S, Du Four S, Vandenbroucke F, et al. Single-center experience with ipilimumab in an expanded access program for patients with pretreated advanced melanoma. *J Immunother*. 2013;36:215–222.
166. Bot I, Blank CU, Boogerd W, Brandsma D. Neurological immune-related adverse events of ipilimumab. *Pract Neurol*. 2013;13:278–280.
167. Phan GQ, Yang JC, Sherry RM, et al. Cancer regression and autoimmunity induced by cytotoxic T lymphocyte-associated antigen 4 blockade in patients with metastatic melanoma. *Proc Natl Acad Sci USA*. 2003;100:8372–8377.
168. Attia P, Phan GQ, Maker AV, et al. Autoimmunity correlates with tumor regression in patients with metastatic melanoma treated with anti-cytotoxic T-lymphocyte antigen-4. *J Clin Oncol*. 2005;23:6043–6053.
169. Maker AV, Phan GQ, Attia P, et al. Tumor regression and autoimmunity in patients treated with cytotoxic T lymphocyte-associated antigen 4 blockade and interleukin 2: a phase I/II study. *Ann Surg Oncol*. 2005;12:1005–1016.
170. Shevach EM, Stephens GL. The GITR-GITRL interaction: co-stimulation or contrasuppression of regulatory activity? *Nat Rev Immunol*. 2006;6:613–618.
171. Peterson AC, Migawa MT, Martin MJ, et al. Evaluation of functionalized tryptophan derivatives and related compounds as competitive inhibitors of indoleamine 2,3-dioxygenase. *Med Chem Res*. 1994;3:531–544.
172. Austin CJ, Mailu BM, Maghzal GJ, et al. Biochemical characteristics and inhibitor selectivity of mouse indoleamine 2,3-dioxygenase-2. *Amino Acids*. 2010;39:565–578.
173. Qian F, Liao J, Villella J, et al. Effects of 1-methyltryptophan stereoisomers on IDO2 enzyme activity and IDO2-mediated arrest of human T cell proliferation. *Cancer Immunol Immunother*. 2012;61:2013–2020.
174. Yuasa HJ, Ball HJ, Austin CJ, Hunt NH. 1-L-methyltryptophan is a more effective inhibitor of vertebrate IDO2 enzymes than 1-D-methyltryptophan. *Comp Biochem Physiol B Biochem Mol Biol*. 2010;157:10–15.

175. Hou DY, Muller AJ, Sharma MD, et al. Inhibition of indoleamine 2,3-dioxygenase in dendritic cells by stereoisomers of 1-methyl-tryptophan correlates with antitumor responses. *Cancer Res.* 2007;67:792–801.
176. Liu X, Shin N, Koblish HK, et al. Selective inhibition of IDO1 effectively regulates mediators of antitumor immunity. *Blood.* 2010;115:3520–3530.
177. Meininger D, Zalameda L, Liu Y, et al. Purification and kinetic characterization of human indoleamine 2,3-dioxygenases 1 and 2 (IDO1 and IDO2) and discovery of selective IDO1 inhibitors. *Biochim Biophys Acta.* 2011;1814:1947–1954.
178. Bakmiwewa SM, Fatokun AA, Tran A, Payne RJ, Hunt NH, Ball HJ. Identification of selective inhibitors of indoleamine 2,3-dioxygenase 2. *Bioorg Med Chem Lett.* 2012;22:7641–7646.

CHAPTER

4

The Role of Myeloid-Derived Suppressor Cells in Immunosuppression in Brain Tumors

K. Gabrusiewicz[1]*, N.A. Colwell*[2]*, A.B. Heimberger*[1]

[1]The University of Texas MD Anderson Cancer Center, Houston, TX, United States; [2]Florida International University Herbert Wertheim College of Medicine, Miami, FL, United States

OUTLINE

Translational Immunotherapy of Brain Tumors
http://dx.doi.org/10.1016/B978-0-12-802420-1.00004-1

BACKGROUND AND DISCOVERY OF THE MYELOID CELL

In healthy individuals, immature myeloid cells (IMCs) form in the bone marrow from hematopoietic stem cells, migrate into peripheral lymphoid organs, and differentiate into mature granulocytes, macrophages, or dendritic cells (DCs). However, under various pathological conditions such as infection,[1] cancer, trauma,[2,3] or inflammatory autoimmune responses,[4,5] IMC expansion can be promoted and differentiation inhibited. A number of tumor-derived factors and host cytokines activate and attract IMCs to the inflammatory site or tumor, thereby resulting in an immunosuppressive immune cell population denoted as myeloid-derived suppressor cells (MDSCs).

MDSCs were first described in 1964 as an infiltrative population within a squamous cell carcinoma.[6] Later, in animal models of either tumor-[7,8] or *Bacillus Calmette–Guerin*[9,10]–induced inflammation, an undefined IMC infiltration was observed that inhibited T cell activity. Initially, these cells were named veto cells, null cells, or natural suppressor cells because they did not express markers for mature T lymphocytes, B lymphocytes, or natural killer (NK) cells.[11–13] After a preliminary characterization, immunologists referred to them as "IMCs"[14,15] or "myeloid suppressor cells";[16] however, the latter designation incorrectly implied that these cells were normal, mature myeloid cells. Consensus designates them as MDSCs,[17] but there still remains some controversy surrounding this nomenclature. Alternative proposed terminology such as "myeloid immunosuppressive cells" or "myeloid immune suppressor cells", would better reflect their function.

MDSCs in Mice

In mice, MDSCs are characterized by the coexpression of the myeloid-cell lineage differentiation antigen Gr-1 and the integrin CD11b. Notably, CD11b is expressed on a variety of immune cells, including monocytes, granulocytes, macrophages, DCs, and microglia (brain resident macrophages). The Gr-1 antigen is a cell-surface protein that belongs to the Ly6 family of proteins and is expressed on monocytes, and granulocytes. In naïve mice, MDSCs account for 20–30% of bone marrow cells and 2–4% of cells in the spleen. In cancer, MDSCs can be detected in the bone marrow, spleen, blood, tumor, and lymph nodes. Subpopulations of MDSCs have been distinguished using two criteria: nuclear morphology and expression of Ly6C and Ly6G. Based on these markers and other supplementary markers such as F4/80, colony-stimulating factor 1 receptor (CSF-1R, CD115), and interleukin 4 receptor alpha (IL-4Rα, CD124), two subpopulations have been identified. These include $CD11b^{+}Ly6G^{+}Ly6C^{low}$ polymorphonuclear granulocytic MDSCs (G-MDSCs), which express high levels of arginase 1 (Arg1), and $CD11b^{+}Ly6G^{-}Ly6C^{high}$ monocytic MDSCs (MO-MDSCs), that preferentially express nitric oxide synthase 2 (iNOS) (Table 4.1). G-MDSCs are the predominant subset of IMCs that accumulate in the bone marrow and spleen of tumor-bearing mice, but this is variable within tumor models.[18,19]

MDSCs in Humans

The role of MDSCs in humans was described in the setting of autologous stem-cell transplantation in cancer patients who had elevated levels of immunosuppressive $CD14^{+}$ monocytes and T cell dysfunction.[20,21] Unlike those in mice, human MDSCs are less phenotypically defined but do express the common myeloid markers CD11b and CD33. In addition, human MDSCs lack or have a low level of human leukocyte antigen DR (HLA-DR).[22–24] Depending on CD14 and CD15 expression, two MDSC subpopulations have been identified: $CD33^{+}/CD11b^{+}/CD14^{+}/CD15^{-}/HLA\text{-}DR^{-}$ monocytic and $CD33^{+}/CD11b^{+}/CD14^{-}/CD15^{+}/HLA\text{-}DR^{-}$ granulocytic MDSCs (Table 4.1). Additional makers are under consideration to describe MDSCs, including high levels of CD66b, low levels of CD62L and CD16, and expression of vascular endothelial growth factor receptor 1 and S100A9.[25,26] Human fibrocytic MDSCs have recently been described as a novel MDSC subset in patients with metastatic cancer.[27,28] Morphologically, this population resembles IMCs and expresses markers that overlap with the previously characterized $CD33^{low/absent}CD11b^{+}/CD15^{+}/CD66b^{+}/IL\text{-}4R\alpha^{+}$ MDSCs but is distinct with regard to high expression of HLA-DR, IL-7Rα, thymic stromal lymphopoietin protein receptor, alpha-smooth muscle actin, and collagens[27] (Table 4.1). Moreover, in contrast to MO-MDSCs and G-MDSCs, fibrocytic MDSC-mediated T cell suppression and regulatory T cell (T_{reg}) expansion is related to production of indoleamine 2,3-dioxygenase.[27,28]

MDSC Activation, Proliferation, Migration, and Expansion

Although initially described in association with cancer, MDSCs are a normal component of the inflammatory response that enhances innate

TABLE 4.1 Phenotype of Murine and Human MDSCs

	Phenotype			
	Murine MDSCs		Human MDSCs	
Subtype	**Main Markers**	**Additional Markers**	**Main Markers**	**Additional Markers**
MO-MDSCs	$CD11b^+$	$IL\text{-}4R\alpha^+$	$CD33^+$	$IL\text{-}4R\alpha^+$
	$Ly6G^-$	$F4/80^+$	$CD11b^+$	$CSF\text{-}1R^+$
	$Ly6C^{high}$	$iNOS^+$	$CD14^+$	$Tie2^+$
			$CD15^-$	
			$HLA\text{-}DR^-$	
G-MDSCs	$CD11b^+$	$IL\text{-}4R\alpha^+$	$CD33^+$	$IL\text{-}4R\alpha^+$
	$Ly6G^+$	$Arg1^+$	$CD11b^+$	$CD66b^+$
	$Ly6C^{low}$		$CD14^-$	
			$CD15^+$	
			$HLA\text{-}DR^-$	
F-MDSCs	NI	NI	$CD33^{low/absent}$	$IL\text{-}4R\alpha^+$
			$CD11b^+$	$CD66b^+$
			$CD15^+$	$IL\text{-}7R\alpha^+$
			$HLA\text{-}DR^+$	$TSLPR^+$
				$\alpha\text{-}SMA^+$
				Collagen I^+/V^+

α-SMA, alpha-smooth muscle actin; *Arg1*, arginase 1; *CSF-1R*, colony-stimulating factor 1 receptor; *F*, fibrocytic; *G*, granulocytic; *HLA-DR*, human leukocyte antigen D-related; *iNOS*, inducible nitric oxide synthase; *MDSC*, myeloid-derived suppressor cell; *MO*, monocytic; *NI*, not identified; *TSLPR*, thymic stromal lymphopoietin receptor.

immune responses through the production of reactive oxygen species (ROS) and inflammatory cytokines. When needed, precursors of MDSCs proliferate in the bone marrow, are mobilized into the systemic circulation, accumulate at sites of inflammation, and differentiate into mature cells in peripheral tissues.[29] During infection, adaptive effector responses secrete proinflammatory factors that recruit the MDSC, which then downregulates the immune response and promotes tissue remodeling and repair.[30] During chronic inflammation, the effector response is unable to clear the antigenic source and proinflammatory cytokines continue to accumulate in spite of MDSCs' attempts to suppress this response. The persistent (albeit low level) activation of MDSCs results in secretion of ROS and their intermediates, with subsequent damage to the DNA of nearby tissue bystander cells.[30] However, persistence of the MDSCs indicates a pathological process such as chronic inflammation and/or tumorigenesis.[30] Physiologic MDSC expansion begins in the bone marrow and is regulated by a variety of cytokines such as granulocyte-macrophage colony-stimulating factor (GM-CSF) and macrophage colony-stimulating factor (M-CSF).[31] Initially, $CD34^+$ stem cells differentiate into a common myeloid progenitor cell and then into IMCs. In healthy subjects, the IMCs leave the bone marrow and accumulate in peripheral tissues where they mature into monocytes, granulocytes, or DCs. However, during pathologic conditions, this final differentiation step is inhibited by interleukins (ILs) such as IL-6 and IL-1β and, as a result, activated MDSCs are formed.[31,32]

MDSC function may operate at two different levels of intensity, rather than simply being active or inactive.[33] MDSC function can also be separated into two different processes: expansion and activation. Although the two processes are distinct, overlapping regulatory signaling pathways are involved. The "two-signal" model may explain why acute inflammation generates only a mild MDSC response in the absence of sustained upregulation of proinflammatory factors and why stimulation of signal transducer and activator of transcription (STAT) 3 and STAT5 during physiologic steady-state hematopoiesis does not induce MDSC accumulation. However, in the presence of significant proinflammatory molecules, STAT3 and STAT5 stimulation results in robust expansion of MDSCs.[33] The process of MDSC activation specifically refers to the induction of immune suppressive function, a defining characteristic of the MDSC phenotype. Bacterial and viral products, tumor stromal cells, and activated T cells are the main drivers of MDSC activation, which occurs via several downstream signaling pathways. Examples include phosphoinositide 3-kinase/Akt, Janus kinase (JAK)/STAT, and nuclear factor kappa-light-chain-enhancer of activated B cells (NF-κB), resulting in the expression of genes like iNOS, IL-10, and Arg1 that confer the immunosuppressive function specific to the MDSC phenotype.[32]

Chemokines are also likely to participate in the process of MDSC activation. For example, S100A8/A9 can activate MDSCs via the NADPH complex, which triggers proliferation and synergy with ROS production via the STAT3 pathway. This mechanism of expansion and activation is commonly associated with tumorigenesis.[29] MDSCs are also activated by IL-6, CXCL1, and hypoxia itself via activation of STAT3, NF-κB, and hypoxia-inducible factor (HIF)-1 alpha. The formation of new blood vessels subsequently promotes activation of the MDSCs, which secrete several proangiogenic factors.[34] STAT3 activation induces antiapoptotic protein expression such as that of Bcl-xL, c-myc, cyclin D1, or survivin, which prevents MDSC apoptosis and differentiation into mature cell types.[33,35] The end result is promotion of the survival of IMCs and inhibition of their further differentiation into mature myeloid cells. In murine models, STAT3 may also be activated by apoptosis inhibitor 6, which causes expansion of MDSCs.[36]

STAT1 and STAT6 pathways are also associated with MDSC activation and contribute to the upregulation of proteins involved in immune suppression, such as Arg1 and transforming growth factor-beta (TGF-β).[29] STAT6 tends to be activated by cytokines such as IL-4 and IL-13, secreted by activated T cells and macrophages in the presence of ongoing inflammation, whereas STAT1 is induced by interferon-gamma (IFN-γ) and IL-1β in the context of tumorigenesis.[29]

Suppressive Mechanisms Mediated by MDSCs

Immunosuppressive activity of MDSCs can be achieved by direct cell-to-cell contact using cell surface receptors and/or through the secretion of short-lived mediators. These mechanisms may be relatively specific to MDSC subtypes. For example, the granulocytic MDSCs express high levels of ROS and low levels of NO, whereas the monocytic population produces less ROS but more NO.[19] Despite having different mechanisms of action, both subsets of MDSCs negatively control T cell response in the tumor microenvironment (TME).

Suppression via Arg1 and iNOS

The suppressive activity of MDSCs is associated with the metabolism of L-arginine. Arg1 and iNOS are enzymes that compete for the same substrate, L-arginine. iNOS converts L-arginine to NO and L-citrulline, whereas Arg1 catalyzes the production of L-ornithine and urea.[37] Arginine metabolism in myeloid cells has been used to distinguish M1 tumor-suppressive from M2 tumor-promoting macrophages. In tumor-supportive M2-polarized macrophages, Arg1 production is induced by IL-4, IL-13, TGF-β, macrophage-stimulating protein, and/or GM-CSF. On the other hand, tumor-suppressive M1-polarized macrophages express iNOS, which is stimulated by IFN-γ (mainly in combination with lipopolysaccharide) and other proinflammatory cytokines such as IL-1, tumor necrosis factor-alpha (TNF-α), IFN-α, or IFN-β. STAT1 is activated in response to signals within M1-polarized macrophages, whereas STAT3 and STAT6 are selectively activated within M2-polarized macrophages.[38,39] MDSCs do not entirely fit into the current dichotomous view of macrophages because they express high levels of Arg1 and iNOS, and both of these enzymes can directly inhibit T cell function[40,41] (Fig. 4.1). MDSCs interact also with tumor-associated microglia/macrophages (TAMs). Through the production of IL-10, MDSCs downregulate IL-12 secretion

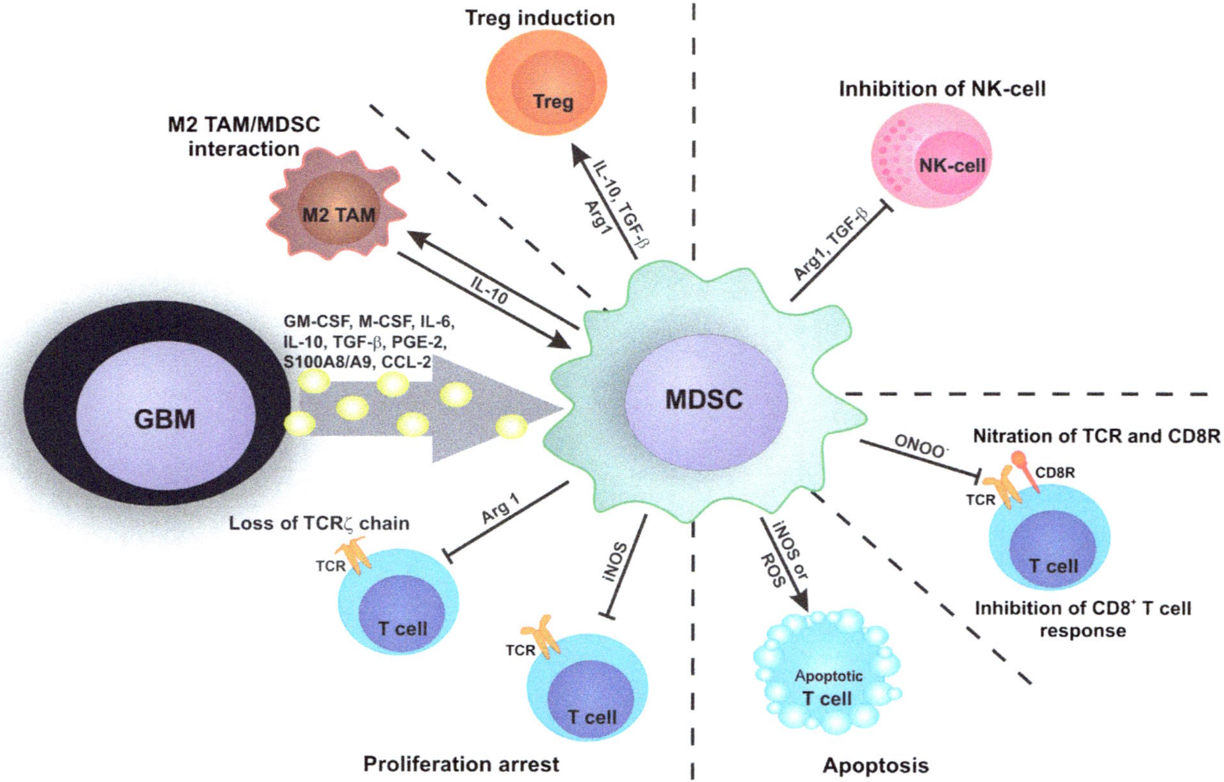

FIGURE 4.1 Immunosuppressive properties of MDSCs. GBM-derived factors (GM-CSF, M-CSF, IL-6, IL-10, TGF-β, PGE-2, S100A8/A9, and CCL-2) block myeloid cell maturation and promote differentiation into MDSCs. As a result, MDSCs through multiple mechanisms contribute to the immunosuppressive microenvironment of GBM. MDSC-dependent T cell suppression and apoptosis are regulated by Arg1/iNOS and ROS, respectively. Secretion of Arg1, TGF-β, and IL-10 promotes Treg development and expansion. MDSC-produced IL-10 stimulates TAMs to polarize toward the tumor-supportive M2 phenotype. NK-cell function is inhibited through Arg1 and TGF-β production by MDSCs. $ONOO^-$ causes nitration of tyrosine residues in the TCR and CD8R, which results in inhibition of $CD8^+$ T cell response. *Arg1*, arginase 1; *GBM*, glioblastoma; *GM-CSF*, granulocyte-macrophage colony-stimulating factor; *IL*, interleukin; *iNOS*, inducible nitric oxide synthase; *M-CSF*, macrophage colony-stimulating factor; *MDSC*, myeloid-derived suppressor cell; *ROS*, reactive oxygen species; *TAM*, tumor-associated microglia/macrophage; *TCR*, T cell receptor; *TGF*, transforming growth factor.

by macrophages, skewing M1 macrophages toward the M2 phenotype.[42] L-arginine catabolism by Arg1 leads to depletion of nonessential amino acids from the microenvironment, thereby inducing T cell dysfunction.[43] T cell suppression is regulated by several distinct mechanisms, including the loss of the T cell receptor CD3ζ chain[44–46] and a decrease in expression of cell cycle regulators cyclin D3 and cyclin-dependent kinase 4. NO inhibits T cell activity by blocking the activation of JAK3, STAT5, extracellular signal-regulated kinases, and Akt.[47] NO has also been reported to down-regulate major histocompatibility complex class II molecule (MHC II)[48] and promote T cell apoptosis.[49]

Suppression via Reactive Oxygen Species

Production of ROS by MDSCs has been reported in both mouse tumor models and in patients with cancer.[19,50–52] Several tumor-derived factors, such as TGF-β, IL-3, IL-6, IL-10, platelet-derived growth factor, and GM-CSF, can induce the production of ROS by MDSCs. MDSC-mediated ROS are also created by increased activity of nicotinamide adenine dinucleotide phosphate-oxidase 2. The lack of NOX2 activity abolished the ability of MDSCs to suppress T cells.[53] Interestingly, ROS sensitize T cells to apoptosis by decreasing expression of Bcl-2.[54]

Suppression via Peroxynitrite

Peroxynitrite ($ONOO^-$) is formed from the reaction between NO and O_2^-. MDSCs are potent producers of peroxynitrite, and high levels of this anion are associated with tumor progression.[55–57] The main role of peroxynitrite is the modification of proteins by oxidation or nitration of amino acids such as tyrosine, cysteine, methionine, and tryptophan. These modifications of proteins can change their physical and chemical properties. For instance, peroxynitrite causes nitration of tyrosine residues in the T cell receptor (TCR) and CD8 coreceptor, leading to changed conformation of TCR chains, thus inhibiting $CD8^+$ T cell–specific response.[58]

Other Immunosuppressive Mechanisms

There is increasing evidence that MDSCs induce Tregs in tumor-bearing hosts that may be dependent on Arg1 produced by MDSCs.[59] Studies carried out by Huang[60] and coworkers demonstrated that MDSC-stimulated Treg expansion required IL-10 and IFN-γ but not NO. Recently, MDSCs isolated from patients with chronic lymphocytic leukemia showed the ability to suppress T cells and promote Treg development.[61] In contrast, other groups have found that expansion of MDSCs is not related to Treg induction.[62,63] Thus further work is required to resolve these discrepancies.

Finally, MDSCs have also been shown to suppress NK cell function via membrane bound TGF-β, STAT5 activity, Arg1, the NKp30 receptor, or downregulation of NKG2D.[64–67]

MDSCs in Glioma Development

MDSCs were first described in murine gliomas in 2008.[41] In GL261 models, MDSCs constituted less than 5% of the $CD11b^+$-tumor infiltrating cells and expressed $CD11c^+/Gr\text{-}1^+/IL\text{-}4R\alpha^+$. These cells had the unique feature of simultaneously expressing markers for proinflammatory M1 cells, such as iNOS, IL-1β, TNF-α, and CXCL10, plus tumor-supportive M2-polarized macrophages, such as Arg1, CCL17, CD206, and CD36, indicating their pleiotropic character. More recently, studies in genetically engineered murine models (GEMMs) of glioma demonstrated that MDSCs constituted about 8% of the total tumor cell count and mostly consisted of the $CD11b^+/Gr\text{-}1^{low}$ monocytic population rather than the $CD11b^+/Gr\text{-}1^{high}$ granulocytic subset.[68] The functional role of MDSCs in gliomagenesis and malignant degeneration is a current area of investigation. Notably, tumor-bearing mice depleted of $CD11b^+Gr1^+$ cells survive longer than those treated with control isotype IgG,[69,70] suggesting that this immune population does contribute to the pathogenesis and prognosis of gliomas.

IL-4Rα expression on MDSCs is known to play a crucial role in their immunosuppressive activity.[71,72] In murine model systems of glioma, and in glioblastoma (GBM) patients, IL-4Rα is upregulated on $CD11b^+/Gr1^+$ cells.[69] GM-CSF secretion by glioma cells induces IL-4Rα expression on myeloid cells, thereby promoting the production of Arg1 via IL-13. IL-4Rα knockout mice developed gliomas with a lower infiltration of MDSCs than was seen in wild-type mice. These studies also revealed that the $CD11b^+Ly6C^+$ MO-MDSCs expressed higher levels of immunosuppressive Arg1 and TGF-β than $CD11b^+Ly6G^+$ granulocytic MDSCs. Subsequent studies in the GL261 murine model demonstrated that the tumors were predominantly infiltrated by $CD11b^+/Gr\text{-}1^{low}$ MO-MDSCs.[73] MDSCs were significantly elevated in the spleen and bone marrow after mice received additional syngeneic $CD11b^+$ monocytes, suggesting that tumor-released factors can direct monocytes toward the immunosuppressive MDSC state.

GBM patients have also been shown to have an elevated number of $CD33^+/HLA\text{-}DR^-$ MDSCs in the peripheral blood relative to normal donors.[74,75] MDSCs found in the blood of GBM patients are mainly composed of the $CD33^+/CD15^+/CD14^-/HLA\text{-}DR^-$ neutrophilic subset, and contain less of the $CD33^+/CD15^-/CD14^-/HLA\text{-}DR^-$ negative lineage and the $CD33^+/CD15^-/CD14^+/HLA\text{-}DR^-$ monocytic subpopulation.[75,76] However, a new study contradicts the previously published results and indicates that the monocytic subset is the most abundant MDSC subpopulation in the blood of GBM patients.[77] Within the GBM tumor mass, MDSCs represent about 5% of the cells and have been characterized predominantly as the $CD33^+/CD15^-/CD14^-/HLA\text{-}DR^-$ negative lineage, followed by $CD33^+/CD15^+/CD14^-/HLA\text{-}DR^-$ neutrophilic and $CD33^+/CD15^-/CD14^+/HLA\text{-}DR^-$ monocytic subtypes.[68] Contrary to these findings, Gielsen and coworkers[77] have recently shown that the neutrophilic MDSCs are the predominant population in GBM tumors. Other studies on freshly resected tumor tissue from GBM

TABLE 4.2 MDSCs as a Target for Glioblastoma Therapy

Drugs	Mechanism of Action	Preclinical and Clinical Outcome
Anti-CCL2	Decreases chemotaxis of MDSCs and TAMs	Activity in GL261 glioma model No clinical trial
COX-2 inhibitors (acetylsalicylic acid, celecoxib)	Prevents production of PGE2, decreases CCL2-dependent chemotaxis of MDSCs, increases infiltration $CD8^+$ T cells by enhancing expression of CXCL10	Activity in murine glioma model No significant clinical activity
Galectin-1 knock down	Decreases accumulation of MDSCs and TAMs	Activity in GL261 glioma model No clinical therapeutic approach available
13-Cis-retinoic acid	Induces differentiation of MDSCs	Not tested in glioma model No clinical benefits
IL-4Rα aptamer	Induces apoptosis of MDSCs by STAT6 suppression	Not tested in glioma model No clinical trial
Arg1 inhibitor (nor-NOHA)	Prevents L-arginine depletion in TME, restores T cell function	Not tested in glioma model No clinical trial
L-arginine supplementation	Compensates L-arginine depletion in TME, restores T cell function	Not tested in glioma model phase I study with oral L-arginine (NCT02017249)
CD200R antagonist peptide (A26059)	Blocks expansion of MDSCs, reduces Arg1 secretion, activates $CD8^+$ T cell response	Activity in GL261 glioma model No clinical trial
Anti-FGL2 antibody	Reduces MDSC number, TAMs, $CD39^+$ Tregs, and PD-1	Activity in GL261 glioma model No clinical trial
Anti–TGF-β antibody	Enhances iNOS production while decreasing Arg1 expression in MDSCs	No activity in GL261 glioma model or U87MG xenograft model No significant clinical benefits

Continued

TABLE 4.2 MDSCs as a Target for Glioblastoma Therapy—cont'd

Drugs	Mechanism of Action	Preclinical and Clinical Outcome
IL-12 immunotherapy	Decreases MDSC number, increases expression of CD80 and MHC II on MDSCs	Activity in GL261 model phase I study with adenovirus expressing IL-12 (NCT02026271)
miR-142-3p	Blocks TGFBR1 signaling and induces apoptosis in M2-polarized macrophages	Activity in GEMM of glioma and GL261 glioma model No clinical trial
Anti-VEGF and anti–G-CSF antibodies	Suppresses MDSCs, Bv8, and tumor vasculature	Not tested in glioma model No clinical trial
Sunitinib	Decreases MDSC number, induces T cell function	Activity in GEMM of glioma, no activity in U87MG xenograft model No clinical benefits
WP1066	Blocks STAT3 pathway and M2 polarization	Activity in GEMM of glioma and GL261 glioma model phase I study with WP1066 (NCT01904123)
CSF-1R inhibitors (BLZ945, PLX3397)	Blocks M2-polarized macrophages	Activity in GEMM of glioma and GL261 glioma model No clinical trial

Arg1, arginase 1; *Bv8*, Bombina variegata peptide 8; *COX-2*, cyclooxygenase 2; *CSF-1R*, colony-stimulating factor 1 receptor; *FGL2*, fibrinogen-like protein 2; *G-CSF*, granulocyte colony-stimulating factor; *GEMM*, genetically engineered mouse model; *IL*, interleukin; *iNOS*, inducible nitric oxide synthase; *MDSC*, myeloid-derived suppressor cell; *MHC II*, major histocompatibility complex class II; *PD-1*, programmed cell death protein 1; *PGE2*, prostaglandin E2; *STAT*, signal transducer and activator of transcription; *TAM*, tumor-associated macrophage; *T cell*, T lymphocyte; *TGF-β*, transforming growth factor beta; *TGFBR1*, transforming growth factor beta receptor 1; *TME*, tumor microenvironment; *Tregs*, regulatory T cells; *VEGF*, vascular endothelial growth factor.

patients have shown that tumor-infiltrating MDSCs highly express HLA-DR.[76] Thus it remains unclear whether HLA-DR is expressed by GBM-associated MDSCs and which subtype occupies the tumor most frequently.

The accumulation of the MDSCs seems to correlate with glioma grade,[78] but to date, no studies have specifically addressed the prognostic influence of the MDSC on GBM patient outcome. MDSCs do appear to be induced by the presence of glioma cells. Healthy donor human $CD14^+$ cells exposed to GBM cells acquired properties of MDSCs manifested by increased production of IL-10, TGF-β, and B7-H1 (PD-L1) expression. These cells also had diminished phagocytic ability and increased ability to induce apoptosis in T cells.[74,79]

PRECLINICAL DATA: TARGETING THE MDSC POPULATION

Chemokine and Cytokine Approaches

MDSCs are specifically recruited to the hypoxic TME via chemotactic factors such as CXCL1, CXCL12, CXCL5, CCL2, CXCL2, and S100A8/A9. In syngeneic and xenograft murine glioma model systems, administration of an anti-CCL2 antibody can decrease the number of MDSCs and TAMs in the TME, leading to prolonged survival of tumor-bearing mice[80] (Table 4.2). Of note, chemokines can be redundant, and therefore upregulation of alternative chemokine pathways is conceivable in the setting of focal targeting. However, upregulation of several of these chemokines is specifically mediated by STAT3,[33] rendering it a viable therapeutic target. Cancer cells also secrete factors like prostaglandin E2 (PGE2) and CXCL12 (also known as stromal cell-derived factor 1) that participate in the recruitment of MDSCs to the tumor site, separate from hypoxia-induced mechanisms.[32] Cyclooxygenase-2 (COX-2) inhibitors such as acetylsalicylic acid or celecoxib can prevent production of PGE2, reduce CCL2-mediated accumulation of $CD11b^{+}Ly6G^{+}$ granulocytic MDSCs in both bone marrow and the TME, and increase $CD8^{+}$ T cells in a CXCL10-dependent manner.[70] These mechanisms are by no means specific to the MDSC population and have not shown significant clinical activity.[81–83] Silencing glioma-derived galectin-1 significantly prolonged the survival of glioma-bearing mice by decreasing the accumulation of glioma-infiltrating microglia/macrophages and MDSCs,[84] but currently there is no viable therapeutic approach for accomplishing this in human subjects.

Differentiation

Although a key feature of MDSCs is the inhibition of their peripheral differentiation into mature myeloid cells, it appears that MDSCs can be induced to differentiate into pathologic TAMs at the tumor site.[31] This plasticity could be harnessed therapeutically, as MDSCs have been induced to differentiate into physiologic, mature myeloid cells via the use of retinoic acid in both mouse and human studies.[85] However, clinical trials of these agents in GBM patients have not demonstrated clinical benefit.[86]

IL-4Rα Targeting

Roth et al.[87] designed a highly efficient RNA aptamer targeting mouse and human IL-4Rα, a cell-surface receptor known to be overexpressed on MDSCs in tumor-bearing mice[69] and cancer patients.[69,72] This aptamer preferentially affected MDSCs and TAMs, inducing their apoptosis, promoting greater T cell infiltration and tumor regression. The anti–IL-4Rα aptamer prevented IL-13 from binding to IL-4Rα on myeloid cells followed by STAT6 phosphorylation, demonstrating the antagonist properties of this molecule. Despite high efficacy of the anti–IL-4Rα aptamer on MDSCs in a mammary carcinoma model, the oligonucleotide has not yet been tested in a rodent glioma model.

Arginase 1

Significant elevations of Arg1 activity and granulocyte colony-stimulating factor production have been detected in the serum of GBM patients.[75,88] MDSCs suppress T cell function via this mechanism, and this process can be reversed by pharmacologic Arg1 inhibition or with arginine supplementation.[88] A clinical trial is currently underway in GBM patients to assess immune function in subjects treated with oral arginine (NCT02017249). Immunosuppression in gliomas is also regulated by tumor-derived CD200 and its receptor on myeloid cells. A CD200R antagonist peptide (A26059) has been shown to block the expansion of MDSCs, reduce secretion of Arg1, and to activate the $CD8^+$ T cell response.[89] The preclinical testing was conducted in a clonotypic murine model, and whether this approach can exert a therapeutic response in a heterogeneous GEMM of glioma is unknown but nonetheless would represent a promising therapeutic direction.

Fibrinogen-Like Protein 2 Targeting

Most recently, we have shown that fibrinogen-like protein 2 (FGL2) plays a role in MDSC glioma accumulation. FGL2 was shown to be upregulated in human GBM, correlated with glioma grade,[90] and promoted tumor growth. FGL2 induces CD39 expression in glioma-associated lymphocytes. CD39 converts ATP to adenosine[91,92] which then suppresses T cell effector function,[93] enhances M2 activity,[94] and promotes the tumor supportive role of MDSCs.[95] The reduction of CD39 is associated with diminished M2 and MDSC accumulation in glioma. The use of an anti-FGL2 antibody in GL261-bearing mice reduced not only the number of MDSCs but also the levels of Tregs and TAMs. A humanized anti-FGL2 antibody is in the process of being developed.

M1 Polarization or M2 Blockade

The concept of blocking M2 polarization has been building momentum. Initially, this was attempted by blocking TGF-β in the GL261 glioma model,[41] but thus far anti-TGF-β therapeutics have not demonstrated significant clinical activity. IL-12 immunotherapy can alter the number of tumor-infiltrating

MDSCs and push them toward an M1 antigen-presenting cell phenotype, overexpressing CD80 and MHC II.[96] Another approach includes using miR-142-3p to block TGFBR1 M2 autocrine-dependent signaling, resulting in preferential M2 apoptosis that correlates with glioma growth suppression.[97] This strategy will require further formulation development prior to clinical trial implementation.

Recent studies suggest that MDSCs contribute to tumor recurrence after anti-VEGF therapies.[98,99] Treatment of refractory tumors with the combination of anti-VEGF and anti–G-CSF dramatically suppressed circulating or tumor-associated MDSCs, contributing to reduced tumor growth.[100] In GEMM of glioma, sunitinib, a tyrosine kinase inhibitor with antiangiogenic and anti-invasive properties, has also been shown to limit both granulocytic and MO-MDSCs in the circulation and in the tumor.[68] Sunitinib is also a nonspecific STAT3 blockade agent.[101] STAT3 has been shown to be a key driver for M2 conversion.[102,103] Our group has shown that a small molecule inhibitor of STAT3, WP1066, can block M2 polarization,[79] which is related to the MDSC immunosuppressive function. Notably, STAT3 blockade approaches are not specific to the MDSC population but exert their effects on a variety of immune suppressive mechanisms that contribute to their therapeutic activity.[104–106] Clinical implementation of WP1066 was confounded by formulation issues that have ultimately been resolved by using nanoparticles, and the clinical trial is anticipated to be open by 2016 (NCT01904123). But perhaps the strategy that would be most ready for clinical trial applications would be one targeting CSF-1R. Blocking M2 polarization with the inhibitor CSF-1R has been shown to suppress glioma growth.[107] Currently, there are open clinical trials of agents targeting CSF-1R in patients with solid malignancies (NCT01346358, NCT02452424) but none yet in glioma patients.

CLINICAL APPROACHES TO TARGETING MDSCs

In contradistinction to other immune suppressive targets and immune suppressive populations, the development of targeted approaches to MDSCs is relatively new (Table 4.2). In some cases, a therapeutic has shown a decrease in $CD11b^+/Gr\text{-}1^+$ MDSCs in the TME,[108] such as in the case of adoptive immunotherapy, but it is unclear whether this represents a direct effect or a surrogate of therapeutic activity. There is a rationale behind targeting immunosuppressive populations such as MDSCs, given their role in tumor-mediated immune suppression and association with tumor progression. Targeting MDSCs can be achieved by depleting them, or by inhibiting their differentiation or function. There are a variety of methods that could deplete MDSCs, including use of monoclonal antibodies and chemotherapeutics, and a variety of novel strategies exist that can block polarization that are starting to make their way into clinical application. Although there are efforts underway to suppress MDSCs within the context of anticancer immunotherapy, there are always possible unintended consequences for immune surveillance and antimicrobial activity. MDSC-targeted therapeutic approaches may have a preferential beneficial effect on patients with the mesenchymal subtype of GBM,[109] as this group demonstrates an enrichment of MDSC-associated markers.[110]

References

1. Zeng QL, Yang B, Sun HQ, et al. Myeloid-derived suppressor cells are associated with viral persistence and downregulation of TCR zeta chain expression on $CD8^+$ T cells in chronic hepatitis C patients. *Mol Cells.* 2014;37(1):66–73.
2. Huang H, Cao K, Malik S, et al. Combination of 12-O-tetradecanoylphorbol-13-acetate with diethyldithiocarbamate markedly inhibits pancreatic cancer cell growth in 3D culture and in immunodeficient mice. *Int J Mol Med.* 2015;35(6):1617–1624.
3. Ruan X, Darwiche SS, Cai C, Scott MJ, Pape HC, Billiar TR. Anti-HMGB1 monoclonal antibody ameliorates immunosuppression after peripheral tissue trauma: attenuated T-lymphocyte response and increased splenic $CD11b^+$ $Gr\text{-}1^+$ myeloid-derived suppressor cells require HMGB1. *Mediators Inflamm.* 2015;2015:458626.
4. Crook KR, Jin M, Weeks MF, et al. Myeloid-derived suppressor cells regulate T cell and B cell responses during autoimmune disease. *J Leukoc Biol.* 2015;97(3):573–582.
5. Ioannou M, Alissafi T, Lazaridis I, et al. Crucial role of granulocytic myeloid-derived suppressor cells in the regulation of central nervous system autoimmune disease. *J Immunol.* 2012;188(3):1136–1146.
6. Lappat EJ, Cawein M. A study of the leukemoid response to transplantable A-280 tumor in mice. *Cancer Res.* 1964;24:302–311.
7. Lee MY, Rosse C. Depletion of lymphocyte subpopulations in primary and secondary lymphoid organs of mice by a transplanted granulocytosis-inducing mammary carcinoma. *Cancer Res.* 1982;42(4):1255–1260.
8. Tsuchiya Y, Igarashi M, Suzuki R, Kumagai K. Production of colony-stimulating factor by tumor cells and the factor-mediated induction of suppressor cells. *J Immunol.* 1988;141(2):699–708.
9. Bennett JA, Mitchell MS. Induction of suppressor cells by intravenous administration of Bacillus Calmette-Guerin and its modulation by cyclophosphamide. *Biochem Pharmacol.* 1979;28(12):1947–1952.
10. Wren SM, Wepsic HT, Larson CH, De Silva MA, Mizushima Y. Inhibition of the graft-versus-host response by BCGcw-induced suppressor cells or prostaglandin E1. *Cell Immunol.* 1983;76(2):361–371.
11. Duwe AK, Singhal SK. The immunoregulatory role of bone marrow. I. Suppression of the induction of antibody responses to T-dependent and T-independent antigens by cells in the bone marrow. *Cell Immunol.* 1979;43(2):362–371.
12. Oseroff A, Okada S, Strober S. Natural suppressor (NS) cells found in the spleen of neonatal mice and adult mice given total lymphoid irradiation (TLI) express the null surface phenotype. *J Immunol.* 1984;132(1):101–110.
13. Slavin S, Strober S. Induction of allograft tolerance after total lymphoid irradiation (TLI): development of suppressor cells of the mixed leukocyte reaction (MLR). *J Immunol.* 1979;123(2):942–946.

14. Deng ZB, Liu Y, Liu C, et al. Immature myeloid cells induced by a high-fat diet contribute to liver inflammation. *Hepatology*. 2009;50(5):1412–1420.
15. Kusmartsev S, Gabrilovich DI. Role of immature myeloid cells in mechanisms of immune evasion in cancer. *Cancer Immunol Immunother*. 2006;55(3):237–245.
16. Yang R, Cai Z, Zhang Y, Yutzy WH, Roby KF, Roden RB. CD80 in immune suppression by mouse ovarian carcinoma-associated Gr-1$^+$CD11b$^+$ myeloid cells. *Cancer Res*. 2006;66(13):6807–6815.
17. Gabrilovich DI, Bronte V, Chen SH, et al. The terminology issue for myeloid-derived suppressor cells. *Cancer Res*. 2007;67(1):425. author reply 426.
18. Haverkamp JM, Smith AM, Weinlich R, et al. Myeloid-derived suppressor activity is mediated by monocytic lineages maintained by continuous inhibition of extrinsic and intrinsic death pathways. *Immunity*. 2014;41(6):947–959.
19. Youn JI, Nagaraj S, Collazo M, Gabrilovich DI. Subsets of myeloid-derived suppressor cells in tumor-bearing mice. *J Immunol*. 2008;181(8):5797–5802.
20. Mielcarek M, Martin PJ, Torok-Storb B. Suppression of alloantigen-induced T-cell proliferation by CD14$^+$ cells derived from granulocyte colony-stimulating factor-mobilized peripheral blood mononuclear cells. *Blood*. 1997;89(5):1629–1634.
21. Singh RK, Varney ML, Buyukberber S, et al. Fas-FasL-mediated CD4$^+$ T-cell apoptosis following stem cell transplantation. *Cancer Res*. 1999;59(13):3107–3111.
22. Almand B, Clark JI, Nikitina E, et al. Increased production of immature myeloid cells in cancer patients: a mechanism of immunosuppression in cancer. *J Immunol*. 2001;166(1):678–689.
23. Hoechst B, Ormandy LA, Ballmaier M, et al. A new population of myeloid-derived suppressor cells in hepatocellular carcinoma patients induces CD4$^+$CD25$^+$Foxp3$^+$ T cells. *Gastroenterology*. 2008;135(1):234–243.
24. Filipazzi P, Valenti R, Huber V, et al. Identification of a new subset of myeloid suppressor cells in peripheral blood of melanoma patients with modulation by a granulocyte-macrophase colony-stimulation factor-based antitumor vaccine. *J Clin Oncol*. 2007;25(18):2546–2553.
25. Rodriguez PC, Ernstoff MS, Hernandez C, et al. Arginase I-producing myeloid-derived suppressor cells in renal cell carcinoma are a subpopulation of activated granulocytes. *Cancer Res*. 2009;69(4):1553–1560.
26. Zhao F, Hoechst B, Duffy A, et al. S100A9 a new marker for monocytic human myeloid-derived suppressor cells. *Immunology*. 2012;136(2):176–183.
27. Zhang H, Maric I, DiPrima MJ, et al. Fibrocytes represent a novel MDSC subset circulating in patients with metastatic cancer. *Blood*. 2013;122(7):1105–1113.
28. Zoso A, Mazza EM, Bicciato S, et al. Human fibrocytic myeloid-derived suppressor cells express IDO and promote tolerance via Treg-cell expansion. *Eur J Immunol*. 2014;44(11):3307–3319.
29. Gabrilovich DI, Nagaraj S. Myeloid-derived suppressor cells as regulators of the immune system. *Nat Rev Immunol*. 2009;9(3):162–174.
30. Serafini P. Myeloid derived suppressor cells in physiological and pathological conditions: the good, the bad, and the ugly. *Immunol Res*. 2013;57(1–3):172–184.
31. Youn JI, Gabrilovich DI. The biology of myeloid-derived suppressor cells: the blessing and the curse of morphological and functional heterogeneity. *Eur J Immunol*. 2010;40(11):2969–2975.
32. Trikha P, Carson 3rd WE. Signaling pathways involved in MDSC regulation. *Biochim Biophys Acta*. 2014;1846(1):55–65.
33. Condamine T, Gabrilovich DI. Molecular mechanisms regulating myeloid-derived suppressor cell differentiation and function. *Trends Immunol*. 2011;32(1):19–25.
34. Condamine T, Ramachandran I, Youn JI, Gabrilovich DI. Regulation of tumor metastasis by myeloid-derived suppressor cells. *Annu Rev Med*. 2015;66:97–110.
35. Yu H, Pardoll D, Jove R. STATs in cancer inflammation and immunity: a leading role for STAT3. *Nat Rev Cancer*. 2009;9(11):798–809.

36. Qu P, Du H, Li Y, Yan C. Myeloid-specific expression of Api6/AIM/Sp alpha induces systemic inflammation and adenocarcinoma in the lung. *J Immunol*. 2009;182(3):1648–1659.
37. Bronte V, Zanovello P. Regulation of immune responses by L-arginine metabolism. *Nat Rev Immunol*. 2005;5(8):641–654.
38. Mantovani A, Sica A, Allavena P, Garlanda C, Locati M. Tumor-associated macrophages and the related myeloid-derived suppressor cells as a paradigm of the diversity of macrophage activation. *Hum Immunol*. 2009;70(5):325–330.
39. Qian BZ, Pollard JW. Macrophage diversity enhances tumor progression and metastasis. *Cell*. 2010;141(1):39–51.
40. Jiao ZJ, Gao JJ, Hua SH, et al. Correlation between circulating myeloid-derived suppressor cells and Th17 cells in esophageal cancer. *World J Gastroenterol*. 2012;18(38):5454–5461.
41. Umemura N, Saio M, Suwa T, et al. Tumor-infiltrating myeloid-derived suppressor cells are pleiotropic-inflamed monocytes/macrophages that bear M1- and M2-type characteristics. *J Leukoc Biol*. 2008;83(5):1136–1144.
42. Sinha P, Clements VK, Bunt SK, Albelda SM, Ostrand-Rosenberg S. Cross-talk between myeloid-derived suppressor cells and macrophages subverts tumor immunity toward a type 2 response. *J Immunol*. 2007;179(2):977–983.
43. Bronte V, Serafini P, De Santo C, et al. IL-4-induced arginase 1 suppresses alloreactive T cells in tumor-bearing mice. *J Immunol*. 2003;170(1):270–278.
44. Ezernitchi AV, Vaknin I, Cohen-Daniel L, et al. TCR zeta down-regulation under chronic inflammation is mediated by myeloid suppressor cells differentially distributed between various lymphatic organs. *J Immunol*. 2006;177(7):4763–4772.
45. Rodriguez PC, Zea AH, Culotta KS, Zabaleta J, Ochoa JB, Ochoa AC. Regulation of T cell receptor CD3zeta chain expression by L-arginine. *J Biol Chem*. 2002;277(24):21123–21129.
46. Rodriguez PC, Zea AH, DeSalvo J, et al. L-arginine consumption by macrophages modulates the expression of CD3ζ chain in T lymphocytes. *J Immunol*. 2003;171(3):1232–1239.
47. Mazzoni A, Bronte V, Visintin A, et al. Myeloid suppressor lines inhibit T cell responses by an NO-dependent mechanism. *J Immunol*. 2002;168(2):689–695.
48. Harari O, Liao JK. Inhibition of MHC II gene transcription by nitric oxide and antioxidants. *Curr Pharm Des*. 2004;10(8):893–898.
49. Rivoltini L, Carrabba M, Huber V, et al. Immunity to cancer: attack and escape in T lymphocyte-tumor cell interaction. *Immunol Rev*. 2002;188:97–113.
50. Kusmartsev S, Nefedova Y, Yoder D, Gabrilovich DI. Antigen-specific inhibition of $CD8^+$ T cell response by immature myeloid cells in cancer is mediated by reactive oxygen species. *J Immunol*. 2004;172(2):989–999.
51. Schmielau J, Finn OJ. Activated granulocytes and granulocyte-derived hydrogen peroxide are the underlying mechanism of suppression of T-cell function in advanced cancer patients. *Cancer Res*. 2001;61(12):4756–4760.
52. Szuster-Ciesielska A, Hryciuk-Umer E, Stepulak A, Kupisz K, Kandefer-Szerszen M. Reactive oxygen species production by blood neutrophils of patients with laryngeal carcinoma and antioxidative enzyme activity in their blood. *Acta Oncol*. 2004;43(3):252–258.
53. Corzo CA, Cotter MJ, Cheng P, et al. Mechanism regulating reactive oxygen species in tumor-induced myeloid-derived suppressor cells. *J Immunol*. 2009;182(9):5693–5701.
54. Hildeman DA, Mitchell T, Aronow B, Wojciechowski S, Kappler J, Marrack P. Control of Bcl-2 expression by reactive oxygen species. *Proc Natl Acad Sci USA*. 2003;100(25):15035–15040.
55. Dairou J, Atmane N, Rodrigues-Lima F, Dupret JM. Peroxynitrite irreversibly inactivates the human xenobiotic-metabolizing enzyme arylamine N-acetyltransferase 1 (NAT1) in human breast cancer cells: a cellular and mechanistic study. *J Biol Chem*. 2004;279(9):7708–7714.
56. Nakamura Y, Yasuoka H, Tsujimoto M, et al. Nitric oxide in breast cancer: induction of vascular endothelial growth factor-C and correlation with metastasis and poor prognosis. *Clin Cancer Res*. 2006;12(4):1201–1207.

57. Vickers SM, MacMillan-Crow LA, Green M, Ellis C, Thompson JA. Association of increased immunostaining for inducible nitric oxide synthase and nitrotyrosine with fibroblast growth factor transformation in pancreatic cancer. *Arch Surg.* 1999;134(3):245–251.
58. Nagaraj S, Gupta K, Pisarev V, et al. Altered recognition of antigen is a mechanism of $CD8^+$ T cell tolerance in cancer. *Nat Med.* 2007;13(7):828–835.
59. Serafini P, Mgebroff S, Noonan K, Borrello I. Myeloid-derived suppressor cells promote cross-tolerance in B-cell lymphoma by expanding regulatory T cells. *Cancer Res.* 2008;68(13):5439–5449.
60. Huang B, Pan PY, Li Q, et al. $Gr\text{-}1^+CD115^+$ immature myeloid suppressor cells mediate the development of tumor-induced T regulatory cells and T-cell anergy in tumor-bearing host. *Cancer Res.* 2006;66(2):1123–1131.
61. Jitschin R, Braun M, Buttner M, et al. CLL-cells induce IDO^{hi} $CD14^+HLA\text{-}DR^{lo}$ myeloid-derived suppressor cells that inhibit T-cell responses and promote T_{Regs}. *Blood.* 2014;124(5):750–760.
62. Dugast AS, Haudebourg T, Coulon F, et al. Myeloid-derived suppressor cells accumulate in kidney allograft tolerance and specifically suppress effector T cell expansion. *J Immunol.* 2008;180(12):7898–7906.
63. Movahedi K, Guilliams M, Van den Bossche J, et al. Identification of discrete tumor-induced myeloid-derived suppressor cell subpopulations with distinct T cell-suppressive activity. *Blood.* 2008;111(8):4233–4244.
64. Hoechst B, Voigtlaender T, Ormandy L, et al. Myeloid derived suppressor cells inhibit natural killer cells in patients with hepatocellular carcinoma via the NKp30 receptor. *Hepatology.* 2009;50(3):799–807.
65. Li H, Han Y, Guo Q, Zhang M, Cao X. Cancer-expanded myeloid-derived suppressor cells induce anergy of NK cells through membrane-bound TGF-β1. *J Immunol.* 2009;182(1):240–249.
66. Liu C, Yu S, Kappes J, et al. Expansion of spleen myeloid suppressor cells represses NK cell cytotoxicity in tumor-bearing host. *Blood.* 2007;109(10):4336–4342.
67. Oberlies J, Watzl C, Giese T, et al. Regulation of NK cell function by human granulocyte arginase. *J Immunol.* 2009;182(9):5259–5267.
68. Raychaudhuri B, Rayman P, Huang P, et al. Myeloid derived suppressor cell infiltration of murine and human gliomas is associated with reduction of tumor infiltrating lymphocytes. *J Neurooncol.* 2015;122(2):293–301.
69. Kohanbash G, McKaveney K, Sakaki M, et al. GM-CSF promotes the immunosuppressive activity of glioma-infiltrating myeloid cells through interleukin-4 receptor-α. *Cancer Res.* 2013;73(21):6413–6423.
70. Fujita M, Kohanbash G, Fellows-Mayle W, et al. COX-2 blockade suppresses gliomagenesis by inhibiting myeloid-derived suppressor cells. *Cancer Res.* 2011;71(7):2664–2674.
71. Gallina G, Dolcetti L, Serafini P, et al. Tumors induce a subset of inflammatory monocytes with immunosuppressive activity on $CD8^+$ T cells. *J Clin Invest.* 2006;116(10):2777–2790.
72. Mandruzzato S, Solito S, Falisi E, et al. $IL4R\alpha^+$ myeloid-derived suppressor cell expansion in cancer patients. *J Immunol.* 2009;182(10):6562–6568.
73. Chae M, Peterson TE, Balgeman A, et al. Increasing glioma-associated monocytes leads to increased intratumoral and systemic myeloid-derived suppressor cells in a murine model. *Neuro Oncol.* 2015;17(7):978–991.
74. Rodrigues JC, Gonzalez GC, Zhang L, et al. Normal human monocytes exposed to glioma cells acquire myeloid-derived suppressor cell-like properties. *Neuro Oncol.* 2010;12(4):351–365.
75. Raychaudhuri B, Rayman P, Ireland J, et al. Myeloid-derived suppressor cell accumulation and function in patients with newly diagnosed glioblastoma. *Neuro Oncol.* 2011;13(6):591–599.

76. Grauer OM, Wölfer J, Hasselblatt M, et al. Granulocytic myeloid-derived cells obtained from blood mainly suppress T-cell proliferation in patients with glioblastoma. In: *Paper presented at: 11th Congress of the European Association of Neuro-Oncology, 2014, Turin, Italy.* ; 2011.
77. Gielen PR, Schulte BM, Kers-Rebel ED, et al. Increase in both CD14-positive and CD15-positive myeloid-derived suppressor cell subpopulations in the blood of patients with glioma but predominance of CD15-positive myeloid-derived suppressor cells in glioma tissue. *J Neuropathol Exp Neurol*. 2015;74(5):390–400.
78. Prosniak M, Harshyne LA, Andrews DW, et al. Glioma grade is associated with the accumulation and activity of cells bearing M2 monocyte markers. *Clin Cancer Res*. 2013;19(14):3776–3786.
79. Wu A, Wei J, Kong LY, et al. Glioma cancer stem cells induce immunosuppressive macrophages/microglia. *Neuro Oncol*. 2010;12(11):1113–1125.
80. Zhu X, Fujita M, Snyder LA, Okada H. Systemic delivery of neutralizing antibody targeting CCL2 for glioma therapy. *J Neurooncol*. 2011;104(1):83–92.
81. Welzel G, Gehweiler J, Brehmer S, et al. Metronomic chemotherapy with daily low-dose temozolomide and celecoxib in elderly patients with newly diagnosed glioblastoma multiforme: a retrospective analysis. *J Neurooncol*. 2015. June 5 (ePub ahead of print).
82. Penas-Prado M, Hess KR, Fisch MJ, et al. Randomized phase II adjuvant factorial study of dose-dense temozolomide alone and in combination with isotretinoin, celecoxib, and/or thalidomide for glioblastoma. *Neuro Oncol*. 2015;17(2):266–273.
83. Kesari S, Schiff D, Henson JW, et al. Phase II study of temozolomide, thalidomide, and celecoxib for newly diagnosed glioblastoma in adults. *Neuro Oncol*. 2008;10(3):300–308.
84. Verschuere T, Toelen J, Maes W, et al. Glioma-derived galectin-1 regulates innate and adaptive antitumor immunity. *Int J Cancer*. 2014;134(4):873–884.
85. Talmadge JE, Gabrilovich DI. History of myeloid-derived suppressor cells. *Nat Rev Cancer*. 2013;13(10):739–752.
86. Butowski N, Prados MD, Lamborn KR, et al. A phase II study of concurrent temozolomide and cis-retinoic acid with radiation for adult patients with newly diagnosed supratentorial glioblastoma. *Int J Radiat Oncol Biol Phys*. 2005;61(5):1454–1459.
87. Roth F, De La Fuente AC, Vella JL, Zoso A, Inverardi L, Serafini P. Aptamer-mediated blockade of IL4Rα triggers apoptosis of MDSCs and limits tumor progression. *Cancer Res*. 2012;72(6):1373–1383.
88. Sippel TR, White J, Nag K, et al. Neutrophil degranulation and immunosuppression in patients with GBM: restoration of cellular immune function by targeting arginase I. *Clin Cancer Res*. 2011;17(22):6992–7002.
89. Moertel CL, Xia J, LaRue R, et al. CD200 in CNS tumor-induced immunosuppression: the role for CD200 pathway blockade in targeted immunotherapy. *J Immunother Cancer*. 2014;2(1):46.
90. Yan J, Kong LY, Hu J, et al. FGL2 as a multimodality regulator of tumor-mediated immune suppression and therapeutic target in gliomas. *J Natl Cancer Inst*. 2015;107(8).
91. Regateiro FS, Cobbold SP, Waldmann H. CD73 and adenosine generation in the creation of regulatory microenvironments. *Clin Exp Immunol*. 2013;171(1):1–7.
92. Xu S, Shao QQ, Sun JT, et al. Synergy between the ectoenzymes CD39 and CD73 contributes to adenosinergic immunosuppression in human malignant gliomas. *Neuro Oncol*. 2013;15(9):1160–1172.
93. Ohta A, Sitkovsky M. Extracellular adenosine-mediated modulation of regulatory T cells. *Front Immunol*. 2014;5:304.
94. Dwyer KM, Deaglio S, Gao W, Friedman D, Strom TB, Robson SC. CD39 and control of cellular immune responses. *Purinergic Signal*. 2007;3(1–2):171–180.
95. Antonioli L, Pacher P, Vizi ES, Hasko G. CD39 and CD73 in immunity and inflammation. *Trends Mol Med*. 2013;19(6):355–367.

96. Thaci B, Ahmed AU, Ulasov IV, et al. Depletion of myeloid-derived suppressor cells during interleukin-12 immunogene therapy does not confer a survival advantage in experimental malignant glioma. *Cancer Gene Ther*. 2014;21(1):38–44.
97. Xu S, Wei J, Wang F, et al. Effect of miR-142-3p on the M2 macrophage and therapeutic efficacy against murine glioblastoma. *J Natl Cancer Inst*. 2014;106(8):dju162.
98. Piao Y, Liang J, Holmes L, et al. Glioblastoma resistance to anti-VEGF therapy is associated with myeloid cell infiltration, stem cell accumulation, and a mesenchymal phenotype. *Neuro Oncol*. 2012;14(11):1379–1392.
99. Shojaei F, Wu X, Malik AK, et al. Tumor refractoriness to anti-VEGF treatment is mediated by $CD11b^{+}Gr1^{+}$ myeloid cells. *Nat Biotechnol*. 2007;25(8):911–920.
100. Shojaei F, Wu X, Qu X, et al. G-CSF-initiated myeloid cell mobilization and angiogenesis mediated tumor refractoriness to anti-VEGF therapy in mouse models. *Proc Natl Acad Sci USA*. 2009;106(16):6742–6747.
101. Xin H, Zhang C, Herrmann A, Du Y, Figlin R, Yu H. Sunitinib inhibition of Stat3 induces renal cell carcinoma tumor cell apoptosis and reduces immunosuppressive cells. *Cancer Res*. 2009;69(6):2506–2513.
102. Sica A, Bronte V. Altered macrophage differentiation and immune dysfunction in tumor development. *J Clin Invest*. 2007;117(5):1155–1166.
103. Qin H, Holdbrooks AT, Liu Y, Reynolds SL, Yanagisawa LL, Benveniste EN. SOCS3 deficiency promotes M1 macrophase polarization and inflammation. *J Immunol*. 2012;189(7):3439–3448.
104. Kong LY, Wu AS, Doucette T, et al. Intratumoral mediated immunosuppression is prognostic in genetically engineered murine models of glioma and correlates to immunotherapeutic responses. *Clin Cancer Res*. 2010;16(23):5722–5733.
105. Wei J, Barr J, Kong LY, et al. Glioblastoma cancer-initiating cells inhibit T-cell proliferation and effector responses by the signal transducers and activators of transcription 3 pathway. *Mol Cancer Ther*. 2010;9(1):67–78.
106. Hussain SF, Kong LY, Jordan J, et al. A novel small molecule inhibitor of signal transducers and activators of transcription 3 reverses immune tolerance in malignant glioma patients. *Cancer Res*. 2007;67(20):9630–9636.
107. Pyonteck SM, Akkari L, Schuhmacher AJ, et al. CSF-1R inhibition alters macrophage polarization and blocks glioma progression. *Nat Med*. 2013;19(10):1264–1272.
108. Zhang Y, Luo F, Li A, et al. Systemic injection of TLR1/2 agonist improves adiptive antigen-specific T cell therapy in glioma-bearing mice. *Clin Immunol*. 2014;154(1):26–36.
109. Verhaak RG, Hoadley KA, Purdom E, et al. Integrated genomic analysis identifies clinically relevant subtypes of glioblastoma characterized by abnormalities in PDGFRA, IDH1, EGFR, and NF1. *Cancer Cell*. 2010;17(1):98–110.
110. Doucette TA, Rao G, Rao A, et al. Immune heterogeneity of glioblastoma subtypes: extrapolation from the cancer genome atlas. *Cancer Immunol Res*. 2013;1(2):112–122.

CHAPTER

5

Tumor-Specific Mutations in Gliomas and Their Implications for Immunotherapy

C.J. Pirozzi, A.B. Carpenter, T. Hennika, O.J. Becher, H. Yan

Duke University Medical Center, Durham, NC, United States

OUTLINE

http://dx.doi.org/10.1016/B978-0-12-802420-1.00005-3

INTRODUCTION

The culmination of the Human Genome Project in April 2003 marked the beginning of the postgenome era. Taking 13 years and approximately $2.7 billion to complete, this endeavor set the stage for tremendous advancements in understanding the molecular underpinnings of variation within humans as well as the genetic basis for disease. As it is now possible to sequence a genome within a few days at a cost close to $1000, efforts are being put forth to sequence and characterize the personalized genomic aberrations present in human tumors on an individual basis. Initiatives, including The Cancer Genome Atlas (TCGA), are well underway to sequence cancers, especially those cancers exhibiting poor prognosis with a high overall public health impact. To date, TCGA has successfully and completely sequenced 31 cancers, with data for all but four publicly available.

Due to the dismal prognosis and an inability to improve survival despite the best of efforts, glioblastoma (GBM) was one of the first cancers to be sequenced by the TCGA.[1] The ultimate goals of the project were to acquire a baseline knowledge of GBM genetics and to determine the genetic alterations in GBM for application toward diagnostics, prognostics, and elucidation of therapeutically targetable components. Several important findings resulted, including the discovery of mutations in core components of the retinoblastoma tumor suppressor, TP53 tumor suppressor, and receptor tyrosine kinase (RTK) pathways.[1,2] Concurrent studies have identified several pathways and genes previously unknown to be altered in GBM.[2] This chapter will present those mutations identified in these sequencing studies and explore their immunotherapeutic potential. Specifically, mutations in isocitrate dehydrogenase 1 (IDH1) and epidermal growth factor receptor (EGFR) identified in adult gliomas, and mutations in *H3F3A* identified in pediatric high grade gliomas will be discussed.

MUTATIONS AS IMMUNOTHERAPEUTIC TARGETS

Treating and curing gliomas are incredibly difficult. For the past several decades, the median survival has remained stagnant at approximately 14.7 months.[15] Any aberrant growth within the confined space of the skull inevitably leads to compression, increased pressure, and ultimately neurological symptoms and dysfunction. Subsequently, tumor growth in this region is likely to interfere with vital structures, and for this reason, surgical resection is usually incomplete. Current standard of care therapy for glioma relies on surgical resection, followed by adjuvant external beam radiation and chemotherapy. This regimen can be highly toxic and cause nonspecific damage to a healthy brain.[15,16] Additionally, these tumors are highly infiltrative by nature, conferring a high rate of recurrence.[4] Considering these difficulties, harnessing the power of the patient's immune system to target tumor-specific mutations offers an ideal means of treating gliomas with minimal invasiveness and off-target collateral damage.

Glioma immunotherapy functions by augmenting the host antitumor immune response against tumor antigens and offers a promising alternative to the current standard of care. It can be highly specific, minimally toxic, and is capable of killing tumor cells while sparing normal tissue. Additionally, this modality can activate immune surveillance mechanisms, offering long-term control over this disease. One area of concern surrounding cancer immunotherapy is that many of the antigens used to elicit an immune response may be present in both tumor cells and in normal cells. However, in light of next-generation sequencing and the ability to rapidly and affordably assess a cancer patient's genome, exome, epigenome, and transcriptome, tumor-specific antigens can be determined on an individual and tumor-specific basis. These tumor-specific antigens are in contrast to tumor-associated antigens, which may display more promiscuous expression among both normal and tumor tissues. Mining the data provided by sequencing of patient tumors will permit the identification of unique therapeutic targets functioning as neoantigens.[17] Since these mutated gene products will be expressed in a tumor-specific manner, resulting in tumor-specific cytotoxicity, the risk and subsequent adverse events of targeting normal tissue will be abrogated.[16,18]

ISOCITRATE DEHYDROGENASE 1 MUTATIONS IN GLIOMAS

The most frequently mutated gene identified in gliomas is *IDH1*, which was found to be mutated in as high as 80% of the progressive gliomas (tumors that have an inherent propensity for progressing from the Grade II and Grade III gliomas to the secondary Grade IV GBMs).[3]

The hotspot mutation lies at arginine 132 of *IDH1* and is most commonly mutated to histidine. In its most recently completed cancer genome study, the TCGA has corroborated the findings that *IDH1* is frequently mutated in this subset of gliomas and is an early hit during oncogenic transformation, preceding other genetic aberrations.[4,5]

IDH1 acts in the tricarboxylic acid cycle, where it functions in the oxidative decarboxylation of isocitrate to α-ketoglutarate (α-KG) in a nicotinamide adenine dinucleotide phosphate ($NADP^+$)-dependent manner.[6] The functional consequences of the R132H hotspot mutation have been determined to be a gain of enzymatic function in the NADPH-dependent reduction of α-KG to a new metabolic product, 2-hydroxyglutarate (2-HG).[7] This was detected in tumors at levels 100-fold higher in *IDH1*-mutant tumors than in *IDH1*-wild-type tumors. 2-HG, now considered an "oncometabolite," has been shown to have widespread effects within mutant cells. For instance, 2-HG promotes the glioma-CpG island methylator phenotype, blocks the normal differentiation patterns of the cells in which it is mutated, and dysregulates families of α-KG-dependent dioxygenases, further leading to restructuring of the epigenome.[8–14]

IDH1 MUTATIONS AS AN IMMUNOTHERAPEUTIC TARGET

Mutations in IDH1 are early events in tumorigenesis and are present in nearly all tumor cells. Furthermore, immunohistochemical staining of IDH1-mutant tumors reveals that it is highly tumor specific. Targeting this mutation through immunotherapy will likely eliminate not only tumor bulk but also microscopic foci that would otherwise go unresected and, in some instances, untreated, thereby reducing the high probability of recurrence. These features, coupled with a mutation frequency as high as 80% in the progressive gliomas, make mutant IDH1 a particularly appealing and ideal candidate for immunotherapy-related modalities.

PRECLINICAL INVESTIGATIONS: IDH1-R132H PEPTIDE VACCINES

Since mutant IDH1 is exclusively expressed in tumors, is an early hit in tumor development, and is a likely oncogenic driver, it is a particularly desirable target for peptide vaccines. Peptide vaccination is a form of non-cell-based, active immunotherapy that stimulates adaptive immunity against tumor-associated and/or tumor-specific antigens.[3,5,19,20] Peptide vaccines are comprised of 8–25 amino acids that encompass an epitope within an antigenic target.[18] To enhance their immunogenicity, they are often conjugated to carrier proteins. In contrast to dendritic cell (DC) vaccines, peptide vaccines are relatively easy to manufacture and do not require laborious preparations.[18] In most instances of peptide vaccination, tumor-associated antigens are being targeted. Because these antigens are expressed in normal tissues as well, T cells directed against these antigens are subject to negative selection in the thymus.[20,21] However, tumor-specific neoantigens, like mutant IDH1, are not subject to this negative selection; rather, mutant IDH1 neoantigens have higher specificity, reduced off-target cytotoxicity, and can elicit highly potent immune responses.[19,22]

Several preclinical studies have been described followed by the initiation of several clinical trials seeking to determine the potential of therapeutically treating IDH1-mutant tumors through the use of mutant-IDH1 peptide vaccines.[23,24] Among the first such preclinical investigations was a study by Schumacher et al. that identified an IDH1-R132H-containing immunogenic epitope suitable for mutation-specific vaccination.[23] Major histocompatibility complex (MHC) binding prediction algorithms and T2 binding assays showed no evidence of peptide binding to the common MHC class I human leukocyte antigen (HLA) allele, HLA-A*0201. However, there was evidence of peptide binding to the common MHC class II allele, HLA-DRB1*0101. In fact, additional studies from the same group investigated whether an IDH1-R132H epitope was presented on HLA-DR in human glioma tissue through an adaptation of the proximity ligation assay typically used to visualize protein–protein interactions.[25] Here, they identified colocalization of IDH1-R132H peptides with HLA-DR in glioma tissue, lending credence that IDH1 mutant peptides bind to MHC class II molecules. Subsequent studies were performed in transgenic MHC-humanized mice, A2.DR1 (HLA-A*0201 HLA-DRA*0101 HLA-DRB*0101), devoid of mouse MHC class I and II alleles. When these animals were vaccinated using a peptide containing amino acids 123–142 of IDH1, a robust interferon (IFN)-γ T-cell response resulted only when the R132H-containing peptide was used. This response was restricted to class II $CD4^+$ T cells and elicited a T_H1 phenotype. IDH1-R132H-specific antibodies were detected in the serum of animals immunized with this peptide. Ensuring the clinical relevance of this finding, human samples were screened for similar responses. Four out of 25 patients with mutations in *IDH1*, but 0 out of 29 patients with wild-type tumors, displayed IFN-γ-producing T cells specific for IDH1-R132H. Similar to the animal studies, when an HLA-DR-blocking antibody was used, the effects were abrogated, lending further support that this is an MHC class II-restricted response. These antibodies were neither present in tumors that were IDH1-wild-type nor in healthy donors. This suggested for the first time that IDH1-R132H is endogenously processed in tumor-bearing patients and is presented on MHC class II molecules to $CD4^+$ T-cells, inducing a T_H1 response with subsequent, mutation-specific antibody production.

Utilizing a chemically-induced sarcoma model overexpressing either wild-type or mutant IDH1, Schumacher et al. sought to assess the effects of the p123-142 IDH1-R132H peptide vaccine on tumor growth. Two approaches were taken. First, a preventive vaccination, which entailed immunization with the p123-142 peptide containing IDH1-R132H, was assessed. Following administration of the vaccine, IDH1-R132H sarcoma cells were transplanted. Excitingly, suppression of tumor growth was observed in the vaccinated animals. Similarly, when vaccination followed IDH1-wild-type and IDH1-R132H tumor establishment, suppression of tumor growth was observed only in those tumors of animals vaccinated with the IDH1-R132H peptide. In both instances, tumor growth was unaffected when the reciprocal experiments were performed with either the mutant peptide vaccine being used against a wild-type tumor, or a wild-type vaccine being used against a mutant tumor. Scrutiny of the immune response yielding this abatement of tumor progression showed that IDH1-R132H sarcomas were infiltrated with the IDH1-R132H HLA-DRB1*0101 T cells, and that $CD4^+$ T cells and not $CD8^+$ T cells were antigen specific. To further strengthen the significance of $CD4^+$ T cells in tumor suppression, $CD4^+$ T cells were depleted in the context of the IDH1-R132H peptide vaccine and tumor burden. Subsequently, the therapeutic effects of the IDH1-R132H peptide vaccine were abrogated, indicating therapeutic efficacy is dependent on $CD4^+$ T cells. A similar effect was observed when B cells were depleted, suggesting B-cell contribution to the vaccine efficacy.

Tumor suppression as a consequence of the mutant IDH1 peptide vaccination is enlightening, especially considering that antitumor immunity resultant from antigen-specific $CD4^+$ T cells in the absence of an MHC class I-restricted $CD8^+$ T-cell response is controversial, though not completely undescribed. In fact, one study showed that when HLA class II-restricted epitopes of tumor-associated antigens were isolated and expanded in vitro, followed by infusion into a patient with metastatic melanoma, long-term complete remission occurred.[26] Similarly, another study showed that transfer of small numbers of tumor antigen-specific $CD4^+$ T cells into irradiated recipients was capable of expansion and differentiation into cytotoxic $CD4^+$ T cells, leading to elimination of the melanoma.[27] Interestingly, one study even determined the "mutanome," the unique set of mutations that a tumor possesses, in several murine tumor models. Here, they found that there is a considerable fraction of nonsynonymous cancer mutations that are immunogenic, and the majority of these immunogenic mutations are recognized by $CD4^+$ T cells.[28] In fact, 95%, 80%, and 70% of the mutation-specific immunogenic responses were conferred by $CD4^+$ T cells when three common tumor models, a melanoma, a colon carcinoma, and a mammary carcinoma line, respectively, were analyzed. These findings are significantly higher than what was previously reported, which was approximately 0.5% of mutated peptides in melanoma conferring CD4+ T-cell reactivity.[29] This optimistic finding suggests that there may be a much larger repertoire of epitopes to take advantage of considering the strong antitumor activity and even complete regression of aggressive tumors that resulted upon vaccination with these CD4+ immunogenic mutations.[28] This discovery indicates a strong foundation for the use of CD4+ T cells in eliciting a cytotoxic, tumor-specific response, though there still have not been studies investigating this in the context of the IDH1 mutation.

Another preclinical study was recently published displaying varying results compared to the previous study. In this study, the GL261 line, which is a carcinogen-induced murine glioma line that has become a common tool for assessing experimental glioma therapeutic modalities, was utilized.[24,30–32] Through overexpression of the IDH1-R132H mutation in the GL261 line, the group generated a syngeneic intracranial injection model, permitting the evaluation of the mutant IDH1 peptide vaccine's efficacy. Nine, 16, and 23 days following tumor implantation, mice were vaccinated with either four short peptides (two 9-mers and two 10-mers) or one 16-mer peptide, all of which spanned the region encoding the mutation. These vaccines were shown to significantly increase survival compared to all control cohorts, with 25% of immunized animals displaying no signs of tumor burden. In vitro proliferation assays were used to assess immunogenicity and functionality of the peptides. Here, they showed splenocyte proliferation in the presence of mutant IDH1 peptides, but not control peptides, suggesting an ability of the peptides to induce an immunologically relevant response. Interestingly, and in contrast to the previous preclinical study, a CD8+ T-cell response in the spleens of immunized animals was observed with a specific cytotoxicity against IDH1-mutant GL261 cells. Complementary to the CD8+ T-cell response, IDH1-R132H-specific IgG in the serum of those immunized, tumor-bearing animals was observed, suggesting a CD4+ T-cell response. Interestingly, tumors derived from the mutant IDH1-R132H GL261 cells following immunization appeared histologically similar to those tumors derived from the parental GL261 cell line, suggesting that vaccination nullified the phenotypic effects of the mutation. This included a reduction in the number and size of blood vessels and a reduction in Hypoxia-inducible factor 1 (HIF-1α) and VEGF expression. Interestingly, when the single 16-mer peptide vaccine was used, an induction of a CD8+ T-cell response and a significant increase in survival compared to control animals were observed. However, this survival was blunted compared to the cohort immunized with the four shorter peptides. The difference in survival may be due to an inability of the 16-mer to elicit an antibody response, and may warrant further investigation into a peptide vaccine eliciting both a cytotoxic and antibody response for maximum tumor control.

CLINICAL INVESTIGATIONS: IDH1-R132H PEPTIDE VACCINES

As a result of these promising preclinical studies, initiation of two clinical trials has commenced, with patients being actively recruited. The two trials, one sponsored by the National Center for Tumor Diseases in Heidelberg, Germany, and the other by Duke University in the United States of America, both rely on peptide vaccination targeting IDH1-R132H (NCT02454634 and NCT02193347, respectively). Though the aims of both Phase I trials are to evaluate safety, tolerability, and immunogenicity of the peptide vaccines, they vary subtly. Where the German study will utilize a 20-mer peptide administered with topical imiquimod (Aldara) in patients with World Health Organization (WHO) Grade III and IV gliomas, the Duke University trial will utilize a 25-mer peptide in patients with recurrent WHO Grade II gliomas administered with granulocyte macrophage colony-stimulating factor (GM-CSF). Both trials will administer the vaccine only in those patients testing positive for the IDH1-R132H mutation and will incorporate experimental arms evaluating the combined response of the peptide vaccine with temozolomide (TMZ). Additionally, Duke University's trial will incorporate another experimental arm with preconditioning of the vaccine site with a single dose of tetanus and diphtheria toxoids (Td). This is in light of recent findings showing that preconditioning of the vaccine site with a potent recall antigen can significantly improve the efficacy of tumor-antigen specific DCs.[33] Similarly, this study will investigate whether preconditioning will elicit a more efficient immune response following peptide vaccination.

Despite a prior meta-analysis of the tumor-response rate following peptide vaccination showing that only 2.6% (of 440 patients) responded to vaccination treatment, current advances in the determination of tumor-specific antigens and their subsequent targeting have proven to be efficacious with minimal toxicity and prolonged, progression-free survival (PFS).[34,35] With the ever-growing repertoire of tumor-specific mutations resulting from sequencing of human tumors, coupled with a requirement for multiple therapeutic hits against the tumor to ensure prevention of tumor evasion, individualized peptide vaccine cocktails containing patients' tumor-specific antigens are being developed. One such initiative is the Glioma Actively Personalized Vaccine Consortium, sponsored by the European Union, which currently has active clinical trials. This is among the first clinical projects engaging fully active personalized medicine with patients being treated based on their specific mutation spectrum. In these trials, patients receive a series of two vaccines: the first contains a cocktail of peptides specific for highly overexpressed tumor-associated antigens, and the second contains a cocktail of tumor specific peptides, thus targeting as many tumor-related antigens as possible and reducing the risk of tumor evasion.

ROADBLOCKS AND FUTURE DIRECTIONS FOR TARGETING IDH1 MUTATIONS

Currently, the only preclinical data and clinical trials pertaining to IDH1-mutated gliomas are peptide-based vaccinations. However, there is the opportunity for antibody-based therapies and T-cell-based therapies. There has been some conflicting data regarding $CD4^+$ T-cell and $CD8^+$ T-cell responses in the context of mutant IDH1. This could be due to different tumor models being used (sarcoma versus glioma models) or different immune-competent mouse strains being used (MHC-humanized mouse versus C57Bl/6).[23,24] Though the clinical trials are founded on strong studies, the gap between preclinical models and treating patients must be overcome to instill unwavering confidence in the safety, efficacy, and ultimate success of these treatments. As each of the described preclinical models has limitations, the importance of establishing genetically faithful and immunologically relevant models has never been more imperative toward reaching this goal. The next logical steps for preclinical evaluation would be the generation of immune-competent genetic mouse models that spontaneously form gliomas and faithfully recapitulate the histological and biological features characteristic of WHO Grade II, III, and IV gliomas. Mice that meet these criteria and contain those tumor-specific mutations are difficult and expensive to generate; moreover, it is still impossible to reproduce a patient's tumor burden within the model, making these studies high risk. The generation of orthotopic intracranial injection models with tumorigenic mouse glioma lines of known background and mutation status would offer similar benefits and allow for a more time-efficient assessment of tumor-inducing mutations. Last, patient-derived human xenografts into humanized animals would be another avenue to pursue, though primary IDH1-mutant tumors have been historically difficult to culture and establish reliable xenografts in mice.[36,37] With representative model systems, additional immunotherapeutic modalities can be investigated.

EPIDERMAL GROWTH FACTOR RECEPTOR AND VARIANT III IN GLIOMAS

Another promising target for primary GBM is EGFRvIII, the most commonly mutated form of the epidermal growth factor receptor. EGFRvIII is expressed in approximately 20–30% of GBMs, and typically occurs concurrently with the overexpression of wild-type EGFR, which is present in approximately 50% of GBMs. It is estimated that within a given tumor sample, EGFRvIII is expressed in 37–86% of cancer cells.[38] Despite the heterogeneity of expression, EGFRvIII is a tumor-specific epitope not found on normal tissue, deeming it an ideal candidate for targeted therapy. It was originally thought that EGFR amplification exclusively promoted cancer, but it is now clear that EGFRvIII also facilitates tumorigenesis and plays a vital role in tumor progression, proliferation, and inhibition of apoptosis.[40]

EGFR is a 170-kDa transmembrane glycoprotein and is a member of the ErbB family (also known as the Her family) of receptors, which has four members, all of which have been identified as being amplified, overexpressed, or altered in a range of human malignancies.[40,41] All four members share a similar structure consisting of (1) a ligand binding extracellular domain (ECD) with two cysteine rich regions, (2) a single membrane spanning region, and (3) a cytoplasmic domain containing multiple tyrosine residues that are phosphorylated upon ligand binding and receptor activation. EGFR is activated by members of the epidermal growth factor family, including EGF, amphiregulin, TGF alpha, heparin-binding EGF-like (HB-EGF), and epiregulin.[42] Binding of these factors results in receptor dimerization, inducing a phosphorylation cascade and downstream pathway signaling involved in cell division, survival, and cell death.

The EGFRvIII mutation yields the loss of 801 base pairs between exons 2 and 7 in the wild-type EGFR gene, resulting in an in-frame truncation of amino acids 6–273 in the ECD of the protein that is involved in dimerization and ligand binding.[43] The deletion allows two otherwise distant and independent portions of the molecule to bind, creating an antigenic junction characterized by a novel glycine residue, flanked by amino acid sequences that are not typically adjacent in the wild-type receptor.[40] This mutation and its downstream effects have created a tumor-specific epitope that is present on the surface of cancer cells, yet is absent from normal adult tissue. This is an important feature in avoiding potential autoimmunity and makes EGFRvIII a particularly favorable candidate target for therapy, as normal brain parenchyma is intolerant to collateral damage or toxicity.[44]

In normal cells, EGFR has a role in cell proliferation via ligand-induced receptor dimerization and tyrosine autophosphorylation. EGFR signaling also plays an important role in cellular proliferation in GBM. As such, ligand binding by EGF results in the activation of the RTK/RAS/phosphoinositide 3-kinase (PI3K) pathways by receptor phosphorylation, which promotes proliferation, angiogenesis, and increased local tissue invasion.[39] Ligand binding to RTKs simultaneously activates PI3K, which phosphorylates phosphatidylinositol 4,5 to phosphatidylinositol 3,4,5 (PIP2 to PIP3). This results in further activation of AKT and mTOR, a protein which exists as two complexes: mammalian target of rapamycin complex 1 (mTORC1) and mammalian target of rapamycin complex 2 (mTORC2). In addition to inhibition of catabolic activity, activation of mTORC1 promotes cellular growth by biosynthesis of proteins, lipids, and organelles. mTORC2 activation results in phosphorylation and subsequent activation of several molecules, including AKT, that are involved in cell survival, metabolism, and proliferation. phosphatase and tensin homolog (PTEN) is a major inhibitor of this pathway, preventing phosphorylation of PIP2 to PIP3. Loss of PTEN, which is observed in 36% of gliomas, results in a dramatic upregulation of this pathway and could be a source of resistance to EGFR therapies.

The altered ECD of EGFRvIII results in changes in the functional characteristics of the receptor compared to wild-type EGFR. Variant III shows constitutive tyrosine kinase activity, despite its ligand independence.[41] The signaling intensity also differs between wild-type EGFR and EGFRvIII. Wild-type EGFR, when stimulated by EGF, transmits a high intensity and short duration signal that is 5–10 times greater in magnitude than EGFRvIII, which produces a low intensity and continuous signal, suggesting a differing downstream cascade.[46] Despite having remarkably lower activity, the variant receptor is known to exhibit increased stability.[43] This is thought to be important for EGFRvIII's role in oncogenesis. It is hypothesized that the signaling potency of EGFRvIII is increased by its ability to signal for a prolonged time as a result of the impaired endocytosis and degradation of the receptor due to inefficient ubiquitination and rapid cycling, thereby prolonging kinase activity and downstream signaling.[41]

Interestingly, it has also been shown that cooperation exists between EGFR and EGFRvIII in transformation studies in vitro. Since EGFRvIII is unable to bind or be activated by the EGF ligand, the effect of the cooperative activity on EGFRvIII was unexpected. In cells expressing both EGFR and EGFRvIII, the addition of the EGF ligand results in increased phosphorylation of EGFR and EGFRvIII, suggesting that ligand-dependent activation of EGFR promotes cross-phosphorylation of EGFRvIII. It was also discovered that the presence of EGFR and EGFRvIII increases phosphorylation of STAT3 and STAT5, inducing malignancy greater than either isoform alone by promoting an immunosuppressive microenvironment.[47] This suggests that in addition to EGFRvIII's known roles in cancer stem cell self-renewal, proliferation, and invasion, EGFRvIII also promotes immune suppression.[39]

These aforementioned features of EGFRvIII qualify it as an ideal candidate for immunotherapeutic targeting. EGFRvIII is expressed exclusively on tumor cells, has a unique signaling mechanism, and has an immunogenic epitope capable of producing an EGFRvIII-specific immune response. However, despite the prominence of EGFRvIII as a potential immunotherapeutic target, its expression is heterogeneous, representing 37–86% of tumor cells in a given tissue sample.[38] Nevertheless, EGFRvIII has yielded encouraging preclinical and clinical data and is a prime target for GBM immunotherapy, particularly when combined with other modalities.

EGFR MUTATIONS AS AN IMMUNOTHERAPEUTIC VACCINATION TARGET

There are a number of active and passive immunotherapeutic vaccinations targeting EGFRvIII in gliomas.[40] Active vaccination relies on provoking the immune system to mount physiological responses against specific antigens, promoting long-term acquired immunity. This is done through administering an immunogen with an adjuvant to direct antigen-specific lymphocytes.[18] Passive vaccination, on the other hand, involves either the direct infusion of antibodies or the adoptive transfer of antigen-specific T lymphocytes.[40] Induced antitumor immunity against EGFRvIII has been demonstrated by both DC and peptide vaccination approaches, which are both active methods.[48] Due to the relatively high cost and technical difficulty of DC vaccinations, however, the most viable vaccinations to date are peptide vaccines.

Rindopepimut (PEPvIII-KLH [keyhole limpet hemocyanin], CDX-100 of Celldex Therapeutics, Phillipsburg, NJ) is a peptide vaccine derived from the novel fusion junction amino acid sequence of EGFRvIII and consists of a 14-mer peptide (H-Leu-Glu-Glu-Lys-Lys-Gln-Ans-Tyr-Val-Val-Thr-Asp-His-Cys-OH). This peptide vaccine has been shown to elicit both humoral and cellular immune responses as well as EGFRvIII-specific immunity when coupled to the KLH carrier.[40,49] This active vaccination is well characterized and is now being investigated in an international Phase III clinical trial for newly diagnosed GBM patients at 200 locations worldwide.[18]

PRECLINICAL INVESTIGATIONS: EGFRvIII VACCINATION

Administration of EGFRvIII-specific monoclonal antibodies demonstrates significant antitumor activity both in vitro and in vivo. Passive administration of Y10 and L8A4 (IgG_{2a} and IgG_1, respectively) leads to significant inhibition of tumor growth in subcutaneous murine melanoma models using syngeneic tumors transfected with a murine homolog of the variant receptor (msEGFRvIII). In vitro studies suggest that Y10 mediates a range of functions in cells expressing msEGFRvIII, such as inhibition of DNA synthesis, cellular proliferation, and antibody-dependent cell-mediated cytotoxicity. Two in vivo experiments were completed to test the efficacy of the antibodies against challenges from EGFRvIII-positive melanoma cell lines. In the first experiment, mice that were pretreated with Y10 and then challenged with melanoma tumor cells never developed palpable tumors, whereas the control, nonimmunized cohort developed tumors. Mice were then challenged with intracranial melanoma tumors and were administered antibody therapy. Intraperitoneal administration inhibited tumor growth but did not increase the median survival of mice. However, when mice were treated intratumorally with Y10, they exhibited a longer than usual tumor-free survival rate, even after treatment was discontinued. Intratumoral injection of Y10 increased survival by 286%, with 26% of mice becoming long-term survivors ($P < .001$).[40,49]

Active vaccinations have also been effective against intracranial msEGFRvIII-expressing tumors. C3H mice received an intraperitoneal immunization with DCs pulsed with PEPvIII-KLH, which resulted in a 600% increase in median survival, with 63% becoming long-term survivors. Furthermore, mice survived re-challenges with tumors, suggesting that immunization was sufficient to create long-term immunological memory against msEGFRvIII. A follow-up study was done in which C3H mice were treated with a single vaccination of PEPvIII-KLH in Freund's adjuvant as opposed to the DC vaccine. This protocol resulted in a 26% increase in median survival, with 40% of mice surviving long term.[50] Mice that failed to exhibit a response to PEPvIII-KLH were found to have downregulated EGFRvIII expression or completely lacked EGFRvIII expression, as observed through immunohistochemistry, indicating that antigen escape may be a mechanism of treatment failure.[40,49]

CLINICAL INVESTIGATIONS: EGFRvIII VACCINATION

Phase I: VICTORI

Due to the success of PEPvIII-KLH in preclinical studies, a Phase I clinical trial, VICTORI, was designed and conducted at Duke University Medical Center. Fifteen patients with newly diagnosed GBMs were enrolled to evaluate the toxicity, immunogenicity, and potential efficacy of PEPvIII-based vaccination, 12 of which underwent leukapheresis to obtain peripheral blood mononuclear cells for the generation of DCs.[18,40,48,49] Prior to intradermal inoculation, DCs were pulsed with PEPvIII-KLH. Patients received three total immunizations of 1.1×10^8 DCs every 2 weeks. The pulsed DCs were generally well-tolerated with minimal adverse effects, and no tested dose exceeded Grade II toxicity as measured through the National Cancer Institute Common Toxicity Criteria. Furthermore, 56% of patients showed EGFRvIII antigen-specific cellular and humoral immune responses as detected ex vivo. The median survival for patients was 22.8 months after histological diagnosis. Though the outcomes were not statistically significant, 9 of the 12 patients actually surpassed outcome expectations according to Curran's recursive partition analysis. The results might have been biased since EGFRvIII expression was not inclusion criteria for the Phase I toxicity trial. Nevertheless, the outcomes of the DC-based PEPvIII-KLH vaccine were encouraging and warranted further investigation.

Phase II: ACTIVATE, ACT II, and ACT III

The promising results from the Phase I trial led to the initiation and completion of several Phase II trials, including ACTIVATE, ACT II, and ACT III. Due to the expenses and variable nature of DC vaccines that were described in VICTORI, PEPvIII-KLH was alternatively administered as a peptide vaccine in these trials.[40] Nineteen adults with newly diagnosed EGFRvIII-expressing primary GBMs were enrolled in ACTIVATE. After gross total resection (>95%) and standard of care radiation therapy with concurrent TMZ, patients received intradermal injections of PEPvIII-KLH with GM-CSF as an adjuvant (rindopepimut). Similar to the observations from VICTORI, patients did not experience significant adverse events. Both time to progression (TTP) and overall survival (OS) were prolonged, with TTP improving from 7.1 (n=29) months to 12 (n=12) months (P=.0058) and OS extending up to 26 months compared to controls. Additionally, a robust anti-EGFRvIII immune response was detected, conferring significantly longer OS compared to patients who did not develop a humoral immune response (47.7 months compared to 22.8 months, P=.025). Several patients (3/17) developed a cell-mediated response and displayed a significantly longer PFS and OS (P=.03), with median OS extending beyond 50 months compared to 23.1 months.[51] As seen in prior in vivo experiments and human trials, the majority (82%) of recurrent tumors had lost EGFRvIII expression.[40]

Prior investigations utilizing both animal and human samples suggest that lymphopenia may support antitumor immunity by eliminating the effect of Treg cells and homeostatic mechanisms, inducing T-cell tolerance.[52] ACT II sought to test the hypothesis that TMZ-induced lymphopenia enhances immune responses to the PEPvIII-KLH vaccine, and 22 patients with newly diagnosed GBMs were enrolled. A similar regimen to ACTIVATE was used, with the inclusion of two different TMZ dose schedules. In the first group, 12 patients received TMZ at a dose of 200 mg/m^2 for 5 days of a 28-day cycle. In the second group, 10 patients were given a dose-intensified regimen (DI), receiving 100 mg/m^2 for 21 days of a 28-day cycle. Immunization with rindopepimut was performed within 6 weeks of radiotherapy on day 21 of each 28-day TMZ cycle. The grade of lymphopenia as well as the proportion of Tregs were greater and sustained in the DI group compared to the standard TMZ dose group. All patients developed an EGFRvIII-specific humoral response and, interestingly, despite inducing more severe lymphopenia, treatment with DI TMZ enhanced both humoral and cellular immune responses following vaccination compared to the standard TMZ group. In addition, more patients in the DI group developed delayed-type hypersensitivity reactions to PEPvIII compared to the standard dose group. After 6 months, PFS was found in nine out of 12 patients in the standard group and nine out of 10 patients in the DI group, whereas 10 out of 12 patients in the standard control group and nine out of 10 patients in the DI group were alive at 12 months. The median PFS from histological diagnosis for all patients was 15.2 months and OS for all patients was 23.6 months. Though these values are significant improvements compared to historical controls, after adjusting for Karnofsky Performance Status and age, there were no significant advantages between groups. As seen previously, 21/23 (91%) of recurrent tumors lost EGFRvIII expression.[51]

A third clinical trial, ACT III, was established to investigate the efficacy and safety of the rindopepimut vaccine.[38] ACT III included 65 patients from 31 institutions with newly diagnosed primary, EGFRvIII-positive GBMs who underwent gross total resection with no signs of progression after standard radiotherapy and TMZ treatment. The treatment regimen consisted of TMZ maintenance therapy along with a series of vaccinations. Rindopepimut was once again well-tolerated with minor injection site reactions (ISRs), which were all resolved without intervention. Notably, there were no signs of cumulative toxicity throughout the course of the study; however, two patients were removed from the study due to adverse events which were potentially related to rindopepimut. These included a grade II hypersensitivity reaction and a grade III toxic epidermal necrolysis. Eighty-five percent of patients were found to have developed anti-EGFRvIII antibody titers, which increased throughout the duration of the study. Both PFS and OS were found to be significantly higher than historical controls at 12.3 and 24.6 months, respectively ($P=.0088$ and $<.0001$). Additionally, patient outcomes were analyzed based on O6 methylguanine methyltransferase (MGMT) promoter methylation status. ACT III revealed that rindopepimut improved both PFS and OS in patients with both methylated and unmethylated MGMT promoters compared to respective matched historical controls.[51] As in ACTIVATE (82%) and ACT II (91%), loss of EGFRvIII expression was seen in four out of six (67%) patients with recurrent disease.[51]

Phase II: ReACT

A Phase II trial known as ReACT was established to evaluate the efficacy of rindopepimut in combination with bevacizumab in patients with recurrent EGFRvIII-positive GBMs.[38,49] Seventy-two EGFRvIII-positive patients in this study were split into two cohorts. The first cohort was randomized to receive bevacizumab plus rindopepimut/GM-CSF (35 patients) while the second cohort received bevacizumab plus KLH control (37 patients).[53] The patients were either at their first or second recurrence and did not have a restriction limit based on tumor size, though it was required for patients to be on less than 4 mg of dexamethazone for the trial. The primary outcome measure was to determine the PFS at 6 months, while the secondary outcomes pertained to safety, tolerability, antitumor activity, and an EGFRvIII-specific immune response. As previous trials have demonstrated, rindopepimut continued to produce consistent and impressive results in ReACT, in which 27% remained progression-free compared to 11% on the control arm (P = .0048). Furthermore, patients treated with rindopepimut also demonstrated a 3.2-month improvement in median OS compared to the controls (12.0 versus 8.8 months, hazard ratio 0.47; P = .0208).[53] Although not a primary outcome measure, rindopepimut recipients also had a higher rate of overall radiographic response of 24% compared to 17% in the control arm. The prolonged survival of patients receiving the vaccination was associated with a high anti-EGFRvIII humoral immune response. Eighty-six percent of patients demonstrated a greater than or equal to fourfold increase in anti-EGFRvIII antibody titers compared with a pretreatment baseline.[53] Fourteen patients who had anti-EGFRvIII antibody titers of greater than 12,800 had a median OS of 20.8 months, whereas those who achieved titers less than 12,800 had a survival of 10.4 months, suggesting that the anti-EGFRvIII antibody titer may be a potential biomarker of an antitumor benefit. The toxicity of rindopepimut was again very well tolerated with few adverse events, which were limited to ISR, fatigue, rash, nausea, and pruritus. ReACT, in summary, demonstrated a statistically significant, clinically meaningful OS benefit in addition to an improved quality of life due to minimal side effects and a reduced need for steroids.

Phase III: ACT IV

ACT IV, a phase III clinical trial, was initiated to investigate the safety and efficacy of rindopepimut in newly diagnosed GBM patients.[38] This trial was the most comprehensive study of patients with EGFRvIII-expressing GBM to date, being conducted at more than 200 clinical trial sites and enrolling 745 patients. This randomized, double-blind trial compared rindopepimut (PEPvIII-KLK) and TMZ treatment with KLH alone and TMZ. Unfortunately, based on a preplanned interim analysis, it was established that the overall survival among patients with minimal residual disease, the primary endpoint, was not statistically significant between the rindopepimut-treated arm (20.4 months) and the control arm of the study (21.1 months). Although the results for the rindopepimut-treated arm were consistent with prior phase II studies (21.8 months), the control arm in this study significantly outperformed expectations based on the prior phase II trial (21.1 and 16.0 months, respectively). This study was subsequently discontinued; however, it does stress the importance of continued investigation on combinatorial therapies, including the use of this peptide vaccine in combination with other drugs, such as the angiogenesis inhibitor, bevacizumab (as in ReACT), as well as establishment of comprehensive, large-scale, multi-institutional investigations.

ROADBLOCKS FOR TARGETING EGFRvIII

Although rindopepimut has been demonstrated to be effective against EGFRvIII-positive, primary GBMs, there are several limitations of the vaccination therapy. First, EGFRvIII-positive tumors are in a small subset (20–30%) of GBM patients; therefore, rindopepimut cannot be used for the majority of patients.[38] Furthermore, patients must be screened for EGFRvIII before they can be considered for vaccination. This excludes rindopepimut from being used in the majority of secondary GBM patients since EGFRvIII is seldom seen in secondary GBMs.[49] Additionally, recurrent tumors from the ACTIVATE, ACT II, and ACT III trials were found to have lost EGFR expression, indicating that the tumor is able to escape through the selection and proliferation of EGFR-negative cells.[38] Even for EGFRvIII-positive tumors, expression is heterogeneous and not expressed on all tumor cells. One strategy to overcome this may be to vaccinate against multiple antigens simultaneously or combine vaccinations with other adjuvant therapies with the aim of decreasing tumor evasion.

Another issue within the studies is that the OS might have been biased as a result of the study design. First, all patients did not receive salvage therapy uniformly for tumor progression because the use of additional therapeutic agents was at the discretion of the patient's neuro-oncologist. The presence of IDH1 and 2 mutations was not exclusion criteria for enrollment, and overrepresentation of these mutations in the study cohort might have artificially improved OS as compared to historical controls.[49] In addition to small sample sizes of the studies, only patients with a high performance status and those not having signs of disease progression were enrolled. Finally, the efficacy of rindopepimut on residual tumor burden is unclear. This is because even though a significant portion of GBM patients do not receive gross total resection, all patients in this particular study underwent image-verified gross total resection.

Nevertheless, for EGFRvIII-positive, primary GBM patients, rindopepimut has trended toward being an effective immunotherapy. Rindopepimut has improved both PFS and OS in Phase I and Phase II clinical trials. The larger Phase III international trial addressed many of the shortcomings of the smaller previous studies. While the OS was similar for those treated with rindopepimut between trials, the OS of patients in the control arm outperformed those treated with rindopepimut in this trial, leading to its discontinuation. The data from this large-scale study is still being analyzed to understand these differences and, once complete, should shed light on the full effects of rindopepimut. GBM is a devastating disease and rindopepimut may still provide a significant improvement to many patients by supplementing the current standard of care, which will be the basis for future investigations.

HISTONE MUTATION BACKGROUND

The final mutations identified in recent sequencing studies that display immunotherapeutic potential for gliomas are mutations in histone 3 (H3). Histones are involved in the basic packaging of DNA, allowing 2m of DNA to fit inside the nucleus of a single cell. Histones are the primary protein components of chromatin, acting as spools around which DNA winds.[55] There are four core histones: H2A, H2B, H3, and H4. DNA in the nucleus is wrapped around a nucleosome, comprised of two of each of the core histones.[54] Nucleosomal arrays were observed with electron microscopy as a series of "beads on a string," the "beads" being individual nucleosomes and the "string" being the linker DNA. Linker histones, such as histone H1, and other nonhistone proteins can interact with the nucleosomes to form higher-order chromatin structures.[56] The amino-terminal tails of each of the core histones protrude from the nucleosome and can be altered by a myriad of posttranslational modifications (PTMs).[54] For each posttranslational modification of histones, enzymes exist, which either modify a specific residue on the histone tail (sometimes referred to as writers) or remove it (sometimes referred to as erasers). The following are some basic classes of enzymes that are known to modify histones: histone acetyltransferases, which acetylate lysines on the histone tails and induce chromatin decondensation; histone deacetylases, which remove the acetyl groups from the lysines and promote a tighter binding of histones to DNA; histone methyltransferases (HMTs), which add methyl groups to lysines and thus promote or inhibit transcription depending on the target histone residue; and histone demethylases, which counteract the activity of HMTs.[56–58] Increasing evidence suggests that histone-modifying enzymes are deregulated in human cancers, which indicates that changes in the expression of histone-modifying enzymes have important and tumor-specific roles in cancer development.[56] The histone code—PTMs of specific histone residues—regulates virtually all processes that act on or depend on DNA, including replication and repair, regulation of gene expression, and maintenance of centromeres and telomeres.[59,60] Posttranslational modifications of histones create an epigenetic mechanism for the regulation of a variety of normal and disease-related processes. Therefore, given the fundamental roles of histone modifications, it is not surprising that aberrations in histone modifications, including the histones themselves, are revealed in cancer.[56,61]

The human histone H3 family consists of a number of related proteins: H3.1 and H3.2 (commonly referred to as "canonical" H3), histone variant H3.3, the centromere-specific variant centromere protein (CENP)-A/CenH3, the testes-specific H3t, and the testes-specific H3.5.[62,63] Histone H3.1 differs from H3.2 by a single amino acid (Ser^{96} in H3.2), and H3.3 is distinguished by four additional amino acid substitutions (Ser^{31}, Ala^{87}, Ile^{89}, Gly^{90}).[64] These clustered amino acids that differ between H3.1 and H3.3 have been linked to the different binding of chaperones,[65–68] with specifically G90 of H3.3 promoting binding to the death-domain-associated protein.[69,70] Histones H3.1 and H3.2 are synthesized during S phase,[71] are incorporated de novo into newly replicated chromatin as well as during DNA repair, and are thus termed "DNA synthesis-coupled." In contrast, the "replacement histone," H3.3, is expressed throughout the cell cycle, as well as in quiescent cells,[70,72] and is a universal, replication-independent histone 3, which is predominantly incorporated into transcription sites and associated with active and open chromatin.[73–75] The canonical histones are encoded by unusual transcripts that lack introns and polyadenylation signals: histone H3.2 is encoded by three genes (*HIST2H3A*, *HIST2H3C*, and *HIST2H3D*), whereas 10 genes encode H3.1 clustered on chromosome 6. The DNA synthesis-independent H3.3 is expressed from only two genes, *H3F3A* on chromosome 1 and *H3F3B* on chromosome 17. These genes produce identical proteins even though they have distinct regulatory sequences and yield distinct polyadenylated transcripts with unusually long 5′ and 3′ untranslated regions (UTRs).[70,76,77]

HISTONE MUTATIONS IN PEDIATRIC GLIOMAS

All reported H3 mutations identified in human gliomas lead to single codon changes within the N-terminal tail of the H3 protein, a region enriched in PTMs.[63] The missense mutations were predominantly found in the genes encoding H3.3, *H3F3A*, and, to a lesser extent, in H3.1 genes, *HIST1H3B* and *HIST1H3C*.[70] The first studies identified H3.3 mutations encoding a K27M substitution, in which a key lysine residue on the histone H3 tail is substituted for a methionine, in addition to less frequent mutations encoding a glycine 34 to arginine or valine (G34R/V) substitution.[60,63,78–80] H3K27 is a critical residue that, when trimethylated, is involved in transcriptional repression via polycomb repressive complexes 1 and 2 (PRC1 and PRC2).[54] The K27M mutation is a gain of function mutation, which competitively inhibits the methyltransferase activity of Enhancer of zeste homolog 2—the active enzyme subunit in PRC2—and thus abolishes polycomb-mediated repression of numerous genes.[54,81] In the absence of PRC2, genes that should be silent are expressed, which is thought to drive cell transformation.[81–84]

The K27M mutation occurs mainly in younger patients (median age, 10–11 years), and is most often found in *H3F3A* (>70%), with a few occurrences in *HIST1H3B* (~20%) and rarely in *HIST1H3C*.[60,70,79,80,85–87] In all tumor samples analyzed thus far, the histone mutations are heterozygous, with one allele being wild type.[60] K27M and G34R/V mutations are mutually exclusive in tumors and show distinct gene expression profiles and DNA methylation patterns.[60,63,80] Mutated tumors display different patterns of localization within the central nervous system. K27M tumors are primarily restricted to midline locations (spinal cord, thalamus, pons, and brainstem) and G34R/V tumors to cerebral hemispheres.[80,88] Additionally, both mutations correspond to different patient age ranges, with K27M mutations more prevalent in younger patients (range, 5–29 years) and G34R/V mutations occurring in slightly older patients (range, 9–42 years).[60,63,80]

H3.3 MUTATIONS AS AN IMMUNOTHERAPEUTIC TARGET

Immunotherapy continues to gain momentum as a potentially promising treatment for childhood High grade gliomas (HGGs) with a variety of ongoing clinical trials, for example, the use of checkpoint inhibitors such as programmed death-1 (PD1) pathway inhibitors, convection-enhanced delivery with a radioactive antibody (clinicaltrials.gov #NCT01502917), and vaccines. Pollack et al.[89] recently published a study of subcutaneous vaccinations with glioma-associated antigen epitope peptides in HLA-A2-positive children with newly diagnosed brainstem gliomas and high-grade gliomas. They report preliminary evidence of immunologic and clinical responses. The vaccination was generally well tolerated, and the authors note that careful monitoring and management of pseudo-progression is essential.[89]

Thus far there have not been any publications on immunotherapeutic targeting of the histone mutations. There are several reasons why these mutations may be good targets for immunotherapy. First, these are thought to be initiating mutations, and so are found in all tumor cells. Second, the amino acid substitutions as a result of the mutations in H3 are almost always the same and have a limited repertoire (K27M, G34R, or G34V). In midline locations, the amino acid substitution is almost always K27M and is found in up to 80% of gliomas in the pons or thalamus. In the cerebral cortex location, the amino acid substitution is either G34R or G34V. Of course, several scientific questions remain, such as are the mutant peptides generated by these histone mutations presented by MHC molecules? What immunosuppressive mechanisms are present in tumors harboring histone mutations? What immunotherapeutic approach will be most efficacious in successfully targeting these mutations? Research over the next few years will shed light on these exciting questions.

ROADBLOCKS AND FUTURE DIRECTIONS FOR TARGETING H3 MUTATIONS

Clinical trials in children with HGGs are a challenge given the small patient population. In addition, to target tumor-specific mutations with immunological therapies, a brain tumor biopsy must be performed to determine whether or not an H3 histone mutation is present. However, in the management of Diffuse Intrinsic Pontine Glioma for example, biopsies are not the standard of care in the United States, partly because of its location within the pons, a part of the brainstem, which is critical for survival. More recently, specialized centers have demonstrated these biopsies to be safe and feasible, and biopsies have been incorporated into several prospective clinical trials successfully. Furthermore, the sample obtained from a biopsy is small and may not accurately represent the entire tumor because of the inherent heterogeneity of genetic alterations in different tumor cells.[90] Despite these roadblocks, there is great excitement for the generation of H3 mutant-specific vaccinations and their subsequent evaluation in both preclinical genetic mouse glioma models and in clinical trials.

Ultimately, the process of targeting tumor-specific or tumor-associated antigens for immunotherapeutic purposes shows great promise of inducing specific killing of tumor cells while limiting damage to surrounding tissue. However, intense investigation is still required to harness the power and full potential of the immune system in a safe way, and to ensure immune responses continue to be effective over time, thereby preventing immune escape. For this reason, as is usually the case, combination therapy, both in the form of targeting several antigens, or in the form of multimodal therapeutics, will indubitably prove most effective.

References

1. Comprehensive genomic characterization defines human glioblastoma genes and core pathways. *Nature*. 2008;455(7216):1061–1068.
2. Parsons DW, Jones S, Zhang X, et al. An integrated genomic analysis of human glioblastoma multiforme. *Science*. 2008;321(5897):1807–1812.
3. Yan H, Parsons DW, Jin G, et al. IDH1 and IDH2 mutations in gliomas. *N Engl J Med*. 2009;360(8):765–773.
4. Brat DJ, Verhaak RG, Aldape KD, et al. Comprehensive, integrative genomic analysis of diffuse lower-grade gliomas. *N Engl J Med*. 2015;372(26):2481–2498.
5. Watanabe T, Nobusawa S, Kleihues P, Ohgaki H. IDH1 mutations are early events in the development of astrocytomas and oligodendrogliomas. *Am J Pathol*. 2009;174(4):1149–1153.
6. Xu X, Zhao J, Xu Z, et al. Structures of human cytosolic NADP-dependent isocitrate dehydrogenase reveal a novel self-regulatory mechanism of activity. *J Biol Chem*. 2004;279(32):33946–33957.
7. Dang L, White DW, Gross S, et al. Cancer-associated IDH1 mutations produce 2-hydroxyglutarate. *Nature*. 2009;462(7274):739–744.

8. Noushmehr H, Weisenberger DJ, Diefes K, et al. Identification of a CpG island methylator phenotype that defines a distinct subgroup of glioma. *Cancer Cell*. 2010;17(5):510–522.
9. Turcan S, Rohle D, Goenka A, et al. IDH1 mutation is sufficient to establish the glioma hypermethylator phenotype. *Nature*. 2012;483(7390):479–483.
10. Lu C, Ward PS, Kapoor GS, et al. IDH mutation impairs histone demethylation and results in a block to cell differentiation. *Nature*. 2012;483(7390):474–478.
11. Duncan CG, Barwick BG, Jin G, et al. A heterozygous IDH1R132H/WT mutation induces genome-wide alterations in DNA methylation. *Genome Res*. 2012;22(12):2339–2355.
12. Xu W, Yang H, Liu Y, et al. Oncometabolite 2-hydroxyglutarate is a competitive inhibitor of alpha-ketoglutarate-dependent dioxygenases. *Cancer Cell*. 2011;19(1):17–30.
13. Sasaki M, Knobbe CB, Munger JC, et al. IDH1(R132H) mutation increases murine haematopoietic progenitors and alters epigenetics. *Nature*. 2012;488(7413):656–659.
14. Figueroa ME, Abdel-Wahab O, Lu C, et al. Leukemic IDH1 and IDH2 mutations result in a hypermethylation phenotype, disrupt TET2 function, and impair hematopoietic differentiation. *Cancer Cell*. 2010;18(6):553–567.
15. Stupp R, Mason WP, van den Bent MJ, et al. Radiotherapy plus concomitant and adjuvant temozolomide for glioblastoma. *N Engl J Med*. 2005;352(10):987–996.
16. Iacob G, Dinca EB. Current data and strategy in glioblastoma multiforme. *J Med Life*. 2009;2(4):386–393.
17. Overwijk WW, Wang E, Marincola FM, Rammensee HG, Restifo NP. Mining the mutanome: developing highly personalized Immunotherapies based on mutational analysis of tumors. *J Immunother Cancer*. 2013;1:11.
18. Swartz AM, Batich KA, Fecci PE, Sampson JH. Peptide vaccines for the treatment of glioblastoma. *J Neuro-oncol*. 2015;123(3):433–440.
19. Jackson C, Ruzevick J, Brem H, Lim M. Vaccine strategies for glioblastoma: progress and future directions. *Immunotherapy*. 2013;5(2):155–167.
20. Sayegh ET, Oh T, Fakurnejad S, Bloch O, Parsa AT. Vaccine therapies for patients with glioblastoma. *J Neuro-oncol*. 2014;119(3):531–546.
21. Swartz AM, Batich KA, Fecci PE, Sampson JH. Peptide vaccines for the treatment of glioblastoma. *J Neuro-oncol*. 2014.
22. Reardon DA, Freeman G, Wu C, et al. Immunotherapy advances for glioblastoma. *Neuro-oncology*. 2014;16(11):1441–1458.
23. Schumacher T, Bunse L, Pusch S, et al. A vaccine targeting mutant IDH1 induces antitumour immunity. *Nature*. 2014;512(7514):324–327.
24. Pellegatta S, Valletta L, Corbetta C, et al. Effective immuno-targeting of the IDH1 mutation R132H in a murine model of intracranial glioma. *Acta Neuropathol Commun*. 2015;3:4.
25. Bunse L, Schumacher T, Sahm F, et al. Proximity ligation assay evaluates IDH1R132H presentation in gliomas. *J Clin Invest*. 2015;125(2):593–606.
26. Hunder NN, Wallen H, Cao J, et al. Treatment of metastatic melanoma with autologous $CD4^+$ T cells against NY-ESO-1. *N Engl J Med*. 2008;358(25):2698–2703.
27. Quezada SA, Simpson TR, Peggs KS, et al. Tumor-reactive $CD4^+$ T cells develop cytotoxic activity and eradicate large established melanoma after transfer into lymphopenic hosts. *J Exp Med*. 2010;207(3):637–650.
28. Kreiter S, Vormehr M, van de Roemer N, et al. Mutant MHC class II epitopes drive therapeutic immune responses to cancer. *Nature*. 2015;520(7549):692–696.
29. Linnemann C, van Buuren MM, Bies L, et al. High-throughput epitope discovery reveals frequent recognition of neo-antigens by $CD4^+$ T cells in human melanoma. *Nat Med*. 2015;21(1):81–85.
30. Szatmari T, Lumniczky K, Desaknai S, et al. Detailed characterization of the mouse glioma 261 tumor model for experimental glioblastoma therapy. *Cancer Sci*. 2006;97(6):546–553.
31. Maes W, Van Gool SW. Experimental immunotherapy for malignant glioma: lessons from two decades of research in the GL261 model. *Cancer Immunol Immunother*. 2011;60(2):153–160.

32. Oh T, Fakurnejad S, Sayegh ET, et al. Immunocompetent murine models for the study of glioblastoma immunotherapy. *J Transl Med*. 2014;12:107.
33. Mitchell DA, Batich KA, Gunn MD, et al. Tetanus toxoid and CCL3 improve dendritic cell vaccines in mice and glioblastoma patients. *Nature*. 2015;519(7543):366–369.
34. Rosenberg SA, Yang JC, Restifo NP. Cancer immunotherapy: moving beyond current vaccines. *Nat Med*. 2004;10(9):909–915.
35. Sampson JH, Archer GE, Mitchell DA, et al. An epidermal growth factor receptor variant III-targeted vaccine is safe and immunogenic in patients with glioblastoma multiforme. *Mol Cancer Ther*. 2009;8(10):2773–2779.
36. Bralten LB, Kloosterhof NK, Balvers R, et al. IDH1 R132H decreases proliferation of glioma cell lines in vitro and in vivo. *Ann Neurol*. 2011;69(3):455–463.
37. Piaskowski S, Bienkowski M, Stoczynska-Fidelus E, et al. Glioma cells showing IDH1 mutation cannot be propagated in standard cell culture conditions. *Br J Cancer*. 2011;104(6):968–970.
38. Paff M, Alexandru-Abrams D, Hsu FP, Bota DA. The evolution of the EGFRvIII (rindopepimut) immunotherapy for glioblastoma multiforme patients. *Hum Vaccin Immunother*. 2014;10(11):3322–3331.
39. Padfield E, Ellis HP, Kurian KM. Current therapeutic advances targeting EGFR and EGFRvIII in glioblastoma. *Front Oncol*. 2015;5:5.
40. Choi BD, Archer GE, Mitchell DA, et al. EGFRvIII-targeted vaccination therapy of malignant glioma. *Brain Pathol*. 2009;19(4):713–723.
41. Gan HK, Kaye AH, Luwor RB. The EGFRvIII variant in glioblastoma multiforme. *J Clin Neurosci*. 2009;16(6):748–754.
42. Erdem-Eraslan L, Gao Y, Kloosterhof NK, et al. Mutation specific functions of EGFR result in a mutation-specific downstream pathway activation. *Eur J Cancer*. 2015;51(7):893–903.
43. Zadeh G, Bhat KP, Aldape K. EGFR and EGFRvIII in glioblastoma: partners in crime. *Cancer Cell*. 2013;24(4):403–404.
44. Suryadevara CM, Verla T, Sanchez-Perez L, et al. Immunotherapy for malignant glioma. *Surg Neurol Int*. 2015;6(suppl 1):S68–S77.
45. Deleted in review.
46. Chumbalkar V, Latha K, Hwang Y, et al. Analysis of phosphotyrosine signaling in glioblastoma identifies STAT5 as a novel downstream target of ΔEGFR. *J Proteome Res*. 2011;10(3):1343–1352.
47. Fan QW, Cheng CK, Gustafson WC, et al. EGFR phosphorylates tumor-derived EGFRvIII driving STAT3/5 and progression in glioblastoma. *Cancer Cell*. 2013;24(4):438–449.
48. Congdon KL, Gedeon PC, Suryadevara CM, et al. Epidermal growth factor receptor and variant III targeted immunotherapy. *Neuro-oncology*. 2014;16(suppl 8):viii20–25.
49. Babu R, Adamson DC. Rindopepimut: an evidence-based review of its therapeutic potential in the treatment of EGFRvIII-positive glioblastoma. *Core Evid*. 2012;7:93–103.
50. Heimberger AB, Crotty LE, Archer GE, et al. Epidermal growth factor receptor VIII peptide vaccination is efficacious against established intracerebral tumors. *Clin Cancer Res*. 2003;9(11):4247–4254.
51. Schuster J, Lai RK, Recht LD, et al. A phase II, multicenter trial of rindopepimut (CDX-110) in newly diagnosed glioblastoma: the ACT III study. *Neuro-oncology*. 2015;17(6):854–861.
52. Sampson JH, Aldape KD, Archer GE, et al. Greater chemotherapy-induced lymphopenia enhances tumor-specific immune responses that eliminate EGFRvIII-expressing tumor cells in patients with glioblastoma. *Neuro-oncology*. 2011;13(3):324–333.
53. Neagu MR, Reardon DA. Rindopepimut vaccine and bevacizumab combination therapy: improving survival rates in relapsed glioblastoma patients? *Immunotherapy*. 2015;7(6):603–606.
54. Liu X, McEachron TA, Schwartzentruber J, Wu G. Histone H3 mutations in pediatric brain tumors. *Cold Spring Harb Perspect Biol*. 2014;6(4):a018689.

55. Luger K, Rechsteiner TJ, Flaus AJ, Waye MM, Richmond TJ. Characterization of nucleosome core particles containing histone proteins made in bacteria. *J Mol Biol*. 1997;272(3):301–311.
56. Fullgrabe J, Kavanagh E, Joseph B. Histone onco-modifications. *Oncogene*. 2011;30(31):3391–3403.
57. Allis CD, Berger SL, Cote J, et al. New nomenclature for chromatin-modifying enzymes. *Cell*. 2007;131(4):633–636.
58. Fullgrabe J, Hajji N, Joseph B. Cracking the death code: apoptosis-related histone modifications. *Cell Death Differ*. 2010;17(8):1238–1243.
59. Chi P, Allis CD, Wang GG. Covalent histone modifications: miswritten, misinterpreted and mis-erased in human cancers. *Nat Rev Cancer*. 2010;10(7):457–469.
60. Schwartzentruber J, Korshunov A, Liu XY, et al. Driver mutations in histone H3.3 and chromatin remodelling genes in paediatric glioblastoma. *Nature*. 2012;482(7384):226–231.
61. Sharma S, Kelly TK, Jones PA. Epigenetics in cancer. *Carcinogenesis*. 2010;31(1):27–36.
62. Szenker E, Ray-Gallet D, Almouzni G. The double face of the histone variant H3.3. *Cell Res*. 2011;21(3):421–434.
63. Yuen BT, Knoepfler PS. Histone H3.3 mutations: a variant path to cancer. *Cancer cell*. 2013;24(5):567–574.
64. Franklin SG, Zweidler A. Non-allelic variants of histones 2a, 2b and 3 in mammals. *Nature*. 1977;266(5599):273–275.
65. Tagami H, Ray-Gallet D, Almouzni G, Nakatani Y. Histone H3.1 and H3.3 complexes mediate nucleosome assembly pathways dependent or independent of DNA synthesis. *Cell*. 2004;116(1):51–61.
66. Drane P, Ouararhni K, Depaux A, Shuaib M, Hamiche A. The death-associated protein DAXX is a novel histone chaperone involved in the replication-independent deposition of H3.3. *Genes Dev*. 2010;24(12):1253–1265.
67. Lewis PW, Elsaesser SJ, Noh KM, Stadler SC, Allis CD. Daxx is an H3.3-specific histone chaperone and cooperates with ATRX in replication-independent chromatin assembly at telomeres. *Proc Natl Acad Sci USA*. 2010;107(32):14075–14080.
68. Wong LH, McGhie JD, Sim M, et al. ATRX interacts with H3.3 in maintaining telomere structural integrity in pluripotent embryonic stem cells. *Genome Res*. 2010;20(3):351–360.
69. Elsasser SJ, Huang H, Lewis PW, Chin JW, Allis CD, Patel DJ. DAXX envelops a histone H3.3-H4 dimer for H3.3-specific recognition. *Nature*. 2012;491(7425):560–565.
70. Kallappagoudar S, Yadav RK, Lowe BR, Partridge JF. Histone H3 mutations – a special role for H3.3 in tumorigenesis? *Chromosoma*. 2015;124(2):177–189.
71. Osley MA. The regulation of histone synthesis in the cell cycle. *Annu Rev Biochem*. 1991;60:827–861.
72. Wu RS, Tsai S, Bonner WM. Patterns of histone variant synthesis can distinguish G0 from G1 cells. *Cell*. 1982;31(2 Pt 1):367–374.
73. Elsaesser SJ, Goldberg AD, Allis CD. New functions for an old variant: no substitute for histone H3.3. *Curr Opin Genet Dev*. 2010;20(2):110–117.
74. Talbert PB, Henikoff S. Histone variants – ancient wrap artists of the epigenome. *Nat Rev Mol Cell Biol*. 2010;11(4):264–275.
75. Fontebasso AM, Liu XY, Sturm D, Jabado N. Chromatin remodeling defects in pediatric and young adult glioblastoma: a tale of a variant histone 3 tail. *Brain Pathol*. 2013;23(2):210–216.
76. Wells D, Kedes L. Structure of a human histone cDNA: evidence that basally expressed histone genes have intervening sequences and encode polyadenylylated mRNAs. *Proc Natl Acad Sci USA*. 1985;82(9):2834–2838.
77. Wells D, Hoffman D, Kedes L. Unusual structure, evolutionary conservation of non-coding sequences and numerous pseudogenes characterize the human H3.3 histone multigene family. *Nucl Acids Res*. 1987;15(7):2871–2889.

78. Castel D, Philippe C, Calmon R, et al. Histone H3F3A and HIST1H3B K27M mutations define two subgroups of diffuse intrinsic pontine gliomas with different prognosis and phenotypes. *Acta Neuropathol.* 2015.
79. Wu G, Broniscer A, McEachron TA, et al. Somatic histone H3 alterations in pediatric diffuse intrinsic pontine gliomas and non-brainstem glioblastomas. *Nat Genet.* 2012;44(3):251–253.
80. Sturm D, Witt H, Hovestadt V, et al. Hotspot mutations in H3F3A and IDH1 define distinct epigenetic and biological subgroups of glioblastoma. *Cancer Cell.* 2012;22(4):425–437.
81. Lewis PW, Muller MM, Koletsky MS, et al. Inhibition of PRC2 activity by a gain-of-function H3 mutation found in pediatric glioblastoma. *Science.* 2013;340(6134):857–861.
82. Bender S, Tang Y, Lindroth AM, et al. Reduced H3K27me3 and DNA hypomethylation are major drivers of gene expression in K27M mutant pediatric high-grade gliomas. *Cancer Cell.* 2013;24(5):660–672.
83. Becher OJ, Wechsler-Reya RJ. Cancer. For pediatric glioma, leave no histone unturned. *Science.* 2014;346(6216):1458–1459.
84. Chan KM, Fang D, Gan H, et al. The histone H3.3K27M mutation in pediatric glioma reprograms H3K27 methylation and gene expression. *Genes Dev.* 2013;27(9):985–990.
85. Khuong-Quang DA, Buczkowicz P, Rakopoulos P, et al. K27M mutation in histone H3.3 defines clinically and biologically distinct subgroups of pediatric diffuse intrinsic pontine gliomas. *Acta Neuropathol.* 2012;124(3):439–447.
86. Wu G, Diaz AK, Paugh BS, et al. The genomic landscape of diffuse intrinsic pontine glioma and pediatric non-brainstem high-grade glioma. *Nat Genet.* 2014;46(5):444–450.
87. Fontebasso AM, Papillon-Cavanagh S, Schwartzentruber J, et al. Recurrent somatic mutations in ACVR1 in pediatric midline high-grade astrocytoma. *Nat Genet.* 2014;46(5):462–466.
88. Bjerke L, Mackay A, Nandhabalan M, et al. Histone H3.3 mutations drive pediatric glioblastoma through upregulation of MYCN. *Cancer Discov.* 2013;3(5):512–519.
89. Pollack IF, Jakacki RI, Butterfield LH, et al. Antigen-specific immune responses and clinical outcome after vaccination with glioma-associated antigen peptides and polyinosinic-polycytidylic acid stabilized by lysine and carboxymethylcellulose in children with newly diagnosed malignant brainstem and nonbrainstem gliomas. *J Clin Oncol.* 2014;32(19):2050–2058.
90. Paugh BS, Broniscer A, Qu C, et al. Genome-wide analyses identify recurrent amplifications of receptor tyrosine kinases and cell-cycle regulatory genes in diffuse intrinsic pontine glioma. *J Clin Oncol.* 2011;29(30):3999–4006.

SECTION II

STUDYING BRAIN TUMOR IMMUNOTHERAPY

CHAPTER

6

Preclinical Immunotherapeutic Animal Models for Brain Tumors

V. Chandramohan, L. Sanchez-Perez, Y. He, C.J. Pirozzi, K.L. Congdon, D.D. Bigner

Duke University Medical Center, Durham, NC, United States

OUTLINE

Translational Immunotherapy of Brain Tumors
http://dx.doi.org/10.1016/B978-0-12-802420-1.00006-5

INTRODUCTION

Malignant gliomas—comprising of glioblastoma multiforme, anaplastic astrocytomas, anaplastic oligodendrogliomas, anaplastic oligoastrocytomas, anaplastic ependymomas, and anaplastic gangliogliomas—are the most common primary malignant brain tumors of the central nervous system (CNS).[1] Glioblastoma (GBM), the most common and most malignant type of glioma, has an annual incidence rate of 3.2 cases/100,000 people in the United States.[2] The current standard of care for patients with GBM involves surgery and concurrent radiotherapy and temozolomide (TMZ), followed by adjuvant TMZ.[3] Despite the multimodal approach, the median survival of patients with GBM is only 15 months, and the 5-year survival rate is 5%.[2,3] Due to the extensive infiltration of therapy-resistant tumor cells into the surrounding brain parenchyma, intratumoral and intertumoral heterogeneity, an immunosuppressive tumor microenvironment and poor delivery of therapeutics across the blood–brain barrier, GBM relapse is universal and is associated with a poor prognosis.[4] Hence, there is a need for the development of novel therapeutic regimens that can overcome the drawbacks associated with conventional therapies and potentiate the complete eradication of GBM tumor cells.

Immunotherapeutics have emerged as potent anticancer agents owing to recent advances in the understanding of tumor-associated antigens and immune checkpoint modulators. Immunotherapeutic approaches involving the body's own immune system not only activate the immune effectors to seek out and destroy tumor cells with exquisite specificity, but also generate a long-term surveillance against cancer cells. The CNS, once considered an immune-privileged site, has emerged as an immunologically active niche, opening up the possibility for brain tumor immunotherapy.[5] Accordingly, tumor vaccines and adoptive cell transfer-based cancer immunotherapy have proven to be safe, prolonged survival, and even generated complete remissions in some GBM patients.[6–9] However, GBM tumors induce and maintain immunosuppressive phenotypes, both systemically and in the tumor microenvironment, which act as barriers for successful immunotherapy.[10] Hence, immunotherapy, while unlikely to be the long awaited panacea for GBM, provides new treatment options that have the potential to improve prognoses for malignant brain tumor patients if incorporated into the current standard of care, or if combined with novel GBM-targeted therapies.

Consequently, the next challenge will be to test novel combinations of promising therapeutics in biologically relevant systems. Immunocompetent animal models of gliomas recapitulate the human tumor scenario by facilitating interactions between the tumor, stroma, immune response, and antiglioma therapeutics, thereby providing a physiologically relevant in vivo system for the investigation of novel antiglioma therapies. Additionally, preclinical testing in immunocompetent animal models will help identify appropriate drugs/targets, determine therapeutic efficacy in the ideal genetic and cellular milieu, and discover biomarkers for subsequent clinical testing. In this chapter, we provide a summary of different immunocompetent animal models of malignant brain tumors currently available for the preclinical testing of novel antiglioma therapeutics.

SPONTANEOUS AND EXPERIMENTALLY-INDUCED AUTOCHTHONOUS BRAIN TUMOR ANIMAL MODELS

Animal models of brain tumors serve as relevant preclinical models for the in vivo testing of potential therapeutic targets and for the identification of appropriate therapy combinations for future clinical testing.

A reliable brain tumor animal model should effectively recapitulate the clinical characteristics of malignant brain tumors for the therapeutic results to be successfully translated clinically. The ideal brain tumor model should

1. be comprised of tumor cells of glial origin;
2. be able to grow in culture and be amenable for genetic modifications;
3. exhibit cellular and molecular characteristics that resemble the human disease;
4. grow in the orthotopic microenvironment, replicate the tumor–host interactions, and exhibit appropriate latency with a predictable survival time;
5. display glioma histopathological features, including intraparenchymal growth, infiltration, and the ability to form neovasculature;
6. grow in a syngeneic host that is small, inexpensive, and easily generated/maintained for the reliable testing of therapeutics in a large number of animals;

7. exhibit a therapeutic responsiveness that reproduces the clinical results; and
8. allow for the monitoring of tumor growth through noninvasive imaging techniques.

The current brain tumor models satisfy some of the attributes specified earlier; however, none of the models are able to fulfill all of the requisites.

Spontaneous Brain Tumor Models

The ideal animal model of brain tumors should consist of spontaneous tumors with characteristics identical to those of its human counterpart. Spontaneous tumors have been observed in other species, including dogs and cats, but rarely in primates.[11–15] Compared to dogs and cats, only a few cases of spontaneous brain tumors have been described in mice. Spontaneous gliomas (astrocytomas) in 32/4000 inbred VM mice were discovered by Fraser during the examination of over 10,000 mouse brains used in scrapie research.[16] Intracerebral transplantable tumor cell lines have been established from these spontaneous astrocytomas and have been utilized for the development of brain tumor models for therapeutic studies.[17–19] In comparison, there was a higher incidence of spontaneous CNS tumors in rats (>1%) than in mice (>0.001%).[20] The low incidence of spontaneous tumors in laboratory animals makes their applicability for therapeutic studies impractical.

Chemical Carcinogen-Induced Brain Tumor Models

The history of experimental carcinogenesis dates back to 1915, when Yamagiwa and Ichikawa succeeded in inducing folliculoepitheliomata (papillomas) by chronically (30–100 days) applying coal tar on rabbits' ears.[21] Seligman and Shear successfully induced brain tumors (gliomas) in 11/20 mice by intracerebrally implanting methylcholanthrene (MCA) pellets.[22] The effectiveness of the aromatic hydrocarbons in inducing brain tumors varied depending upon the species as well as on the strain tested. The tumor incidences in mice were similar with both MCA and benzpyrene; however, MCA implantation yielded a higher percentage of gliomas (15/26) as opposed to sarcomas (5/26).[23,24] The aromatic hydrocarbon 6,9,10-trimethyl-1,2-benzanthracene was highly effective in inducing brain tumors in rats (40 gliomas/66 tumors).[25] However, aromatic hydrocarbon-induced tumors in animals had long latency periods of 155–528 days.[25]

Systemic testing for oncogenesis led Druckrey et al. to identify nitrosoureas, a new class of synthetic carcinogens that are greatly successful in inducing CNS tumors in rats and rabbits.[25] Among the different *N*-nitrosourea derivatives, *N*-methyl-*N*-nitrosourea (MNU) induced tumors exclusively in the brain and spinal cord, whereas *N,N′,N′*-trimethyl-*N*-nitrosourea (TMNU) induced tumors frequently in the peripheral nervous system.[26,27] The MNU-induced tumors were detected 97–99 days after intravenous and intraperitoneal administration of the carcinogen. Despite a lengthy observation period, 8–16 doses of 20 mg/kg of MNU delivered intravenously at 4-week intervals failed to induce tumors in rhesus monkeys.[25]

Druckrey et al. found that ethylnitrosurea (ENU) was highly efficient in inducing CNS tumors in younger rats (3 months of age) compared to older rats.[28] This discovery prompted a series of transplacental CNS tumor induction studies in pregnant rats through the intravenous injection of ENU.[29] A single intravenous injection of ENU on the 22nd day of pregnancy led to the development of nervous system tumors in 100% (142/142) of the offspring.[30] In contrast to adult rats, MNU demonstrated lesser activity during transplacental administration. Despite the reliable induction of tumors of neural origin by carcinogens, the extensive variability in tumor type, location, latency period, and extraneural tumor induction are major drawbacks that preclude carcinogen use for brain tumor therapeutic studies.

Virus-Induced Brain Tumor Models

Brain tumors have been successfully induced in experimental animal models using different groups of oncogenic viruses.[31] Adenovirus and papovavirus are the two DNA viruses that have been successfully utilized for the induction of brain tumors in animal models. Human adenovirus 12 has been demonstrated to induce neuroblastic and retinoblastic cellular tumors in neonatal rats, mice, hamsters, and natal multimammate mice, with a tumor incidence rate of 8–100% and a latency period of 31–235 days.[31] Transplantable rat and mouse brain tumor cell lines have been established from human adenovirus 12–induced tumors. Intracerebral injection of Simian adenovirus 7 led to the development of fibrosarcomas, malignant ependymomas, and choroid plexus papillomas, with a tumor incidence rate of 22–93%.[31] The inoculation of Simian virus 20 (SV20) and avian adenovirus induced undifferentiated intracranial tumors and intraventricular tumors, respectively, in neonatal hamsters.[32,33] Human papovaviruses (Mad-1 to -4) isolated from progressive multifocal leukoencephalopathy brain tumor tissue induced cerebellar medulloblastomas, undifferentiated thalamic gliomas, intraventricular papillary ependymomas, pineocytomas, and a meningeal tumor in new-born hamsters postintracerebral inoculation.[34] Human papovavirus embryonic kidney cells induced choroid plexus papillomas and fibrosarcomas of the scalp when injected intracerebrally into new-born hamsters. Bovine papilloma virus is a nonhuman papovavirus that can induce both benign and malignant tumors. It has been known to cause anything from meningiomas to invasive meningeal sarcomas in the hamster, and fibromatous noninvasive, well-differentiated sarcomas to fibrosarcomas in the calf.[31] Intracerebral inoculation of SV40 in new-born Syrian hamsters or Natal multimammate mice induced the development of ventricular tumors that were classified as ependymomas, choroid plexus papillomas, and meningeal sarcomas.

The ASV, murine sarcoma virus (MSV), and simian sarcoma virus (SSV) are RNA viruses (retroviruses) that have been known to induce CNS tumors in animals.[31] ASV is the most widely used retrovirus for the induction of brain tumors in experimental animal models.[31] Rabotti and Raine successfully induced gliomas by intracerebrally inoculating new-born hamsters with the Schmidt–Ruppin strain, a variant of the chicken rous sarcoma virus.[35] Brain tumors have also been successfully induced in mice, rats, cats, dogs, hamsters, gerbils, and nonhuman primates by intracerebral inoculation of different ASV strains, such as Bryan, Schmidt-Ruppin, CT-559, and Bratislava-77.[31] Development of brain tumor models in dogs is advantageous due to their larger brain size, the possibility for surgical intervention, and the availability of larger tumor volumes for evaluation. However, the expense and the lack of inbred strains preclude the usage of dog models. Contrarily, the ASV-induced rat brain tumor models are more advantageous because there is a higher incidence of gliomas, a shorter induction time (three months), an availability of inbred rat strains, an ability to establish in vitro cultures, and the potential for intracranial and subcutaneous transplantations.

Moloney MSV, Kirsten MSV, and Harvey MSV induced tumors in rats and mice when inoculated intracerebrally. The different brain tumor types induced by Kirsten MSV were GBMs, gemistocytic astrocytomas, oligodendrogliomas, and hemangioblastomas in rats and sarcomas in mice.[31,36] Injections of SSV directly into the brains of new-born marmosets generated brain tumors that were morphologically similar to human GBMs.

SYNGENEIC TRANSPLANTATION BRAIN TUMOR MODELS

Syngeneic transplantation tumor systems involve serial transplantation of tumor tissues derived from spontaneous-, chemical-, or virus-induced tumors passed intracranially or subcutaneously in animals that have the same genetic backgrounds as the original animal tumor strains. The advantages of transplantable tumor models include the high degree of morphologic uniformity between tumors and a short latency due to rapid tumor growth posttransplantation. The animals used most commonly to establish transplantable brain tumor models are rats and mice.

Because of the larger size of their brains (a rat's brain is ~1200 mg and a mouse's brain is ~400 mg), rat brain tumor models are advantageous as they allow for accurate stereotactic implantation of tumor cells, better in vivo localization and imaging studies posttherapy, and administration of larger volumes of the therapeutic agent by convection enhanced delivery. Contrarily, the potential for easier genetic engineering/manipulation in mice provides an opportunity to investigate the function of genetic factors, signaling pathways, cell types, and the tumor microenvironment during therapeutic interventions in mouse brain tumor models. Additionally, there are a larger number of monoclonal antibodies (mAbs) available for the therapeutic targeting of mouse molecules (immune cell receptors and chemokines) compared to rat molecules. Therefore, both model systems provide a unique set of advantages for immunotherapy studies. Over the past several decades, several rat and mouse glioma lines have been established, cultured ex vivo, and serially transplanted into immunocompetent recipient animals, permitting large-scale tumor induction studies, which allowed for comprehensive investigations of tumor biology, tumor immunology, and immunotherapeutics.[37] The following models have played a critical role in understanding gliomas as well as how best to utilize immunotherapy for this devastating disease.

Rat Brain Tumor Models

T9/9L Gliosarcoma Model

The T9/9L gliosarcoma tumor was induced by weekly tail vein injections of 5 mg/kg of MNU into CD Fischer rats for 36 weeks.[38,39] Twenty-six weeks posttermination of MNU injections, a grade III astrocytoma tumor, designated as tumor 9 (T9), developed in one of these rats. A cell line established from the grade III astrocytoma T9 was subsequently grown in the flanks of adult Fischer 344 rats and was later designated as tumor 9L. The morphological feature of the 9L tumors was different from that of the original tumor and was similar to that of a GBM and sarcoma mix or gliosarcoma. The 9L tumors exhibited pleomorphic features with abundant giant cells, mitoses, and pseudopalisading necrosis.[40] Molecular characterization of 9L gliosarcoma identified the presence of a mutant *p53* gene and a wild-type *p16/cyclin-dependent kinase inhibitor 2A (Cdkn2a)/INK4a* locus.[41,42] Additionally, when compared to SD rat stem cell–derived astrocytes, 9L tumors revealed an increased expression of epidermal growth factor receptor (*EGFR*) and its ligand, transforming growth factor alpha (*TGF*α), a decreased expression of fibroblast growth factor (*FGF*)-*9* and platelet-derived growth factor receptor beta, and a loss of expression of *FGF-2* and *FGFR-1* genes.[43] The 9L gliosarcoma cell line exhibits features identical to cancer stem-like cells (CSLCs) when grown in serum-free media supplemented with growth factors.[44] The 9L gliosarcoma CSLCs express the neural stem cell markers nestin and Sox2, and have the capacity to self-renew and differentiate into different CNS cell types (neuron and glial cells). When implanted intracranially, the 9L gliosarcoma CSLCs form tumors that are more aggressive than the original 9L tumors.[44]

The 9L gliosarcoma-Fischer rat tumor model has been used for the investigation of adoptive cell therapy, cytokine therapy, dendritic cell (DC) vaccination, radiosurgery, and oligonucleotide vaccination strategies. Systemic chemotherapy (cyclophosphamide), along with the intracranial infusion of lymphokine-activated killer cells and interleukin-2 (IL-2) combination therapy (chemoadoptive immunotherapy), resulted in 23% of long-term survivors (9/39), with no evidence of tumors in 18% (7/39) of Fischer rats bearing intracranial 9L gliosarcoma tumors.[45] Similarly, local delivery of IL-12 into the rat brain by 9L gliosarcoma cells genetically engineered to express IL-12 (9L-IL12 cells) significantly prolonged the survival time of animals coimplanted intracranially with the parental 9L tumor cells.[46] Continuous subcutaneous delivery of a cytokine cocktail consisting of either granulocyte-macrophage colony-stimulating factor (GM-CSF) alone, GM-CSF with IL-2, or GM-CSF with IL-12, along with irradiated 9L cells injected subcutaneously for each group generated a delayed-type hypersensitivity (DTH) response and led to an increased survival rate in rats with intracranial 9L tumors when compared to untreated animals.[47,48] Sequential vaccination of Fischer 344 rats harboring intracranial 9L tumors with DCs pulsed with apoptotic 9L glioma cells generated significantly more long-term survivors [75% (6/8) long-term survivors (>90 days)] compared to the vehicle control group (0/16).[49] Interestingly, CpG-containing oligonucleotides (ODN)—which are known to induce local inflammatory responses through $CD8^+$ T cell infiltration and major histocompatibility complex (MHC) class II upregulation in normal rat brains and decrease tumor volumes of 9L subcutaneous tumors—caused an increase in the size of the intracranial 9L tumors compared to control groups. Ginzkey et al. underscored the possibility that therapeutics would display differential effects against subcutaneous versus intracranial tumors and suggested that CpG-ODN treatment in glioma patients might not generate beneficial outcomes.[50,51] Finally, reversal of glioma immunosuppression through TGF-β gene therapy (9L gliosarcoma cells genetically modified to inhibit TGF-β expression with an antisense plasmid vector or a truncated, soluble form of the TGF-β receptor) resulted in an improved survival of animals bearing intracranial 9L tumors.[52,53]

Thus, several immunotherapy studies have demonstrated impressive antitumor efficacy, including apparent cures of animals bearing 9L gliosarcoma tumors. However, it must be emphasized that 9L gliosarcoma tumors have been shown to be highly immunogenic. Animals immunized subcutaneously with irradiated 9L cells were resistant to intracranial tumor implantation, compared to 100% tumor takes in immunologically naive animals.[54,55] Therefore, the immunogenicity of the 9L glioma must be taken into consideration when utilizing this model to evaluate the efficacy of novel immunotherapeutic strategies. In addition to Fischer rats, 9L gliosarcoma cells can also develop circumscribed, noninfiltrative intracranial tumors in allogeneic Wistar rats.[56] Immunostaining of the 9L tumors in Wistar rats revealed the presence of Ki-67-positive cells, the absence of glial fibrillary acidic proteins (GFAP)-positive astrocytic cells, and the presence of both activated ED1-positive macrophages/microglia and T lymphocytes.[56]

Rat Glioma 2/D74 Glioma Model

The rat glioma 2 (RG2), also known as D74-RG2 or D74, is a well-differentiated glioma cell line. The RG2 cell line was established from brain tumors induced by a single intravenous injection of 50 mg/kg body weight of ENU into a pregnant CD Fischer rat on the 20th day of gestation.[57,58] Implantation of 1×10^5 RG2 cells into the brains of syngeneic rats produced large anaplastic tumors within 9 days in the right basal ganglia.[59] Histological examination of the tumors revealed frequent mitoses, giant cells, scarcely distributed blood vessels, focal necrosis, and infiltrative growth into the adjacent brain, making it an attractive model for brain tumors.[59] Molecular characterization of the RG2 glioma identified the presence of a wild-type *p53* gene and the deletion of the *p16/Cdkn2a/INK4a* locus.[42] Gene expression studies demonstrated increased expression of the *PDGFβ, IGF-I, Ras, ErbB3/human epidermal growth factor receptor 3 (HER3)* precursor messenger RNA (mRNA), and *cyclin D2* genes in RG2 gliomas.[43]

The RG2 glioma is nonimmunogenic in syngeneic Fischer rats and expresses low basal levels of MHC class I, which is upregulated upon treatment with interferon-gamma (IFN-γ) in vitro.[60,61] The RG2 glioma tumor model has been used for the investigation of oligonucleotide and cytokine vaccination strategies, regulatory $CD4^+CD25^+$ T cell (Treg) depletion studies, and cytokine and tumor-targeted cytotoxic agent combination therapies. Schartner et al. utilized the RG2 glioma model to demonstrate that compared to normal brains, MHC class II induction by CpG-ODN, IFN-γ, and IFN-γ/lipopolysaccharide was significantly impaired in tumor microglia/macrophages.[62] In a subcutaneous model of the RG2 glioma, low-dose metronomic TMZ regimens (0.5 and 2 mg/kg for 21 days) induced Treg depletion among the splenocyte and tumor-infiltrating lymphocyte populations.[63] Compared to untreated controls, animals treated with 0.5 mg/kg of TMZ demonstrated a slight decrease in tumor growth; however, the decrease in tumor growth was not statistically significant.[63] The immune-stimulating cytokine fms-like tyrosine kinase-3 ligand (Ad-Flt3L) and the conditionally cytotoxic herpes simplex type 1-thymidine kinase (Ad-TK) combination therapy synergized with either the overexpression of IFN-γ or IL-2 or the inhibition of the nuclear factor kappa-B signaling pathway, leading to enhanced cytotoxic T lymphocyte–mediated immune responses and improved survival of rats bearing intracranial RG2 gliomas.[64] Additionally, the Ad-Flt3L and Ad-TK/ganciclovir combination immunotherapy also synergized with rapamycin (inhibitor of the mammalian target of rapamycin) treatment, leading to an enhanced cytotoxic activity of $CD8^+$ T cells and a development of $CD8^+$ memory T cell response, which improved the survival of RG2 tumor-bearing animals from 35 to 60 days ($P < .001$).[65] Analogous to human GBM, the RG2 glioma is resistant to most common therapeutic modalities, including radiation, chemotherapy, and immunotherapy, making it an attractive preclinical model for the testing of novel GBM therapeutics.[59,60]

F98/D98 Glioma Model

The F98, also known as D98, is an undifferentiated glioma cell line that was established from brain tumors induced in CD Fischer rats by a single transplacental exposure of 50 mg/kg body weight of ENU to fetuses on the 20th day of gestation.[57,58] Orthotopic implantation of 1×10^4 F98 glioma cells resulted in 100% tumor take, with a median survival of 26 days.[66] Magnetic resonance imaging studies revealed that F98 glioma cells form aggressive tumors by infiltrating the surrounding normal brain tissue during tumor growth and progression and disseminating small necrotic areas throughout the tumor.[67] Immunohistochemical staining of F98 tumors revealed positive staining for GFAP and vimentin.[66] Molecular characterization of the F98 glioma identified the presence of a wild type *p53* gene and deletion of the *p16/Cdkn2a/INK4a* locus.[42] Gene expression profiling of F98 gliomas (relative to SD rat astrocytes) revealed an increased expression of the *PDGFβ, Rb, Ras, EGFR, cyclin D1*, and *cyclin D2* genes.[43]

The low immunogenicity of F98 gliomas was established by tumor rechallenge studies, wherein CD Fischer rats were immunized with irradiated F98 tumor cells and challenged with intracerebral implants of 2×10^4 F98 cells. The median survival time was 22 days for immunized animals, compared to 18 days for nonimmunized controls.[61] However, vaccination of Fischer rats with mitomycin C-treated F98 or F98/B7.1 F98 glioma cells transfected with the B7.1 (CD80) costimulatory molecule did not improve the survival time of animals bearing F98/B7.1 intracranial tumors when compared to unvaccinated or wild-type F98 vaccinated animals.[68] Similarly, intracerebral implantation of F98 cells, followed by subcutaneous vaccinations with irradiated F98 cells in the presence or absence of GM-CSF, failed to demonstrate significant differences in survival times between vaccinated and untreated animals.[69] Adoptive immunotherapy with adherent lymphokine-activated killer (A-LAK) cells, injected intracranially together with 4000 F98 glioma cells, increased the mean survival time from 22.3 days for the control group to 46.1 days for the rats treated with A-LAK cells.[70] However, increasing the number of implanted F98 cells to 12,500 cells only produced a moderate improvement in the survival time of A-LAK-treated animals (27.8 days) compared to untreated controls (20.8 days).[70] A previous study involving a glioma cell line overexpressing the human wild-type EGFR protein concluded that a total of 1×10^5 cells or greater is required to evoke a xeno-immune response against human EGFR proteins in rats.[71] Along similar lines, F98 cell lines overexpressing human wild-type EGFR ($F98_{EGFR}$) and EGFRvIII ($F98_{EGFRvIII}$) proteins have been established. Ciesielski et al. utilized the $F98_{EGFRvIII}$ tumor model to evaluate the efficacy of EGFRvIII peptide vaccination.[72] Fischer rats were either unvaccinated or administered a total of four vaccinations of either EGFRvIII multiple antigenic peptide [a peptide containing multiple copies of the unique EGFRvIII epitope (MAP)], GM-CSF, or a combination of MAP/GM-CSF. Two weeks after the final booster vaccination, rats underwent intracerebral implantation of 1×10^4 $F98_{EGFRvIII}$ glioma cells. In this model, the median survival times of unvaccinated, GM-CSF-vaccinated, MAP-vaccinated, and MAP/GM-CSF-vaccinated rats were 17.5 days, 18.5 days, 21 days, and 30.5 days, respectively.[72] A MAP/GM-CSF combination vaccination strategy generated a 72% increase in the median survival of intracerebral $F98_{EGFRvIII}$ tumor-bearing rats. Volovitz et al. defined a location-based immunotherapeutic phenomenon, termed "split immunity," using the F98 glioma model.[73] The authors demonstrated that F98 tumors that thrive intracranially (in an immune-privileged site) can be inhibited by injecting live, unmodified F98 tumor cells subcutaneously (in a site that is not immune-privileged), thereby generating protective immunity that spreads back to the privileged site.[73] Because F98 gliomas reproduce critical characteristics of human GBMs, such as a highly infiltrative growth pattern, an increased expression of *EGFR*, a positive staining for GFAP, and a low immunogenicity, they have been used as preclinical models of brain tumors to evaluate the efficacy of a variety of experimental therapeutics. Bioluminescent cell lines for the 9L, RG2, and F98 glioma tumor models have been established by stably transfecting the different rat tumor cell lines with the luciferase gene.[74–76] These tumor cell lines will aid in the monitoring of the size of intracranially implanted tumors through the measurement of bioluminescence. These models should allow for the rapid, noninvasive imaging of intracranial tumor growth to evaluate the effectiveness of novel therapeutic modalities.

CNS-1 Glioma Model

The CNS-1 glioma cell line was developed by weekly intravenous injections of MNU for a period of 6 months into inbred Lewis rats.[77] Intracranial implantation of CNS-1 tumors in the syngeneic host demonstrated extensive leptomeningeal, perivascular, and periventricular spread, as well as individual tumor cell permeation of parenchyma.[77] The mean survival times of Lewis rats implanted intracranially with an increasing number of

CNS-1 cells (5×10^4–1×10^6 cells) ranged between 20.5 and 30.2 days. CNS-1 tumor cells demonstrated positive reactivity for GFAP, S100, and vimentin by immunohistochemistry (IHC) analysis.[77] Therefore, the CNS-1 glioma provides an excellent tumor model for the investigation of a variety of immunotherapeutic modalities.

An injection of immunostimulatory CpG-ODNs into the tumor bed potentiated a rejection of established intracranial CNS-1 tumors and induced intratumoral infiltration of macrophages/microglia, CD8, and natural killer cells.[78] CNS-1 tumor-bearing Lewis rats that were cured by CpG-ODN injection were also protected against CNS-1 tumor rechallenge.[78] The CNS-1 glioma model was used to evaluate the applicability of gene therapy to mitigate TGF-β-mediated immunosuppression.[79] The therapeutic efficacy of decorin, a small proteoglycan which can bind to and inactivate TGF-β, was studied using recombinant adenoviruses (RAds) expressing the human decorin gene in the rat CNS-1 glioma model. Intratumoral delivery of 8×10^7 infectious units of RAd/hCMV/decorin had a small (but significant) effect in prolonging the survival of CNS-1 tumor-bearing rats; however, decorin expression had to be present in all tumor cells for the therapy to be effective.[79] The efficacy of various immunostimulatory therapies [RAdFlt3L (human Flt3L), RAdCD40L (mouse CD40L), RAdIL-12], in combination with conditional cytotoxicity (RAdTK), was tested in an orthotopic model of the CNS-1 glioma.[80] Only the combined delivery of RAdFlt3L+RAdTK significantly prolonged the survival of large CNS-1 tumor-bearing animals (≥70%), and the prolongation of survival was primarily mediated through functional macrophages and $CD4^+$ cells.[80] Despite demonstrating significant efficacy in syngeneic rat models with the RAds-mediated gene therapy, future studies should take into consideration the fact that the authors in these studies used human and mouse proteins as immunostimulatory molecules. In addition, the CNS-1 glioma model has also been utilized to demonstrate the antitumor efficacy of the systemic delivery of propentofylline, an atypical methylxanthine with CNS glial modulating and antiinflammatory actions. Propentofylline targeted TROY, a novel signaling molecule upregulated in infiltrating microglia, and caused a significant decrease in tumor volume.[81,82] Propentofylline specifically decreased matrix metallopeptidase 9 (MMP-9) expression and microglial migration toward CNS-1 tumor cells and did not affect peripheral macrophages.[81,82] The antitumor efficacy of parenterally injected imidazoquinoline-based toll-like receptor 7/8 (TLR7/8) small molecule agonist resiquimod (R848) was investigated in a subcutaneous CNS-1 glioma model. The antitumor effect of R848 was primarily mediated through the activation of immune cells and was independent of any direct effect on CNS-1 tumor cells. Significantly, R848 monotherapy was able to reject smaller established CNS-1 tumors as well as develop an immunological memory against tumor rechallenge.

C6 Glioma Model

Brain tumors were established by weekly tail vein injections of 5mg/kg of MNU to randomly bred Wistar rats over a period of 36 weeks.[38,39] Tumors were excised from animals exhibiting neurological symptoms and placed in primary culture. A cell line established from a tumor designated as "#6" was subsequently cloned and renamed "C6". The C6 glioma exhibited an astrocytic pattern with moderate cellular pleomorphism, increased mitotic activity, and conspicuous palisading of tumor cells around areas of necrosis.[38,39] The C6 rat glioma model has been widely used in experimental neuro-oncology to evaluate the efficacy of a variety of tumor therapies.[83] Because the original tumor was established in an outbred Wistar rat, there is no syngeneic host for the propagation of C6 gliomas. Different rat strains, including inbred rat strain IX (BDIX), inbred rat strain X (BDX), SD, and Wistar have been used as "syngeneic" hosts for the C6 glioma, and it has been demonstrated to be immunogenic in Wistar rats.[84,85] Since C6 glioma cells are allogeneic in all inbred strains, investigators should be cautious about using this tumor model for their immunotherapy research.

BT4C Model

Brain cell suspensions were established from BD IX-rat fetuses excised at 20–90h posttransplacental injection of 75μg/g of ENU on the 18th day of gestation.[86] The brain cells became tumorigenic after propagation for ~200days in culture. Seven malignant neurogenic cell lines (BT1C-BT7C) were established by this procedure.[87] When reimplanted subcutaneously into BD IX rats, the BT cell lines had a mean latency period of 48±13days. The malignant BT cell lines generated neurinoma-like, glioma- or GBM-like, and undifferentiated, pleomorphic neoplasms.[87] Histological diagnosis was confirmed only for the neurinoma-like tumors; hence, glioma- or GBM-like descriptions for the BT cell line-derived tumors should be considered with caution.[87] The BT4C tumor has served as a useful model to test novel chemotherapies, gene therapy, antiangiogenic therapy, and radiotherapy, but it has not been utilized in immunotherapy studies.[83]

Mouse Brain Tumor Models

SMA-560 Anaplastic Astrocytoma

Spontaneously arising brain tumors are rare and often have a prolonged onset with low penetrance (0.054%).[88] In a series of experiments looking at a variety of mouse strains, the VM/Dk inbred mouse strain displayed a tumor incidence of approximately 1–5% by 500days.[16] Based on cellular morphology, these tumors were determined to be astrocytomas and upon serial transplantation into the CNS of the recipient VM/Dk mouse strain, a spontaneous mouse astrocytoma (SMA) line was generated. While using approximately 1×10^4 cells, survival times ranged from 145 to 176 days at the first passage, and decreased to 50–60 days at the second passage. Several lines derived from SMA have been generated that have both in vitro and in vivo growth capabilities while retaining their tumorigenic potential and astrocytic histological features.[18] SMA-560 was one line that most closely fit the criteria required of a reliable brain tumor model; it is of an autochthonous origin, it has a capacity for in vitro growth, it has the ability to be transplanted into a comparable location (giving rise to reproducible cell types), and it gives rise to tumors in a timely manner while maintaining appropriate tumor histology.[89] Further characterization of SMA-560 through IHC, electron microscopy, and its response to differentiation cues, considered jointly with its astrocytoma-specific expression profile, which includes low S100 expression, high GFAP, and glutamine synthetase expression, has implicated that this line is formative of anaplastic astrocytomas.[18,37,90] Furthermore, additional characterization of SMA-560 revealed that it is immunologically relevant because it expresses the immunosuppressive protein TGF-β.[91]

SMA-560 has been used in a variety of ways to investigate different immunotherapeutic modalities. Expanding on the role of TGF-β in immunosuppression, SMA-560 was utilized to show the potential of targeting TGF-β. In one study, an inhibitor of TGF-βRI kinase was used to treat tumor-bearing animals. Blocking TGF-β led to a trend of increased apoptosis as well as an increase of $CD3^+$ T cell infiltration among treated tumors.[92] Additionally, surface mobilization of CD107a on T cells—an accepted measure of tumor-specific cytolytic activity—was shown to be elevated twofold in treated animals.[92] Abrogation of the TGF-β-induced immune-suppression enhanced the tumor-specific cytolytic function of $CD8^+$ T cells and restored antitumor immune activity, increasing the overall survival of the animals.

SMA-560 has also been used to investigate DC-based therapies. One study in particular incorporated the most common genetic aberration in human GBMs, EGFRvIII, into SMA-560. DCs pulsed with homogenate from this engineered line were used to vaccinate animals prior to tumor implantation.[93] Although unpulsed DCs did not protect against CNS challenge, mice treated with the pulsed DCs displayed a 160% increase in median survival. Following tumor rechallenge, those animals previously immunized with pulsed DCs survived significantly longer than their control counterparts. This sustained response was likely due to both the induced tumor-specific cytotoxic T lymphocyte (CTL) response and the tumor-specific humoral immunity, which developed in response to the homogenate-pulsed DCs. This study demonstrated the efficacy of DCs pulsed with glioma tumor homogenate. It also showed that in these cases, there were no detrimental effects to the normal brain tissue, emphasizing the tumor-specific nature of immunotherapy, an aspect imperative to the appeal and overall success of glioma immunotherapy.

Patients with malignant gliomas exhibit cellular immune defects with a concomitant increase in the number of Treg cells among $CD4^+$ T cell populations. Similarly, SMA-560 tumor-bearing syngeneic VM/Dk mice demonstrated CD4 lymphopenia, in which the majority of peripheral blood $CD4^+$ T cells were comprised of Treg cells.[94] Inhibition of Treg cell function in this model through systemic anti-CD25 administration resulted in enhanced lymphocyte proliferation and IFN-γ responses. Additionally, a combination therapy involving anti-CD25 and a DC vaccine against intracranial SMA-560 tumors resulted in 100% long-term survivors.[94]

SMA-560 has also been used in the assessment of the efficacy of chimeric antigen receptor (CAR) T cells. Specifically, third generation, EGFRvIII-specific murine CAR T cells were generated and tested. Following implantation of SMA-560 expressing EGFRvIII, EGFRvIII CAR T cells were administered. Significantly prolonged survival was only observed in the context of lymphodepletion.[95] Not only were EGFRvIII CAR T cells capable of curing tumor-bearing mice, but an immune response and tumor rejection ensued following rechallenge with an EGFRvIII-negative cell line. This suggests that treatment with EGFRvIII CAR T cells could elicit de novo priming against additional tumor antigens. This is particularly important for gliomas, specifically EGFRvIII gliomas, where immune escape is prevalent and tumors can often recur that do not harbor the initially targeted mutation. The combination of being histologically, biologically, immunologically, and clinically relevant makes SMA-560 an ideal glioma line for the investigation of immunotherapeutic modalities directed toward the cure of malignant gliomas.

Three additional spontaneous VM/Dk tumor lines, VM-M2, VM-M3, and VM-NM1, were established by Dr. Seyfried's group at Boston College.[96] Upon intracranial implantation, the VM-M2 and VM-M3 cell lines form highly invasive tumors, whereas the malignant VM-NM1 cell line fails to form invasive tumors. Gene expression analysis and phagocytosis assay revealed that both the VM-M2 and the VM-M3 tumors exhibit properties identical to macrophages and the VM-NM1 tumor has properties similar to neural stem/progenitor cells, indicating that spontaneous VM brain tumors have different cell type origins. These spontaneous VM/Dk tumor cell lines are yet to be utilized for immunotherapy studies.[96]

CT-2A Glioma Model

The CT-2A line was originally generated through implantation of the chemical carcinogen 20-MCA into the cerebral cortex of C57Bl/6 mice.[97] Four hundred and fifty days after transplantation, a tumor designated as CT-2 arose, and following serial subcutaneous passages in the flank, a soft, noncohesive, and mostly hemorrhagic tissue, referred to as CT-2A, was derived. Intracranial injection of CT-2A led to features of high-grade astrocytomas, including a high mitotic index, microvascular proliferation, pseudopalisading necrosis, and intratumoral heterogeneity.[98] These tumors followed blood vessel tracks, infiltrated into surrounding brain parenchyma, and expressed astrocytic markers, including GFAP, recapitulating the human disease. Additionally, CT-2A was shown to express the brain tumor stem cell marker CD133, as well as other stem cell markers, including Oct4, Nanog, and Nestin.[99] Considering CD133-positive glioma cells have been shown to confer glioma radiochemoresistance and promote recurrence, the stem-like characteristics and expression profile of CT-2A make it an ideal candidate for investigating how to best eradicate those brain tumor stem cells that promote resistance to current standard of care therapies.[98–102] Although not many immunotherapy-related studies have taken advantage of CT-2A just yet, investigating the immunogenicity of CD133-positive CT-2A cells and their mechanisms of immune evasion could potentially be beneficial.

GL261 Glioma Model

The GL261 glioma line is one of the most utilized syngeneic immunocompetent models to date and has been particularly amenable to immunotherapeutic investigation. This line was originally generated through placement of 20-MCA pellets into the right cerebral hemisphere of C57Bl/6 mice,[103] where it gave rise to what was considered as an ependymoblastoma.[104] However, recent publications have shown this tumor to be poorly differentiated, lacking ependymal differentiation and lacking expression of glial molecular markers.[105] Biologically, this tumor does mimic the behavior of GBM due to its rapid infiltration into the brain, resulting in death 25 days following injection of 1×10^5 cells.[22,37,104,106] The development of GL261 tumors has been thoroughly studied and has been shown to progress through five stages including implantation, perivascular organization, proliferation, apoptosis, and involution of vasculature, and angiogenesis.[107] Genetic characterization of this line showed that it carries point mutations in *K-ras* and *p53*. Additionally, in its normal state, GL261 expressed MHC-I, but not MHC-II, although MHC-II could be induced following transduction with IFN-γ-encoding adenovirus.[106] Importantly, this line is moderately immunogenic in C57Bl/6 animals. Immunogenicity of GL261 was demonstrated by the lack of tumor development in C57Bl/6 animals when they were vaccinated with irradiated GL261 cells prior to tumor implantation. When the tumor-free animals were rechallenged, they still failed to develop tumors, suggesting that there was long-lasting immunity as a result of the irradiated GL261 vaccination.[106] This immunity was dependent on the number of irradiated cells used for vaccination.

Despite the caveats surrounding GL261, the growth features, tumor penetrance, presence of clinically relevant genetic mutations, and expression of MHC-I have made GL261 a leading model for immunotherapeutic investigations, ranging from adoptive cell transfer therapy, treatment with mAbs, immunization with DCs, and gene therapy, among others.[108]

Adoptive transfer of tumor-reactive T lymphocytes was demonstrated to induce tumor regression in the GL261 brain tumor model.[109] In one study, GL261 was injected into the hind flank to stimulate an immune response in C57Bl/6 animals. After 9 days, draining lymph node cells were isolated and activated via anti-CD3 and low-dose IL-2. When these activated tumor-draining lymph node cells were intravenously administered to GL261 tumor-bearing animals following sublethal whole body irradiation, 100% of animals displayed long-term survival.[110] Adoptive transfer of tumor-draining lymph node cells was shown to confer a specific protective immunity in the recipients, indicating an effective memory response.

DC tumor vaccination, an active and specific form of immunotherapy, has also been evaluated using GL261.[111,112] For these studies, DCs were isolated from C57Bl/6 animals and pulsed with GL261 extract. Pulsed DCs were administered intraperitoneally into mice bearing GL261 gliomas. After 100 days, 60% of the animals treated with pulsed DCs were still alive, compared to 25% and 30% for the animals treated with unpulsed DCs and saline controls, respectively.[111] A DTH test was administered, in which irradiated GL261 cells were applied to each animal's footpad. The animals that were treated with pulsed DCs and survived the therapy also displayed a swelling of the footpad, suggesting that a memory immune response specific to GL261 had been established as a result of the vaccine. The data surrounding the efficacy of tumor lysate-pulsed DCs against gliomas have been corroborated, and is the foundation for several DC-based clinical trials.[112]

Although it has not been observed in the studies referenced, DCs pulsed with whole tumor cell antigens run the risk of generating autoimmune responses. Pulsing DCs with tumor-specific antigens reduce this risk and has yielded comparable and promising results. This has been described on several accounts in which GL261 was utilized. Survivin is one tumor-associated antigen that is rarely detectable in normal adult tissues, but is highly expressed in malignant gliomas, as well as in many other tumor types. In one study, bone marrow DCs were transfected with full-length copies of survivin and were used to vaccinate animals. Survivin-specific CTLs were detected and survival was prolonged in those animals with brain tumor burden.[113] Similar results were obtained when DCs were pulsed with other tumor-associated antigens, including gp100, tyrosinase-related protein-2 (TRP-2), and EphA2, among others.[114,115]

GL26 Glioma

With similar growth rates and effects on tumor-burdened animals, the GL26 line is very similar to GL261, though it is used far less frequently. Despite also being diagnosed as an ependymoblastoma, there are several differences that exist between the two lines, including GL26's development of greater necrosis and vascularity, as well as GL26's tendency to give rise to more hemorrhagic tumors.[37,104]

Interestingly, GL26 shares many of the same tumor-associated antigens as melanoma, making it particularly useful in investigations of DCs pulsed with melanoma-associated antigens and GL26-associated antigens. Similar to GL261, animals treated with these tumor antigen-pulsed DCs responded well with prolonged survival.[115]

GL26 has also been used to demonstrate the potential of using IL-12 to bridge the gap between the innate and specific immune responses through activating natural killer (NK) cells, promoting CTL maturation, and biasing the $CD4^+$ T cells toward Th1 differentiation, thereby enhancing the antitumor immune response.[116] Here, they expressed IL-12 in DCs via adenoviral-mediated gene transfer and pulsed the DCs with GL26 tumor cell lysate.[116] The IL-12-expressing DCs pulsed with tumor cell lysate displayed antitumor immunity, with tumor growth inhibition and prolonged survival when GL26 was injected both subcutaneously and intracranially.

4C8 Glioma Model

The 4C8 glioma line was originally established to model oligodendrogliomas in B6D2F1 mice. To this end, Hayes et al. generated a transgenic animal with the *neu* oncogene (*c-neu*) expressed under control of the myelin basic protein (MBP) promoter.[117] Spontaneous tumors developed; however, they were histologically and immunohistochemically more comparable to GBM rather than oligodendrogliomas. MOCH-1 was one of the original clones established from the spontaneous tumor and was shown to have astrocytic properties when cultured in high serum, but oligodendroglial properties when cultured in low serum, suggesting these cells display a baseline level of plasticity with properties dictated by environmental conditions.[117] The 4C8 line was a subclone of MOCH-1 and was shown to differentiate into mature, membrane sheet-bearing oligodendrocyte-like cells that expressed major myelin markers at levels similar to primary oligodendrocytes.[118] Morphological changes were induced when cultured in the presence of serum, generating large somas and star-shaped processes that expressed GFAP.[105,118] When the 4C8 line was cultured in the presence of serum and intracranially injected into B6D2F1 mice, dense cellular masses containing giant cells developed. These lesions were surrounded by nodules, contained pseudopalisading necrosis, exhibited GFAP expression, and lacked MBP expression, suggesting that this line forms GBM-like tumors.[105] When injected intracranially with a dose of 1×10^5 cells, median survival of this tumor model was 62 days.[119]

Several immunotherapy-related studies have been performed using the 4C8 glioma line. One study sought to evaluate novel type 1 herpes simplex virus (HSV) therapies for malignant gliomas by treating established tumors with engineered HSVs.[119] An additional arm to this study included treatment with HSVs that expressed IL-12 to assess whether an antitumor immune response by HSV-mediated expression of proinflammatory cytokines is more or less robust than direct oncolysis alone. Intratumoral injection of HSVs expressing IL-12 into 4C8 tumors led to a significant prolongation of survival, with a median survival greater than 108 days. This was in stark contrast to animals treated with saline or HSV alone, which had median survivals of 75 and 71 days, respectively. For those animals treated with HSVs expressing IL-12, 40% were long-term survivors, compared to 20% in the HSV alone treatment group. All long-term survivors were deemed tumor free through histological analysis.[119]

The repertoire of immune cells present after tumor infiltration was also investigated. Tumor-bearing animals treated with HSV alone showed increased $CD4^+$ T cells, $CD8^+$ T cells, and NK cells; however, there was a much larger increase in these cells when treated with HSV expressing IL-12. This suggests that although both viruses elicit an immune response, the expression of IL-12 promoted greater immune cell infiltration into the tumor. This study was the foundation for additional preclinical investigations in which HSV expressing IL-12 was assessed for efficacy, safety, biodistribution, and capability of activating primate lymphocytes.[120] Excitingly, there was no evidence of toxicity following direct inoculation of HSV expressing IL-12 into the right frontal lobes of *Aotus nancymae* and there was no evidence of pathology 5 years following the injection. Additionally, it was shown that murine IL-12 produced by HSV was capable of activating primate lymphocytes in a lymphocyte activation assay.[120] These promising results laid the foundation for use of genetically engineered HSV-1 expressing IL-12 in GBM patients in the NCT02062827 clinical trial. This trial is currently ongoing, though not recruiting participants.

GENETICALLY ENGINEERED MOUSE MODELS

Despite their obvious value in basic and translational research, including testing novel immunotherapeutic strategies, there is a major limitation with the spontaneously or chemically induced brain tumor models. Since these tumors occur spontaneously or due to random mutagenesis, there is no control on the genetic alterations they possess, a pitfall that is manifested in the current genomic era when major cancer-causing mutations have been well-documented and therapies targeting specific mutations have become an increasingly promising approach.

With the advance of molecular genetic technologies, genetically engineered mouse models (GEMMs), with either germline or somatic cell genetic modifications, have become an important tool. Although a GEMM is challenging to obtain, once available, it provides several advantages over currently available brain tumor models. First, due to modification of well-studied cancer genes, including oncogenes such as *EGFR* or tumor suppressor genes such as *tumor protein 53 (TP53)* and *phosphatase and tensin homolog (PTEN)*, a GEMM allows faithful modeling of a human glioma, and can be readily achieved via transgene or gene knockout. Second, gene modifications can be temporally and spatially controlled, thus a more precise simulation of human cancer development can be mimicked. This is typically achieved using the Cre–Lox recombination system, in which a tissue-specific, promoter-driven expression of Cre recombinase induces genetic alterations, such as the deletion of tumor suppressor genes or activation of oncogenes. The most commonly used tissue-specific promoters for gliomagenesis studies include the nestin promoter, which drives Cre expression in neural progenitor cells, and the *GFAP* promoter, which drives Cre expression in the astrocyte lineage and subventricular zone progenitors. In some cases, a CreER variant can also be used so that the Cre's activity can be further controlled posttranscriptionally by tamoxifen. Finally, unlike well-established tumor cell lines, tumor cells derived from GEMMs do not have the problem of extensive passage/proliferation in vitro and therefore better maintain their original genotype and phenotype.

Oncogene Activation

EGFR is one of the most important oncogenes in gliomas, with frequent gain-of-function genetic alterations, including gene amplification and/or point mutations that lead to constitutively active, oncogenic variants (e.g., *EGFRvIII*).[121,122] Initially, the generation of *EGFR* or oncogenic *EGFRvIII*-based GEMMs was attempted primarily via cell type–specific somatic gene transfer or transgene introduction into the germline under the control of a cell type–specific promoter. In the first approach, the replication-competent avian leucosis virus splice acceptor viral system was used to deliver the constitutively active EGFR into glial progenitors via a lineage specific promoter-driven expression of a viral receptor.[123,124] This approach demonstrated that oncogenic *EGFR* alone was insufficient in driving gliomagenesis; instead, a second genetic alteration, such as the loss of the tumor suppressor gene *INK4a/alternative-reading-frame protein (ARF)*, is necessary for fully driving glioma progression that resembles human tumors. Reiterating this finding, subsequent studies using germline transgene and oncogenic *EGFR* expression in specific lineages, including astrocytes via the *GFAP* promoter and oligodendrocytes via the *S100β* promoter, also found that oncogenic *EGFR* cooperates with oncogenic *Ras* or loss of tumor suppressor pathways in driving the development of high grade gliomas.[125–127] Alternatively, GEMMs can also be generated through retroviral delivery of the gene of interest—frequently an oncogene for a gain of function study (e.g., *EGFR*)—into neural progenitors in ex vivo cultures, and subsequent implantation of these transduced cells into the mouse intracranial compartment.

Genetic engineering approaches (described earlier) have also been utilized for the generation of GEMMs involving other GBM-related genetic alterations, including *PDGF*, *Ras*, and *Akt*, by either modifying each oncogene individually or in combination with other oncogenes or tumor suppressor genes.[128–130] For example, direct injections of a retrovirus expressing external *PDGFβ* into mouse brains, which targeted more than one specific lineage of cells, induced gliomas in a portion of the host mice.[131] In most studies, however, successful GEMMs were frequently developed using the combination of one of these genetic alterations with a loss of a tumor suppressor pathway. For example, mice with transgenic expression of *PDGFβ* under the control of the *GFAP* promoter had no phenotype; however, in a *TP53*-null background, most mice develop brain tumors with features mimicking human GBM.[132] Further supporting the cooperative role between PDGF signaling and loss of tumor suppressors, direct injections of a retrovirus expressing external *PDGFβ* into $INK4a^{-/-}$ mouse brains induced gliomas, from which stable tumor cell lines could be established for further mechanistic studies.[133] Supporting the oncogenic role of the PDGF signaling pathway, a separate study found that an overexpression of *PDGFα*, plus a loss of *INK4a/ARF*, achieved by retroviral expression of *PDGFα* in ex vivo cultured progenitor cells from *INK4a/ARF* knockout mice, induced glioma in an in vivo mouse tumor model.[134] Due to a relatively high penetrance and short latency, *PDGF*-driven GEMMs have been elegantly used to dissect cooperating molecular pathways and reveal distinct roles of tumor suppressors in gliomagenesis.[132,134,135]

Similarly, the transgenic expression of V12-*Ras* under *GFAP*-driven Cre induced astrocytomas, in which the penetrance, latency, and tumor grade were dependent on the dose of expression.[136,137] One of these studies also revealed spontaneous TP53 mutations in tumors induced by the V12-*Ras* transgene, suggesting a selection pressure for tumor-driving genetic alterations in mice—a resemblance that also occurs in human cancers.[137]

Loss of Tumor Suppressors

TP53 inactivation is one of the most common genetic alterations in human cancers, including both primary and secondary GBMs. For example, in high-grade GBMs, mutations in *TP53* and other genes encoding the TP53 pathway regulators, including mouse double minute 2 (*MDM2*), *MDM4*, and *CDKN2A*, occur in about 90% of patients.[121,138] As such, it is also one of the most commonly studied genes in cancer modeling in mice. Experiments using GEMMs found that a deletion of individual tumor suppressors or a combination of two or more tumor suppressor genes in distinct cell lineages led to the development of different types of brain tumors.[139,140] For example, a conditional *TP53* knockout mouse line, in which gene knockout was achieved by an inducible *GFAP*-CreER, found that *TP53* deletion in adult neural progenitors and astrocytes led to high grade astrocytoma with low frequency and long latency.[140] Underscoring the unique tumor suppressor role of TP53, the loss of two other tumor suppressors individually—*PTEN* or *Rb*—failed to generate tumors. However, the loss of either one of these two tumor suppressors in the context of *TP53* deletion increased the frequency of tumor formation, and the knockout of both *PTEN* and *Rb* in conjunction with *TP53* deletion significantly shortened the latency of high-grade tumors, further supporting the cooperative effects of different genetic alterations.[140]

Another common tumor suppressor gene frequently mutated in GBM is *NF1*. *NF1* is associated with the genetically heritable disease neurofibromatosis type I, in which brain tumor occurrence is a key feature.[121,141–143] Although *NF1* deletion alone was incapable of inducing GBM, a heterozygous knockout of both *NF1* and *TP53* efficiently induced different types of tumors in certain mouse strains, including brain tumors that resembled secondary GBMs in humans with a penetrance of close to 100%.[144] Several cell lines (KR129, KR130, and KR158) have been established from brain tumors obtained from *NF1* and *TP53* double knockout mice.[144] The invasive mouse glioma model KR158 (K-Luc) exhibited positive staining for S100 by IHC and has been successfully used for immunotherapy studies involving hematopoietic stem cells[145] or CpG-ODN[146] in C57Bl/6 mice.

Since they demonstrate that the loss of tumor suppressors in the absence of the activation of a well-established oncogene(s) is sufficient to drive tumorigenesis, these studies are especially important. This is an interesting notion as a large body of genetic findings in humans to date indeed confirm that a lot more tumor suppressors than oncogenes are involved in human cancers and oncogene alterations are not necessarily required for tumorigenesis.[147]

Collectively, studies using GEMMs have provided strong evidence to support the in vivo role of the most important oncogenes and tumor suppressors. These studies have demonstrated the importance of a cell's origin in determining which type of tumors will arise. Furthermore, they have found that in most cases, cooperation between two or more oncogenic alterations are necessary to generate tumors resembling those in humans. These studies have clearly illustrated the power and faithfulness of GEMMs in brain tumor research. GEMMs provide valuable tools for studying mechanisms underlying oncogene-driven gliomagenesis and cooperating oncogenic alterations, as well as for evaluating novel preclinical therapies. Although GEMMs have yet to be broadly used in immunotherapy studies, their syngeneic nature and well-defined genotypes suggest that they can become uniquely useful for this approach, especially if models can be efficiently generated and the heterogeneity of tumor cells can be better characterized.

XENOGRAFT MODELS OF GLIOBLASTOMA

Syngeneic murine brain tumor models offer us an opportunity to develop novel therapies for histologically similar human tumors. However, differences between the biology of the mouse and human species preclude the evaluation of novel therapies specifically aimed at targeting human diseases.[148,149] Therefore, to overcome this obvious need, xenograft (tissue from another species) models have been developed.[150] Xenograft models are established in immunocompromised hosts capable of accepting human tissue, including tumors, without prompting the elimination of the transplanted human sample. The utilization of immunocompromised animal models capable of accepting tumor xenografts has been invaluable to the scientific community. These models have been used to obtain critical data that have led to the approval of radiotherapy regimens, chemotherapeutics, and most recently, biologic agents for the treatment of brain tumors.[151,152] Herein, we describe the most common xenograft models utilized in scientific research, and how these models can further be exploited for the development of novel immunotherapies.

The discovery of the nude (*NU/NU*) mouse in the 1960s was critical for the initiation of studies evaluating human tissue in mice,[153] including brain tumors. The nude mouse is an immunocompromised strain, which lacks hair due to a spontaneous mutation in the *FOXN1* gene, but also lacks a thymus, resulting in a limited T cell development.[154] The absence of this compartment allows the mice to accept what would normally be an incompatible tissue from another species, in this case human tumor tissue. Nude mice were originally the preferred strain for performing studies with human tumors as well as other tissues; however, these mice possess B cells capable of producing antibodies in a T cell independent manner, and contain a normal innate immune system with increased NK cell activity. Due to these immune alterations, nude mice are still capable of eliminating human tumors.[155–157] Therefore, to overcome these deficits, a different immunocompromised model that carries the protein kinase, DNA-activated, catalytic polypeptide severe combined immunodeficiency mutation (*Prkdc*^{scid}) was developed in CB17 mice.[158] *Prkdc*^{scid} mutant mice has a limited capacity to correct double-strand breaks in the genome,[159] such as those required for T and B cell generation, resulting in a lack of mature T and B cells.[158] In addition, mice with the *Prkdc*^{scid} mutation have a significant decrease in NK cell activity, allowing them to accept human tumors and normal tissue at a higher rate than nude mice.[160,161]

The nude and *Prkdc*^{scid} mice were the strains used most frequently to evaluate chemotherapeutics, radiation, and biologics against brain tumors.[152,162] Although these mice were of great utility for several decades, there was still a relatively high incidence of spontaneous human tissue elimination. This was generally attributed to the presence of a normal innate myeloid compartment and a low but significant number of T cells.[163] In the early 2000s, however, the development of novel immunocompromised strains led investigators to reassess the utilization of nude and *Prkdc*^{scid} mice. Therefore, a new strain of mice, termed nonobese-diabetic (NOD)-*Scid* IL-2R$\gamma_C^{-/-}$ (NOG or NSG mice), was developed.[164] This strain is based on the NOD background, which has genetic polymorphisms in immunoregulatory genes.[165] This strain was then bred with mice containing the *Prkdc*^{scid} mutation to generate the NOD-*scid* mouse, possessing both attenuated innate and adaptive immune responses.[166] This strain was subsequently bred with mice deficient for the IL-2 receptor γ_{Common} gene (IL-2Rγ_C, also known as CD132). The absence of the IL-2R$\gamma_C^{-/-}$ eliminates the signaling of homeostatic cytokines IL-2, IL-4, IL-7, IL-9, IL-15, and IL-21, which results in the absence of T cells, B cells, and NK cells, as well as a significant reduction in innate myeloid components.[164] The resultant NSG strain is a severely immunocompromised mouse model capable of routinely accepting human tumor tissue. Furthermore, GBM xenografts in this model have been shown to be histologically accurate.[167] Cumulatively, these qualities enable researchers to model the interaction of novel chemotherapies, radiotherapies, and biologics in vivo against human tumors.

HUMANIZED MOUSE MODELS

Since biologic agents aimed at modulating immunity could not be tested due to the absence of a competent human immune system, the evaluation of novel therapies on immunocompromised models was initially limited to chemotherapy and radiotherapy. Largely due to the activity of the remaining NK cells and other innate immune system components, early studies utilizing the nude,[168] *Prkdc*^{scid},[169] and NOD-*Scid*[170] strains failed to obtain consistent engraftment of human immune cells.[150] Out of these three models, the NOD-*Scid* allowed for a higher, although inconsistent, rate of hematopoietic engraftment. However, engraftment-required host irradiation[171] and the incidence of thymic lymphomas,[166] as well as the induction of graft versus host disease (GvHD),[172,173] limited the lifespan of these mice and the available study time.

In contrast, the NOD-*Scid*/IL-2R$\gamma_C^{-/-}$ mice presented the unique opportunity to reliably develop a long-term human immune system with both innate and adaptive cellular components.[174] The initial strategy used to model a human system in NOD-*Scid*/IL-2R$\gamma_C^{-/-}$ mice required the transfer and engraftment of human peripheral blood lymphocytes (PBLs). This model, known as the Hu-PBL, is widely used to study human NK cell, T cell, and B cell biology, but unfortunately does not allow for studies aimed at examining the role of other immune cells of myeloid origin.[175,176] To overcome this limitation, a more robust model of the human immune system was developed, in which human hematopoietic stem cells (HSCs) are transferred rather than PBLs. This model has been termed human-*scid* repopulating cells (Hu-SRC)[174,177] and is generated by the administration of enriched CD34$^+$ HSCs to newborn NOD-*Scid*/IL-2R$\gamma_C^{-/-}$ mice, permitting the reconstitution of the host bone marrow and the circulation with human immune cells. This humanized model allows for the successful differentiation of HSCs into both myeloid and lymphoid lineages, including the generation of human monocytes, neutrophils, basophils, DCs, T cells, B cells, platelets, and red blood cells.[174,177,178] The Hu-SRC mouse model in the NOD-*Scid*/IL-2R$\gamma_C^{-/-}$ strain allows for the presence of the human immune system with a relatively low incidence of GvHD, permitting a longer time for analyzing than the NOD-*Scid* strain.[174] In addition, the T cell and B cell repertoire tends to be broader in the Hu-PBLs model and possesses a phenotype similar to the human adaptive system.[179,180] Since they are capable of mounting humoral and cellular responses to vaccines[177] and infections[181,182] and can reject allogeneic grafts,[183] Hu-SRC mice are immune competent. However, despite the advantages of the NOD-*Scid*/IL-2R$\gamma_C^{-/-}$ mouse model over the nude, *Prkdc*^{scid}, and NOD-*Scid* strains, this model is still constrained due to its high radiosensitivity from the *Prkdc*^{scid} mutation.[184]

Due to the demonstrated importance of IL-2R$\gamma_C^{-/-}$ deficiency, other mouse models maintaining IL-2R$\gamma_C^{-/-}$ deficiency but substituting the *Prkdc*^{scid} mutation for the recombinant activated gene (RAG) deficiency were developed.[185] The loss of RAG activity also leads to T and B cell deficiency and models such as NOD/RAG$^{-/-}$IL-2R$\gamma_C^{-/-}$ [185] and the RAG$^{-/-}$IL-2R$\gamma_C^{-/-}$[186] double knockout share similar phenotypes. Furthermore, the NOD/RAG$^{-/-}$IL-2R$\gamma_C^{-/-}$ mouse model permits robust human HSC engraftment while being radiation resistant,[184,186] allowing for the evaluation of radiation-based therapies. Therefore, the use of the Hu-SRC system within the NOD-*Scid*/IL-2R$\gamma_C^{-/-\ -/-}$, NOD/RAG$^{-/-}$IL-2R$\gamma_C^{-/-}$, and RAG$^{-/-}$ IL-2R$\gamma_C^{-/-}$ strains has become a powerful tool for scientists. Patient-derived xenografts can now be implanted in mice containing patient-derived immune systems with the potential for long-term study, providing ideal models for evaluating novel therapies, including radiation, chemotherapy, and biologic agents.

The advantages of the NOD-*Scid*/IL-2R$\gamma_C^{-/-}$, NOD/RAG$^{-/-}$IL-2R$\gamma_C^{-/-}$, and RAG$^{-/-}$ IL-2R$\gamma_C^{-/-}$ mouse models over prior strains have made these "the standard models" to utilize in studies involving engraftment of human tumors or immune cells. However, limitations still exist in these models preventing the completely identical recapitulation of a human immune system in a mouse.[150,176] Murine growth factors and cytokines responsible for the differentiation and survival of HSCs and descendant immune cells have a lack of cross-reactivity for human hematopoietic cells; this has a detrimental impact on the developing human immune system within the mouse.[150,176] To overcome this, exogenous administration of cytokines or growth factors as well as novel transgenic models are being developed to include expression of thrombopoietin,[187] IL-15,[188] Flt-3L,[189] and CSF-1[190] to promote differentiation and survival of human HSCs, NK cells, DCs, and macrophages, respectively. Although these modulations benefit the development of a healthy human innate compartment, limitations on the cellular and humoral responses from the lymphoid compartment still exist in these models. The NOD-*Scid*/IL-2R$\gamma_C^{-/-}$, NOD/RAG$^{-/-}$IL-2R$\gamma_C^{-/-}$, and RAG$^{-/-}$IL-2R$\gamma_C^{-/-}$mouse Hu-SRC models have relatively low numbers of circulating mature T cells specific against human tissue.[191] This insufficiency is believed to be driven by the lack of human MHC expression within the thymus epithelium, which is responsible for the positive selection of human-restricted T cells.[192] To overcome this deficiency, transgenic NOD-*Scid*/IL-2R$\gamma_C^{-/-}$, NOD/RAG$^{-/-}$IL-2R$\gamma_C^{-/-}$, and RAG$^{-/-}$IL-2R$\gamma_C^{-/-}$ models, including the human MHC molecules human leukocyte antigen (HLA)-A2 and HLA-DR1 class I and class II molecules,[193,194] are being developed to enhance the human specificity of the arising T cell response. Because T cells are now capable of directly interacting with human B cells and are providing critical signals required for optimal responses,[195] improved T cell recognition of the human MHC molecules has also increased humoral B cell responses. Finally, in addition to these transgenic models, a triple xenograft transplantation model including bone marrow CD34$^+$ HSCs, fetal liver, and thymus (cumulatively termed BLT) is currently being evaluated for its capacity to develop normal counts of T and B lymphocytes with specificity against human HLAs.[196,197] If successful, the BLT model system could overcome the need for complicated breeding with transgenic HLA mice, although it does rely heavily on technically challenging surgical techniques. As these models demonstrate their utility and their ability to overcome the deficits of currently employed humanized mouse models, it is expected that these models will become critical for the evaluation of novel immunotherapies against cancer, including malignant brain tumors.

CONCLUSION

For research regarding experimental therapeutics, tumor models in animals have been indispensable. Therefore, considerable efforts have been expended for developing representative animal models of brain tumors. Even though it has been difficult to establish animal models that reflect all the characteristics of human tumors, there are a number of transplantable tumor models in syngeneic mice that have demonstrated considerable promise. Each of these models has contributed invaluable insight toward the development of novel brain tumor therapies currently being tested in the clinic, and will continue to aid in the evaluation of novel therapeutic agents. Alternatively, humanized mouse models are emerging as a promising platform for evaluating novel immunotherapeutics against human tumors in the presence of an active human immune system, and also represent a valuable option for future preclinical studies.

Despite the improvements in preclinical murine models, it is important to note that although all animal models have their strengths, they also have deficiencies that place limitations on their use, and knowledge of the advantages and inadequacies is critical for acquiring preclinical results that are beneficial for clinical translation. Given the promising effects immunotherapy has recently had on patients with GBM, in vivo models will continue to prove invaluable in preclinical settings. The increased development and availability of cell line-derived models, GEMMs, xenografts, and humanized models for GBMs emblematize great promise in improving treatment for this deadly disease.

Acknowledgments

This work was supported by the following NIH grants: P01 (5P01CA154291) and the Outstanding Investigator Award (1R35CA197264). We thank Jenna Lewis for the editorial review of the chapter.

References

1. Louis DN, Ohgaki H, Wiestler OD, Cavenee WK. *WHO Classification of Tumours of the Central Nervous System*. 4th ed. International Agency for Research on Cancer; 2007.
2. Ostrom QT, Gittleman H, Fulop J, et al. CBTRUS Statistical Report: primary brain and central nervous system tumors diagnosed in the United States in 2008–2012. *Neuro Oncol*. 2015;17(suppl 4):iv1–iv62.
3. Stupp R, Mason WP, van den Bent MJ, et al. Radiotherapy plus concomitant and adjuvant temozolomide for glioblastoma. *N Engl J Med*. 2005;352(10):987–996.
4. Claes A, Idema AJ, Wesseling P. Diffuse glioma growth: a guerilla war. *Acta Neuropathol*. 2007;114(5):443–458.
5. Weller RO, Engelhardt B, Phillips MJ. Lymphocyte targeting of the central nervous system: a review of afferent and efferent CNS-immune pathways. *Brain Pathol*. 1996;6(3):275–288.

6. Finocchiaro G, Pellegatta S. Perspectives for immunotherapy in glioblastoma treatment. *Curr Opin Oncol*. 2014;26(6):608–614.
7. Krebs S, Rodriguez-Cruz TG, Derenzo C, Gottschalk S. Genetically modified T cells to target glioblastoma. *Front Oncol*. 2013;3:322.
8. Mitchell DA, Batich KA, Gunn MD, et al. Tetanus toxoid and CCL3 improve dendritic cell vaccines in mice and glioblastoma patients. *Nature*. 2015;519(7543):366–369.
9. Reardon DA, Freeman G, Wu C, et al. Immunotherapy advances for glioblastoma. *Neuro Oncol*. 2014;16(11):1441–1458.
10. Perng P, Lim M. Immunosuppressive mechanisms of malignant gliomas: parallels at non-CNS sites. *Front Oncol*. 2015;5:153.
11. Hayes HM, Priester Jr WA, Pendergrass TW. Occurrence of nervous-tissue tumors in cattle, horses, cats and dogs. *Int J Cancer*. 1975;15(1):39–47.
12. Snyder JM, Shofer FS, Van Winkle TJ, Massicotte C. Canine intracranial primary neoplasia: 173 cases (1986–2003). *J Vet Intern Med*. 2006;20(3):669–675.
13. Fankhauser R, Luginbuhl H. Tumors of the CNS. In: 3rd ed. Dobberstein J, Pallaske G, Stunzi H, eds. *Special Pathological Anatomy of Domestic Animals, by E. Joest*. vol. 3. Berlin-Hamburg: Paul Parey; 1968:366–423 (Ger).
14. Troxel MT, Vite CH, Van Winkle TJ, et al. Feline intracranial neoplasia: retrospective review of 160 cases (1985–2001). *J Vet Intern Med*. 2003;17(6):850–859.
15. Lapin BA, Yakovleva LA. *Comparitive Pathology in Monkeys*. Jena: Fischer; 1964 (Ger).
16. Fraser H. Astrocytomas in an inbred mouse strain. *J Pathol*. 1971;103(4):266–270.
17. Fraser H, McConnell I. Letter: experimental brain tumours. *Lancet*. 1975;1(7897):44.
18. Serano RD, Pegram CN, Bigner DD. Tumorigenic cell culture lines from a spontaneous VM/Dk murine astrocytoma (SMA). *Acta Neuropathol*. 1980;51(1):53–64.
19. Shelton LM, Mukherjee P, Huysentruyt LC, Urits I, Rosenberg JA, Seyfried TN. A novel pre-clinical in vivo mouse model for malignant brain tumor growth and invasion. *J Neurooncol*. 2010;99(2):165–176.
20. Krinke GJ, Kaufmann W, Mahrous AT, Schaetti P. Morphologic characterization of spontaneous nervous system tumors in mice and rats. *Toxicol Pathol*. 2000;28(1):178–192.
21. Yamagiwa K, Ichikawa K. Experimental study of the pathogenesis of carcinoma. *CA Cancer J Clin*. 1977;27(3):174–181.
22. Seligman AM, Shear MJ. Studies in carcinogenesis. VIII. Experimental production of brain tumors in mice with methylcholanthrene. *Am J Cancer*. 1939;37:364–395.
23. Zimmerman HM, Arnold H. Experimental brain tumors: I. Tumors produced with methylcholanthrene. *Cancer Res*. 1941:919–938.
24. Zimmerman HM, Arnold H. Experimental brain tumors: II. Tumors produced with benzpyrene. *Am J Pathol*. 1943;19(6):939–955.
25. Jainisch W, Schreiber D. In: Bigner DD, Swenberg JA, eds. *Experimental Tumors of the Central Nervous System*. 1st English ed. Kalamazoo, Mich: Upjohn; 1977:18–30.
26. Ivankovic S. Selective induction of tumors in the brain and the nervous system. *Cancer Res Cancer Control*. 1967;6:97–101 (Ger).
27. Ivankovic S, Druckrey H, Preussmann R. The induction of tumros in the PNS and CNS of rats by trimethyl-nitroso-urea. *Z Krebsforsch*. 1965;66:541–548 (Ger).
28. Druckrey H, Ivankovic S, Preussmann R. Selective induction of malignant tumors in the brain and spinal cord of rats with MNU. *Z Krebsforsch*. 1965;66:389–408 (Ger).
29. Druckrey H, Ivankovic S, Preussmann R. Teratogenic and carcinogenic effects in the offspring after single injection of ethylnitrosourea to pregnant rats. *Nature*. 1966;210(5043):1378–1379.
30. Druckrey H. *Chemical Structure and Action in Transplacental Carcinogenesis and Teratogenesis. Transplacental Carcinogenesis*. Lyon: International Agency for Research on Cancer; 1973:45–58.
31. Bullard DEB, Bigner DD. Animal models and virus induction of tumours. In: Thomas DGT, Graham DI, eds. *Brain Tumours. Scientific Basis, Clinical Investigation and Current Therapy*. Butterworth and Co; 1980:51–84.

32. Mancini LO, Yates VJ, Jasty V, Anderson J. Ependymomas induced in hamsters inoculated with an avian adenovirus (CELO). *Nature*. 1969;222(5189):190–191.
33. Merkow L, Slifkin M, Pardo M, Rapoza NP. Studies on the pathogenesis of Simian adenovirus-induced tumors. 3. The histopathology and ultrastructure of intracranial neoplasms induced by SV20. *J Natl Cancer Inst*. 1968;41(5):1051–1070.
34. Padgett BL, Walker DL, ZuRhein GM, Varakis JN. Differential neurooncogenicity of strains of JC virus, a human polyoma virus, in newborn Syrian hamsters. *Cancer Res*. 1977;37(3):718–720.
35. Rabotti GF, Raine WA. Brain tumours induced in Hamsters inoculated intracerebrally at Birth with Rous Sarcoma Virus. *Nature*. 1964;204:898–899.
36. Duffy PE. Virus induced cerebral sarcoma. *J Neuropathol Exp Neurol*. 1970;29(3):370–391.
37. Oh T, Fakurnejad S, Sayegh ET, et al. Immunocompetent murine models for the study of glioblastoma immunotherapy. *J Trans Med*. 2014;12:107.
38. Benda P, Someda K, Messer J, Sweet WH. Morphological and immunochemical studies of rat glial tumors and clonal strains propagated in culture. *J Neurosurg*. 1971;34(3):310–323.
39. Schmidek HH, Nielsen SL, Schiller AL, Messer J. Morphological studies of rat brain tumors induced by N-nitrosomethylurea. *J Neurosurg*. 1971;34(3):335–340.
40. Weizsaecker M, Deen DF, Rosenblum ML, Hoshino T, Gutin PH, Barker M. The 9L rat brain tumor: description and application of an animal model. *J Neurol*. 1981;224(3):183–192.
41. Asai A, Miyagi Y, Sugiyama A, et al. Negative effects of wild-type p53 and s-Myc on cellular growth and tumorigenicity of glioma cells. Implication of the tumor suppressor genes for gene therapy. *J Neurooncol*. 1994;19(3):259–268.
42. Schlegel J, Piontek G, Kersting M, et al. The p16/Cdkn2a/Ink4a gene is frequently deleted in nitrosourea-induced rat glial tumors. *Pathobiology*. 1999;67(4):202–206.
43. Sibenaller ZA, Etame AB, Ali MM, et al. Genetic characterization of commonly used glioma cell lines in the rat animal model system. *Neurosurg Focus*. 2005;19(4):E1.
44. Ghods AJ, Irvin D, Liu G, et al. Spheres isolated from 9L gliosarcoma rat cell line possess chemoresistant and aggressive cancer stem-like cells. *Stem Cells*. 2007;25(7):1645–1653.
45. Kruse CA, Mitchell DH, Kleinschmidt-DeMasters BK, et al. Systemic chemotherapy combined with local adoptive immunotherapy cures rats bearing 9L gliosarcoma. *J Neurooncol*. 1993;15(2):97–112.
46. DiMeco F, Rhines LD, Hanes J, et al. Paracrine delivery of IL-12 against intracranial 9L gliosarcoma in rats. *J Neurosurg*. 2000;92(3):419–427.
47. Jean WC, Spellman SR, Wallenfriedman MA, et al. Effects of combined granulocyte-macrophage colony-stimulating factor (GM-CSF), interleukin-2, and interleukin-12 based immunotherapy against intracranial glioma in the rat. *J Neurooncol*. 2004;66(1–2):39–49.
48. Wallenfriedman MA, Conrad JA, DelaBarre L, et al. Effects of continuous localized infusion of granulocyte-macrophage colony-stimulating factor and inoculations of irradiated glioma cells on tumor regression. *J Neurosurg*. 1999;90(6):1064–1071.
49. Witham TF, Erff ML, Okada H, Chambers WH, Pollack IF. 7-Hydroxystaurosporine-induced apoptosis in 9L glioma cells provides an effective antigen source for dendritic cells and yields a potent vaccine strategy in an intracranial glioma model. *Neurosurgery*. 2002;50(6):1327–1334. discussion 1334–1325.
50. Ginzkey C, Eicker SO, Marget M, et al. Increase in tumor size following intratumoral injection of immunostimulatory CpG-containing oligonucleotides in a rat glioma model. *Cancer Immunol Immunother*. 2010;59(4):541–551.
51. Auf G, Carpentier AF, Chen L, Le Clanche C, Delattre JY. Implication of macrophages in tumor rejection induced by CpG-oligodeoxynucleotides without antigen. *Clin Cancer Res*. 2001;7(11):3540–3543.
52. Fakhrai H, Dorigo O, Shawler DL, et al. Eradication of established intracranial rat gliomas by transforming growth factor beta antisense gene therapy. *Proc Natl Acad Sci USA*. 1996;93(7):2909–2914.

53. Witham TF, Villa L, Yang T, et al. Expression of a soluble transforming growth factor-beta (TGFbeta) receptor reduces tumorigenicity by regulating natural killer (NK) cell activity against 9L gliosarcoma in vivo. *J Neurooncol*. 2003;64(1–2):63–69.
54. Iwadate Y, Yamaura A, Sakiyama S, Sato Y, Tagawa M. Glioma-specific cytotoxic T cells can be effectively induced by subcutaneous vaccination of irradiated wild-type tumor cells without artificial cytokine production. *Int J Oncol*. 2003;23(2):483–488.
55. Denlinger RH, Axler DA, Koestner A, Liss L. Tumor-specific transplantation immunity to intracerebral challenge with cells from a methylnitrosourea- induced brain tumor. *J Med*. 1975;6(3–4):249–259.
56. Stojiljkovic M, Piperski V, Dacevic M, Rakic L, Ruzdijic S, Kanazir S. Characterization of 9L glioma model of the Wistar rat. *J Neurooncol*. 2003;63(1):1–7.
57. Ko L, Koestner A, Wechsler W. Morphological characterization of nitrosourea-induced glioma cell lines and clones. *Acta Neuropathol*. 1980;51(1):23–31.
58. Kobayashi N, Allen N, Clendenon NR, Ko LW. An improved rat brain-tumor model. *J Neurosurg*. 1980;53(6):808–815.
59. Weizsacker M, Nagamune A, Winkelstroter R, Vieten H, Wechsler W. Radiation and drug response of the rat glioma RG2. *Eur J Cancer Clin Oncol*. 1982;18(9):891–895.
60. Oshiro S, Liu Y, Fukushima T, Asotra K, Black KL. Modified immunoregulation associated with interferon-gamma treatment of rat glioma. *Neurol Res*. 2001;23(4):359–366.
61. Tzeng JJ, Barth RF, Orosz CG, James SM. Phenotype and functional activity of tumor-infiltrating lymphocytes isolated from immunogenic and nonimmunogenic rat brain tumors. *Cancer Res*. 1991;51(9):2373–2378.
62. Schartner JM, Hagar AR, Van Handel M, Zhang L, Nadkarni N, Badie B. Impaired capacity for upregulation of MHC class II in tumor-associated microglia. *Glia*. 2005;51(4):279–285.
63. Banissi C, Ghiringhelli F, Chen L, Carpentier AF. Treg depletion with a low-dose metronomic temozolomide regimen in a rat glioma model. *Cancer Immunol Immunother*. 2009;58(10):1627–1634.
64. Mineharu Y, Muhammad AK, Yagiz K, et al. Gene therapy-mediated reprogramming tumor infiltrating T cells using IL-2 and inhibiting NF-kappaB signaling improves the efficacy of immunotherapy in a brain cancer model. *Neurotherapeutics*. 2012;9(4):827–843.
65. Mineharu Y, Kamran N, Lowenstein PR, Castro MG. Blockade of mTOR signaling via rapamycin combined with immunotherapy augments antiglioma cytotoxic and memory T-cell functions. *Mol Cancer Ther*. 2014;13(12):3024–3036.
66. Mathieu D, Lecomte R, Tsanaclis AM, Larouche A, Fortin D. Standardization and detailed characterization of the syngeneic Fischer/F98 glioma model. *Can J Neurol Sci*. 2007;34(3):296–306.
67. Doblas S, He T, Saunders D, et al. Glioma morphology and tumor-induced vascular alterations revealed in seven rodent glioma models by in vivo magnetic resonance imaging and angiography. *J Magn Reson Imaging*. 2010;32(2):267–275.
68. Paul DB, Barth RF, Yang W, Shen GH, Kim J, Triozzi PL. B7.1 expression by the weakly immunogenic F98 rat glioma does not enhance immunogenicity. *Gene Ther*. 2000;7(12):993–999.
69. Clavreul A, Delhaye M, Jadaud E, Menei P. Effects of syngeneic cellular vaccinations alone or in combination with GM-CSF on the weakly immunogenic F98 glioma model. *J Neurooncol*. 2006;79(1):9–17.
70. Tzeng JJ, Barth RF, Clendenon NR, Gordon WA. Adoptive immunotherapy of a rat glioma using lymphokine-activated killer cells and interleukin 2. *Cancer Res*. 1990;50(14):4338–4343.
71. Fenstermaker RA, Capala J, Barth RF, Hujer A, Kung HJ, Kaetzel Jr DM. The effect of epidermal growth factor receptor (EGFR) expression on in vivo growth of rat C6 glioma cells. *Leukemia*. 1995;9(suppl 1):S106–S112.

72. Ciesielski MJ, Kazim AL, Barth RF, Fenstermaker RA. Cellular antitumor immune response to a branched lysine multiple antigenic peptide containing epitopes of a common tumor-specific antigen in a rat glioma model. *Cancer Immunol Immunother.* 2005;54(2):107–119.
73. Volovitz I, Marmor Y, Azulay M, et al. Split immunity: immune inhibition of rat gliomas by subcutaneous exposure to unmodified live tumor cells. *J Immunol.* 2011;187(10):5452–5462.
74. Alain T, Lun X, Martineau Y, et al. Vesicular stomatitis virus oncolysis is potentiated by impairing mTORC1-dependent type I IFN production. *Proc Natl Acad Sci USA.* 2010;107(4):1576–1581.
75. Bryant MJ, Chuah TL, Luff J, Lavin MF, Walker DG. A novel rat model for glioblastoma multiforme using a bioluminescent F98 cell line. *J Clin Neurosci.* 2008;15(5):545–551.
76. Rehemtulla A, Stegman LD, Cardozo SJ, et al. Rapid and quantitative assessment of cancer treatment response using in vivo bioluminescence imaging. *Neoplasia.* 2000;2(6):491–495.
77. Kruse CA, Molleston MC, Parks EP, Schiltz PM, Kleinschmidt-DeMasters BK, Hickey WF. A rat glioma model, CNS-1, with invasive characteristics similar to those of human gliomas: a comparison to 9L gliosarcoma. *J Neurooncol.* 1994;22(3):191–200.
78. Carpentier AF, Xie J, Mokhtari K, Delattre JY. Successful treatment of intracranial gliomas in rat by oligodeoxynucleotides containing CpG motifs. *Clin Cancer Res.* 2000;6(6):2469–2473.
79. Biglari A, Bataille D, Naumann U, et al. Effects of ectopic decorin in modulating intracranial glioma progression in vivo, in a rat syngeneic model. *Cancer Gene Ther.* 2004;11(11):721–732.
80. Ali S, King GD, Curtin JF, et al. Combined immunostimulation and conditional cytotoxic gene therapy provide long-term survival in a large glioma model. *Cancer Res.* 2005;65(16):7194–7204.
81. Jacobs VL, Landry RP, Liu Y, Romero-Sandoval EA, De Leo JA. Propentofylline decreases tumor growth in a rodent model of glioblastoma multiforme by a direct mechanism on microglia. *Neuro Oncol.* 2012;14(2):119–131.
82. Jacobs VL, Liu Y, De Leo JA. Propentofylline targets TROY, a novel microglial signaling pathway. *PLoS One.* 2012;7(5):e37955.
83. Barth RF, Kaur B. Rat brain tumor models in experimental neuro-oncology: the C6, 9L, T9, RG2, F98, BT4C, RT-2 and CNS-1 gliomas. *J Neurooncol.* 2009;94(3):299–312.
84. Beutler AS, Banck MS, Wedekind D, Hedrich HJ. Tumor gene therapy made easy: allogeneic major histocompatibility complex in the C6 rat glioma model. *Hum Gene Ther.* 1999;10(1):95–101.
85. Parsa AT, Chakrabarti I, Hurley PT, et al. Limitations of the C6/Wistar rat intracerebral glioma model: implications for evaluating immunotherapy. *Neurosurgery.* 2000;47(4):993–999 [discussion 999–1000].
86. Laerum OD, Rajewsky MF. Neoplastic transformation of fetal rat brain cells in culture after exposure to ethylnitrosourea in vivo. *J Natl Cancer Inst.* 1975;55(5):1177–1187.
87. Laerum OD, Rajewsky MF, Schachner M, Stavrou D, Haglid KG, Haugen A. Phenotypic properties of neoplastic cell lines developed from fetal rat brain cells in culture after exposure to ethylnitrosourea in vivo. *Z Krebsforsch Klin Onkol Cancer Res Clin Oncol.* 1977;89(3):273–295.
88. Morgan KT, Frith CH, Swenberg JA, McGrath JT, Zulch KJ, Crowder DM. A morphologic classification of brain tumors found in several strains of mice. *J Natl Cancer Inst.* 1984;72(1):151–160.
89. Crafts D, Wilson CB. Animal models of brain tumors. *National Cancer Institute Monograph.* 1977;46:11–17.
90. Pilkington GJ, Darling JL, Lantos PL, Thomas DG. Cell lines (VMDk) derived from a spontaneous murine astrocytoma. Morphological and immunocytochemical characterization. *J Neurol Sci.* 1983;62(1–3):115–139.

91. Wrann M, Bodmer S, de Martin R, et al. T cell suppressor factor from human glioblastoma cells is a 12.5-kd protein closely related to transforming growth factor-beta. *EMBO J*. 1987;6(6):1633–1636.
92. Tran TT, Uhl M, Ma JY, et al. Inhibiting TGF-beta signaling restores immune surveillance in the SMA-560 glioma model. *Neuro Oncol*. 2007;9(3):259–270.
93. Heimberger AB, Crotty LE, Archer GE, et al. Bone marrow-derived dendritic cells pulsed with tumor homogenate induce immunity against syngeneic intracerebral glioma. *J Neuroimmunol*. 2000;103(1):16–25.
94. Fecci PE, Sweeney AE, Grossi PM, et al. Systemic anti-CD25 monoclonal antibody administration safely enhances immunity in murine glioma without eliminating regulatory T cells. *Clin Cancer Res*. 2006;12(14 Pt 1):4294–4305.
95. Sampson JH, Choi BD, Sanchez-Perez L, et al. EGFRvIII mCAR-modified T-cell therapy cures mice with established intracerebral glioma and generates host immunity against tumor-antigen loss. *Clin Cancer Res*. 2014;20(4):972–984.
96. Huysentruyt LC, Mukherjee P, Banerjee D, Shelton LM, Seyfried TN. Metastatic cancer cells with macrophage properties: evidence from a new murine tumor model. *Int J Cancer*. 2008;123(1):73–84.
97. Seyfried TN, el-Abbadi M, Roy ML. Ganglioside distribution in murine neural tumors. *Mol Chem Neuropathol*. 1992;17(2):147–167.
98. Martinez-Murillo R, Martinez A. Standardization of an orthotopic mouse brain tumor model following transplantation of CT-2A astrocytoma cells. *Histol Histopathol*. 2007;22(12):1309–1326.
99. Binello E, Qadeer ZA, Kothari HP, Emdad L, Germano IM. Stemness of the CT-2A immunocompetent mouse brain tumor model: characterization in vitro. *J Cancer*. 2012;3:166–174.
100. Bao S, Wu Q, McLendon RE, et al. Glioma stem cells promote radioresistance by preferential activation of the DNA damage response. *Nature*. 2006;444(7120):756–760.
101. Pallini R, Ricci-Vitiani L, Montano N, et al. Expression of the stem cell marker CD133 in recurrent glioblastoma and its value for prognosis. *Cancer*. 2011;117(1):162–174.
102. Zeppernick F, Ahmadi R, Campos B, et al. Stem cell marker CD133 affects clinical outcome in glioma patients. *Clin Cancer Res*. 2008;14(1):123–129.
103. Grosser BI. 11-beta-Hydroxysteroid metabolism by mouse brain and glioma 261. *J Neurochem*. 1966;13(6):475–478.
104. Ausman JI, Shapiro WR, Rall DP. Studies on the chemotherapy of experimental brain tumors: development of an experimental model. *Cancer Res*. 1970;30(9):2394–2400.
105. Weiner NE, Pyles RB, Chalk CL, et al. A syngeneic mouse glioma model for study of glioblastoma therapy. *J Neuropathol Exp Neurol*. 1999;58(1):54–60.
106. Szatmari T, Lumniczky K, Desaknai S, et al. Detailed characterization of the mouse glioma 261 tumor model for experimental glioblastoma therapy. *Cancer Sci*. 2006;97(6):546–553.
107. Zagzag D, Amirnovin R, Greco MA, et al. Vascular apoptosis and involution in gliomas precede neovascularization: a novel concept for glioma growth and angiogenesis. *Lab Invest*. 2000;80(6):837–849.
108. Maes W, Van Gool SW. Experimental immunotherapy for malignant glioma: lessons from two decades of research in the GL261 model. *Cancer Immunol Immunother*. 2011;60(2):153–160.
109. Barth Jr RJ, Bock SN, Mule JJ, Rosenberg SA. Unique murine tumor-associated antigens identified by tumor infiltrating lymphocytes. *J Immunol*. 1990;144(4):1531–1537.
110. Plautz GE, Touhalisky JE, Shu S. Treatment of murine gliomas by adoptive transfer of ex vivo activated tumor-draining lymph node cells. *Cell Immunol*. 1997;178(2):101–107.
111. Ni HT, Spellman SR, Jean WC, Hall WA, Low WC. Immunization with dendritic cells pulsed with tumor extract increases survival of mice bearing intracranial gliomas. *J Neurooncol*. 2001;51(1):1–9.

112. Pellegatta S, Poliani PL, Corno D, et al. Dendritic cells pulsed with glioma lysates induce immunity against syngeneic intracranial gliomas and increase survival of tumor-bearing mice. *Neurol Res*. 2006;28(5):527–531.
113. Ciesielski MJ, Apfel L, Barone TA, Castro CA, Weiss TC, Fenstermaker RA. Antitumor effects of a xenogeneic survivin bone marrow derived dendritic cell vaccine against murine GL261 gliomas. *Cancer Immunol Immunother*. 2006;55(12):1491–1503.
114. Hatano M, Kuwashima N, Tatsumi T, et al. Vaccination with EphA2-derived T cell-epitopes promotes immunity against both EphA2-expressing and EphA2-negative tumors. *J Transl Med*. 2004;2(1):40.
115. Prins RM, Odesa SK, Liau LM. Immunotherapeutic targeting of shared melanoma-associated antigens in a murine glioma model. *Cancer Res*. 2003;63(23):8487–8491.
116. Kim CH, Hong MJ, Park SD, et al. Enhancement of anti-tumor immunity specific to murine glioma by vaccination with tumor cell lysate-pulsed dendritic cells engineered to produce interleukin-12. *Cancer Immunol Immunother*. 2006;55(11):1309–1319.
117. Hayes C, Kelly D, Murayama S, Komiyama A, Suzuki K, Popko B. Expression of the neu oncogene under the transcriptional control of the myelin basic protein gene in transgenic mice: generation of transformed glial cells. *J Neurosci Res*. 1992; 31(1):175–187.
118. Dyer CA, Philibotte T. A clone of the MOCH-1 glial tumor in culture: multiple phenotypes expressed under different environmental conditions. *J Neuropathol Exp Neurol*. 1995;54(6):852–863.
119. Hellums EK, Markert JM, Parker JN, et al. Increased efficacy of an interleukin-12-secreting herpes simplex virus in a syngeneic intracranial murine glioma model. *Neuro Oncol*. 2005;7(3):213–224.
120. Markert JM, Cody JJ, Parker JN, et al. Preclinical evaluation of a genetically engineered herpes simplex virus expressing interleukin-12. *J Virol*. 2012;86(9):5304–5313.
121. Cancer Genome Atlas Research N. Comprehensive genomic characterization defines human glioblastoma genes and core pathways. *Nature*. 2008;455(7216):1061–1068.
122. Verhaak RG, Hoadley KA, Purdom E, et al. Integrated genomic analysis identifies clinically relevant subtypes of glioblastoma characterized by abnormalities in PDGFRA, IDH1, EGFR, and NF1. *Cancer Cell*. 2010;17(1):98–110.
123. Fisher GH, Orsulic S, Holland E, et al. Development of a flexible and specific gene delivery system for production of murine tumor models. *Oncogene*. 1999;18(38):5253–5260.
124. Holland EC. A mouse model for glioma: biology, pathology, and therapeutic opportunities. *Toxicol Pathol*. 2000;28(1):171–177.
125. Ding H, Shannon P, Lau N, et al. Oligodendrogliomas result from the expression of an activated mutant epidermal growth factor receptor in a RAS transgenic mouse astrocytoma model. *Cancer Res*. 2003;63(5):1106–1113.
126. Weiss WA, Burns MJ, Hackett C, et al. Genetic determinants of malignancy in a mouse model for oligodendroglioma. *Cancer Res*. 2003;63(7):1589–1595.
127. Zhu H, Acquaviva J, Ramachandran P, et al. Oncogenic EGFR signaling cooperates with loss of tumor suppressor gene functions in gliomagenesis. *Proc Natl Acad Sci USA*. 2009;106(8):2712–2716.
128. Chow LM, Baker SJ. Capturing the molecular and biological diversity of high-grade astrocytoma in genetically engineered mouse models. *Oncotarget*. 2012;3(1):67–77.
129. Rankin SL, Zhu G, Baker SJ. Review: insights gained from modelling high-grade glioma in the mouse. *Neuropathol Appl Neurobiol*. 2012;38(3):254–270.
130. Hambardzumyan D, Parada LF, Holland EC, Charest A. Genetic modeling of gliomas in mice: new tools to tackle old problems. *Glia*. 2011;59(8):1155–1168.
131. Uhrbom L, Hesselager G, Nister M, Westermark B. Induction of brain tumors in mice using a recombinant platelet-derived growth factor B-chain retrovirus. *Cancer Res*. 1998;58(23):5275–5279.

132. Hede SM, Hansson I, Afink GB, et al. GFAP promoter driven transgenic expression of PDGFB in the mouse brain leads to glioblastoma in a Trp53 null background. *Glia*. 2009;57(11):1143–1153.
133. Uhrbom L, Hesselager G, Ostman A, Nister M, Westermark B. Dependence of autocrine growth factor stimulation in platelet-derived growth factor-B-induced mouse brain tumor cells. *Int J Cancer*. 2000;85(3):398–406.
134. Liu KW, Feng H, Bachoo R, et al. SHP-2/PTPN11 mediates gliomagenesis driven by PDGFRA and INK4A/ARF aberrations in mice and humans. *J Clin Invest*. 2011;121(3):905–917.
135. Tchougounova E, Kastemar M, Brasater D, Holland EC, Westermark B, Uhrbom L. Loss of Arf causes tumor progression of PDGFB-induced oligodendroglioma. *Oncogene*. 2007;26(43):6289–6296.
136. Ding H, Roncari L, Shannon P, et al. Astrocyte-specific expression of activated p21-ras results in malignant astrocytoma formation in a transgenic mouse model of human gliomas. *Cancer Res*. 2001;61(9):3826–3836.
137. Shannon P, Sabha N, Lau N, Kamnasaran D, Gutmann DH, Guha A. Pathological and molecular progression of astrocytomas in a GFAP:12 V-Ha-Ras mouse astrocytoma model. *Am J Pathol*. 2005;167(3):859–867.
138. Parsons DW, Jones S, Zhang X, et al. An integrated genomic analysis of human glioblastoma multiforme. *Science*. 2008;321(5897):1807–1812.
139. Jacques TS, Swales A, Brzozowski MJ, et al. Combinations of genetic mutations in the adult neural stem cell compartment determine brain tumour phenotypes. *EMBO J*. 2010;29(1):222–235.
140. Chow LM, Endersby R, Zhu X, et al. Cooperativity within and among Pten, p53, and Rb pathways induces high-grade astrocytoma in adult brain. *Cancer Cell*. 2011;19(3):305–316.
141. Trovo-Marqui AB, Tajara EH. Neurofibromin: a general outlook. *Clin Genet*. 2006;70(1):1–13.
142. Singhal S, Birch JM, Kerr B, Lashford L, Evans DG. Neurofibromatosis type 1 and sporadic optic gliomas. *Arch Dis Child*. 2002;87(1):65–70.
143. Gutmann DH, James CD, Poyhonen M, et al. Molecular analysis of astrocytomas presenting after age 10 in individuals with NF1. *Neurology*. 2003;61(10):1397–1400.
144. Reilly KM, Loisel DA, Bronson RT, McLaughlin ME, Jacks T. Nf1;Trp53 mutant mice develop glioblastoma with evidence of strain-specific effects. *Nat Genet*. 2000;26(1):109–113.
145. Flores C, Pham C, Snyder D, et al. Novel role of hematopoietic stem cells in immunologic rejection of malignant gliomas. *Oncoimmunology*. 2015;4(3):e994374.
146. Ouyang M, White EE, Ren H, et al. Metronomic doses of temozolomide enhance the efficacy of carbon nanotube CpG immunotherapy in an invasive glioma MODEL. *PLoS One*. 2016;11(2):e0148139.
147. Vogelstein B, Papadopoulos N, Velculescu VE, Zhou S, Diaz Jr LA, Kinzler KW. Cancer genome landscapes. *Science*. 2013;339(6127):1546–1558.
148. Mouse Genome Sequencing C, Waterston RH, Lindblad-Toh K, et al. Initial sequencing and comparative analysis of the mouse genome. *Nature*. 2002;420(6915):520–562.
149. Drake AC. Of mice and men: what rodent models don't tell us. *Cell Mol Immunol*. 2013;10(4):284–285.
150. Shultz LD, Ishikawa F, Greiner DL. Humanized mice in translational biomedical research. *Nat Rev Immunol*. 2007;7(2):118–130.
151. Gould SE, Junttila MR, de Sauvage FJ. Translational value of mouse models in oncology drug development. *Nat Med*. 2015;21(5):431–439.
152. McNeill RS, Vitucci M, Wu J, Miller CR. Contemporary murine models in preclinical astrocytoma drug development. *Neuro Oncol*. 2015;17(1):12–28.
153. Flanagan SP. 'Nude', a new hairless gene with pleiotropic effects in the mouse. *Gen Res*. 1966;8(3):295–309.

154. Mecklenburg L, Tychsen B, Paus R. Learning from nudity: lessons from the nude phenotype. *Exp Dermatol*. 2005;14(11):797–810.
155. Warner NL, Woodruff MF, Burton RC. Inhibition of the growth of lymphoid tumours in syngeneic athymic (nude) mice. *Int J Cancer*. 1977;20(1):146–155.
156. Klein AS, Plata F, Jackson MJ, Shin S. Cellular tumorigenicity in nude mice. Role of susceptibility to natural killer cells. *Exp Cell Biol*. 1979;47(6):430–445.
157. Nassiry L, Miller SC. Renewal of natural killer cells in mice having elevated natural killer cell activity. *Nat Immun Cell Growth Regul*. 1987;6(5):250–259.
158. Bosma GC, Custer RP, Bosma MJ. A severe combined immunodeficiency mutation in the mouse. *Nature*. 1983;301(5900):527–530.
159. Fulop GM, Phillips RA. The scid mutation in mice causes a general defect in DNA repair. *Nature*. 1990;347(6292):479–482.
160. Kubota T, Yamaguchi H, Watanabe M, et al. Growth of human tumor xenografts in nude mice and mice with severe combined immunodeficiency (SCID). *Surg Today*. 1993;23(4):375–377.
161. Mosier DE. Adoptive transfer of human lymphoid cells to severely immunodeficient mice: models for normal human immune function, autoimmunity, lymphomagenesis, and AIDS. *Adv Immunol*. 1991;50:303–325.
162. Taghian A, Budach W, Zietman A, Freeman J, Gioioso D, Suit HD. Quantitative comparison between the transplantability of human and murine tumors into the brain of NCr/Sed-nu/nu nude and severe combined immunodeficient mice. *Cancer Res*. 1993;53(20):5018–5021.
163. Bosma GC, Fried M, Custer RP, Carroll A, Gibson DM, Bosma MJ. Evidence of functional lymphocytes in some (leaky) scid mice. *J Exp Med*. 1988;167(3):1016–1033.
164. Ito M, Hiramatsu H, Kobayashi K, et al. NOD/SCID/gamma(c)(null) mouse: an excellent recipient mouse model for engraftment of human cells. *Blood*. 2002;100(9):3175–3182.
165. Wicker LS, Chamberlain G, Hunter K, et al. Fine mapping, gene content, comparative sequencing, and expression analyses support Ctla4 and Nramp1 as candidates for Idd5.1 and Idd5.2 in the nonobese diabetic mouse. *J Immunol*. 2004;173(1):164–173.
166. Shultz LD, Schweitzer PA, Christianson SW, et al. Multiple defects in innate and adaptive immunologic function in NOD/LtSz-scid mice. *J Immunol*. 1995;154(1):180–191.
167. Joo KM, Kim J, Jin J, et al. Patient-specific orthotopic glioblastoma xenograft models recapitulate the histopathology and biology of human glioblastomas in situ. *Cell Rep*. 2013;3(1):260–273.
168. Lapidot T, Lubin I, Terenzi A, Faktorowich Y, Erlich P, Reisner Y. Enhancement of bone marrow allografts from nude mice into mismatched recipients by T cells void of graft-versus-host activity. *Proc Natl Acad Sci USA*. 1990;87(12):4595–4599.
169. Lapidot T, Pflumio F, Doedens M, Murdoch B, Williams DE, Dick JE. Cytokine stimulation of multilineage hematopoiesis from immature human cells engrafted in SCID mice. *Science*. 1992;255(5048):1137–1141.
170. Hesselton RM, Greiner DL, Mordes JP, Rajan TV, Sullivan JL, Shultz LD. High levels of human peripheral blood mononuclear cell engraftment and enhanced susceptibility to human immunodeficiency virus type 1 infection in NOD/LtSz-scid/scid mice. *J Infect Dis*. 1995;172(4):974–982.
171. Lowry PA, Shultz LD, Greiner DL, et al. Improved engraftment of human cord blood stem cells in NOD/LtSz-scid/scid mice after irradiation or multiple-day injections into unirradiated recipients. *Biol Blood Marrow Transplant*. 1996;2(1):15–23.
172. Verlinden SF, Mulder AH, de Leeuw JP, van Bekkum DW. T lymphocytes determine the development of xeno GVHD and of human hemopoiesis in NOD/SCID mice following human umbilical cord blood transplantation. *Stem Cells*. 1998;16(suppl 1):205–217.
173. Berney T, Molano RD, Pileggi A, et al. Patterns of engraftment in different strains of immunodeficient mice reconstituted with human peripheral blood lymphocytes. *Transplantation*. 2001;72(1):133–140.

174. Shultz LD, Lyons BL, Burzenski LM, et al. Human lymphoid and myeloid cell development in NOD/LtSz-scid IL2R gamma null mice engrafted with mobilized human hemopoietic stem cells. *J Immunol*. 2005;174(10):6477–6489.
175. Alcantar-Orozco EM, Gornall H, Baldan V, Hawkins RE, Gilham DE. Potential limitations of the NSG humanized mouse as a model system to optimize engineered human T cell therapy for cancer. *Hum Gene Ther Methods*. 2013;24(5):310–320.
176. Garcia S, Freitas AA. Humanized mice: current states and perspectives. *Immunol Lett*. 2012;146(1–2):1–7.
177. Ishikawa F, Yasukawa M, Lyons B, et al. Development of functional human blood and immune systems in NOD/SCID/IL2 receptor {gamma} chain(null) mice. *Blood*. 2005;106(5):1565–1573.
178. Tanaka S, Saito Y, Kunisawa J, et al. Development of mature and functional human myeloid subsets in hematopoietic stem cell-engrafted NOD/SCID/IL2rgammaKO mice. *J Immunol*. 2012;188(12):6145–6155.
179. Brugman MH, Wiekmeijer AS, van Eggermond M, et al. Development of a diverse human T-cell repertoire despite stringent restriction of hematopoietic clonality in the thymus. *Proc Natl Acad Sci USA*. 2015;112(44):E6020–E6027.
180. Rossi MI, Medina KL, Garrett K, et al. Relatively normal human lymphopoiesis but rapid turnover of newly formed B cells in transplanted nonobese diabetic/SCID mice. *J Immunol*. 2001;167(6):3033–3042.
181. Strowig T, Gurer C, Ploss A, et al. Priming of protective T cell responses against virus-induced tumors in mice with human immune system components. *J Exp Med*. 2009;206(6):1423–1434.
182. Jaiswal S, Pearson T, Friberg H, et al. Dengue virus infection and virus-specific HLA-A2 restricted immune responses in humanized NOD-scid IL2rgammanull mice. *PLoS One*. 2009;4(10):e7251.
183. Racki WJ, Covassin L, Brehm M, et al. NOD-scid IL2rgamma(null) mouse model of human skin transplantation and allograft rejection. *Transplantation*. 2010;89(5):527–536.
184. Pearson T, Shultz LD, Miller D, et al. Non-obese diabetic-recombination activating gene-1 (NOD-Rag1 null) interleukin (IL)-2 receptor common gamma chain (IL2r gamma null) null mice: a radioresistant model for human lymphohaematopoietic engraftment. *Clin Exp Immunol*. 2008;154(2):270–284.
185. Alt FW, Rathbun G, Oltz E, Taccioli G, Shinkai Y. Function and control of recombination-activating gene activity. *Ann N Y Acad Sci*. 1992;651:277–294.
186. Traggiai E, Chicha L, Mazzucchelli L, et al. Development of a human adaptive immune system in cord blood cell-transplanted mice. *Science*. 2004;304(5667):104–107.
187. Rongvaux A, Willinger T, Takizawa H, et al. Human thrombopoietin knockin mice efficiently support human hematopoiesis in vivo. *Proc Natl Acad Sci USA*. 2011;108(6):2378–2383.
188. Pek EA, Chan T, Reid S, Ashkar AA. Characterization and IL-15 dependence of NK cells in humanized mice. *Immunobiology*. 2011;216(1–2):218–224.
189. Chen Q, Khoury M, Chen J. Expression of human cytokines dramatically improves reconstitution of specific human-blood lineage cells in humanized mice. *Proc Natl Acad Sci USA*. 2009;106(51):21783–21788.
190. Rathinam C, Poueymirou WT, Rojas J, et al. Efficient differentiation and function of human macrophages in humanized CSF-1 mice. *Blood*. 2011;118(11):3119–3128.
191. van Lent AU, Dontje W, Nagasawa M, et al. IL-7 enhances thymic human T cell development in "human immune system" Rag2$^{-/-}$IL-2Rgammac$^{-/-}$ mice without affecting peripheral T cell homeostasis. *J Immunol*. 2009;183(12):7645–7655.
192. Anderson G, Owen JJ, Moore NC, Jenkinson EJ. Thymic epithelial cells provide unique signals for positive selection of CD4+CD8+ thymocytes in vitro. *J Exp Med*. 1994;179(6):2027–2031.

193. Shultz LD, Saito Y, Najima Y, et al. Generation of functional human T-cell subsets with HLA-restricted immune responses in HLA class I expressing NOD/SCID/IL2r gamma(null) humanized mice. *Proc Natl Acad Sci USA*. 2010;107(29):13022–13027.
194. Billerbeck E, Horwitz JA, Labitt RN, et al. Characterization of human antiviral adaptive immune responses during hepatotropic virus infection in HLA-transgenic human immune system mice. *J Immunol*. 2013;191(4):1753–1764.
195. Ito R, Shiina M, Saito Y, Tokuda Y, Kametani Y, Habu S. Antigen-specific antibody production of human B cells in NOG mice reconstituted with the human immune system. *Curr Top Microbiol Immunol*. 2008;324:95–107.
196. Melkus MW, Estes JD, Padgett-Thomas A, et al. Humanized mice mount specific adaptive and innate immune responses to EBV and TSST-1. *Nat Med*. 2006;12(11):1316–1322.
197. Denton PW, Estes JD, Sun Z, et al. Antiretroviral pre-exposure prophylaxis prevents vaginal transmission of HIV-1 in humanized BLT mice. *PLoS Med*. 2008;5(1):e16.

CHAPTER

7

Imaging Studies in Immunotherapy

R.Y. Huang[1], *D.A. Reardon*[2]

[1]Brigham and Women's Hospital, Boston, MA, United States; [2]Dana-Farber Cancer Institute, Boston, MA, United States

OUTLINE

Translational Immunotherapy of Brain Tumors
http://dx.doi.org/10.1016/B978-0-12-802420-1.00007-7

CLINICAL CONSIDERATIONS

The Growing Role of Immunotherapy for Oncology

Although the vision of utilizing the immune system, our body's natural defenses, to attack and eliminate cancer dates back over 120 years to the pioneering work of William Coley who utilized bacterial toxins as cancer therapy,[1] it has only been in recent years that significant breakthroughs with immunotherapeutics have evolved. During this time frame, remarkable advances have evolved regarding the role of immunotherapy for cancer patients. Sipuleucel-T, a vaccine targeting prostatic acid phosphatase, became the first vaccine against nonviral cancers approved by the United States Food and Drug Administration (FDA).[2] In addition, monoclonal antibodies (mAb) capable of inhibiting the immune checkpoint molecules cytotoxic T lymphocyte-4 (CTLA-4; ipilimumab) and programmed death 1 (PD-1; nivolumab and pembroluzimab) achieved unprecedented results for advanced solid and hematologic malignancies.[3–10] Furthermore, durable remissions have been induced by chimeric antigen receptor (CAR) engineered autologous T cells among leukemia patients refractory to conventional therapies including bone marrow transplantation.[11,12]

Immunotherapy offers the potential to generate meaningful long-term antitumor activity by eliciting a two-phase effect. First, antitumor immune responses may contribute to eradicating existing tumors. Second, if effective and specific immune memory responses are generated, future relapse or recurrence may be prevented. Currently, a spectrum of immunotherapeutics with varied mechanisms triggering antitumor immunoreactivity is under clinical development. Active agents such as vaccines against tumor-associated or tumor-specific antigens can act to specifically sensitize immune effector cells against tumor cells. Unarmed or armed antibodies against tumor targets can also attack cancer cells directly. Cellular approaches, such as adoptive T cell transfer or CAR-T cells, are cellular-based strategies that manipulate autologous T cells ex vivo to specifically react against tumor targets. Furthermore, therapeutics that can modulate immune responses such as inhibitory mAbs against immune checkpoints including PD-1 or CTLA-4 have been shown to abrogate immunosuppressive mechanisms exploited by tumors to evade the immune system. In addition, a growing number of studies are incorporating combinations of immunotherapeutics that provide potentially synergistic or complementary mechanisms of action. A current example includes the administration of a tumor specific vaccine in combination with an immune checkpoint inhibitor. A variety of immunotherapeutics are currently undergoing clinical evaluation for neuro-oncology patients, and emerging preliminary results appear encouraging.[13–15]

Although the clinical application of immunotherapies for neuro-oncology patients is in the early stages and much remains to be learned, the immunotherapy response assessment in neuro-oncology (iRANO) criteria were recently drafted to provide guidelines that can be currently applied to consistently assess response across studies and in daily practice. The iRANO criteria are intended to assist treatment decision making among neuro-oncology patients undergoing immunotherapy. In particular, these guidelines are provided to raise awareness of the likelihood of early progressive radiographic change following initiation of immunotherapy. Importantly, they provide an algorithm to guide clinicians confronted with potentially misleading, early progressive imaging changes to decrease the likelihood of inappropriate premature therapy discontinuation. Finally, the iRANO guidelines are expected to be a "work in progress" in that this initial draft will be successively amended as future experience grows, and data regarding their utility accumulate from ongoing immunotherapy trials in neuro-oncology.

Evolution of Radiologic Response Assessment

As the therapeutic arena of oncology has evolved over the past several years, the criteria to assess response have also evolved. Conventional imaging response assessment metrics such as the World Health Organization (WHO),[16] Response Evaluation in Solid Tumors (RECIST),[17] and Macdonald criteria,[18] rose to prominence during the cytotoxic therapy era. These criteria, which are based on percent change of contrast enhancing tumor burden, provide a direct relationship between antitumor benefit and radiographic response for such therapeutics. In this setting, stabilization of tumor growth or actual tumor shrinkage are known to reliably reflect underlying tumor sensitivity, whereas progressive enlargement or new lesions indicate de novo or acquired resistance to treatment. Although cytotoxic agents remain a critical cornerstone of oncology care, a host of additional therapeutics including molecularly targeted agents, antiangiogenics, and most recently, immunotherapeutics, have joined the forefront as standard therapies for cancer care. Nonetheless, the interpretation of the impact of these agents on patient outcome based on imaging parameters has become increasingly challenging and much less straightforward in the modern era. Such difficulties have been particularly noteworthy in the field of neuro-oncology. For example, pseudoprogressive increased enhancement with and without mass effect or clinical decline due to treatment induced inflammation rather than underlying tumor progression have been noted in up to 30% of newly diagnosed glioblastoma (GBM) patients following combined temozolomide chemoradiotherapy.[19] Conversely, the phenomenon of pseudoresponse in which enhancement dramatically improves within a short time frame following initiation of antiangiogenic agents has been observed. In many such cases, these changes appear to be independent of an underlying antitumor effect.[20] Although advanced magnetic resonance imaging (MRI) techniques including diffusion weighted and perfusion sequences continue to be intensively investigated, there is compelling interest to further refine existing and broadly available techniques to assess enhancing and nonenhancing tumor burden, to improve our ability to accurately assess response of our patients undergoing therapy in the modern era. For this reason, a multidisciplinary group of expert investigators developed the RANO[21] for high-grade glioma patients in 2010. Although RANO was formulated on the foundation of response assessment created by the Macdonald criteria,[18] RANO incorporated important new considerations designed to address the phenomena of pseudoprogression as well as pseudoresponse, along with a number of other factors in order to better assess such complexities of imaging assessment. Similar efforts have more recently been developed to apply RANO criteria for the assessment of patients with low-grade glioma,[22] and brain metastases.[23,24]

As mentioned, a critical advance of the RANO criteria is recognition of the phenomenon of radiographic pseudoprogression which may or may not be associated with clinical decline. Although a detailed understanding of the mechanisms of action underlying pseudoprogression remains to be determined, worsened enhancement mimicking tumor growth appears to be due to an inflammatory infiltrate induced by combined modality, temozolomide chemoradiotherapy. In such cases, the progressive imaging changes often plateau eventually and thereafter may ultimately improve. The peak incidence of pseudoprogression in this clinical setting is 3 months although longer and more variable time courses have been described.[25] To not inadvertently classify a patient with pseudoprogression as truly progressive, RANO guidelines preclude classification of progression within a 3-month window following completion of temozolomide radiochemotherapy (the period of greatest pseudoprogression likelihood) unless one of the two following caveats are met: (1) new enhancing disease consistent with tumor outside of the radiation field; or (2) histopathologic confirmation of unequivocal tumor recurrence. In anticipation that other treatments beyond temozolomide chemoradiotherapy may also be associated with pseudoprogressive imaging findings, RANO also includes a general statement that any patient with progressive imaging changes of unclear significance may be permitted to continue current therapy pending follow-up imaging. Importantly, RANO guidelines have been widely adopted by the neuro-oncology community for both daily practice and clinical research. Nonetheless, the RANO criteria alone may not provide sufficient guidance for the accurate assessment of response among patients with brain cancer undergoing immunotherapy.

The Need to Update Imaging Response Assessment Criteria for Immunotherapy

Based on the aggregate, modern era oncology experience to date, the radiographic response assessment for oncology patients treated with immunotherapy has also proven challenging. Given the markedly expanding repertoire of immune-based therapeutics and their widespread application across a multitude of cancer indications, considerable interest has developed to improve understanding of response assessment following administration of these agents. The interpretation of decreased size of an enhancing lesion is straightforward as such changes are felt to indicate a true antitumor effect because immunotherapeutics are not associated with pseudoresponse, as has been observed among GBM patients treated with antiangiogenic therapy.[20] A caveat to the direct relationship between improved imaging findings and antitumor effect however involves corticosteroid administration as discussed in a later section.

In contrast, interpretation of progressive imaging findings following administration of immunotherapeutics can be challenging. In particular, traditional dogma equating worsened enhancement or edema with tumor progression may require careful consideration in the context of immunotherapies. Specifically, some patients with early progressive radiographic changes, including the appearance of new lesions following immunotherapy treatment have subsequently achieved therapeutic benefit including delayed radiographic response or prolonged survival.[26–40]

The ultimate achievement of therapeutic benefit, including improved survival, despite worsened imaging findings early in the course of therapy could be due to one of two possible explanations. First, effective immune responses, that may take several weeks to develop in healthy individuals, may require additional time to evolve in cancer patients who often demonstrate inherent or acquired diminished immunoresponsiveness. In such patients, early imaging may reveal findings indicative of true underlying tumor progression, including the development of new lesions. Nonetheless, with ongoing treatment, tumor stabilization or response may be achieved. Second, pseudoprogressive radiologic features of tumor worsening including increased enhancement and/or edema may evolve in areas of gross or microscopic tumor extension due to the induction of an inflammatory response associated with an effective immunotherapeutic.

Immune-Related Response Criteria

Growing data demonstrate that early radiographic worsening following immunotherapy initiation, due to either real or pseudoprogression, may not preclude subsequent meaningful therapeutic benefit. Cancer immunotherapy studies evaluating cytokines, cancer vaccines, or immune checkpoint inhibitors demonstrate that a subset of patients may achieve a radiographic complete response (CR), partial response (PR), or stable disease (SD), and/or survival benefit after initial imaging findings reveal changes consistent with PD by conventional WHO or RECIST criteria.[26,30–40] For example, among patients undergoing anti-PD-1/PD-L1 therapy, individual spider plots measuring the percent change of target lesion size from baseline over time demonstrate initial enlargement of original tumor burden and/or the development of new lesions early in the course of therapy that can subsequently stabilize and/or regress.[30,32,34]

Among 487 advanced melanoma patients who received treatment with ipilimumab in three phase II studies, radiographic response patterns were classified into four categories.[26] Of note, traditional WHO or RECIST criteria adequately described two of these response patterns including: (1) a conventional radiologic response noted among baseline lesions with no new lesions; and (2) achievement of stable disease initially that subsequently led to gradual steady decrease of overall tumor burden. However, two additional response patterns were described which failed to fit previously established conventional response assessment criteria. Specifically, in one cohort of patients, existing lesions were noted to initially increase to the point of satisfying traditional PD criteria, which was subsequently followed by radiographic response or stable disease without any change or addition in therapy. In another cohort of patients, new noncontiguous foci of tumor were observed early in the course of treatment, a factor which defines progression via traditional response criteria. Nonetheless, these patients then went on to develop stable disease or even radiographic responses without altering treatment.

Additional early studies evaluating immune checkpoint blockade among oncology patients noted similar patterns of delayed response despite early worsening of imaging findings. In each of these examples, patients met traditional criteria that define radiographic progression. Importantly if these criteria had been applied, premature discontinuation of treatment would have occurred that may have lessened the likelihood of ultimate therapeutic benefit. These examples, therefore, highlight a potential disconnect between early radiographic worsening and possible therapeutic gain, including prolongation of survival that can be observed among some patients undergoing treatment with an immunotherapeutic. Among 227 patients who received ipilimumab on a phase II trial, 22 patients with early evidence of progressive disease by conventional WHO imaging criteria later achieved either stable disease (17 patients) or an actual radiographic response (5 patients) with continuation of ipilimumab therapy.[26] Furthermore, there was no difference in overall survival (OS) between these 22 patients with early progression compared to those patients with CR, PR, or SD without early progression. In a separate phase II study of the CTLA-4 blocking mAb tremelimumab, the OS of eight patients who developed early evidence of apparent mixed response, concurrent progression of some existing lesions, or the development of new lesions at the same time that some existing lesions demonstrated radiographic response, was 21–39 months. In contrast, OS for all patients enrolled on the study was only 10 months.[41]

A series of published vaccine studies among brain tumor patients suggest that the incidence of "tumor pseudoprogression" is not infrequent.[42–45] In addition, preliminary results of recently initiated clinical trials evaluating immune checkpoint blocking antibodies among recurrent GBM patients demonstrate that early progressive radiographic changes may subsequently stabilize in some patients (Fig. 7.1).

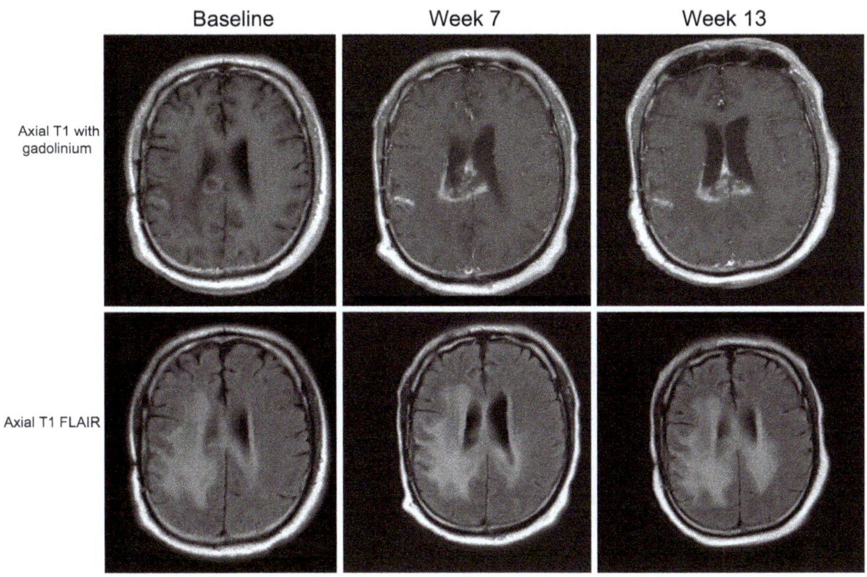

FIGURE 7.1 Imaging findings from a glioblastoma patient demonstrating progressive changes on initial imaging (7 weeks) after immune checkpoint blockade who subsequently achieved stable disease (week 13).

The immune-related response criteria (irRC) were proposed as novel guidelines for response assessment based on the appreciation of frequent challenges associated with the interpretation of progressive imaging findings following immunotherapy noted among patients with a wide array of cancer subtypes.[26,27,46] The irRC include several key considerations including (1) radiographic evidence of clinical activity at the tumor site may take longer to evolve for immunotherapies than for cytotoxic therapies; (2) radiographic responses to immunotherapies may evolve after conventional criteria defining progressive disease have been met; (3) confirmation of progressive disease on follow-up imaging may be appropriate in some cases prior to the discontinuation of immune therapy; (4) "clinically insignificant" progressive disease (e.g., small new lesions despite stable disease or response among other lesions) should not be a sufficient basis to define progressive disease in some cases; and, (5) meaningful clinical benefit may be associated with durable stable disease. In addition, a major component of the irRC, which represents a paradigm shift for response assessment, was an endorsement of immunotherapy continuation pending confirmation of progressive disease beyond initial demonstration of radiographic progression for clinically stable patients unless medically contraindicated.[26] The irRC were therefore issued as guidance to aid the interpretation of imaging changes observed by clinicians among oncology patients undergoing immunotherapy.[26] The overall intent of the irRC was to raise awareness that traditional imaging criteria to define tumor progression may be less reliable and could lead to premature discontinuation of potentially beneficial immunotherapeutics.

The irRC specify that clinically stable patients who exhibit an initial enlargement in lesion size or the development of new radiographic lesions, should not define progressive disease unless additional worsening of imaging findings are confirmed with follow-up imaging. An important caveat to this guidance is that confirmation of radiographic progression is not required for patients with significant clinical decline felt to be attributable to underlying tumor progression. Radiographic confirmation of tumor progression based on follow-up imaging is a unique aspect of the irRC, although confirmation of radiographic response with follow-up imaging has been a long-standing and accepted requisite of most response assessment tools, including RANO. Pending confirmation of progression on follow-up imaging, patients may be allowed to continue an immunotherapeutic. This guideline is included to prevent premature treatment discontinuation and represents an appropriate consideration based on the possibility that subsequent antitumor benefit may be achieved for at least some patients. The premature discontinuation of promising treatment, particularly for malignancies such as GBM, where effective treatment options are rare and durable tumor control remains elusive, is felt to represent an appropriate option for clinically stable patients.[8,26,27]

The primary objective of the iRANO criteria is to refine appropriate response evaluation metrics for brain tumor patients treated with immunotherapy including the appropriate fusion of relevant irRC and RANO factors. Immune responses targeting infiltrative brain tumor cells but not normal brain cells may generate transient enlargement of lesions that may ultimately lead to effective antitumor immune responses and clinical benefit for patients. To better understand the significance of early progressive imaging changes, future radiolographic, histopathologic, and immunologic studies will be required.

iRANO Criteria

A critical factor underlying the development of the RANO criteria was the conundrum associated with the evaluation of patients demonstrating possible temozolomide chemoradiation-induced tumor pseudoprogression. In this regard, RANO represents an acceptable basis to develop guidelines for imaging assessment among neuro-oncology patients being treated with immunotherapy. However, the RANO criteria per se are likely insufficient without further adaptation for immunotherapy for several reasons. First, immunotherapeutics may trigger pseudoprogression based on different mechanism of action compared to temozolomide chemoradiotherapy. Such distinct mechanisms of action may also reflect differences in the kinetics, frequency, and overall impact on patient outcome. For example, pseudoprogression typically peaks within 3 months following the completion of temozolomide chemoradiotherapy; in contrast the peak window for the development of pseudoprogression due to immunotherapy remains unknown. Furthermore, the kinetics of pseudoprogression may substantially differ among different types of immunotherapeutic treatments. Weeks to months may be required for the evolution of inflammatory changes and associated imaging findings related to humoral and/or cellular immune responses triggered by vaccine therapies. In contrast, immunotherapies that utilize ex vivo prepared immune effector cells, such as adoptive T cell therapies, or immunotherapies that effectively block tumor immunosuppressive mechanisms, such as immune checkpoint inhibitors, may generate more rapid antitumor immune responses and associated inflammatory changes on imaging assessment. A second important distinction is that RANO does not allow the continuation of treatment beyond true initial progression. Third, RANO criteria specify that a criterion for progressive disease is the development of any new lesion. Fourth, a repeat scan to confirm progressive disease is not required by the RANO criteria. Based on these important considerations, the iRANO criteria were generated by an independent, multinational panel of multidisciplinary experts to provide a consensus statement of guidance for the interpretation of early progressive imaging changes among neuro-oncology patients treated with immunotherapy reagents.

A distinguishing component of the iRANO criteria that is not present in the irRC is the specification of a timeframe from the start of an immunotherapeutic in which possible pseudoprogression may occur. The temporal window for the occurrence of possible pseudoprogression associated with immunotherapy is set at 6 months by the iRANO criteria. The basis for this cutoff is arbitrary, but reflects the observation that most melanoma or other solid tumor patients achieve evidence of antitumor benefit within 6 months of initiating an immune checkpoint therapeutic, even if early imaging findings suggest possible tumor progression.[8,30,32,34] Although the kinetics of either pseudoprogression or delayed response are not known for brain tumor patients treated with immunotherapy, limited data addressing this event for neuro-oncology suggest that most cases of pseudoprogression occur within 6 months of treatment initiation.[40,42,44] Therefore, iRANO recommends that patients without significant neurologic decline and who develop early radiographic evidence of progression within 6 months of starting an immunotherapeutic agent, should be allowed to pursue further treatment on their immunotherapeutic regimen pending follow-up imaging to confirm progressive disease. On the other hand, if radiographic evidence of progression evolves more than 6 months after initiation of immunotherapy, such patients are expected to have a much higher likelihood of true underlying tumor progression disease and should be classified as progressive disease at that time. For patients outside of the 6-month window from the initiation of immunotherapy, a confirmation of progression on follow-up imaging should not be required and they should be advised to discontinue their current immunotherapy treatment.

Because it is possible that patients who develop radiographic evidence of progressive disease within 6 months of the start of an immunotherapy regimen may still achieve therapeutic benefit, the patient's immunotherapy should be allowed to continue pending follow-up imaging to confirm progressive disease.

Another key distinction between the iRANO criteria and the irRC is the amount of time allowed after initial radiographic progressive disease criteria are met for follow-up imaging to confirm progression. The irRC criteria specify that 4 weeks is sufficient. In contrast, iRANO specifies that 3 months be permitted before confirmation of progression on follow-up imaging is assessed. Although the kinetics of pseudoprogression due to immunotherapy among neuro-oncology patients have not been accurately defined, a 3-month window to confirm progression is similar to the temporal window neuro-oncologists recognize to be the peak period of pseudoprogression following external beam radiotherapy/daily temozolomide for GBM patients.[19,47] Furthermore, 3 months is also the most common temporal window for pseudoprogression observed among patients with advanced melanoma or other solid tumors treated with CTLA-4 or PD-1/PD-L1 immune checkpoint blockade to date.[30,32,34]

Specifically, the iRANO criteria articulate that for patients who are not experiencing significant clinical decline, confirmation of progression on repeat imaging is recommended no earlier than 3 months after initial radiographic progression is detected to diminish the possibility of prematurely defining progression that may in fact represent pseudoprogression. In this case, imaging obtained 3 months after initial radiographic progression is directly compared to the initial scan demonstrating progression and any imaging obtained between these two time points is not relevant, in a manner that reflects the current RANO practice in which imaging obtained 3 months following completion of temozolomide chemoradiotherapy becomes the comparator against which future scans are assessed. Patients with confirmation of radiographic progression relative to comparison with the scan that initially demonstrated evidence of progression, or those who exhibit significant clinical decline during this assessment window, should be classified as having progressive disease with the date of disease progression retroactive to the date of the initial scan that met criteria for radiographic progression. It is recommended that such patients should be discontinued from their current immunotherapy regimen. On the other hand, should follow-up imaging fail to confirm further progression but instead demonstrates tumor stabilization or regression (in the absence of increased corticosteroid dosing), treatment may be allowed to be continued/resumed.

Despite these guidelines, the determination of true tumor progression compared to pseudoprogression will likely remain challenging for many neuro-oncology patients. For this reason, it may be reasonable to encourage the surgical collection of tumor material via either biopsy or resection. In cases where surgical intervention is deemed safe and feasible, histopathologic sampling may be helpful to definitively distinguish true tumor progression from pseudoprogression. If histopathologic analysis reveals findings consistent with recurrent tumor, the iRANO guidelines specify that the date of progressive disease should be back dated to the initial date imaging revealed progressive disease.

During the 3-month window to confirm tumor progression, most patients should be allowed to continue their underlying immunotherapy treatment unless the treating clinician feels that the associated risk/benefit ratio favors a treatment interruption. For example, patients who are experiencing worsened neurologic deficits, or those who require significant increases in corticosteroid dosing, may be considered for an interruption of immunotherapy treatment pending the acquisition of follow-up imaging. If an interruption of immunotherapy occurs, a resumption of treatment may be considered when systemic corticosteroid requirements are decreased or when worsened neurologic deficits stabilize.

TABLE 7.1 Key Distinctions of iRANO Relative to the RANO and Criteria

	Repeat Scan Required to Confirm PD	Temporal Window for Pseudoprogression	Permit Further Therapy Pending Confirmation of PD	New Lesion Defines PD
iRANO	Yes	3 months	Yes	No
RANO	No	3 months	Not applicable	Yes
irRC	Yes	4 weeks	Yes	No

CR, complete response; *iRANO*, immunotherapy response assessment in neuro-oncology; *irRC*, immune-related response criteria; *NA*, not applicable; *PD*, progressive disease; *PR*, partial response; *SD*, stable disease.

Table 7.1 provides a summary of the key distinctions of iRANO relative to the RANO and irRC criteria.

The following section summarizes the key components and definitions included in the iRANO criteria.[48]

Key Definitions

Evaluable for Toxicity

All patients who receive at least one dose of immunotherapy treatment will be evaluable for toxicity from the time of their first treatment.

Evaluable for Objective Response

Only those patients who have measurable disease (at least 1 cm × 1 cm) at baseline (recommend obtaining within 14 days of treatment initiation) scan and have received at least one dose of immunotherapy will be considered evaluable for response. These patients will have their response classified according to the definitions stated later. Patients who exhibit objective disease progression or die prior to the end of cycle 1 will also be considered evaluable. Unless progression is observed, objective response can only be determined when all measurable and nonmeasurable lesions are assessed.

Measurable Disease

For contrast-enhancing tumors, measurable disease is defined as bidimensionally, contrast enhancing, lesions with clearly defined margins by CT or MRI scan, a minimal diameter of 1 cm, and visible on two slices that are at least 5 mm apart with 0 mm skip. Measurement of tumor around a cyst or surgical cavity, if necessary, requires a minimum thickness of 3 mm. If there are too many measurable lesions to measure at each evaluation, the investigator must choose the largest two to be followed. The remaining lesions will be considered nonmeasurable for the purpose of objective response determination.

For noncontrast-enhancing tumors, measurable disease is defined as the T2 or fluid-attenuated inversion recovery (FLAIR) lesions with a minimal diameter of 1 cm, and visible on two slices that are at least 5 mm apart with 0 mm skip.

Nonmeasurable Evaluable Disease

Such tumors include unidimensionally measurable lesions, masses with margins not clearly defined, and/or lesions with maximal diameter <1 cm.

Methods for Evaluation of Measurable Disease

All measurements should be taken and recorded in metric notation, using a ruler, calipers, or digital measurement tool. All baseline evaluations should be performed as closely as possible to the beginning of treatment and never more than 14 days before the beginning of the treatment.

The same method of assessment and the same technique should be used to characterize each identified and reported lesion at baseline and during follow-up.

Response/Progression Categories

Based on discussions earlier, we propose the following definitions for categories of response/progression.

Complete Response

For patients with *enhancing tumors*, all of the following criteria must be met:

a. Complete disappearance of all enhancing measurable and nonmeasurable disease compared to baseline and sustained for at least 4 weeks. In the absence of a confirming scan 4 weeks later, this scan will be considered stable disease.
b. Stable or improved T2/FLAIR changes associated with underlying tumor.
c. No new lesions.
d. All measurable and nonmeasurable lesions must be assessed using the same techniques as baseline.
e. Patients must be on no corticosteroids or physiologic replacement doses only.
f. Stable or improved clinically, for clinical signs and symptoms present at baseline and recorded to be disease related.

For patients with *nonenhancing tumors*, criteria c-f apply and criteria a and b are replaced with the following: complete disappearance of all tumor associated T2/FLAIR changes sustained for at least 4 weeks.

Patients with residual nonmeasurable disease cannot have a complete response. The best response possible for such patients is stable disease.

Partial Response

For patients with *enhancing tumors,* all of the following criteria must be met:

a. Greater than or equal to 50% decrease compared to baseline in the sum of products of perpendicular diameters of all measurable lesions sustained for at least 4 weeks. In the absence of a confirming scan 4 weeks later, this scan will be considered only stable disease.
b. Stable or decreased tumor associated T2/FLAIR changes.
c. No progression of nonmeasurable disease.
d. All measurable and nonmeasurable lesions must be assessed using the same techniques as baseline.
e. The steroid dose at the time of the scan evaluation should be no greater than the dose at time of baseline scan.
f. Stable or improved, for clinical signs and symptoms present at baseline and recorded to be disease related clinically.

For patients with nonenhancing tumors, criteria c–f apply and criteria a and b are replaced with the following: greater than or equal to 50% decrease in the product of the perpendicular diameters of tumor associated T2/FLAIR changes sustained for at least 4 weeks.

Patients with nonmeasurable disease cannot have a partial response. The best response possible for such patients is stable disease.

Progressive Disease

Of note, confirmation of progression is recommended as described in the preceding section for patients who are not experiencing significant clinical decline and who are within 6 months of immunotherapy initiation. In this situation, confirmation of progression is based on comparison with the scan initially demonstrating progression and if progression is confirmed, the date of actual progression should be back dated to the date progressive disease was initially demonstrated. For patients who are more than 6 months from initiation of immunotherapy, progression should be determined based on comparison to the overall best response.

For patients with *enhancing tumors,* any one of the below listed criteria must be met:

a. >25% increase in sum of the products of perpendicular diameters of measurable lesions on stable or increasing doses of corticosteroids.
b. Any new measurable enhancing or nonenhancing lesion.
c. Clear clinical deterioration not attributable to other causes apart from the tumor (e.g., seizures, medication side effects, complications of

therapy, cerebrovascular events, infection). The definition of clinical deterioration is left to the discretion of the treating physician.

d. Failure to return for evaluation due to death or deteriorating condition.

For patients with *nonenhancing tumors*, criteria b–d apply but criterion a is replaced with the following: greater than or equal to 25% increase in the product of the perpendicular diameters of tumor associated T2/FLAIR changes not attributable to radiation effect or comorbid event on stable or increasing doses of corticosteroids.

Stable Disease

All of the following criteria must be met:

a. Does not qualify for CR, PR, or progression.
b. All measurable and nonmeasurable sites must be assessed using the same techniques as baseline.
c. Stable clinically.

In the event that corticosteroid dosing is increased for new symptoms and signs without confirmation of disease progression on neuroimaging, and subsequent follow-up imaging shows that this increase in corticosteroids was required because of disease progression, the last scan considered to show stable disease will be the scan obtained when the corticosteroid dose was equivalent to the baseline dose (Table 7.2).

Unknown Response Status

Progressive disease has not been documented and one or more measurable or nonmeasurable lesions have not been assessed (Table 7.3).

Evaluation of Best Response

The best overall response is the maximum response recorded from the start of treatment until disease progression (taking as reference for progressive disease the smallest measurements recorded since treatment started). If a response recorded at one scheduled MRI does not persist at the next regular scheduled MRI, the response will still be recorded based on the prior scan, but will be designated as a nonsustained response. If the response is sustained, i.e., still present on the subsequent MRI at least 4 weeks later, it will be recorded as a sustained response, lasting until the time of tumor progression. Patients without measurable disease may only achieve SD or PD as their best "response."

Corticosteroids

The iRANO criteria incorporate changes in systemic corticosteroids in the assessment of response as recommended by the RANO criteria.[21]

TABLE 7.2 Summary of the iRANO Response Criteria for *Enhancing Tumors*

	CR[a]	PR[a]	SD	PD[b]
T1-Gd+	None	≥50% decrease	<50% decrease to <25% increase	≥25% increase*
T2/FLAIR	Stable or improved	Stable or improved	Stable or improved	NA
New enhancing or nonenhancing measurable lesion	None	None	None	Present*
Corticosteroids	None	Stable or decrease	Stable or decrease	Stable or increasing
Clinical status	Stable or improved	Stable or improved	Stable or improved	Significantly worse*
Requirement for response	All	All	All	Any*

CR, complete response; *FLAIR*, fluid-attenuated inversion recovery; *iRANO*, immunotherapy response assessment in neuro-oncology; *NA*, not applicable; *PD*, progressive disease; *PR*, partial response; *SD*, stable disease.
*Progression occurs when any of the criteria with * is present.
Percent changes refer to largest bidimensional product measurements.
Increase in corticosteroids alone will not be taken into account in determining progression in the absence of persistent clinical deterioration.
[a] *CR or PR are based on comparison to baseline scans and must be sustained for at least 4 weeks.*
[b] *PD defined following confirmation scan for patients without significant clinical decline if less than 6 months following initiation of immunotherapy.*

TABLE 7.3 Summary of the iRANO Response Criteria for *Nonenhancing Tumors*

	CR[a]	PR[a]	SD	PD[b]
T2/FLAIR	None	≥50% decrease	<50% decrease to <25% increase	≥25% increase*
New enhancing or nonenhancing measurable lesion	None	None	None	Present*
Corticosteroids	None	Stable or decrease	Stable or decrease	Stable or increasing
Clinical status	Stable or improved	Stable or improved	Stable or improved	Significantly worse*
Requirement for response	All	All	All	Any*

CR, complete response; *FLAIR*, fluid-attenuated inversion recovery; *iRANO*, immunotherapy response assessment in neuro-oncology; *NA*, not applicable; *PD*, progressive disease; *PR*, partial response; *SD*, stable disease.
*Progression occurs when any of the criteria with * is present.
Percent changes refer to largest bidimensional product measurements.
Radiologic confirmation of progression is permitted as described below.
Increase in corticosteroids alone will not be taken into account in determining progression in the absence of persistent clinical deterioration.
[a] *CR or PR are based on comparison to baseline scans and must be sustained for at least 4 weeks.*
[b] *PD defined following confirmation scan for patients without significant clinical decline if less than 6 months following initiation of immunotherapy.*

Specifically, classifications of CR, PR, or SD cannot be assigned if patients require increased corticosteroids within 2 weeks of MRI assessment (relative to the dose taken at the time of the prior assessment). Such assessments should be classified as nonevaluable at that time point. Conversely, PD cannot be assigned if patients have worsened findings associated with a decrease in corticosteroid dosing within 2 weeks of MRI assessment (relative to the dose taken at the time of the prior assessment). Again, such assessments should be classified as nonevaluable at that time point.

ADVANCED IMAGING ASSESSMENTS

Advanced Imaging Techniques in Immunotherapy

Although standard imaging methods are currently limited in their ability to distinguish treatment-related changes from progressive tumor following immunotherapy, advanced imaging techniques based on physiological or metabolic properties of tissues can potentially provide more insightful information reflecting the status of tumor. Following initiation of immunotherapy, effective treatment is expected to result in a decrease in tumor burden while generating necrotic tissue with physiologic and metabolic characteristics different from those of viable tumor cells. This outcome is expected to be analogous to tissue necrosis induced by cytotoxic therapy, although the mechanism for tumor cell death during immunotherapy is not entirely clear and may be different. Although findings from studies of pseudoprogression in the setting of cytotoxic therapy can provide guidance for how advanced imaging techniques differentiate necrosis from tumor progression during immunotherapy, the diagnostic thresholds for imaging parameters may be different. Furthermore, inflammatory reactions associated with immune therapy such as hyperemia or immune cell infiltration can have overlapping range of measurements on one or more imaging parameters, thereby lowering diagnostic accuracy. Finally, imaging of patients with delayed treatment response will initially show findings of progressive tumor.

Several advanced MRI techniques have been developed for imaging of brain tumors, including perfusion-weighted imaging (PWI), diffusion-weighted imaging (DWI), MR spectroscopy (MRS), and positron emission tomography (PET). The following sections will present the basic concepts of these imaging techniques as well as up-to-date evidence supporting their roles in evaluating pseudoprogression.

Magnetic Resonance Perfusion-Weighted Imaging

High-grade gliomas recruit greater amount of blood flow compared to normal brain tissue due to angiogenesis. In contrast, nonviable, necrotic tissue resulting from treatment is not expected to derive blood flow, providing a basis for distinguishing such necrotic tissue from recurrent tumor using imaging methods that can quantify tissue blood flow, i.e., PWI. The most widely available PWI technique in clinical use is dynamic susceptibility contrast (DSC) MRI which depends on signal intensity change during a first-pass bolus of paramagnetic contrast agents.[49,50] The DSC-MRI technique has been increasingly incorporated into standard imaging protocols at brain tumor centers due to its short imaging acquisition time and availability of postprocessing software packages. Several hemodynamic parameters can be calculated based on the time–intensity curve derived from DSC-MRI, including relative cerebral blood volume (rCBV), relative peak height (rPH), and percentage of signal intensity recovery (PSR).[51] When applied to distinguish pseudoprogression from tumor progression in GBM following chemoradiation therapy, rPH and rCBV have been shown to be greater in progressive/recurrent tumor, while PSR was significantly lower.[52] When combined with conventional imaging sequences in this treatment setting, DSC-MRI can provide an accuracy of about 90%.[53] From the point of view of clinical utility, there is evidence that MR-PWI aids decision making during standard chemoradiation treatment of GBM.[54]

Dynamic contrast enhanced (DCE-MRI) perfusion technique measures vascular permeability by quantifying movement of contrast agents crossing the blood–brain barrier (BBB) using pharmacokinetic parameters, including Ktrans (transfer coefficient between the intra- and extravascular spaces), Ve (fractional volume of extravascular, extracellular space), and Kep (transfer constant from the extracellular, extravascular space into the plasma).[55–57] The ability to quantify contrast permeability allows more accurate calculation of cerebral blood volume compared to DSC-MRI. A small number of retrospective series have reported DCE-MRI being accurate in distinguishing pseudoprogression from progression following temozolomide chemoradiation treatment.[58,59] However, when the DSC-MRI and DCE-MRI techniques were directly compared in a retrospective analysis of 169 patients, there was no significant difference in the degree of improvement for diagnostic accuracy.[53]

To date there are only a small number of studies that examined the diagnostic value of PWI in immunotherapy of GBM. In a small retrospective analysis of patients with recurrent GBM treated with dendritic cell immune therapy, the maximum rCBV ratios in the contrast-enhancing area could differentiate immune therapy-induced inflammatory response from recurrent tumor.[60] Based on a preclinical rat model of GBM treated with natural killer cellular immunotherapy targeting the neural/glial antigen 2 (NG2) proteoglycan, fractional volume of extravascular, extracellular space (Ve) was associated with treatment response.[61] These preliminary results suggest that changes related to specific forms of immunotherapy may be sufficiently different from tumor tissue, yet these results need further validation.

There are a number of technical challenges that need to be considered when performing and interpreting studies with MR-PWI. Susceptibility artifacts can attenuate imaging signal and are commonly observed near the bone or air interfaces, as well as in the presence of blood products or surgical hardware. Furthermore, there can be variability among different imaging acquisition parameters and postprocessing methods that may impact the diagnostic accuracy of this technique. For example, calculation of rCBV can be either over or underestimated due to BBB disruption resulting in contrast leakage to interstitial space,[62] requiring additional adjustment to imaging acquisition and postprocessing.[63]

Magnetic Resonance Diffusion Weighted Imaging

MR diffusion imaging (DWI-MRI) characterizes tissue by probing motion characteristics of water molecules under influence of surrounding microenvironment. This noninvasive imaging technique is available on most clinical MRI scanners and can help differentiate brain tumor subtypes and tumor grades when combined with conventional MRI sequences.[64–68] The most widely used quantitative parameter in DWI-MRI is apparent diffusion coefficient (ADC), which is lower when water movement is restricted within highly cellular tumors.[67,69,70] Among patients with glioma following chemoradiation treatment, progressive tumors demonstrate lower ADC values as compared to normal brain tissues.[71–73] When this technique was applied to patients with recurrent GBM treated with dendritic cell immune therapy, minimum ADC values in the contrast-enhancing area could differentiate immune therapy-induced inflammatory response from recurrent GBM tumor.[60]

Although DWI is sensitive to treatment-induced alterations, it can also be affected by other brain pathologies including infarction, infection, inflammation, and hemorrhage. To distinguish these conditions from brain tumors or necrosis, DWI is usually interpreted alongside conventional MRI sequences. Similarly, to derive quantitative markers from this imaging technique, lesion of interests often need to be selected or outlined by imaging experts. Finally, quantitative analysis of DWI parameters including calculation of ADC values can be affected by variations in MRI equipment and acquisition parameters, and development and validation of diagnostic markers based on this technique will require standardization of imaging protocols.[74]

Magnetic Resonance Spectroscopy

MRS can simultaneously measure concentrations of several metabolites within both tumor and brain tissues, allowing for diagnosis, grading, and posttreatment assessment of brain.[75] For high-grade gliomas, several metabolites including N-acetyl aspartate (NAA), choline, creatine, and lactate exhibit different concentrations when comparing recurrent tumor from treatment-related necrosis.[76–79] Based on a recent meta-analysis, the diagnostic accuracy in differentiating recurrent glioma from radiation necrosis using choline to NAA ratio has a sensitivity and specificity of 0.88 and 0.86, respectively.[80]

Inflammatory lesions, based on characterization of patients with active demyelination disease, showed an elevated choline to creatine ratio as well as normal or reduced NAA to creatine ratio.[81,82] These findings suggest that treatment-induced response from immunotherapy may overlap with MRS finding of recurrent tumor, although MRS has not been tested in the setting of immunotherapy. Nonetheless, treatment-related MRS changes may be sufficiently different from those of tumor tissues to allow differentiation by this technique. One strategy is to identify tumor or inflammation specific metabolic markers. For example, MRS has been recently applied to detect gliomas carrying isocitrate dehydrogenase mutations using elevated level of 2-hydroxyglutarate within tumor which is normally not present in healthy brain tissue.[83,84] Simultaneously assessing multiple metabolic peaks to generate a combined model is another strategy that has been applied to diagnosing gliomas,[85,86] and this approach may enhance the accuracy of differentiating immune response from tumor.

Comparing to conventional MR sequences, MRS has a lower spatial resolution that limits its ability to assess small lesions, particularly ones with lesion volume less than 1.5 cm^3. Furthermore, this MR technique requires experienced operators to define region-of-interests during imaging acquisition, making it challenging to implement clinically with high reproducibility. Quantitative measurements of metabolic concentrations can also be affected by variations in MR equipment, pulse sequences, and data postprocessing methods.

Positron Emission Tomography

PET is a functional imaging technique that measures active cellular uptake of 18F-fluorodeoxyglucose (18F-FDG), a radiolabeled glucose analogue.[87] Due to greater level of uptake in many cancer cells compared to normal brain tissue, this technique is wildly used in clinical cancer imaging for staging and treatment response assessment.[88] For gliomas, 18F-FDG uptake on PET is associated with tumor grade[89,90] and patient survival.[91–97] Although FDG-PET imaging is currently widely available in most cancer treatment centers, the diagnostic accuracy in distinguishing radiation necrosis from tumor progression in gliomas is only modest.[98–102]

There are a number of challenges using FDG-PET limiting the use of this technique in diagnosing tumor following treatment. First, normal brain tissue typically exhibits high metabolism, making measurement of FDG uptake within lesions near brain cortices difficult. This is further limited by the relative lower resolution of PET imaging, currently at about 5 mm. Finally, treatment induced tumor necrosis can exhibit inflammatory response that results in increased glucose metabolism,[103] making FDG uptake measurement less specific for progressive tumor. In the setting of immunotherapy, nontumor related FDG uptake on PET imaging is expected to be present, as evident on prior studies of immunotherapy for systemic solid tumors.[104,105]

A number of other radiotracers with greater tumor-to-background uptake have been investigated. Malignant brain tumors have enhanced proliferative activities that manifest as increased transportation of amino acids.[106–109] In contrast, normal brain has a relative low level of amino acid uptake, thus a high lesion-to-background ratio can be achieved using amino acid based radiotracers. L-methyl-11C-methionine has greater accuracy in detecting tumor progression following cytotoxic therapy when compared to FDG-PET.[110–112] The clinical application of this radiotracer is limited by the relatively short half-life of 11C requiring an on-site cyclotron. Other amino acid radiotracers with longer half-life have been developed, including 3′-fluoro-3′-deoxy-L-thymidine (18F-FLT), O-2-18F-fluoroethyl-L-tyrosine (18F-FET), and 3,4-dihydroxy-6-[18F]-fluoro-L-phenylalanine. Preliminary evaluation of these radiotracers has shown their prognostic value as early markers of treatment response and survival.[113–116] The diagnostic accuracy of differentiating tumor from treatment-induced necrosis can be affected by tracer accumulation due to BBB leakage that can mimic tracer uptake in growing tumors. In a study using FET PET to monitor the effects of intracavitary radioimmunotherapy in patients with malignant glioma, increasing FET uptake around the tumor cavity can be observed up to 18 months after treatment, although recurrent tumors exhibited a higher degree of FET uptake.[117] Furthermore, kinetic modeling of tracer accumulation allows differentiation of passive tracer accumulation in necrosis from active tracer uptake in tumors,[118,119] and thus may improve the ability of PET imaging techniques to diagnose pseudoprogression.

Detection of Temporal Alternation of Tumor Tissues by Advanced Imaging Techniques

High-grade gliomas are characterized by their high genetic heterogeneity that can result in markedly variable imaging appearance within individual tumors. Comparison with pretreatment imaging can be helpful to account for intertumor variations when defining a diagnostic threshold for progressive tumor versus treatment effect. For example, analysis of PWI comparing post- and pretreatment studies for glioma patients has been shown to enhance diagnostic specificity.[120] A number of studies have characterized treatment-related alterations immediately postradiation compared to baseline studies in DWI and PWI imaging and demonstrate that such changes are associated with patient outcome.[121–124] In a study using a peptide vaccine in patients with high-grade glioma, serial measurements of maximum ^{11}C-methionine level within tumor versus background brain predicted treatment response and patient survival with greater accuracy as compared to enhanced MRI.[125] This type of approach requires incorporation of advanced imaging protocols at the time of pretreatment study. This strategy can be problematic if patients are referred for treatment following imaging studies at outside facilities. Furthermore, while DWI and PWI are increasingly performed as part of the routine imaging protocol for brain tumor patients treated at major cancer centers, it is important to confirm the uniformity of the imaging protocols used at different time points of assessment.

The use of temporal changes as a marker of tumor progression or treatment effect has also been evaluated by performing serial imaging following treatment until time of suspected necrosis. In 21 children with diffuse intrinsic pontine gliomas treated with peptide-based vaccination following radiation therapy, serial parametric response mapping of ADC, performed at multiple time points of therapy, was able to differentiate subjects with pseudoprogression from those without pseudoprogression.[126] Serial MRS can also detect metabolic changes associated with posttreatment necrosis, including a transient increase in choline following radiation therapy and decreased NAA concentrations over time.[76,78,127–129]

Spatial Heterogeneity of Tumor and Treatment Response

Differentiating tumor progression versus treatment-related necrosis, whether using imaging or histopathology methods, can be challenging when the two tissue types coexist. Imaging approaches based on whole tumor rather than regional or hot-spot type analysis can account for the potential spatial heterogeneity. For example,[130] calculated fractional tumor burden demonstrated that this parameter correlated with the relative histologic fraction of viable tumor and was also predictive of OS.[130] One commonly employed strategy to represent data from whole tumor is by summarizing imaging measurements using histograms. This type of approach has been applied to analysis of whole tumor rCBV and ADC to differentiate pseudoprogression from tumor progression.[131–133] For MRS, a multivoxel acquisition approach (chemical shift imaging) can collect spatially registered data within heterogeneous tissues and appears to improve diagnostic accuracy for detecting tumor.[127,134–138]

Clinical implementation of whole tumor evaluation remains challenging due to the need of delineating tumor volumes. To date, an increasing number of automated or semiautomated volume segmentation software programs are becoming available and some of these are currently being validated for clinical use.

CONCLUSION

Based on available preliminary data, the advanced imaging techniques discussed in this chapter hold great promise as tools to supplement conventional imaging in diagnosing pseudoprogression. However, the diagnostic accuracy of these imaging techniques in the setting of immunotherapy is largely unknown due to the current lack of validation studies. Furthermore, there are a number of challenges related to clinical implementation of these techniques as outlined above, requiring standardization of imaging protocols to maximize their reproducibility.

The iRANO criteria have been put forth for the evaluation of neuro-oncology patients undergoing immunotherapy. These recommendations integrate relevant aspects of the RANO and irRC criteria to provide consensus guidelines to aide clinical decision making in every day practice and within the context of clinical trials. Specifically the iRANO guidelines address the interpretation of initially progressive imaging findings among neuro-oncology patients with a goal of lessening the possibility of prematurely stopping potentially beneficial therapies while prioritizing the maintenance of optimal patient safety. The iRANO criteria apply to all neuro-oncology patient populations including those with primary and metastatic tumors and are meant to integrate into the existing RANO working group guidelines for patients with malignant glioma,[21] low-grade glioma,[22] and brain metastases.[24,139] It is fully anticipated that the original iRANO guidelines will benefit from future amendments including possible incorporation of advanced imaging techniques, once sufficient experience and expertise have accumulated.

References

1. Coley WB. The treatment of malignant tumors by repeated inoculations of erysipelas, with a report of ten original cases. *Am J Med Sci*. 1893;105:487–511.
2. Kantoff PW, Higano CS, Shore ND, Berger ER, Small EJ, Penson DF, et al. Sipuleucel-T immunotherapy for castration-resistant prostate cancer. *N Engl J Med*. 2010;363(5):411–422.
3. Le DT, Uram JN, Wang H, Bartlett BR, Kemberling H, Eyring AD, et al. PD-1 blockade in tumors with mismatch-repair deficiency. *N Engl J Med*. 2015;372(26):2509–2520.

4. Ansell SM, Lesokhin AM, Borrello I, Halwani A, Scott EC, Gutierrez M, et al. PD-1 blockade with nivolumab in relapsed or refractory Hodgkin's lymphoma. *N Engl J Med.* 2015;372:311–319.
5. McDermott DF, Drake CG, Sznol M, Choueiri TK, Powderly JD, Smith DC, et al. Survival, durable response, and long-term safety in patients with previously treated advanced renal cell carcinoma receiving nivolumab. *J Clin Oncol.* 2015;33(18):2013–2020.
6. Gettinger SN, Horn L, Gandhi L, Spigel DR, Antonia SJ, Rizvi NA, et al. Overall survival and long-term safety of nivolumab (anti-programmed death 1 antibody, BMS-936558, ONO-4538) in patients with previously treated advanced non-small-cell lung cancer. *J Clin Oncol.* 2015;33(18):2004–2012.
7. Garon EB, Rizvi NA, Hui R, Leighl N, Balmanoukian AS, Eder JP, et al. Pembrolizumab for the treatment of non-small-cell lung cancer. *N Engl J Med.* 2015;372:2018–2028.
8. Hodi FS, O'Day SJ, McDermott DF, Weber RW, Sosman JA, Haanen JB, et al. Improved survival with ipilimumab in patients with metastatic melanoma. *N Engl J Med.* 2010;363(8):711–723.
9. Robert C, Ribas A, Wolchok JD, Hodi FS, Hamid O, Kefford R, et al. Anti-programmed-death-receptor-1 treatment with pembrolizumab in ipilimumab-refractory advanced melanoma: a randomised dose-comparison cohort of a phase 1 trial. *Lancet.* 2014;384(9948):1109–1117.
10. Rizvi NA, Mazieres J, Planchard D, Stinchcombe TE, Dy GK, Antonia SJ, et al. Activity and safety of nivolumab, an anti-PD-1 immune checkpoint inhibitor, for patients with advanced, refractory squamous non-small-cell lung cancer (CheckMate 063): a phase 2, single-arm trial. *Lancet Oncol.* 2015;16(3):257–265.
11. Maude SL, Frey N, Shaw PA, Aplenc R, Barrett DM, Bunin NJ, et al. Chimeric antigen receptor T cells for sustained remissions in leukemia. *N Engl J Med.* 2014;371(16):1507–1517.
12. Grupp SA, Kalos M, Barrett D, Aplenc R, Porter DL, Rheingold SR, et al. Chimeric antigen receptor-modified T cells for acute lymphoid leukemia. *N Engl J Med.* 2013;368(16):1509–1518.
13. Reardon DA, Freeman G, Wu C, Chiocca EA, Wucherpfennig KW, Wen PY, et al. Immunotherapy advances for glioblastoma. *Neuro Oncol.* 2014;16(11):1441–1458.
14. Wainwright DA, Nigam P, Thaci B, Dey M, Lesniak MS. Recent developments on immunotherapy for brain cancer. *Expert Opin Emerg Drugs.* 2012;17(2):181–202.
15. Jackson CM, Lim M, Drake CG. Immunotherapy for brain cancer: recent progress and future promise. *Clin Cancer Res.* 2014;20(14):3651–3659.
16. Miller AB, Hoogstraten B, Staquet M, Winkler A. Reporting results of cancer treatment. *Cancer.* 1981;47(1):207–214.
17. Therasse P, Arbuck SG, Eisenhauer EA, Wanders J, Kaplan RS, Rubinstein L, et al. New guidelines to evaluate the response to treatment in solid tumors. European Organization for Research and Treatment of Cancer, National Cancer Institute of the United States, National Cancer Institute of Canada. *J Natl Cancer Inst.* 2000;92(3):205–216.
18. Macdonald DR, Cascino TL, Schold Jr SC, Cairncross JG. Response criteria for phase II studies of supratentorial malignant glioma. *J Clin Oncol.* 1990;8(7):1277–1280.
19. Brandsma D, Stalpers L, Taal W, Sminia P, van den Bent MJ. Clinical features, mechanisms, and management of pseudoprogression in malignant gliomas. *Lancet Oncol.* 2008;9(5):453–461.
20. Chinot OL, Macdonald DR, Abrey LE, Zahlmann G, Kerloeguen Y, Cloughesy TF. Response assessment criteria for glioblastoma: practical adaptation and implementation in clinical trials of antiangiogenic therapy. *Curr Neurol Neurosci Rep.* 2013;13(5):347.
21. Wen PY, Macdonald DR, Reardon DA, Cloughesy TF, Sorensen AG, Galanis E, et al. Updated response assessment criteria for high-grade gliomas: response assessment in neuro-oncology working group. *J Clin Oncol.* 2010;28(11):1963–1972.

22. van den Bent MJ, Wefel JS, Schiff D, Taphoorn MJ, Jaeckle K, Junck L, et al. Response assessment in neuro-oncology (a report of the RANO group): assessment of outcome in trials of diffuse low-grade gliomas. *Lancet Oncol.* 2011;12(6):583–593.
23. Lin NU, Lee EQ, Aoyama H, Barani IJ, Barboriak DP, Baumert BG, et al. Proposed response assessment criteria for brain metastases: response assessment in neuro-oncology (RANO) working group. *Lancet Oncol.* 2015;16(6):e270–e278.
24. Lin NU, Wefel JS, Lee EQ, Schiff D, van den Bent MJ, Soffietti R, et al. Challenges relating to solid tumour brain metastases in clinical trials, part 2: neurocognitive, neurological, and quality-of-life outcomes. A report from the RANO group. *Lancet Oncol.* 2013;14(10):e407–16.
25. Radbruch A, Fladt J, Kickingereder P, Wiestler B, Nowosielski M, Baumer P, et al. Pseudoprogression in patients with glioblastoma: clinical relevance despite low incidence. *Neuro Oncol.* 2015;17(1):151–159.
26. Wolchok JD, Hoos A, O'Day S, Weber JS, Hamid O, Lebbe C, et al. Guidelines for the evaluation of immune therapy activity in solid tumors: immune-related response criteria. *Clin Cancer Res.* 2009;15(23):7412–7420.
27. Hoos A, Eggermont AM, Janetzki S, Hodi FS, Ibrahim R, Anderson A, et al. Improved endpoints for cancer immunotherapy trials. *J Natl Cancer Inst.* 2010;102(18):1388–1397.
28. Hoos A. Evolution of end points for cancer immunotherapy trials. *Ann Oncol.* 2012;23(suppl 8). viii47–52.
29. Okada H, Pollack IF. Do we need novel radiologic response criteria for brain tumor immunotherapy? *Expert Rev Neurother.* 2011;11(5):619–622.
30. Topalian SL, Hodi FS, Brahmer JR, Gettinger SN, Smith DC, McDermott DF, et al. Safety, activity, and immune correlates of anti-PD-1 antibody in cancer. *N Engl J Med.* 2012;366(26):2443–2454.
31. Topalian SL, Sznol M, McDermott DF, Kluger HM, Carvajal RD, Sharfman WH, et al. Survival, durable tumor remission, and long-term safety in patients with advanced melanoma receiving nivolumab. *J Clin Oncol.* 2014;32(10):1020–1030.
32. Hamid O, Robert C, Daud A, Hodi FS, Hwu WJ, Kefford R, et al. Safety and tumor responses with lambrolizumab (anti-PD-1) in melanoma. *N Engl J Med.* 2013;369(2):134–144.
33. Wolchok JD, Kluger H, Callahan MK, Postow MA, Rizvi NA, Lesokhin AM, et al. Nivolumab plus ipilimumab in advanced melanoma. *N Engl J Med.* 2013;369(2):122–133.
34. Brahmer JR, Tykodi SS, Chow LQ, Hwu WJ, Topalian SL, Hwu P, et al. Safety and activity of anti-PD-L1 antibody in patients with advanced cancer. *N Engl J Med.* 2012;366(26):2455–2465.
35. Little RF, Pluda JM, Wyvill KM, Rodriguez-Chavez IR, Tosato G, Catanzaro AT, et al. Activity of subcutaneous interleukin-12 in AIDS-related Kaposi sarcoma. *Blood.* 2006;107(12):4650–4657.
36. van Baren N, Bonnet MC, Dreno B, Khammari A, Dorval T, Piperno-Neumann S, et al. Tumoral and immunologic response after vaccination of melanoma patients with an ALVAC virus encoding MAGE antigens recognized by T cells. *J Clin Oncol.* 2005;23(35):9008–9021.
37. Kruit WH, van Ojik HH, Brichard VG, Escudier B, Dorval T, Dreno B, et al. Phase 1/2 study of subcutaneous and intradermal immunization with a recombinant MAGE-3 protein in patients with detectable metastatic melanoma. *Int J Cancer.* 2005;117(4):596–604.
38. Di Giacomo AM, Danielli R, Guidoboni M, Calabro L, Carlucci D, Miracco C, et al. Therapeutic efficacy of ipilimumab, an anti-CTLA-4 monoclonal antibody, in patients with metastatic melanoma unresponsive to prior systemic treatments: clinical and immunological evidence from three patient cases. *Cancer Immunol Immunother.* 2009;58(8):1297–1306.

39. Hodi FS, Lawrence D, Lezcano C, Wu X, Zhou J, Sasada T, et al. Bevacizumab plus ipilimumab in patients with metastatic melanoma. *Cancer Immunol Res*. 2014;2(7):632–642.
40. Okada H, Kalinski P, Ueda R, Hoji A, Kohanbash G, Donegan TE, et al. Induction of CD8+ T-cell responses against novel glioma-associated antigen peptides and clinical activity by vaccinations with {alpha}-type 1 polarized dendritic cells and polyinosinic-polycytidylic acid stabilized by lysine and carboxymethylcellulose in patients with recurrent malignant glioma. *J Clin Oncol*. 2011;29(3):330–336.
41. Kirkwood JM, Lorigan P, Hersey P, Hauschild A, Robert C, McDermott D, et al. Phase II trial of tremelimumab (CP-675,206) in patients with advanced refractory or relapsed melanoma. *Clin Cancer Res*. 2010;16(3):1042–1048.
42. Pollack IF, Jakacki RI, Butterfield LH, Hamilton RL, Panigrahy A, Potter DM, et al. Antigen-specific immune responses and clinical outcome after vaccination with glioma-associated antigen peptides and polyinosinic-polycytidylic acid stabilized by lysine and carboxymethylcellulose in children with newly diagnosed malignant brainstem and nonbrainstem gliomas. *J Clin Oncol*. 2014;32(19):2050–2058.
43. Okada H, Kalinski P, Ueda R, Hoji A, Kohanbash G, Donegan TE, et al. Induction of CD8+ T-cell responses against novel glioma-associated antigen peptides and clinical activity by vaccinations with {alpha}-type 1 polarized dendritic cells and polyinosinic-polycytidylic acid stabilized by lysine and carboxymethylcellulose in patients with recurrent malignant glioma. *J Clin Oncol*. 2011;29(3):330–336.
44. Sampson JH, Heimberger AB, Archer GE, Aldape KD, Friedman AH, Friedman HS, et al. Immunologic escape after prolonged progression-free survival with epidermal growth factor receptor variant III peptide vaccination in patients with newly diagnosed glioblastoma. *J Clin Oncol*. 2010;28(31):4722–4729.
45. Chiocca EA, Aguilar LK, Bell SD, Kaur B, Hardcastle J, Cavaliere R, et al. Phase IB study of gene-mediated cytotoxic immunotherapy adjuvant to up-front surgery and intensive timing radiation for malignant glioma. *J Clin Oncol*. 2011;29(27):3611–3619.
46. Hoos A, Parmiani G, Hege K, Sznol M, Loibner H, Eggermont A, et al. A clinical development paradigm for cancer vaccines and related biologics. *J Immunother*. 2007;30(1):1–15.
47. Brandes AA, Tosoni A, Spagnolli F, Frezza G, Leonardi M, Calbucci F, et al. Disease progression or pseudoprogression after concomitant radiochemotherapy treatment: pitfalls in neurooncology. *Neuro Oncol*. 2008;10(3):361–367.
48. Okada H, Weller M, Huang R, Finocchiaro G, Gilbert MR, Wick W, et al. Immunotherapy response assessment in neuro-oncology (iRANO): a report of the RANO working group. *Lancet Oncol*. 2015.
49. Rosen BR, Belliveau JW, Vevea JM, Brady TJ. Perfusion imaging with NMR contrast agents. *Magn Reson Med*. 1990;14(2):249–265.
50. Villringer A, Rosen BR, Belliveau JW, Ackerman JL, Lauffer RB, Buxton RB, et al. Dynamic imaging with lanthanide chelates in normal brain: contrast due to magnetic susceptibility effects. *Magn Reson Med*. 1988;6(2):164–174.
51. Rosen BR, Belliveau JW, Buchbinder BR, McKinstry RC, Porkka LM, Kennedy DN, et al. Contrast agents and cerebral hemodynamics. *Magn Reson Med*. 1991;19(2):285–292.
52. Barajas Jr RF, Chang JS, Segal MR, Parsa AT, McDermott MW, Berger MS, et al. Differentiation of recurrent glioblastoma multiforme from radiation necrosis after external beam radiation therapy with dynamic susceptibility-weighted contrast-enhanced perfusion MR imaging. *Radiology*. 2009;253(2):486–496.
53. Kim HS, Goh MJ, Kim N, Choi CG, Kim SJ, Kim JH. Which combination of MR imaging modalities is best for predicting recurrent glioblastoma? Study of diagnostic accuracy and reproducibility. *Radiology*. 2014;273(3):831–843.
54. Geer CP, Simonds J, Anvery A, Chen MY, Burdette JH, Zapadka ME, et al. Does MR perfusion imaging impact management decisions for patients with brain tumors? A prospective study. *Am J Neuroradiol*. 2012;33(3):556–562.

55. Tofts PS. Modeling tracer kinetics in dynamic Gd-DTPA MR imaging. *J Magn Reson Imaging*. 1997;7(1):91–101.
56. Tofts PS, Kermode AG. Measurement of the blood–brain barrier permeability and leakage space using dynamic MR imaging. 1. Fundamental concepts. *Magn Reson Med*. 1991;17(2):357–367.
57. Tofts PS, Brix G, Buckley DL, Evelhoch JL, Henderson E, Knopp MV, et al. Estimating kinetic parameters from dynamic contrast-enhanced T(1)-weighted MRI of a diffusable tracer: standardized quantities and symbols. *J Magn Reson Imaging*. 1999;10(3):223–232.
58. Larsen VA, Simonsen HJ, Law I, Larsson HB, Hansen AE. Evaluation of dynamic contrast-enhanced T1-weighted perfusion MRI in the differentiation of tumor recurrence from radiation necrosis. *Neuroradiology*. 2013;55(3):361–369.
59. Bisdas S, Naegele T, Ritz R, Dimostheni A, Pfannenberg C, Reimold M, et al. Distinguishing recurrent high-grade gliomas from radiation injury: a pilot study using dynamic contrast-enhanced MR imaging. *Acad Radiol*. 2011;18(5):575–583.
60. Vrabec M, Van Cauter S, Himmelreich U, Van Gool SW, Sunaert S, De Vleeschouwer S, et al. MR perfusion and diffusion imaging in the follow-up of recurrent glioblastoma treated with dendritic cell immunotherapy: a pilot study. *Neuroradiology*. 2011;53(10):721–731.
61. Rygh CB, Wang J, Thuen M, Gras Navarro A, Huuse EM, Thorsen F, et al. Dynamic contrast enhanced MRI detects early response to adoptive NK cellular immunotherapy targeting the NG2 proteoglycan in a rat model of glioblastoma. *PLoS One*. 2014;9(9):e108414.
62. Paulson ES, Schmainda KM. Comparison of dynamic susceptibility-weighted contrast-enhanced MR methods: recommendations for measuring relative cerebral blood volume in brain tumors. *Radiology*. 2008;249(2):601–613.
63. Huang RY, Neagu MR, Reardon DA, Wen PY. Pitfalls in the neuroimaging of glioblastoma in the era of antiangiogenic and immuno/targeted therapy – detecting illusive disease, defining response. *Front Neurol*. 2015;6:33.
64. Yamasaki F, Kurisu K, Satoh K, Arita K, Sugiyama K, Ohtaki M, et al. Apparent diffusion coefficient of human brain tumors at MR imaging. *Radiology*. 2005;235(3):985–991.
65. Guo AC, Cummings TJ, Dash RC, Provenzale JM. Lymphomas and high-grade astrocytomas: comparison of water diffusibility and histologic characteristics. *Radiology*. 2002;224(1):177–183.
66. Dorenbeck U, Grunwald IQ, Schlaier J, Feuerbach S. Diffusion-weighted imaging with calculated apparent diffusion coefficient of enhancing extra-axial masses. *J Neuroimaging*. 2005;15(4):341–347.
67. Sugahara T, Korogi Y, Kochi M, Ikushima I, Shigematu Y, Hirai T, et al. Usefulness of diffusion-weighted MRI with echo-planar technique in the evaluation of cellularity in gliomas. *J Magn Reson Imaging*. 1999;9(1):53–60.
68. Murakami R, Hirai T, Sugahara T, Fukuoka H, Toya R, Nishimura S, et al. Grading astrocytic tumors by using apparent diffusion coefficient parameters: superiority of a one- versus two-parameter pilot method. *Radiology*. 2009;251(3):838–845.
69. Ellingson BM, Malkin MG, Rand SD, Connelly JM, Quinsey C, LaViolette PS, et al. Validation of functional diffusion maps (fDMs) as a biomarker for human glioma cellularity. *J Magn Reson Imaging*. 2010;31(3):538–548.
70. Hayashida Y, Hirai T, Morishita S, Kitajima M, Murakami R, Korogi Y, et al. Diffusion-weighted imaging of metastatic brain tumors: comparison with histologic type and tumor cellularity. *Am J Neuroradiol*. 2006;27(7):1419–1425.
71. Hein PA, Eskey CJ, Dunn JF, Hug EB. Diffusion-weighted imaging in the follow-up of treated high-grade gliomas: tumor recurrence versus radiation injury. *Am J Neuroradiol*. 2004;25(2):201–209.

72. Asao C, Korogi Y, Kitajima M, Hirai T, Baba Y, Makino K, et al. Diffusion-weighted imaging of radiation-induced brain injury for differentiation from tumor recurrence. *Am J Neuroradiol*. 2005;26(6):1455–1460.
73. Sundgren PC, Fan X, Weybright P, Welsh RC, Carlos RC, Petrou M, et al. Differentiation of recurrent brain tumor versus radiation injury using diffusion tensor imaging in patients with new contrast-enhancing lesions. *Magn Reson Imaging*. 2006;24(9):1131–1142.
74. Padhani AR, Liu G, Koh DM, Chenevert TL, Thoeny HC, Takahara T, et al. Diffusion-weighted magnetic resonance imaging as a cancer biomarker: consensus and recommendations. *Neoplasia*. 2009;11(2):102–125.
75. Oz G, Alger JR, Barker PB, Bartha R, Bizzi A, Boesch C, et al. Clinical proton MR spectroscopy in central nervous system disorders. *Radiology*. 2014;270(3):658–679.
76. Schlemmer HP, Bachert P, Herfarth KK, Zuna I, Debus J, van Kaick G. Proton MR spectroscopic evaluation of suspicious brain lesions after stereotactic radiotherapy. *Am J Neuroradiol*. 2001;22(7):1316–1324.
77. Dowling C, Bollen AW, Noworolski SM, McDermott MW, Barbaro NM, Day MR, et al. Preoperative proton MR spectroscopic imaging of brain tumors: correlation with histopathologic analysis of resection specimens. *Am J Neuroradiol*. 2001;22(4):604–612.
78. Rabinov JD, Lee PL, Barker FG, Louis DN, Harsh GR, Cosgrove GR, et al. In vivo 3-T MR spectroscopy in the distinction of recurrent glioma versus radiation effects: initial experience. *Radiology*. 2002;225(3):871–879.
79. Pratt WB, Ruddon RW, Ensminger WD, et al. *The anticancer drugs*. Oxford University Press; 1994:3–16.
80. Zhang H, Ma L, Wang Q, Zheng X, Wu C, Xu BN. Role of magnetic resonance spectroscopy for the differentiation of recurrent glioma from radiation necrosis: a systematic review and meta-analysis. *Eur J Radiol*. 2014;83(12):2181–2189.
81. Richards TL. Proton MR spectroscopy in multiple sclerosis: value in establishing diagnosis, monitoring progression, and evaluating therapy. *Am J Roentgenol*. 1991;157(5):1073–1078.
82. Chang L, Munsaka SM, Kraft-Terry S, Ernst T. Magnetic resonance spectroscopy to assess neuroinflammation and neuropathic pain. *J Neuroimmune Pharmacol*. 2013;8(3):576–593.
83. Andronesi OC, Rapalino O, Gerstner E, Chi A, Batchelor TT, Cahill DP, et al. Detection of oncogenic IDH1 mutations using magnetic resonance spectroscopy of 2-hydroxyglutarate. *J Clin Invest*. 2013;123(9):3659–3663.
84. Choi C, Ganji SK, DeBerardinis RJ, Hatanpaa KJ, Rakheja D, Kovacs Z, et al. 2-hydroxyglutarate detection by magnetic resonance spectroscopy in IDH-mutated patients with gliomas. *Nat Med*. 2012;18(4):624–629.
85. Imani F, Boada FE, Lieberman FS, Davis DK, Mountz JM. Molecular and metabolic pattern classification for detection of brain glioma progression. *Eur J Radiol*. 2014;83(2):e100–e105.
86. Ranjith G, Parvathy R, Vikas V, Chandrasekharan K, Nair S. Machine learning methods for the classification of gliomas: initial results using features extracted from MR spectroscopy. *Neuroradiol J*. 2015;28(2):106–111.
87. Di Chiro G, DeLaPaz RL, Brooks RA, Sokoloff L, Kornblith PL, Smith BH, et al. Glucose utilization of cerebral gliomas measured by [^{18}F] fluorodeoxyglucose and positron emission tomography. *Neurology*. 1982;32(12):1323–1329.
88. Kelloff GJ, Hoffman JM, Johnson B, Scher HI, Siegel BA, Cheng EY, et al. Progress and promise of FDG-PET imaging for cancer patient management and oncologic drug development. *Clin Cancer Res*. 2005;11(8):2785–2808.
89. Kincaid PK, El-Saden SM, Park SH, Goy BW. Cerebral gangliogliomas: preoperative grading using FDG-PET and 201Tl-SPECT. *Am J Neuroradiol*. 1998;19(5):801–806.
90. Delbeke D, Meyerowitz C, Lapidus RL, Maciunas RJ, Jennings MT, Moots PL, et al. Optimal cutoff levels of F-18 fluorodeoxyglucose uptake in the differentiation of low-grade from high-grade brain tumors with PET. *Radiology*. 1995;195(1):47–52.

91. Di Chiro G. Positron emission tomography using [^{18}F] fluorodeoxyglucose in brain tumors. A powerful diagnostic and prognostic tool. *Invest Radiol.* 1987;22(5):360–371.
92. Hustinx R, Smith RJ, Benard F, Bhatnagar A, Alavi A. Can the standardized uptake value characterize primary brain tumors on FDG-PET? *Eur J Nucl Med.* 1999;26(11):1501–1509.
93. Kosaka N, Tsuchida T, Uematsu H, Kimura H, Okazawa H, Itoh H. ^{18}F-FDG PET of common enhancing malignant brain tumors. *Am J Roentgenol.* 2008;190(6):W365–W369.
94. De Witte O, Lefranc F, Levivier M, Salmon I, Brotchi J, Goldman S. FDG-PET as a prognostic factor in high-grade astrocytoma. *J Neurooncol.* 2000;49(2):157–163.
95. Pardo FS, Aronen HJ, Fitzek M, Kennedy DN, Efird J, Rosen BR, et al. Correlation of FDG-PET interpretation with survival in a cohort of glioma patients. *Anticancer Res.* 2004;24(4):2359–2365.
96. Tralins KS, Douglas JG, Stelzer KJ, Mankoff DA, Silbergeld DL, Rostomily RC, et al. Volumetric analysis of ^{18}F-FDG PET in glioblastoma multiforme: prognostic information and possible role in definition of target volumes in radiation dose escalation. *J Nucl Med.* 2002;43(12):1667–1673.
97. Spence AM, Muzi M, Graham MM, O'Sullivan F, Link JM, Lewellen TK, et al. 2-[^{18}F] Fluoro-2-deoxyglucose and glucose uptake in malignant gliomas before and after radiotherapy: correlation with outcome. *Clin Cancer Res.* 2002;8(4):971–979.
98. Kim YH, Oh SW, Lim YJ, Park CK, Lee SH, Kang KW, et al. Differentiating radiation necrosis from tumor recurrence in high-grade gliomas: assessing the efficacy of ^{18}F-FDG PET, ^{11}C-methionine PET and perfusion MRI. *Clin Neurol Neurosurg.* 2010;112(9):758–765.
99. Gomez-Rio M, Rodriguez-Fernandez A, Ramos-Font C, Lopez-Ramirez E, Llamas-Elvira JM. Diagnostic accuracy of 201Thallium-SPECT and ^{18}F-FDG-PET in the clinical assessment of glioma recurrence. *Eur J Nucl Med Mol Imaging.* 2008;35(5):966–975.
100. Ricci PE, Karis JP, Heiserman JE, Fram EK, Bice AN, Drayer BP. Differentiating recurrent tumor from radiation necrosis: time for re-evaluation of positron emission tomography? *Am J Neuroradiol.* 1998;19(3):407–413.
101. Di Chiro G, Oldfield E, Wright DC, De Michele D, Katz DA, Patronas NJ, et al. Cerebral necrosis after radiotherapy and/or intraarterial chemotherapy for brain tumors: PET and neuropathologic studies. *Am J Roentgenol.* 1988;150(1):189–197.
102. Valk PE, Budinger TF, Levin VA, Silver P, Gutin PH, Doyle WK. PET of malignant cerebral tumors after interstitial brachytherapy. Demonstration of metabolic activity and correlation with clinical outcome. *J Neurosurg.* 1988;69(6):830–838.
103. Kubota R, Yamada S, Kubota K, Ishiwata K, Tamahashi N, Ido T. Intratumoral distribution of fluorine-18-fluorodeoxyglucose in vivo: high accumulation in macrophages and granulation tissues studied by microautoradiography. *J Nucl Med.* 1992;33(11):1972–1980.
104. Engell-Noerregaard L, Hendel HW, Johannesen HH, Alslev L, Svane IM. FDG PET scans as evaluation of clinical response to dendritic cell vaccination in patients with malignant melanoma. *Cancer Immunol Immunother.* 2013;62(1):17–25.
105. Gilles R, de Geus-Oei LF, Mulders PF, Oyen WJ. Immunotherapy response evaluation with ^{18}F-FDG-PET in patients with advanced stage renal cell carcinoma. *World J Urol.* 2013;31(4):841–846.
106. Isselbacher KJ. Sugar and amino acid transport by cells in culture–differences between normal and malignant cells. *N Engl J Med.* 1972;286(17):929–933.
107. Busch H, Davis JR, Honig GR, Anderson DC, Nair PV, Nyhan WL. The uptake of a variety of amino acids into nuclear proteins of tumors and other tissues. *Cancer Res.* 1959;19:1030–1039.
108. Kato T, Shinoda J, Oka N, Miwa K, Nakayama N, Yano H, et al. Analysis of 11C-methionine uptake in low-grade gliomas and correlation with proliferative activity. *Am J Neuroradiol.* 2008;29(10):1867–1871.
109. Sato N, Suzuki M, Kuwata N, Kuroda K, Wada T, Beppu T, et al. Evaluation of the malignancy of glioma using 11C-methionine positron emission tomography and proliferating cell nuclear antigen staining. *Neurosurg Rev.* 1999;22(4):210–214.

110. Li DL, Xu YK, Wang QS, Wu HB, Li HS. ^{11}C-methionine and ^{18}F-fluorodeoxyglucose positron emission tomography/CT in the evaluation of patients with suspected primary and residual/recurrent gliomas. *Chin Med J*. 2012;125(1):91–96.
111. Van Laere K, Ceyssens S, Van Calenbergh F, de Groot T, Menten J, Flamen P, et al. Direct comparison of ^{18}F-FDG and ^{11}C-methionine PET in suspected recurrence of glioma: sensitivity, inter-observer variability and prognostic value. *Eur J Nucl Med Mol Imaging*. 2005;32(1):39–51.
112. Chung JK, Kim YK, Kim SK, Lee YJ, Paek S, Yeo JS, et al. Usefulness of ^{11}C-methionine PET in the evaluation of brain lesions that are hypo- or isometabolic on ^{18}F-FDG PET. *Eur J Nucl Med Mol Imaging*. 2002;29(2):176–182.
113. Chen W, Silverman DH, Delaloye S, Czernin J, Kamdar N, Pope W, et al. ^{18}F-FDOPA PET imaging of brain tumors: comparison study with ^{18}F-FDG PET and evaluation of diagnostic accuracy. *J Nucl Med*. 2006;47(6):904–911.
114. Oborski MJ, Demirci E, Laymon CM, Lieberman FS, Mountz JM. Assessment of early therapy response with ^{18}F-FLT PET in glioblastoma multiforme. *Clin Nucl Med*. 2014;39(10):e431–e432.
115. Galldiks N, Langen KJ, Holy R, Pinkawa M, Stoffels G, Nolte KW, et al. Assessment of treatment response in patients with glioblastoma using O-(2-^{18}F-fluoroethyl)-L-tyrosine PET in comparison to MRI. *J Nucl Med*. 2012;53(7):1048–1057.
116. Piroth MD, Pinkawa M, Holy R, Klotz J, Nussen S, Stoffels G, et al. Prognostic value of early [^{18}F]fluoroethyltyrosine positron emission tomography after radiochemotherapy in glioblastoma multiforme. *Int J Radiat Oncol Biol Phys*. 2011;80(1):176–184.
117. Popperl G, Gotz C, Rachinger W, Schnell O, Gildehaus FJ, Tonn JC, et al. Serial O-(2-[^{18}F]fluoroethyl)-L: -tyrosine PET for monitoring the effects of intracavitary radioimmunotherapy in patients with malignant glioma. *Eur J Nucl Med Mol Imaging*. 2006;33(7):792–800.
118. Muzi M, Spence AM, O'Sullivan F, Mankoff DA, Wells JM, Grierson JR, et al. Kinetic analysis of 3′-deoxy-3′-^{18}F-fluorothymidine in patients with gliomas. *J Nucl Med*. 2006;47(10):1612–1621.
119. Spence AM, Muzi M, Link JM, O'Sullivan F, Eary JF, Hoffman JM, et al. NCI-sponsored trial for the evaluation of safety and preliminary efficacy of 3′-deoxy-3′-[^{18}F]fluorothymidine (FLT) as a marker of proliferation in patients with recurrent gliomas: preliminary efficacy studies. *Mol Imaging Biol*. 2009;11(5):343–355.
120. Tsien C, Galban CJ, Chenevert TL, Johnson TD, Hamstra DA, Sundgren PC, et al. Parametric response map as an imaging biomarker to distinguish progression from pseudoprogression in high-grade glioma. *J Clin Oncol*. 2010;28(13):2293–2299.
121. Cao Y, Shen Z, Chenevert TL, Ewing JR. Estimate of vascular permeability and cerebral blood volume using Gd-DTPA contrast enhancement and dynamic T2*-weighted MRI. *J Magn Reson Imaging*. 2006;24(2):288–296.
122. Mangla R, Singh G, Ziegelitz D, Milano MT, Korones DN, Zhong J, et al. Changes in relative cerebral blood volume 1 month after radiation-temozolomide therapy can help predict overall survival in patients with glioblastoma. *Radiology*. 2010;256(2):575–584.
123. Chenevert TL, Stegman LD, Taylor JM, Robertson PL, Greenberg HS, Rehemtulla A, et al. Diffusion magnetic resonance imaging: an early surrogate marker of therapeutic efficacy in brain tumors. *J Natl Cancer Inst*. 2000;92(24):2029–2036.
124. Chenevert TL, McKeever PE, Ross BD. Monitoring early response of experimental brain tumors to therapy using diffusion magnetic resonance imaging. *Clin Cancer Res*. 1997;3(9):1457–1466.
125. Chiba Y, Kinoshita M, Okita Y, Tsuboi A, Isohashi K, Kagawa N, et al. Use of ^{11}C-methionine PET parametric response map for monitoring WT1 immunotherapy response in recurrent malignant glioma. *J Neurosurg*. 2012;116(4):835–842.

126. Ceschin R, Kurland BF, Abberbock SR, Ellingson BM, Okada H, Jakacki RI, et al. Parametric response mapping of apparent diffusion coefficient as an imaging biomarker to distinguish pseudoprogression from true tumor progression in peptide-based vaccine therapy for pediatric diffuse intrinsic pontine glioma. *Am J Neuroradiol*. 2015.
127. Rock JP, Scarpace L, Hearshen D, Gutierrez J, Fisher JL, Rosenblum M, et al. Associations among magnetic resonance spectroscopy, apparent diffusion coefficients, and image-guided histopathology with special attention to radiation necrosis. *Neurosurgery*. 2004;54(5):1111–1117. [discussion 7–9].
128. Esteve F, Rubin C, Grand S, Kolodie H, Le Bas JF. Transient metabolic changes observed with proton MR spectroscopy in normal human brain after radiation therapy. *Int J Radiat Oncol Biol Phys*. 1998;40(2):279–286.
129. Kaminaga T, Shirai K. Radiation-induced brain metabolic changes in the acute and early delayed phase detected with quantitative proton magnetic resonance spectroscopy. *J Comput Assist Tomogr*. 2005;29(3):293–297.
130. Hu LS, Eschbacher JM, Heiserman JE, Dueck AC, Shapiro WR, Liu S, et al. Reevaluating the imaging definition of tumor progression: perfusion MRI quantifies recurrent glioblastoma tumor fraction, pseudoprogression, and radiation necrosis to predict survival. *Neuro Oncol*. 2012;14(7):919–930.
131. Baek HJ, Kim HS, Kim N, Choi YJ, Kim YJ. Percent change of perfusion skewness and kurtosis: a potential imaging biomarker for early treatment response in patients with newly diagnosed glioblastomas. *Radiology*. 2012;264(3):834–843.
132. Ellingson BM, Cloughesy TF, Lai A, Nghiemphu PL, Liau LM, Pope WB. Quantitative probabilistic functional diffusion mapping in newly diagnosed glioblastoma treated with radiochemotherapy. *Neuro Oncol*. 2013;15(3):382–390.
133. Ellingson BM, Cloughesy TF, Zaw T, Lai A, Nghiemphu PL, Harris R, et al. Functional diffusion maps (fDMs) evaluated before and after radiochemotherapy predict progression-free and overall survival in newly diagnosed glioblastoma. *Neuro Oncol*. 2012;14(3):333–343.
134. McKnight TR, von dem Bussche MH, Vigneron DB, Lu Y, Berger MS, McDermott MW, et al. Histopathological validation of a three-dimensional magnetic resonance spectroscopy index as a predictor of tumor presence. *J Neurosurg*. 2002;97(4):794–802.
135. Yang I, Huh NG, Smith ZA, Han SJ, Parsa AT. Distinguishing glioma recurrence from treatment effect after radiochemotherapy and immunotherapy. *Neurosurg Clin N Am*. 2010;21(1):181–186.
136. Weybright P, Sundgren PC, Maly P, Hassan DG, Nan B, Rohrer S, et al. Differentiation between brain tumor recurrence and radiation injury using MR spectroscopy. *Am J Roentgenol*. 2005;185(6):1471–1476.
137. Zeng QS, Li CF, Zhang K, Liu H, Kang XS, Zhen JH. Multivoxel 3D proton MR spectroscopy in the distinction of recurrent glioma from radiation injury. *J Neurooncol*. 2007;84(1):63–69.
138. Smith EA, Carlos RC, Junck LR, Tsien CI, Elias A, Sundgren PC. Developing a clinical decision model: MR spectroscopy to differentiate between recurrent tumor and radiation change in patients with new contrast-enhancing lesions. *Am J Roentgenol*. 2009;192(2):W45–W52.
139. Lin NU, Lee EQ, Aoyama H, Barani IJ, Baumert BG, Brown PD, et al. Challenges relating to solid tumour brain metastases in clinical trials, part 1: patient population, response, and progression. A report from the RANO group. *Lancet Oncol*. 2013;14(10):e396–406.

CHAPTER

8

Immunotherapy Clinical Trials in Neuro-Oncology

S. Khagi, G. Vlahovic

Duke University Medical Center, Durham, NC, United States

OUTLINE

Translational Immunotherapy of Brain Tumors
http://dx.doi.org/10.1016/B978-0-12-802420-1.00008-9

INTRODUCTION

Immunotherapy is a broad term and, when applied loosely in oncology, can encompass a multitude of modalities. In fact, common cytotoxic chemotherapy can trigger an immune response due to cancer DNA instability and the development of neoantigens.[1] Ionizing radiotherapy has also been demonstrated to produce novel neoantigens that are recognized and acted upon by the immune system.[2] Thus, given the broad context and for the purposes of this chapter, we will narrow our definition of immunotherapy. Within each subsection of this chapter, we will begin our discussion examining seminal immunotherapy studies in systemic oncology and then focus the majority of our attention on adult high-grade glioma clinical trials assessing the direct stimulation, activation, and maintenance of effector cells using various biotherapeutic modalities.

EARLY IMMUNOTHERAPY

The use of agents to stimulate the immune system against a malignant process had demonstrated efficacy as far back as the 1890s. Serendipitously coming upon a 30-year-old case report revealing tumor regression after bouts of erysipelas, a surgeon by the name of William B. Coley devised a mixture of *Streptococcus pyogenes* and *Serratia marcescens*. Formulating this mixture into an injection, he successfully treated a young man suffering from an abdominal mass with multiple administrations of his bacterial brew.[3]

In the mid-20th century, as the field of oncology continued to grow, our understanding of the interplay between the immune system and cancer began to take shape. The field progressed beyond the early anecdotal case reports of bacterial and viral infections causing complete or partial responses to a more nuanced understanding of immunobiology and cancer. The Australian virologist Frank MacFarlane Burnet posited the theory of immunosurveillance.[4] The idea grew out of his seminal work describing "self" and "nonself," as well as the concept of immunotolerance.[5] The theory was applied to tumorigenesis and demonstrated the immune system's unique ability to recognize cancer-related peptides as "nonself." With the abrogation of immune competence in the setting of HIV, immunosurveillance was further corroborated by reports of increased rates of cancer in those patients affected with acquired immunodeficiency syndrome in the 1980s.[6]

Building on early preclinical and clinical studies assessing immunostimulation using the Bacillus Calmette–Guerin vaccine and granulocyte-macrophage colony-stimulating factor (GM-CSF) combined with manipulated prostate tumor cells, studies in malignant melanoma were some of the first that targeted tumor–associated antigens.[7–10] This was due, in part, to the observation that cytotoxic T lymphocytes could be found infiltrating melanoma biopsy specimens. Building on the inherent immunogenic potential of melanoma, research into tumor-associated proteins (i.e., gp100, MAGE-1, MAGE-3, MART-1, and melan A) was shown to induce a T cell response, leading to the development of vaccines. Among the best studied vaccines was gp100; however, studies did not prove convincing enough to implement this approach into clinical practice.[11]

IMMUNOTHERAPY IN THE 21ST CENTURY

The early 21st century has borne witness to a remarkable groundswell of research that has led to the utilization of immunotherapy in the clinic. Although this is beyond the scope of this chapter, much of what will be discussed owes credit to these early preclinical and clinical endeavors that have paved the way to harnessing the immune system as a means of treatment and prevention of cancer. For example, the three-dimensional structure of cytotoxic T lymphocyte–associated protein 4 (CTLA-4) was elucidated and its function described by Schwartz and Stamper;[12,13] cytotoxic T lymphocytes were shown to be highly active after vaccination and portend a survival advantage in patients with melanoma;[14] Riddell, Greenberg, and Jensen demonstrated that genetically modified T lymphocytes engineered to express chimeric antigens have clinically meaningful activity in lymphoma;[15] and the human papillomavirus quadrivalent recombinant vaccine (Gardasil), the first vaccine approved to prevent a subtype of cancer, demonstrated efficacy in triggering a humoral response using components of a viral capsid and, in turn, the inhibition of oncogenic human papillomavirus variants that cause squamous cell carcinoma of the cervix.[16] In all, the first decade of the 21st century has firmly set in place the foundation for many of the immunotherapeutic modalities that are being investigated and used in the field of oncology today. The following sections will discuss these varied modalities, with each subsection divided into an overview of important trials in systemic oncology followed by an appraisal of current trials in neuro-oncology.

CHECKPOINT INHIBITORS

Systemic Oncology

Lessons learned from early trials in solid tumors eventually led the way to more successful endeavors. In 2010, Hodi et al. published their seminal work assessing the efficacy of an anti-CTLA-4 monoclonal antibody called ipilimumab in patients with advanced melanoma.[17] A total of 676 patients were randomized in a 3:1:1 fashion to receive ipilimumab with gp100, ipilimumab alone, or gp100 alone. With overall survival (OS) being the primary endpoint, the trial demonstrated that ipilimumab, with or without gp100, had a median survival of about 10 months. These results fundamentally changed the practice of treating advanced melanoma, with ipilimumab becoming the first checkpoint inhibitor approved for the treatment of cancer; however, it also revealed the intrinsic complexity of assessing the immunotherapeutic response, as well as managing syndromes of autoimmunity elicited by these new agents.

Blocking programmed cell death ligand 1 (PD-1), another autoimmune checkpoint receptor on T lymphocytes exploited by cancer cells has also demonstrated robust clinical data leading to the approval of multiple agents. Several phase III trials have demonstrated the efficacy of PD-1 inhibition using monoclonal antibodies. These results led to the approval of nivolumab for melanoma, non–small cell carcinoma and renal cell carcinoma and in turn, demonstrated the potential to elicit a vigorous immunological response leading to manifestations of autoimmunity in a subset of patients. CHECKMATE 057 was a large international, open-label, randomized trial comparing nivolumab to docetaxel in 582 patients with advanced nonsquamous, non–small cell lung carcinoma who had progressed after first line platinum-doublet therapy. With the primary endpoint of OS, Borghaei et al. demonstrated that nivolumab had an improved median outcome compared to docetaxel [12.2 months vs. 9.4 months; hazard ratio (HR) 0.73; 96% confidence interval (CI) 0.59–0.89; $P = 0.002$]. Immune-related events represented the minority of overall complications, with a relatively small percentage of events leading to discontinuation of treatment compared to the control arm.[18] Similarly, CHECKMATE 017 compared nivolumab to docetaxel in recurrent squamous cell non–small cell lung carcinoma, and demonstrated an OS of slightly more than 3 months (9.2 months vs. 6 months; HR 0.59; 95% CI, 0.44–0.79; $P < 0.001$).[19] In CHECKMATE 025, a randomized trial of 802 patients with advanced renal cell carcinoma who failed at least one antiangiogenic agent, Motzer et al. showed that nivolumab was superior to everolimus, with an OS of over 2 years favoring the experimental arm.[20] Thus nivolumab became the first Food and Drug Administration (FDA)-approved checkpoint inhibitor in recurrent renal cell carcinoma.

Similar to nivolumab, pembrolizumab has been granted accelerated approval for the treatment of advanced non–small cell lung carcinoma in patients who are positive for PD-L1 expression via immunohistochemistry (IHC) and progressed after standard platinum doublet therapy. In KEYNOTE-010, a large international and randomized phase III trial, the investigators demonstrated that PD-L1 expression of greater than 1% yielded higher response rates among the pembrolizumab arms compared to docetaxel. Again, immune-related events were noted and occurred at a rate of about 20% in both pembrolizumab arms.[21] Pembrolizumab has gone on to receive expanded approval in advanced melanoma both in the first line and recurrent settings following ipilimumab and, if BRAF V600 positive, after small molecule inhibitors targeting the said pathway.[22]

In an effort to improve on the response rate of individual checkpoint inhibitors and assess the risk of toxicity with a combined regimen, recent studies have looked at coupling PD-1 and CTLA-4 inhibitors in the treatment of melanoma. Among these was a large, randomized, double-blind phase III study in which Larkin and colleagues assigned patients to nivolumab plus ipilimumab, nivolumab plus placebo, or ipilimumab plus placebo. The results demonstrated that the combination of nivolumab and ipilimumab resulted in a progression-free survival (PFS) increase compared to nivolumab alone and ipilimumab alone (11.5 months vs. 6.9 months vs. 2.9 months, respectively). Results were statistically significant between groups. Furthermore, in a subgroup analysis, patients with PD-L1 negative tumors demonstrated greater benefit from the combined regimen compared to either agent alone. Consequently, these results led to the conditional accelerated approval of this combined regimen as granted by the FDA in 2015.[23]

Neuro-Oncology

The evaluation of checkpoint inhibitors for the treatment of high–grade glioma is well underway, with a focus on ipilimumab, nivolumab, and pembrolizumab. At this point, the use of checkpoint inhibitors has yet to be approved by the US FDA in primary brain malignancies. Both newly diagnosed patients and those with recurrent disease are potentially eligible for trial enrollment comparing checkpoint inhibitors with or without the standard of care. In this section, we will examine select clinical trials that are currently underway.

In an intricately designed trial, nivolumab is being compared within multiple patient cohorts. A subset of these cohorts (1, 1b, 1c, 1d) fall into a phase I safety lead-in analysis prior to the accrual of patients into cohort 2. The latter cohort will be included in a phase III randomized, open-label study designed to assess nivolumab versus bevacizumab. In brief, the phase I portion of the study aims to demonstrate the safety and tolerability of nivolumab alone or in combination with ipilimumab. Cohorts will be stratified based on the newly diagnosed or recurrent status of participants, as well as on the *O*-6-methylguanine-DNA methyltransferase (MGMT) promotor methylation status of newly diagnosed patients. With the completed accrual of cohort 2, safety data are now being evaluated from earlier cohorts, as well as OS, PFS, and overall response rate (ORR). Correlative biomarkers will also be assessed, which include PD-L1 expression, the mutational status of epidermal growth factor receptor (EGFR) and KRAS, as well as immunomonitoring of various systemic T lymphocyte populations.[24]

For the treatment of recurrent glioblastoma (GBM), Dana-Farber Cancer Institute and its partner institutions have opened a randomized, open-label phase II trial (NCT02337491) comparing the efficacy of pembrolizumab with or without bevacizumab. The primary outcome of this trial is to assess 6-month PFS in these two cohorts of patients. Additionally, along with PFS, the cohort receiving pembrolizumab and bevacizumab will be monitored in such a way that will establish the maximum tolerated dose (MTD) of the former.[25]

Taken together, the aforementioned trials have distinct similarities and differences concerning their eligibility criteria for patient enrollment. Most notably, patients with active autoimmune disease are excluded from participation.[24,25] This is not surprising given that trials in systemic malignancies (discussed earlier) yielded serious immunologic adverse reactions. However, the CHECKMATE 143 study goes beyond active autoimmunity as a potential excluding factor and omits patients who have a history of or are suspected of having an autoimmune disorder.[24] Given that CHECKMATE 143 utilizes CTLA-4 and PD-1 inhibition, it is apparent why extra care must be taken when combining these immunotherapeutic modalities. The MGMT promoter methylation status of patients is another point of distinction between CHECKMATE 143 and NCT02337491. Given that a newly diagnosed groups is among the cohorts being assessed in CHECKMATE 143, it is reasonable to assess the utility of a nonmethylated MGMT promotor as a prognostic and predictive biomarker when comparing these experimental regimens to the standard approach.[24,26]

THERAPEUTIC CANCER VACCINES

Systemic Oncology

Therapeutic vaccines in cancer are a broad subset of immunotherapeutic agents. The study of these modalities has encompassed the use of active vaccine immunotherapy, adoptive cell transfer, tumor-associated and tumor-specific antigen stimulation of effector cells, as well as the use of nucleic acid transfer via an assortment of vectors to stimulate the immune system against a variety of malignancies. The overall endpoint of using these agents is to stimulate effector cells against cancer antigens, sustain that response against disease elsewhere in the body, and develop immunologic memory.[27]

Sipuleucel-T was approved by the FDA in 2010 for the treatment of advanced castrate-resistant prostate cancer.[28] Sipuleucel-T is a dendritic cell (DC)-based vaccine and represents an example of active vaccine immunotherapy. Its technology utilizes leukapheresis and segregation of peripheral blood mononuclear cells (including DCs), followed by ex vivo stimulation. Autologous peripheral blood mononuclear cells are then reintroduced into the patient.[29] In a seminal study, Kantoff and colleagues reported on a randomized, double-blind, placebo-controlled phase III trial using sipuleucel-T or a placebo. The results demonstrated a relative reduction of death of 22% in the cohort that received sipuleucel-T compared to the placebo arm (HR, 0.78; 95% CI, 0.61–0.98; $P=0.03$). However, the study revealed that asymptomatic patients with only a small burden of disease derived the most benefit, as the experimental arm was not associated with significant tumor regression but did demonstrate a decrease in prostate specific antigen levels.

Neuro-Oncology

Although no single agent has yet to gain approval, therapeutic cancer vaccines and adoptive cell transfer strategies for brain tumors have been an active area of research and clinical investigation. Active vaccine immunotherapy and adoptive transfer of immune effector cells have made up a significant proportion of trials that have progressed furthest in development; modalities that include DCs and engineered chimeric antigen receptor (CAR) T lymphocytes, respectively. Novel vaccine strategies focused on targeting tumor-specific antigens, glioma stem cells, isocitrate dehydrogenase (IDH) mutated tumors, and heat–shock protein (HSP)-directed mechanisms have also been incorporated into various clinical trial formats. Within this section of the chapter we will discuss select trials that represent promising avenues and novel means of targeting high-grade glioma.

Currently undergoing a phase III trial, DCVax represents an active immunotherapy vaccine, by which autologous DCs are exposed to patient tumor lysate in an ex vivo setting, and then intradermally reintroduced to the patient.[30] As described in a phase I study published by Prins and colleagues, autologous monocyte-derived DCs were stimulated in culture with GM-CSF and interleukin (IL)-4. One day prior to its intradermal administration, the product was exposed to the patient's cell-free, protein-enriched tumor lysate. The trial included both newly diagnosed patients and those with recurrent GBM.[31] The vaccine was well tolerated. Among the most salient findings, the investigators concluded that patients with tumors exhibiting a mesenchymal phenotype had the longest duration of benefit from the vaccine (discussed further in Study Design section). Not unexpectedly, data suggested a greater benefit in newly diagnosed patients compared to patients with recurrent disease (median OS 35.9 months vs. 17.9 months, $P = 0.03$).[31] It was unclear if the use of topical imiquimod and polyinosinic-polycytidylic acid [poly(I:C)] stabilized by lysine and carboxymethylcellulose (poly-ICLC) (a toll-like receptor agonist), utilized in the study as a means to enhance the immune response, played a significant role in improving survival. A higher dose of the vaccine did not correlate to an increase in survival, toxicity, or differences in immunological parameters.[31] As mentioned earlier, an international, multicenter, randomized, double-blind, placebo-controlled phase III study for newly diagnosed patients with GBM is currently underway.[30]

Specifically focusing on glioma-associated antigens (GAA), a broad assortment of epitopes that help distinguish tumor from most normal tissue, has provided further fodder for vaccine development. Honing in on a combination of unique tumor- and glioma stem cell–associated antigens is ICT-107. In an early phase study of 21 patients with GBM, Phuphanich et al. administered the autologous DC vaccine pulsed with either major histocompatibility complex (MHC) subtype human leukocyte antigen A1 (HLA-A1) or HLA-A2–binding epitopes directed against antigens human epidermal growth factor receptor 2 (HER2), tyrosinase-related protein 2 (TRP-2), gp100, MAGE-1, IL-13Rα2, and AIM-2.[32] The newly diagnosed cohort was treated with standard chemoradiotherapy prior to receiving the vaccine. Patients with an HLA-A2 subtype demonstrated a trend toward a better response. Furthermore, postvaccine sampling of a subset of patient tumors at the time of recurrence revealed a loss of CD133, a known stem cell marker. The latter finding suggested that the stem cell compartment may be preferentially targeted. Newly diagnosed patients demonstrated a PFS of 16.9 months and an OS of 38.4 months.[32] Subsequently, phase II data from a double-blind placebo-controlled trial of ICT-107 in 124 newly diagnosed patients with GBM randomized 2:1 to either receive the vaccine or autologous unpulsed DCs, revealed an improved PFS compared to the placebo arm (24.1 months vs. 8.5 months; HR 0.26; 95% CI 0.09–0.70; $P = 0.004$). However, this PFS benefit did not translate into an OS advantage in both the intention-to-treat and the per protocol cohorts ($P = 0.64$, HR 0.89; and $P = 0.48$, HR = 0.84; respectively). A pre-specified subgroup analysis suggested that patients expressing HLA-A2 with and without MGMT methylation derived the most benefit.[33] A phase III, randomized, double-blind, placebo-controlled trial evaluating newly diagnosed GBM patients enriched for HLA-A2 is currently underway.[34]

Similarly, a study by Pollack et al. demonstrated that a glioma antigen–specific response can be elicited by using a peptide vaccine in pediatric patients, without the need for autologous DCs primed against GAA.[35] Twenty-six HLA-A2–positive patients with brain stem glioma or high-grade glioma not otherwise found in the brain stem were treated with a combinatorial vaccine using poly-ICLC and peptides against epitopes of IL-13 receptor α2 (IL-13Rα2), ephrin type-A receptor 2, and survivin. In this small pilot study, median OS for patients was 13.3 months (95% CI, 10.9–20.3 months). Compared to historical data, these data represented only a marginal improvement in survival. However, immune assays demonstrated that 50% of patients were able to mount an immune response to at least one antigen. The authors note that the lack of prolonged benefit from the vaccine may have been due to immune tolerance or immunogenic escape of subclones not expressing the aforementioned target antigens.

Another important cell surface antigen is the EGFR. It can be expressed on various normal tissue, but can be constitutively active and heavily amplified in multiple tumor subtypes, including GBM. However, EGFR variant III (EGFRvIII), which has been described to have an in-frame deletion of 801 base pairs, represents a unique tumor-specific antigen.[36] Its presence has been reported in approximately 30–50% of GBM and has been deemed a glioma-specific antigen.[37,38] Capitalizing on the specificity of this antigen, rindopepimut is an altered EGFRvIII peptide conjugated to keyhole limpet hemocyanin (KLH), which has been shown to preferentially target cells that express EGFRvIII and induce immunological memory against the antigen.[39] In a phase II trial assessing the efficacy of rindopepimut in newly diagnosed patients with EGFRvIII-positive GBM, 65 patients were administered the vaccine with adjuvant temozolomide after having undergone standard resection followed by chemoradiotherapy.[40] With the primary endpoint of PFS at 5.5 months, the study demonstrated that 66% (95% CI: 55–76%, $P=0.0168$) reached the predefined threshold for significance. Median PFS was 9.2 months (95% CI: 7.4–11.3 months) and median OS was 21.8 months (95% CI: 17.9–26.5 months).[40]

Given that bevacizumab plays a central role in the management of recurrent GBM and with preclinical models suggesting synergy when immunotherapy is combined with bevacizumab, Reardon et al. designed a phase II trial assessing the addition of rindopepimut in patients with recurrent GBM who were bevacizumab naïve. A total of 73 patients were subdivided in a 1:1 fashion to receive bevacizumab with either a control or rindopepimut. The combination arm showed a 2-month median OS advantage compared to the control arm (11.3 months vs. 9.3 months, HR 0.53; $P=0.013$).[41,42] The study also demonstrated that patients who generated the highest antibody titers and those that generated these titers quickly had an association with a prolonged OS. At 24 months, 25% of the patients in the combination arm were still alive compared to 0% in the control arm.[42]

With strong signals of clinical efficacy in early phase studies, a large, multicenter, international, double-blind, randomized phase III trial was initiated to determine the impact of rindopepimut on survival parameters in patients with newly diagnosed GBM. A total of 745 patients were randomized to receive either rindopepimut or KLH. Both arms included GM-CSF, in addition to receiving adjuvant temozolomide. Despite showing signs of similar efficacy to prior studies, an interim analysis revealed that rindopepimut did not outperform KLH with respect to median OS (20.4 months vs. 21.4 months; HR = 0.99).[43] Subsequently, the trial was stopped early.

Other peptide-focused vaccines are currently in various stages of clinical development. A prime example is the HSP vaccine. The basis for this approach centers around the intrinsic role of HSP, particularly HSP-96 (as defined by its molecular weight), in acting as a chaperone to deliver tumor-associated antigens to DCs.[44] In a single-arm, multicenter, phase II trial assessing the impact of an autologous tumor-derived HSP-96 vaccine on OS in newly diagnosed GBM patients receiving standard of care therapy, Bloch and colleagues evaluated 46 patients and measured PD-L1 expression as an outcome correlate. Median PFS was 17.8 months (95% CI: 11.3–21.6 months) and median OS was 23.8 months (95% CI: 19.8–30.2 months). Interestingly, in those patients that had a high level of PD-L1 expression, median OS was 18 months (95% CI: 10–23.3 months). This is in stark contrast to those with low PD-L1 expression, who exhibited a median OS of 44.7 months (95% CI: not calculable). Along with MGMT, the expression levels of PD-L1 were determined to be an independent marker for survival.[44] Currently, a randomized, open-label phase II trial is underway to determine the impact of bevacizumab with or without the HSP-96 vaccine in the recurrent setting.[45]

Another example of a promising vaccine strategy is the use of cytomegalovirus (CMV) antigens to stimulate autologous immune effector cells. Evidence for the role of CMV in glioma tumorigenesis has been previously described.[46] An antigen unique to CMV is pp65. Preclinical models have suggested that autologous T lymphocytes activated by CMV pp65-RNA–pulsed DCs resulted in activity against GBM cells in vitro.[47] In an effort to improve the clinical efficacy of DC activation and migration to the target area in vivo, Mitchell et al. developed a small, randomized trial in which 12 patients were randomized to receive preconditioning with either mature, unpulsed, autologous DCs, or tetanus/diphtheria toxoid (Td) prior to administration of CMV pp65–pulsed autologous DCs. Patients who received Td preconditioning demonstrated a more robust CMV pp65–pulsed DC accumulation at the site of vaccination. These results also correlated to a survival advantage as patients who received Td had a longer median OS, with three of six patients surviving ≥36.6 months.[48] Built on the aforementioned potential of eliciting an immune response against CMV-specific antigens inherent to high-grade glioma, a novel early phase trial is currently underway that will assess the safety and efficacy of a CMV pp65–pulsed DC vaccine in combination with nivolumab.[49]

Mutations in the isocitrate dehydrogenase 1 (*IDH1*) gene have been found to be inherent to multiple tumor subtypes, particularly gliomas. This oncogenic metabolic gene is mutated early in the process of tumorigenesis, with a majority of low-grade gliomas and a subset of transformed high-grade gliomas exhibiting the IDH1 product.[50] This alteration occurs commonly as a missense mutation of arginine to histidine (R132H).[51] A subset of patients who have IDH1 (R132H)–mutated tumors can have $CD4^+$ lymphocytes and antibodies that are specifically directed against this tumor-specific antigen. Transgenic mouse data demonstrated that the R132H-containing epitope is expressed via an MHC class II–dependent manner, and when vaccinated using a R132H-directed peptide vaccine, a T cell mediated, IgG-directed response demonstrated a survival advantage compared to mice who lacked the *IDH1* mutation.[52] Another preclinical modal has identified a promising 25 amino acid peptide vaccine, PEPIDH1M, which is homologous to the *IDH1* R132H mutation in humans.[53] A phase I trial is ongoing to assess the safety and tolerability of PEPIDH1M combined with GM-CSF when administered intradermally with Td preconditioning, followed by surgical resection and chemotherapy.[54] Additionally, a phase I trial out of Germany is currently evaluating the safety and efficacy of an intradermally administered 20 amino acid peptide vaccine targeting IDH1 R132H in combination with topical imiquimod in patients with newly diagnosed high-grade gliomas after having undergone resection followed by chemoradiotherapy.[55]

CHIMERIC ANTIGEN RECEPTOR T CELLS

Systemic Oncology

An exciting example of adoptive cell transfer is CAR T cells. In brief, present technology involves leukapheresing unselected autologous T lymphocytes, that are then transfected ex vivo using a viral vector encoding a chimeric T cell receptor that targets a specific antigen and possess unique costimulatory domains to potentiate the response.[56] Granted breakthrough designation by the FDA in 2014, CAR T cells engineered against CD19 (CTL019) demonstrated dramatic responses in children with relapsed acute lymphoblastic leukemia (ALL). Maude et al. reported on 30 patients with relapsed or refractory CD19–positive ALL treated with CTL019. Twenty-seven out of 30 patients achieved a complete response and had an OS rate of 78% (95% CI, 65–95%). Immunomonitoring revealed that the majority of responders had high levels of circulating CTL019-positive cells. The majority of patients developed B cell aplasia, as well as cytokine release syndrome. The latter necessitating treatment with tocilizumab, an anti–IL-6 monoclonal antibody.[57]

Neuro-Oncology

As in systemic malignancies, autologous engineered CAR T lymphocytes are being explored as a novel means of adoptive cell transfer in the treatment of glioma. Utilizing EGFRvIII as a unique tumor-specific antigen, preclinical data suggest that retrovirally transfected CAR T cells can be engineered to preferentially target EGFRvIII-mutated cells, but also impart global immunity against intracranially implanted malignant glioma in a syngeneic, immunocompetent mouse model.[58] A group from the University of Pennsylvania also generates an EGFRvIII-directed CAR T cell construct; however, the approach utilized a lentiviral transfection, with the lead candidate having the lowest affinity for EGFRvIII among the other CAR T cell constructs that were assessed.[59] As noted by the authors, the rationale for this approach was to limit cross-reactivity with the Erb-family of receptors (i.e., EGFR, HER2, etc.) expressed on normal tissue, potentially mitigating toxicity. Given the promising aforementioned preclinical data, an early phase clinical trial sponsored by the National Cancer Institute is currently assessing the safety, 6-month PFS, and immunodynamics of EGFRvIII-directed CAR T cells in combination with IL-2 and a nonmyeloablative lymphocyte depleting regimen.[60] Additionally, a phase I clinical trial based on the latter preclinical EGFRvIII-directed CAR T cell model has been accruing patients with either newly diagnosed or recurrent EGFRvIII-positive GBM.[61]

In the same vein as EGFRvIII, IL-13 receptor α2 (IL13R α2) represents a tumor-specific antigen that can be targeted via CAR T cell–directed technology. Studies have demonstrated that this antigen is closely related to a mesenchymal gene signature (a "proinflammatory" signature), can be found in approximately half of GBM patients, and may portend a poor survival.[62–64] Given the potential for therapeutic targeting of this receptor, Jensen and colleagues developed a genetically modified IL-13 chimeric T cell antigen receptor, which utilized an IL-13 E13Y recombinant protein capable of selectively binding IL13Rα2-positive GBM cells.[65] This approach proved promising in preclinical murine models, in which GBM orthotopic xenografts demonstrated regression after exposure to IL13Rα2 CAR T cells.[65] In a small, first-in-human pilot study, Brown et al. instilled intracavitary autologous IL13Rα2 CAR T cells into three patients with recurrent GBM who had undergone a gross total resection. An anti-IL13Rα2 response was observed and histologically confirmed within one patient, with other patients demonstrating radiographic evidence of an inflammatory response to residual disease.[66] Using these data, an expanded phase I trial is currently underway to evaluate an optimized version of IL13Rα2 CAR T cells in patients with recurrent high-grade gliomas.[67]

VIRAL THERAPY IN SOLID TUMORS

Systemic Oncology

Oncolytic viral therapy has been at the forefront of the immunotherapy community's consciousness for well over a decade. Indeed, the potential benefit of using a viral vector to elicit a response against cancer has been anecdotally demonstrated for over a century, even before the causative source was fully understood. For example, in 1896, George Dock published a case of a patient with leukemia who had concomitantly contracted influenza. The patient's spleen size and total white blood cell count improved dramatically; however, the response was short-lived, with the patient succumbing to her disease in a matter of months.[68] Throughout the early 20th century, there remained a steady trickle of reports documenting responses in patients afflicted with hematological malignancies after exposure to a wide variety of viruses; however, durable responses remained elusive and toxicity was prevalent.[69]

The advent of genetic recombination technology led the way to utilizing viruses more effectively against cancer and decreasing potential virulence to the host.[70] Although there have been a myriad of early phase clinical work using various modalities of administration in a variety of malignancies, there has only been a single agent approved by the FDA thus far.

Talimogene laherparepvec (T-VEC) became the first oncolytic virus approved by the FDA.[71] The virus was shown to be effective in advanced melanoma and is based on a genetically modified herpes simplex virus (HSV) type-1 encoding GM-CSF; consequently, this causes the infected cancer cell to express GM-CSF, leading to a localized immune response.[72] The approval of T-VEC was not based on its OS data, but rather on its durable response rate. In a phase III, randomized, open-label design, patients with advanced melanoma were assigned in a 2:1 fashion to either receive intralesional T-VEC or subcutaneous GM-CSF. The primary endpoint was durable response rate, defined as an objective response (complete or partial) lasting greater than 6 months. Secondary endpoints included ORR and OS. The results demonstrated that intralesional injections of T-VEC produced a durable response in 16.3% compared to 2.1% of patients receiving subcutaneous injections of GM-CSF ($P<0.0001$). Median OS data trended toward significance, with 23.3 months in the T-VEC arm versus 18.9 months in the GM-CSF arm (HR, 0.79; 95% CI, 0.62–1.00; $P=0.051$). The most common adverse events in the experimental cohort were fatigue, chills, and fever.[72]

Neuro-Oncology

Viral therapy for malignant glioma has recently garnered a fair amount of attention. Various attenuated, genetically modified viral constructs have been developed, and at least seven have made their way into a clinical trial.[73] Mechanisms are as varied as the viruses themselves; however, the overarching mode of action is that of viral infection followed by tumor cell lysis and concurrent activation of the innate immune system to clear virally infected tumor cells.[73] In 2000, among the first iterations of an oncolytic virus had been demonstrated in an early phase clinical trial assessing the safety of a modified HSV that had been attenuated for virulence and preferential replication, and instilled intracranially via a stereotactic approach. Despite its limitations and variable clinical outcomes, the results suggested that intracranial administration of a live virus is feasible and safe.[74]

In a phase I study, Lang et al. assessed the safety, efficacy, and biological activity of an oncolytic adenovirus, known as Delta-24-RGD (DNX-2401).[75] One arm of the study focused on clinical endpoints and dose limiting toxicity, with patients receiving a single intratumoral injection of the virus. Another group of patients received a single intratumoral injection, followed by resection of the lesion and further administration of virus into the cavity. The study revealed that the virus was safe and an MTD was not reached. Median OS was 11 months, with three patients (12%) having a complete response. The investigators observed that IL-12p70 (a potent inducer of cell-mediated immunity) may be an important biomarker of response, as all three responders had levels approaching from 10-fold to 1000-fold over the norm.

The development of a chimeric poliovirus fused with a portion of a rhinovirus genome (PVS-RIPO) demonstrated remarkable tropism for CD155 receptors on GBM cells and a lack of neurotoxicity in animal models.[76,77] Subsequently, investigators at Duke University have opened an early phase clinical trial assessing the safety and dose tolerability of intracranially instilled PVS-RIPO for recurrent and surgically resectable GBM. As of June 2015, 20 patients had been treated, with 12 patients who are reported to be alive. Two patients demonstrated a survival of over 30 months. The study is slated to move into phase II.[78]

Another approach to targeting high-grade glioma is using a modified virus to hone in on glioma-associated endothelial cells. In a phase I/II multicenter trial, Brenner et al. used an attenuated adenoviral construct carrying a proapoptotic Fas-chimera transgene with a modified endothelin promotor region, also known as ofranergene obadenovec (VB-111), to infect patients with recurrent GBM.[79] With the transgene being preferentially expressed in rapidly dividing glioma-associated endothelial cells, the chimeric cytokine receptor-Fas transmembrane protein elicits an immune response that leads to endothelial cell apoptosis.[80] The study demonstrated that the addition of continuously dosed VB-111 to bevacizumab in patients with recurrent GBM is safe and well tolerated. In fact, patients in the continuously dosed cohort had an improved median OS compared to patients in the cohort that progressed after receiving virus for a limited time followed by sequential bevacizumab (15 months vs. 8 months; $P = 0.048$).[79] These data demonstrate that an immune-mediated antiangiogenic effect via a virus can be enhanced with the use of a monoclonal antibody targeting vascular EGF.

BISPECIFIC T CELL ENGAGERS

Systemic Oncology

Bispecific T cell engagers (BiTEs) are an engineered class of fused, divalent monoclonal antibody constructs, which interact with separate cell surface receptors. The present technology involves the interaction of one end of the antibody fragment with an immune cell, and the other fragment binding to a target cell antigen. This leads to activation of a cytotoxic cell mediated response against the target cell.[81]

One current example of a BiTE antibody is blinatumomab. Preclinical data demonstrated that blinatumomab had potent binding of CD3-positive cytotoxic T lymphocytes and CD19-positive B lymphocytes, the former exerting a cytotoxic effect on the latter.[82] These data led to small early phase clinical trials showing signs of efficacy against CD19-positive B cell leukemias and lymphomas. Most notably was a multicenter, single-arm, open-label phase II trial evaluating 189 adult patients with Philadelphia chromosome negative, relapsed or refractory, B-precursor ALL.[83] The primary endpoints were complete remission or complete hematological remission. The study demonstrated that after two cycles of treatment, 81 patients (43%, 95% CI, 36–50) met the primary end, with 40% of these patients subsequently undergoing an allogeneic hematopoietic stem cell transplant. Complications included cytopenias and cytokine release syndrome. Overall, this led to the accelerated approval of blinatumomab by the FDA. A phase III trial is currently ongoing.[84]

Neuro-Oncology

Although the utilization of BiTEs has yet to make it into an active clinical trial, the preclinical data suggest significant potential of redirecting immune effector cells against glioma-specific antigen using these divalent antibody fragments. For example, as reported by Choi and colleagues, a BiTE directed against EGFRvIII and CD3 demonstrated in vitro cytotoxicity via enriched $CD4^+/CD25^+$ T lymphocytes against EGFRvIII-positive glioma cells in a granzyme-dependent manner.[85] Given that $CD4^+/CD25^+$ T lymphocytes (e.g., T regulatory cells) have been classically associated with tumor-promoting immunosuppression, this study represents a paradigm shift with respect to the potential of repurposing these cells in the presence of cancer.[86]

IMMUNOTHERAPY CLINICAL TRIAL DESIGN IN NEURO-ONCOLOGY

There are currently no approved immunotherapeutic modalities for the treatment of glial neoplasms. As evidenced earlier in this chapter, trials in neuro-oncology are well underway to make these drugs a clinical reality. Given the intricacies of targeting lesions within the central nervous system, these studies have been designed to take into account each immunotherapeutic modality's unique mechanism of action. Within this segment of the chapter we will examine unique characteristics that pertain to designing effective immunotherapy trials for brain tumor patients.

STUDY DESIGN

The overall incidence of adult malignant glial neoplasms remains low compared to the incidence of malignancies involving the breast, lung, and gastrointestinal tract.[87] However, these tumors represent a high-relative mortality compared to these more common malignancies, necessitating the need for improved therapeutics.[88] Given the rarity of these cancers, it is difficult to accrue large numbers of patients into clinical trials in a reasonably short period of time. Therefore cooperative groups focused on brain tumors and industry-sponsored partnerships have helped foster a more robust accrual process to demonstrate statistically and clinically significant results.

The vast majority of clinical trials dedicated to brain tumors follow the classic phased model of drug development. Phase I trials are designed with a small number of patients and are typically used to determine the MTD of a study drug. Phase II trials increase the number of subjects and attempt to demonstrate efficacy within single or multiple disease states. At times, phase I and II studies may be combined, and include a cohort from the former in the efficacy analysis. Phase III trials are designed to confirm efficacy (e.g., OS) within a larger population and are usually randomized and controlled, with arms comparing the study drug to a placebo or the standard of care. Phase II and III trials can also be combined. After years of early phase studies and the culmination of positive phase III data, a study drug can be taken to market via the FDA new drug application process and proceed with postmarketing analysis. Rarely, phase II trials can also lead to an accelerated FDA approval; albeit, this comes with preconditions set forth by the FDA that (1) surrogate endpoints are met that demonstrate strong signals of efficacy, (2) the agent meets an urgent need for new therapeutics in a disease that has few options, and (3) conditional approval is based on the need for further studies that may corroborate positive results demonstrated in earlier studies.[89]

Currently approved immunotherapy agents have undergone a phased approval process as discussed in the previous section of this chapter. These trials uniformly used statistical endpoints that represented paradigms set forth decades ago in trials evaluating cytotoxic chemotherapy in solid tumors. As in previous chemotherapy studies, the use of Response Evaluation Criteria in Solid Tumors (RECIST) criteria was consistently utilized when evaluating radiographic response as a categorical variable. In other words, patients were labeled as either complete, partial, or nonresponders. Since the approval of ipilimumab, there has been a need to implement criteria that more accurately recapitulate the natural immune response as it relates to the radiographic monitoring of patients being treated with immunotherapy (discussed later).

Immunotherapy drug trials may also contain correlative studies that use imaging or patient-derived biological markers. These biomarkers can be designated as surrogate endpoints based on preclinical or early phase clinical studies, and can potentially predict predetermined outcomes (e.g., OS), suggesting a response to treatment. In contrast to pharmacodynamics or pharmacokinetic measurements found in many phase I trials, correlative biomarkers are not a direct measure of the biochemical properties of a compound interacting with the patient. Extensive validation and standardization of assays are required to translate correlative biomarkers into patient-defining criteria for treatment; in other words, personalized immunotherapy.

Correlative biomarkers have played an important role in immunotherapy research. PD-L1 has been used extensively in trials evaluating PD-1 inhibitors. For example, in an early phase trial assessing the efficacy of BMS-936558 (nivolumab), Topalian et al. demonstrated that patients with PD-L1 negative tumors (<5% expression by IHC) had a poor objective response to PD-1 inhibition. This is in contrast to 36% of patients with ≥5% expression of PD-L1 who had an objective response.[90] The KEYNOTE-010 trial went a step further and enrolled patients with non–small cell lung carcinoma with PD-L1 expression of ≥1%, and stratified patients based on expression levels of 1–49% and ≥50% who were then treated with pembrolizumab 2 mg/kg, pembrolizumab 10 mg/kg, or docetaxel.[21] As described previously, the results suggested statistical OS equivalence between the PD-L1 groups. The FDA went on to approve pembrolizumab for patients with PD-L1 expressing non–small cell lung carcinoma, making it the first immunotherapeutic agent to be used on the basis of a patient-specific biomarker.[91]

Immunomonitoring of T lymphocyte populations and cytokine levels has also demonstrated biological relevance as surrogates of response. This is especially important in adoptive cell transfer and vaccine studies. For example, CAR T cell technology relies on ex vivo viral vector transfer of a chimeric T cell receptor gene into autologous patient-derived T lymphocytes. Upon reintroduction of these genetically modified T cells, correlative studies help account for population expansion and maintenance over time. As mentioned previously, Maude et al. used flow cytometry to monitor CTL019 and correlate these results with patient response. The data revealed that 27 out of the 30 patients with ALL who had an objective response had a median of 38.9% circulating CTL019 cells, whereas the remaining three nonresponding patients had significantly lower levels.[57] Furthermore, all patients in the aforementioned trial had some degree of cytokine release syndrome (CRS), a well-established consequence of CAR T cell therapy. The study authors carefully monitored the levels of C-reactive protein, ferritin, and IL-6. These biomarkers were demonstrated to be at higher levels in patients who exhibited more severe symptoms of CRS.[57] Ongoing early phase studies assessing the safety and efficacy of CAR T cell technology in patients with malignant glial neoplasms are monitoring lymphocyte population assays and inflammatory cytokine levels for correlative and safety purposes.[60]

Molecular characterization may also be a biomarker for immunotherapy. As defined by The Cancer Genome Atlas, GBM can be divided into various molecular subtypes based on gene-expression profiling.[92] As discussed in Prins et al., the mesenchymal subtype may portend an improved response to immunotherapy based on observations that patients exposed to an experimental DC vaccine had a more favorable response compared to patients with a proneural subtype. The investigators' conclusion was supported by the finding that tumor-infiltrating lymphocytes were more prone to aggregate in mesenchymal tumors before and after DC vaccination.[31]

Given the difference in its ability to elicit a particular kind of immune response, the MHC HLA-A2 cell surface receptor can be considered another example of a molecular biomarker that can be used to enrich for a specific patient population that may derive a potential benefit from a vaccine-driven approach.[33,34] Future studies may focus on optimizing the efficacy of vaccine development to take into account the inherent genotypic differences in MHC antigen presentation and immune response potentiation.

A potentially new avenue in neuro-oncologic trials is applying adaptive designs to phased studies by using a biomarker-driven methodology and continuous reassessment of response. From immunomonitoring to molecular subtyping of tumors, biomarkers can serve as correlates of response. Using small early phase hypothesis-generating trial data, subsequent studies can be enriched for populations that harbor biomarkers that may portend a positive response to therapy. Multiarm protocols can be designed in such a way where multiple biomarkers can be examined, and predefined interim analyses performed to determine the fate of cohorts based on response to treatment. In doing so, study accrual can be expedited and statistical power maintained; therefore increasing trial efficiency and hastening the drug approval process.[93]

ASSESSING THE CANDIDATE

Selecting suitable candidates for enrollment into adult immunotherapy trials is of paramount importance. Within this section of the chapter we will examine common inclusion and exclusion criteria delineating candidate enrollment into neuro-oncologic immunotherapy trials.

With respect to high-grade glioma, it is important to first define and confirm the disease that is being targeted as delineated by the World Health Organization histologic and molecular characterization.[94] Additionally, enrollment eligibility based on the stage of therapy needs to be defined (e.g., newly diagnosed or recurrent disease). Typically, most neuro-oncologic immunotherapy trials define newly diagnosed patients as those that had their tumor resected but have yet to start standard radiotherapy with concomitant chemotherapy.[95] With respect to recurrent disease, patients must have completed at least 6 weeks of chemoradiation and demonstrated progressive disease thereafter. Patients progressing on or after the completion of adjuvant cytotoxic therapy may also be considered recurrent. Second-line antiangiogenic therapy may, at times, complicate enrollment given that many trials use these agents to offset the need for corticosteroids or as a direct comparator arm in the setting of a randomized trial for recurrent disease.

Performance status is among the most important elements that make up trials in oncology. As is standard within most neuro-oncology clinical studies, patients must possess a preserved performance status, as typically defined by a cut off of ≥70% on the Karnofsky Performance Status scale. Hematological, renal, and hepatic laboratory parameters need to be within suitable levels for the study drug to be administered. These levels can vary depending on the immunotherapy being employed; however, typical values are as follows: total white count of ≥2000 per microliter (μL), hemoglobin ≥8 g per deciliter (dL), platelets >100,000 per μL, creatinine clearance ≥40 mL per minute, hepatic transaminases not greater than three times the upper limit of normal (xULN), and bilirubin <1.5 xULN. Other biological parameters may need to be considered depending on the mechanism of action of the drug (e.g., distinct lymphocyte populations).

Determining the neurological baseline for a clinical trial candidate is of particular importance. Given that some trials may require surgical debulking to decrease the tumor burden, optimize tumor infiltration via effector cells, or implant catheters for direct infusion of an agent, patients need to be screened to determine the potential for progressive postoperative neurological decline. Even if surgery is not part of a protocol, lesions within eloquent areas of the brain (e.g., thalamus, brain stem, cerebellum, etc.) may become inflamed when exposed to the study drug, leading to cerebral edema and potentially significant morbidity. Thus patients need to be carefully evaluated to determine the degree of baseline neurological impairment secondary to disease, its overall stability, and/or the potential for complications while on trial.

Preexisting autoimmunity is another common exclusion criterion. Notable examples of trials that exclude patients with autoimmunity are those involving PD-1 and CTLA-4 inhibition.[24,25] As evidenced within these trials, patients can present with severe dermatologic, gastrointestinal, and pulmonary complications due to effector cell stimulation and cytokine release. Moreover, to manage these complications, the use of high-dose systemic corticosteroids could lead to an impaired immune response against the malignancy (discussed later). Adoptive cell transfer, vaccines, and oncolytic viral therapy studies can have variable criteria excluding preexisting autoimmunity depending on the mechanism of action of the agent being employed. Furthermore, active cardiopulmonary comorbidities, severe infectious disease (e.g., HIV, hepatitis C), pregnancy, and other active malignancies are common preexisting conditions that warrant exclusion.

The use of glucocorticoids needs to be taken into account when selecting patients for enrollment. It has been well established that steroids are capable (and culpable) of depleting lymphocyte populations. In fact, data continue to accumulate regarding the potentially detrimental role of prolonged glucocorticoid administration in patients with high-grade glioma.[96] Given the previously mentioned, and depending on the specifics of the study, patients may either need to be off steroids, on a stable dose, or meeting a predefined cutoff (e.g., dexamethasone 4mg daily total dose) prior to enrollment on trial.

ASSESSMENT OF SAFETY

As eluded to in our discussion of exclusion and inclusion criteria, patient safety is a top consideration when it comes to the design and successful execution of a clinical trial. Hypothesis testing can only be fostered when a clinical trial is feasible with respect to protecting patients from undo harm. Furthermore, patient-provided informed consent can only be obtained when the investigator and the research team are both *informed* regarding the potential adverse outcomes of a new therapy. Although not always necessary, mechanisms for mitigation of untoward effects should be considered as part-and-parcel of certain immunotherapy protocols and afforded to patients who present with symptoms of severe intracranial complications.

Determining the safety of an immunotherapy protocol begins by understanding the mechanism of action of the agent in question. These mechanisms are elucidated in preclinical and animal models; however, these models do not fully recapitulate the human immunological environment. Thus phase I clinical trials inform investigators of the tolerability of an agent and the MTD. However, exceptions to the standard phase I mantra include adoptive cell transfer and vaccines, which may never have a defined MTD given inherent biological mechanisms. Two notable examples of this are DC-Vax and DNX-2401.[31,75]

Current immunotherapy protocols use the Common Terminology Criteria for Adverse Events version 4.03 (CTCAE v4.03) to characterize and grade manifestations of autoimmunity, as well as CRS.[97] Given that a brain tumor is being targeted with an immunotherapy, characterization of an adverse reaction by the investigator needs to be considered in the context of neurological complications (e.g., cerebral edema and encephalopathy), as well as in systemic manifestations of toxicity. The CTCAE v4.03 takes this into account; albeit, it is without the connotations that may relate these symptoms to autoimmunity. Thus separate characterizations are necessary to fully describe adverse events in the context of neuro immunotherapeutic studies.

With autoimmunity being a potentially clear and present adverse manifestation of immunotherapy in the treatment of malignant glioma, countermeasures should be built into clinical trials in an effort to (1) mitigate a clinically significant immune response, and (2) to avoid the use of high-dose corticosteroids which may severely attenuate the effector cell population. The selection of agents is based on the mechanism of action of the immunotherapy being used. For example, the phase I results of intratumorally administered PVS-RIPO demonstrated a profound local inflammatory response in select patients. In an effort to offset the degree of inflammation and not exceed 4 mg of dexamethasone (as per protocol), bevacizumab was used. The dose utilized is that often used for radiation necrosis.[98] In the same token, trials assessing CAR T cells have incorporated tocilizumab (anti–IL-6 monoclonal antibody) to offset CRS, and as a steroid-sparing agent.[99]

ASSESSMENT OF RESPONSE

Clinical and radiographic response evaluation in neuro-oncology can be an exercise in objective subjectivity. In other words, it can be as much of an art as it is a science. Distinguishing between what is true progressive disease (PD) versus pseudo progression (PsP) on magnetic resonance imaging (MRI) has been at the crux of determining a response to treatment in neuro-oncology. And radiographic findings and clinical manifestations may not always correlate to one another. MRI perfusion–diffusion imaging, positron emission tomography combined with structural imaging, and magnetic resonance spectroscopy (MRS) have been used as tools to measure treatment response in neuro-oncology; however, none of these methodologies have been prospectively validated for use in immunotherapy clinical trials. For an in depth discussion of neuroimaging techniques in neuro-oncology, please refer to Chapter 7.

Early attempts at determining response criteria of cytotoxic therapy in malignant glial neoplasms stemmed from standards described in systemic malignancies, with incorporation of some factors inherent to neuro-oncology (e.g., edema and mass effect). One such early response tool was the Levin criteria, which used two-dimensional (2D) cross-sectional measurements and accounted for edema and mass effect.[100] However, given the intrinsic variability in characterization of response using the Levin criteria, the MacDonald criteria were introduced to help inform both radiographic and clinical decision-making. The MacDonald criteria took into account the change in size of a lesion over time, new areas of disease, corticosteroid dosage fluctuation, and the overall clinical status of the patient. Clear metrics defined a complete response (CR), partial response (PR), stable disease (SD), and PD.[101] It was a tool that has been uniformly used by radiologists and neuro-oncologists alike.

In 2000, an international collaborative group developed the RECIST to help standardize radiographic response characterization in systemic malignancies.[102] RECIST version 1.1 is the current iteration and improves on previous measures of lesion number, volumetrics, lymph node characterization, and response confirmation for the purposes of clinical trial endpoints.[103] At present, RECIST criteria are the standard means of radiographic evaluation used in clinical trials in oncology.

Not to be outdone, the neuro-oncologic community came together and developed the response assessment for neuro-oncology (RANO) criteria. First published in 2010, The RANO Working Group brought to light the inherent limitations of using (1) 2D analyses (e.g., MacDonald criteria) to measure irregularly shaped lesions and multifocal disease, (2) not accounting for nonenhancing disease on MRI, and (3) the difficulty of measuring enhancement of surgical/cystic cavities and relating these findings to response.[104] Among the most pertinent recommendations that came out of the RANO Working Group were the following: defining areas of measurable disease as being at least 1 × 1 cm in diameter; characterizing irregularly shaped lesions via the sum of the products of the diameters (SPD); and using changes in the relative percent change in SPD to longitudinally define response. Additionally, areas of nonenhancing disease are taken into account, as well as the effect of corticosteroids and antiangiogenics.

A full appraisal of RECIST and RANO criteria is beyond the scope of this chapter; however, in a retrospective analysis, Gallego Perez-Larraya et al. concluded that RECIST, RANO, and the MacDonald criteria appear to be similar with respect to demonstrating radiographic PFS and clinical measures of response with respect to recurrent GBM.[105] Neuro-oncologic trials have been known to use RECIST criteria to characterize response, but have typically relied on MacDonald criteria for a majority of published data; however, more recently, there has been a shift toward using RANO criteria, especially as more trials are incorporating antiangiogenic therapy and its impact on distinguishing PD from PsP on MRI.

In recent years the field of oncology has shifted its gaze upon evaluating the promise of immunotherapy, and with it came the realization that our current standards of response may not be entirely adequate in fully capturing the breadth and variability of immunologic outcomes. This became apparent with the approval of ipilimumab in advanced melanoma. Early phase clinical studies demonstrated that an inflammatory response caused by CTLA-4 inhibition may occasionally result in a transient worsening of disease and/or the appearance of new lesions.[106] By standard RECIST criteria, this increase would be termed PD and discontinuation of treatment would be warranted. Thus new means of evaluation were needed to address these issues. In 2009, using observations from 487 patients with melanoma receiving ipilimumab, Wolchock et al. introduced the immune-related response criteria (irRC) as a potential means of accounting for a transient increase in disease burden.[107] Along with incorporating the unique kinetics of immunotherapy, the criteria allow for "progressive disease" and builds in time to fully assess response in patients. Although not yet universally adopted, these criteria are being recognized as an important tool in response assessment of immunotherapy clinical trials.[108,109]

As evidenced in the first segment of this chapter, clinical trials in neuro-oncologic immunotherapy are quickly coming online. With potentially practice changing results forthcoming, investigators are faced with the challenge of appropriately characterizing radiographic response and correlating it clinically. In anticipation of these challenges, the neuro-oncologic community has recently proposed a means to account for immune-related radiographic changes. The Immunotherapy Response Assessment for Neuro-Oncology (iRANO) guidelines have been formulated and are in the process of being validated. In brief, the framework of these new guidelines is based on the current RANO criteria and the recently developed irRC. Due to the delayed effects of immunotherapy, the iRANO criteria stipulate that patients who have radiographic progression of disease and who have been on immunotherapy for less than 6 months, should be re-imaged in 3 months to confirm disease resistance to immunotherapy. Additionally, the issues that are being addressed are the following: the appearance of new lesions; continuing immunotherapy treatment until confirmation of disease; defining criteria for discontinuation of therapy; and factoring in the immunological relevance of nonenhancing disease on MRI.[110] The applicability of iRANO will hinge on its validation in future clinical trials. Measuring disease response may vary by the specific mechanism of action of the immunotherapeutic employed, with immune biomarkers potentially being used to further guide assessment. Accounting for these facts while developing iRANO could contribute greatly to disease response evaluation in the era of neuro immunotherapeutics.

Although MRS has not been incorporated into RANO or iRANO criteria to assess treatment response, it can potentially be used to correlate response with tumor-derived biomarkers. As mentioned previously, mutations in IDH are an important driver of tumorigenesis in a subset of high-grade gliomas.[50] IDH mutations result in the production of D-2-hydroxygluterate (D2HG), an abnormal enzymatic by-product that functions as an oncometabolite, which leads to epigenetic silencing of various tumor suppressor mechanisms.[111] In vivo modeling suggests that D2HG levels can be measured using 2D correlation spectroscopy, as well as point-resolved spectroscopy.[112] Using the latter technique, Choi and colleagues were able to detect the presence of D2HG in all IDH-positive glioma patients, without eliciting a similar signal in IDH-negative tumors. The presence of the mutation was confirmed using gene sequencing, which had a greater concordance with metabolic imaging compared to standard immunohistochemical techniques used to detect IDH mutations.[113] With various immunotherapeutic methods targeting IDH-mutated brain tumors, one can envision the use of D2HG spectroscopy as a means to detect patients with the mutation and to potentially monitor response to treatment. Prospective validation of the latter is certainly warranted in future clinical trials.

References

1. Kroemer G, Galluzzi L, Kepp O, Zitvogel L. Immunogenic cell death in cancer therapy. *Annu Rev Immunol*. 2013;31:51–72.
2. Corso CD, Ali AN, Diaz R. Radiation-induced tumor neoantigens: imaging and therapeutic implications. *Am J Cancer Res*. 2011;1(3):390–412.
3. Coley WB. The treatment of inoperable sarcoma by bacterial toxins (the mixed toxins of the streptococcus erysipelas and the bacillus prodigiosus). *Proc R Soc Med*. 1910;3(Surg Sect):1–48.
4. Burnet FM. The concept of immunological surveillance. *Prog Exp Tumor Res*. 1970;13:1–27.
5. Park HW. Germs, hosts, and the origin of Frank Macfarlane Burnet's concept of 'self' and 'tolerance', 1936–1949. *J Hist Med Allied Sci*. 2006;61(4):492–534.
6. Beral V, Peterman T, Berkelman R, Jaffe MD. AIDS-associated non-Hodgkin lymphoma. *Lancet*. April 1991;337(8745):805–809.
7. Finke LH, Wentworth K, Blumenstein B, Rudolph NS, Levitsky H, Hoos A. Lessons from randomized phase III studies with active cancer immunotherapies–outcomes from the 2006 meeting of the Cancer Vaccine Consortium (CVC). *Vaccine*. 2007;25(suppl 2):B97–B109.

8. Dranoff G, Jaffee E, Lazenby A, et al. Vaccination with irradiated tumor cells engineered to secrete murine granulocyte-macrophage colony-stimulating factor stimulates potent, specific and long-lasting anti-tumor immunity. *Proc Natl Acad Sci USA*. 1993;90(8):3539–3543.
9. Cormier JN, Salgaller ML, Prevette T, et al. Enhancement of cellular immunity in melanoma patients immunized with a peptide from MART-1/Melan A. *Cancer J Sci Am*. January–February 1997;3(1):37–44.
10. Marchand M, van Baren N, Weynants P, et al. Tumor regression observed in patients with metastatic melanoma treated with an antigenic peptide encoded by gene MAGE-3 and presented HLA-A1. *Int J Cancer*. January 18, 1999;80(2):219–230.
11. Sosman JA, Carrillo C, Urba WJ, et al. Three phase II cytokine working group trials of gp100 (210M) peptide plus high-dose interleukin-2 in patients with HLA-A2-positive advanced melanoma. *J Clin Oncol*. May 10, 2008;26(14):2292–2298.
12. Schwartz J-CD, Zhang X, Fedorov AA, Nathenson SG, Almo SC. Structural basis for co-stimulation by the human CTLA-4/B7-2 complex. *Nature*. March 29, 2001;410:604–608.
13. Stamper CC, Zhang Y, Tobin JF, et al. Crystal structure of the B7-1/CTLA-4 complex that inhibits human immune responses. *Nature*. March 29, 2001;410:608–611.
14. Speiser DE, Lienard D, Rufer N, et al. Rapid and strong human CD8+ T cell responses to vaccination with peptide, IFA, and CpG oligodeoxynucleotide 7909. *J Clin Invest*. March 2005;115:739–746.
15. Till B, Jensen MC, Wang J, et al. Adoptive immunotherapy for indolent non-Hodgkin lymphoma and mantle cell lymphoma using genetically modified autologous CD20-specific T cells. *Blood*. September 15, 2008;112:2261–2271.
16. The FUTURE II Study Group. Quadrivalent vaccine against human papillomavirus to prevent high-grade cervical lesions. *N Engl J Med*. May 10, 2007;356(19):1915–1927.
17. Hodi FS, O'Day SJ, McDermott DF, et al. Improved survival with ipilimumab in patients with metastatic melanoma. *N Engl J Med*. August 19, 2010;363(8):711–723.
18. Borghaei H, Paz-Ares L, Horn L, et al. Nivolumab versus docetaxel in advanced nonsquamous non-small-cell lung cancer. *N Engl J Med*. October 22, 2015;373(17):1627–1639.
19. Brahmer J, Reckamp KL, Baas P, et al. Nivolumab versus docetaxel in advanced squamous-cell non-small-cell lung cancer. *N Engl J Med*. July 09, 2015;373(2):123–135.
20. Motzer RJ, Escudier B, McDermott DF, et al. CheckMate 025 investigators. Nivolumab versus everolimus in advanced renal-cell carcinoma. *N Engl J Med*. November 05, 2015;373(19):1803–1813.
21. Herbst RS, Baas P, Kim DW, et al. Pembrolizumab versus docetaxel for previously treated, PD-L1-positive, advanced non-small-cell lung cancer (KEYNOTE-010): a randomised controlled trial. *Lancet*. December 18, 2015. pii:S0140-6736(15) 01281–7.
22. U.S. Food and Drug Administration. http://www.fda.gov/Drugs/InformationOnDrugs/ApprovedDrugs/ucm478493.htm; December 21, 2015.
23. U.S. Food and Drug Administration. http://www.fda.gov/Drugs/InformationOnDrugs/ApprovedDrugs/ucm465274.htm; October 01, 2015.
24. *A Study of the Efficacy and Safety of Nivolumab versus Bevacizumab and of Nivolumab with or without Ipilimumab in Glioblastoma Patients at Different Stages of Treatment (CheckMate 143)*; January 26, 2016. http://clinicaltrials.gov/ct2/show/study/NCT02017717.
25. *Pembrolizumab +/– Bevacizumab for Recurrent GBM*; January 19, 2016. http://clinicaltrials.gov/ct2/show/NCT02337491.
26. Hegi ME, Diserens AC, Gorlia T, et al. MGMT gene silencing and benefit from temozolomide in glioblastoma. *N Engl J Med*. March 10, 2005;352(10):997–1003.
27. Goldman B, DeFrancesco L. The cancer vaccine roller coaster. *Nat Biotechnol*. February 2009;27(2):129–139.
28. U.S. Food and Drug Administration. http://www.fda.gov/BiologicsBloodVaccines/CellularGeneTherapyProducts/ApprovedProducts/ucm210012.htm; April 15, 2015.

29. Kantoff PW, Higano CS, Shore ND, et al. Sipuleucel-T immunotherapy for castration-resistant prostate cancer. *N Engl J Med*. July 29, 2010;363(5):411–422.
30. *Study of a Drug [DCVax®-L] to Treat Newly Diagnosed GBM Brain Cancer (GBM)*; January 11, 2016. http://clinicaltrials.gov/ct2/show/NCT00045968.
31. Prins RM, Soto H, Konkankit V, et al. Gene expression profile correlates with T cell infiltration and relative survival in glioblastoma patients vaccinated with dendritic cell immunotherapy. *Clin Cancer Res*. March 15, 2011;17(6):1603–1615.
32. Phuphanich S, Wheeler CJ, Rudnick JD, et al. Phase 1 trial of a multi-epitope-pulsed dendritic cell vaccine for patients with newly diagnosed glioblastoma. *Cancer Immunol Immunother*. January 2013;62(1):125–135.
33. Wen P, Reardon D, Phuphanich S, et al. Randomized double blind placebo-controlled phase 2 trial of dendritic cell (DC) vaccine ICT-107 following standard treatment in newly diagnosed patients with GBM. *Neuro Oncol*. 2014;16(suppl 5):v22.
34. *Phase 3 Randomized, Double-Blind, Controlled Study of ICT-107 in Glioblastoma*; February 02, 2016. http://clinicaltrials.gov/ct2/show/study/NCT02546102.
35. Pollack IF, Jakacki RI, Butterfield LH, et al. Antigen-specific immune responses and clinical outcome after vaccination with glioma-associated antigen peptides and poly-inosinic-polycytidylic acid stabilized by lysine and carboxymethylcellulose in children with newly diagnosed malignant brainstem and nonbrainstem gliomas. *J Clin Oncol*. July 01, 2014;32(19):2050–2058.
36. Sampson JH, Crotty LE, Samson L, et al. Unarmed, tumor-specific monoclonal antibody effectively treats brain tumors. *Proc Natl Acad Sci USA*. June 20, 2000; 97(13):7503–7508.
37. Pelloski CE, Ballman KV, Furth AF, et al. Epidermal growth factor receptor variant III status defines clinically distinct subtypes of glioblastoma. *J Clin Oncol*. June 01, 2007;25(16):2288–2294.
38. Humphrey PA, Wong AJ, Vogelstein B, et al. Anti-synthetic peptide antibody reacting at the fusion junction of deletion-mutant epidermal growth factor receptors in human glioblastoma. *Proc Natl Acad Sci USA*. June 1990;87(11):4207–4211.
39. Heimberger AB, Crotty LE, Archer GE, et al. Epidermal growth factor receptor VIII peptide vaccination is efficacious against established intracerebral tumors. *Clin Cancer Res*. September 15, 2003;9(11):4247–4254.
40. Schuster J, Lai RK, Recht LD, et al. A phase II, multicenter trial of rindopepimut (CDX-110) in newly diagnosed glioblastoma: the ACT III Study. *Neuro Oncol*. June 2015;17(6):854–861.
41. Osada T, Chong G, Tansik R, et al. The effect of anti-VEGF therapy on immature myeloid cell and dendritic cells in cancer patients. *Cancer Immunol Immunother*. August 2008;57(8):1115–1124.
42. Reardon DA, Desjardins A, Schuster J, et al. ReACT: long-term survival from a randomized phase II study of rindopepimut (CDX-110) plus bevacizumab in relapsed glioblastoma. In: *The Annual Meeting of the Society of Neuro-Oncology; November 19–22, 2015; San Antonio, TX*; 2015. Abstract# IMCT-08.
43. RINTEGA® (Rindopepimet) – A Phase 3 Immnunotherapy Targeting EGFRvIII-Expressing Glioblastoma (GBM). http://www.celldex.com/pipline/rindopepimut.php; n.d.
44. Rivoltini L, Castelli C, Carrabba M, et al. Human tumor-derived heat shock protein 96 mediates in vitro activation and in vivo expansion of melanoma- and colon carcinoma-specific T cells. *J Immunol*. 2003;171:3467–3474.
45. *Vaccine Therapy with Bevacizumab versus Bevacizumab Alone in Treating Patients with Recurrent Glioblastoma Multiforme that can be Removed by Surgery*; January 26, 2016. http://clinicaltrials.gov/ct2/show/NCT01814813.
46. Schuessler A, Walker DG, Khanna R. Cytomegalovirus as a novel target for immunotherapy of glioblastoma multiforme. *Front Oncol*. 2014;4:275.

47. Nair SK, De Leon G, Boczkowski D, et al. Recognition and killing of autologous, primary glioblastoma tumor cells by human cytomegalovirus pp65-specific cytotoxic T cells. *Clin Cancer Res*. May 15, 2014;20(10):2684–2694.
48. Mitchell DA, Batich KA, Gunn MD, et al. Tetanus toxoid and CCL3 improve DC vaccines in mice and glioblastoma patients. *Nature*. March 19, 2015;519(7543):366–369.
49. *Nivolumab with DC Vaccines for Recurrent Brain Tumors (AVERT)*; February 08, 2016. http://clinicaltrials.gov/ct2/show/NCT02529072.
50. Cohen A, Holmen S, Colman H. IDH1 and IDH2 mutations in gliomas. *Curr Neurol Neurosci Rep*. May 2013;13(5):345.
51. Watanabe T, Vital A, Nobusawa S, Kleihues P, Ohgaki H. Selective acquisition of IDH1 R 132C mutations in astrocytomas associated with Li-Fraumeni syndrome. *Acta Neuropathol*. 2009;117(6):653–656.
52. Schumacher T, Bunse L, Pusch S, et al. A vaccine targeting mutant IDH1 induces antitumor immunity. *Nature*. August 21, 2014;512(7514):324–327.
53. Archer GE, Reap E, Norberg P, et al. IDH1 mutations as immunotherapeutic target for brain tumors. *Neuro Oncol*. 2014;16(suppl 3):iii40.
54. *IDH1 Peptide Vaccine for Recurrent Grade II Glioma (RESIST)*; February 09, 2016. http://clinicaltrials.gov/ct2/show/NCT02193347.
55. *Phase I Trial of IDH1 Peptide Vaccine in IDH1R132H-Mutated Grade III–IV Gliomas (NOA-16)*; October 22, 2015. http://clinicaltrials.gov/ct2/show/NCT02454634.
56. Cartellieri M, Bachmann M, Feldman A, et al. Chimeric antigen receptor-engineered T cells for immunotherapy of cancer. *J Biomed Biotechnol*. 2010;2010:956304.
57. Maude SL, Frey N, Shaw PA, et al. Chimeric antigen receptor T cells for sustained remissions in leukemia. *N Engl J Med*. October 16, 2014;371(16):1507–1517.
58. Sampson JH, Choi BD, Sanchez-Perez L, et al. EGFRvIII mCAR-modified T-cell therapy cures mice with established intracerebral glioma and generates host immunity against tumor-antigen loss. *Clin Cancer Res*. February 15, 2014;20(4):972–984.
59. Johnson LA, Scholler J, Ohkuri T, et al. Rational development and characterization of humanized anti-EGFR variant III chimeric antigen receptor T cells for glioblastoma. *Sci Transl Med*. February 18, 2015;7(275):275ra22.
60. *CAR T Cell Receptor Immunotherapy Targeting EGFRvIII for Patients with Malignant Gliomas Expressing EGFRvIII*; February 10, 2016. http://clinicaltrials.gov/ct2/show/NCT01454596.
61. *Pilot Study of Autologous T Cells Redirected to EGFRVIII-with a Chimeric Antigen Receptor in Patients with EGFRVIII Gliobloastoma*; August 04, 2014. http://clinicaltrials.gov/ct2/show/NCT02209376.
62. Brown CE, Warden CD, Starr R, et al. Glioma IL13Ralpha2 is associated with mesenchymal signature gene expression and poor patient prognosis. *PLoS One*. 2013;8:e77769.
63. Debinski W, Gibo DM, Hulet SW, Connor JR, Gillespie GY. Receptor for interleukin 13 is a marker and therapeutic target for human high-grade gliomas. *Clin Cancer Res*. 1999;5:985–990.
64. Thaci B, Brown CE, Binello E, Werbaneth K, Sampath P, Sengupta S. Significance of interleukin-13 receptor alpha 2-targeted glioblastoma therapy. *Neuro Oncol*. 2014;16:1304–1312.
65. Kahlon KS, Brown C, Cooper LJN, Raubitschek A, Forman SJ, Jensen MC. Specific recognition and killing of glioblastoma multiforme by interleukin 13-zetakine redirected cytolytic T cells. *Cancer Res*. 2004;64(24):9160–9166.
66. Brown CE, Badie B, Barish ME, et al. Bioactivity and safety of IL13Rα2-redirected chimeric antigen receptor CD8+ T cells in patients with recurrent glioblastoma. *Clin Cancer Res*. September 15, 2015;21(18):4062–4072.
67. *Genetically Modified T-Cells in Treating Patients with Recurrent or Refractory Malignant Glioma*; March 14, 2016. https://clinicaltrials.gov/ct2/show/NCT02208362.
68. Dock G. The influence of complicating disease upon leukemia. *Am J Med Sci*. 1904;127:563–592.

69. Kelly E, Russell SJ. History of oncolytic viruses: genesis to genetic engineering. *Mol Ther*. April 2007;15(4):651–659.
70. Alemany R, Balague C, Curiel DT. Replicative adenoviruses for cancer therapy. *Nat Biotechnol*. 2000;18:723–727.
71. U.S. Food and Drug Administration. http://www.fda.gov/NewsEvents/Newsroom/PressAnnouncements/ucm469571.htm; October 27, 2015.
72. Andtback RH, Kaufman HL, Collichio F, et al. Talimogene laherparepvec improves durable response rate in patients with advanced melanoma. *J Clin Oncol*. September 01, 2015;33(25):2780–2788.
73. Wollmann G, Ozduman K, van den Pol AN. Oncoloytic virus therapy of glioblastoma multiforme – concepts and candidates. *Cancer J*. January–February 2012;18(1):69–81.
74. Markert JM, Medlock MD, Rabkin SD, et al. Conditionally replicating herpes simplex virus mutant, G207 for the treatment of malignant glioma: results of a phase I trial. *Gene Ther*. May 2000;7(10):867–874.
75. Lang FF, Conrad C, Gomez-Manzano C, et al. Phase I clinical trial of oncolytic virus Delta-24-RGD (DNX-2401) with biological endpoints: implications for viro-immunotherapy. In: *Presented at the Annual Meeting of the Society for Neuro-Oncology; November 13–14, 2014; Miami, FL*; 2014.
76. Gromeier M, Lachmann S, Rosenfeld MR, Gutin PH, Wimer E. Intergeneric poliovirus recombinants for the treatment of malignant glioma. *Proc Natl Acad Sci USA*. 2000;97:6803–6808.
77. Cello J, Toyoda H, Dejesus N, Dobrikova EY, Gromeier M, Wimmer E. Growth phenotypes and biosafety profiles in poliovirus-receptor transgenic mice of recombinant oncolytic polio/human rhinovirus. *J Med Virol*. 2008;80:352–359.
78. Desjardins A, Sampson JH, Peters KB, et al. Oncolytic polio/rhinovirus recombinant (PVSRIPO) against recurrent glioblastoma (GBM): optimal dose determination. *J Clin Oncol*. 2015;33 (suppl; abstr 2068).
79. Brenner A, Cohen Y, Vredenburgh J, et al. Phase 2 study of VB-111, an anti-cancer gene therapy, as monotherapy followed by combination of VB-111 with bevacizumab, in patients with recurrent glioblastoma. In: *20th Annual Scientific Meeting and Education Day of the Society for Neuro-Oncology; November 19–22, 2015; San Antonio, TX. Neuro-Oncology 17*; 2015:v45–v54. [abstract].
80. Boldin MP, Mett IL, Varfolomeev EE, et al. Self-association of the "death domains" of the p55 tumor necrosis factor (TNF) receptor and Fas/APO1 prompts signaling for TNF and Fas/APO1 effects. *J Bio Chem*. January 06, 1995;270(1):387–391.
81. Weiner GJ. Building better monoclonal antibody-based therapeutics. *Nat Rev Cancer*. June 2015;5(6):361–370.
82. Dreier T, Lorenczewski G, Brandl C, et al. Extremely potent, rapid and costimulation-independent cytotoxic T-cell response against lymphoma cells catalyzed by a single-chain bispecific antibody. *Int J Cancer*. August 20, 2002;100(6):690–697.
83. Topp MS, Gokbuget N, Stein AS, et al. Safety and activity of blinatumomab for adult patients with relapsed or refractory B-precursor acute lymphoblastic leukaemia: a multicentre, single-arm, phase 2 study. *Lancet Oncol*. January 2015;16(1):57–66.
84. Phase 3 trial of blinatumomab vs standard chemotherapy in pediatric subjects with HR first relapse B-precursor ALL. In: ClinicalTrials.gov [internet]. Bethesda, MD: National Library of Medicine (US); NLM Idenifier: NCT02393859. http://clinicaltrials.gov/sow/NCT02393859.
85. Choi BD, Gedeon PC, Herndon 2nd JE, et al. Human regulatory T cells kill tumor cells through granzyme-dependent cytotoxicity upon retargeting with a bispecific antibody. *Cancer Immunol Res*. September 2013;1(3):163.
86. Zou W. Immunosuppressive networks in the tumour environment and their therapeutic relevance. *Nat Rev Cancer*. 2005;5:263–274.

87. Ostrom QT, Gittleman H, Fulop J, et al. CBTRUS statistical report: primary brain and central nervous system tumors diagnosed in the United States in 2008–2012. *Neuro Oncol*. October 2015;17(suppl 4):iv1–iv62.
88. Ostrom QT, Gittleman H, Liao P, et al. CBTRUS statistical report: primary brain and central nervous system tumors diagnosed in the United States in 2007–2011. *Neuro Oncol*. 2014;16(suppl 4):iv1.
89. U.S. Food and Drug Administration. http://www.fda.gov/ForPatients/Approvals/Fast/ucm405447.htm; September 15, 2014.
90. Topalian SL, Hodi FS, Brahmer JR, et al. Safety, activity and immune correlates of anti-PD-1 antibody in cancer. *N Engl J Med*. June 28, 2012;366(26):2443–2454.
91. U.S. Food and Drug Administration. http://www.fda.gov/newsevents/newsroom/pressannouncements/ucm465444.htm; October 02, 2015.
92. The Cancer Genome Atlas Research Network. Comprehensive genomic characterization defines human glioblastoma genes and core pathways. *Nature*. October 23, 2008;455(7216):1061–1068.
93. Trippa L, Lee EQ, Wen PY, et al. Bayesian adaptive randomized trial design for patients with recurrent glioblastoma. *J Clin Oncol*. September 10, 2012;30(26):3258–3263.
94. Louis DN, Ohgaki H, Wiestler OD, et al. The 2007 WHO classification of tumours of the central nervous system. *Acta Neuropathol*. August 2007;114(2):97–109.
95. Stupp R, Mason WP, van den Bent MJ, et al. Radiotherapy plus concomitant and adjuvant temozolomide for glioblastoma. *N Engl J Med*. March 10, 2005;352(10):987–996.
96. Wong ET, Lok E, Gautam S, Swanson KD. Dexamethasone exerts profound immunologic interference on treatment efficacy for recurrent glioblastoma. *Br J Cancer*. July 14, 2015;113(2):232–241.
97. CTCAE. http://nciterms.nci.nih.gov/ncitbrowser/pages/vocabulary.jsf?dictionary=commonterminologycriteriaforadverseevents&version=4.03; April 29, 2010.
98. *Investigators Update PVS-RIPO Data in Glioblastoma*; July 10, 2015. http://www.ascopost.com/issues/july-10-2015/investigators-update-pvs-ripo-data-in-glioblastoma/.
99. Maude SL, Barrett D, Teachey DT, Grupp SA. Managing cytokine release syndrome associated with novel T cell-engaging therapies. *Cancer J*. March–April 2014;20(2):119–122.
100. Levin VA, Crafts DC, Norman DM, Hoffer PB, Spire JP, Wilson CB. Criteria for evaluating patients undergoing chemotherapy for malignant brain tumors. *J Neurosurg*. September 1977;47(4):329–335.
101. Macdonald DR, Cascino TL, Schold Jr SC, Cairncross JG. Response criteria for phase II studies of supratentorial malignant glioma. *J Clin Oncol*. July 1990;8(7):1277–1280.
102. Therasse P, Arbuck SG, Eisenhauer EA, et al. New guidelines to evaluate the response to treatment in solid tumors. *J Natl Cancer Inst*. February 2000;92:205–216.
103. Eisenhauer EA, Therasse P, Bogaerts J, et al. New response evaluation criteria in solid tumours: revised RECIST guideline (version 1.1). *Eur J Cancer*. January 2009;45(2):228–247.
104. Wen PY, Macdonald DR, Reardon DA, et al. Updated response assessment criteria for high-grade gliomas: response assessment in neuro-oncology working group. *J Clin Oncol*. April 10, 2010;28(11):1963–1972.
105. Gallego Perez-Larraya J, Lahutte M, Petrirena G, et al. Response assessment in recurrent glioblastoma treated with irinotecan-bevacizumab: comparative analysis of the Macdonald, RECIST, RANO, and RECIST + F criteria. *Neuro Oncol*. May 2012;14(5):667–673.
106. Wolchok JD, Ibrahim R, DePril V, et al. Antitumor response and new lesions in advanced melanoma patients on ipilimumab treatment. *J Clin Oncol*. 2008;26. [abstract 3020].
107. Wolchok JD, Hoos A, O'Day S, et al. Guidelines for the evaluation of immune therapy activity in solid tumors: immune-related response criteria. *Clin Cancer Res*. December 2009;15(23):7412–7420.

108. *Study of Nivolumab Plus Chemotherapy in Patients with Advanced Cancer (NivoPlus)*; December 10, 2015. http://clinicaltrials.gov/ct2/show/NCT02423954.
109. *A Phase 1 Study of AM0010 in Patients with Advanced Solid Tumors*; January 11, 2016. http://clinicaltrials.gov/ct2/show/NCT02009449.
110. Okada H, Weller M, Huang R, et al. Immunotherapy response assessment in neuro-oncology: a report of the RANO working group. *Lancet Oncol.* November 2015;16(15):e534–e542.
111. Dang L, White DW, Gross S, et al. Cancer-associated IDH1 mutatons produce 2-hydroxygluterate. *Nature.* June 17, 2010;465(7300):966.
112. Andronesi OC, Rapalino O, Gerstner E, et al. Detection of oncogenic IDH1 mutations using magnetic resonance spectroscopy of 2-hydroxyglutarate. *J Clin Invest.* September 03, 2013;123(9):3659–3663.
113. Choi C, Ganji SK, DeBeradinis RJ, et al. 2-hydroxyglutarate detection by magnetic resonance spectroscopy in IDH-mutated patients with gliomas. *Nat Med.* January 26, 2012;18(4):624–629.

SECTION III

EXPERIMENTAL BRAIN TUMOR IMMUNOTHERAPIES

CHAPTER

9

Peptide-Specific Vaccines

P. Roth, M. Weller

University Hospital Zurich and University of Zurich, Zurich, Switzerland

OUTLINE

BACKGROUND AND DISCOVERY

Immunotherapy as a treatment option against various tumor entities including gliomas has gained increasing interest within the last years. Among the group of CNS tumors, particular attention has been given to gliomas, most notably glioblastoma (GBM), which is the most common malignant primary brain tumor in adults.[1] Despite intense standard treatment, involving maximal safe surgical resection, radiation therapy, and alkylating chemotherapy, GBM remains a particular therapeutic challenge and immunotherapy may represent one novel treatment option.[2]

Translational Immunotherapy of Brain Tumors
http://dx.doi.org/10.1016/B978-0-12-802420-1.00009-0

Immunotherapeutic approaches that aim at harnessing the immune system against a tumor take advantage of various key properties of the immune system such as its specificity when targeting selected antigens as well as its memory function, which may allow for long-lasting control of tumor growth and ultimately result in a cure. Vaccination is one immunotherapeutic strategy that may help boosting T cell–mediated immune responses. Considerable progress has been made within the last years and it has become clear that antitumor vaccination can elicit powerful immune responses without causing severe side effects. To induce immune responses of highest specificity, various vaccination approaches have been developed and examined within preclinical models as well as clinical trials. Among these, peptide-based vaccination is one of the most advanced concepts harboring several key advantages such as the induction of immune responses against predefined antigens, simple administration, and rather low cost compared to other emerging immunotherapeutic strategies such as immune checkpoint inhibition.[3] The current knowledge as well as the challenges associated with peptide-specific vaccines is summarized in this chapter.

PRECLINICAL DATA AND PRINCIPLES OF PEPTIDE VACCINES

Harnessing the immune system against cancer has been considered an attractive therapeutic option for decades. Compared to other treatment strategies, peptide-based vaccination offers several advantages and may ultimately result in a highly specific activation of the immune system without damaging side effects. However, despite several frequent genomic alterations, GBM represents a heterogeneous tumor entity. Accordingly, finding a common target for peptide vaccination that suits all tumors and patients may be impossible. Thus tumor antigens that may be exploited for standardized vaccination using a vaccine "from the shelf" are limited.

Major histocompatibility complex (MHC)-based presentation of short peptide fragments is a requirement for the induction of cellular immunity. Accordingly, peptide vaccines aim at boosting primarily cell-based immune responses, whereas humoral immune mechanisms are only partially affected. Still, due to the complex nature of the immune system, the close interaction between cellular and humoral immunity as well as interindividual differences in the responses to peptide vaccines, it remains largely unpredictable whether and to what extent a subject responds to vaccination.[4] The major challenge for the design of peptide vaccines is the choice of appropriate target antigens. Ideally, such targets are exclusively expressed by tumor cells but never by healthy tissue. Furthermore, abundant expression of an antigen may increase the chance of a powerful antitumor immune response. The term tumor-specific antigen (TSA) describes antigens that are expressed exclusively by tumor cells. Although ideal candidate targets for vaccination, TSAs are rare or mainly expressed at low levels that may preclude their use for immunotherapeutic purposes. TSAs are a result of gene mutations or alterations in the process of gene transcription or translation. Gliomas are heterogeneous tumors and a plethora of mutations and genetic alterations on the genomic, transcriptomic, and proteomic level have been described.[5,6] However, there is high interindividual variability making it difficult to find antigens that are suitable candidates as uniform targets. The most prominent example of a TSA is the variant III of the epidermal growth factor receptor (EGFRvIII), which is found in approximately 20–30% of all GBMs.[7–9] In contrast, tumor-associated antigens (TAA) are antigens that are abundantly expressed by tumor cells but may also be present at low levels in healthy tissue. Thus their therapeutic targeting may allow for a stimulation of the immune system but can also be associated with side effects due to autoimmunity with damage of healthy tissue. Wild-type EGFR is frequently overexpressed in GBM and can be considered a TAA.[10] Other TAA are proteins that are otherwise only preset during fetal development or germ-line restricted, e.g., only detected on testis. TAA pose another important challenge. As they may be expressed on healthy cells, they are recognized as "self-epitopes" by the immune system. To avoid autoimmunity, there are several mechanisms that preclude the induction of immune responses against such self-antigens. Lymphocytes recognizing self-antigens are either depleted by clonal selection or become anergic. As a result, vaccination using peptides derived from TAA may face a variety of tolerance mechanisms that need to be taken into account.

A particular challenge for peptide-based vaccination against GBM is the partially insufficient expression of MHC molecules by glioma cells, e.g., the fraction of cells with stem cell properties.[11] Deficits in the antigen-processing machinery of glioma cells may impede the presentation of peptides in the context of MHC.[12,13] Furthermore, the expression of MHC molecules by glioma cells may be hampered due to epigenetic alterations such as hypermethylation of the human leukocyte antigen (HLA) genes or chromatin structure changes in the region of the HLA promoter.[14] Additionally, insufficient or lacking expression of β2-microglobulin is a frequent condition in tumor cells that may also impair the presentation of antigens to immune cells.[15,16] All these alterations may ultimately result in limited recognition of tumor cells by antigen-specific lymphocytes. Finally, peptide-based vaccination, as basically all other therapeutic approaches aiming at inducing immune cell responses against gliomas and other brain tumors, is further confronted with the immunosuppressive environment surrounding gliomas that may inhibit profound antitumor immune cell attacks.[17]

Peptides that are used for vaccination typically consist of 8–25 amino acids that represent an epitope from a TSA or TAA. Peptides alone are hardly immunogenic. Accordingly, they are frequently conjugated to a carrier protein, which increases the immunogenicity of the vaccine. One of the best characterized carrier for peptides is the keyhole limpet hemocyanin (KLH), a large glycosylated protein that acts as an immune stimulant.[18] The advantage of peptide vaccines is their rather simple generation and storage as well as the mostly uncomplicated way of administration. Isolation and processing of blood cells, e.g., dendritic cells (DC), is not required unless peptides are used for pulsing of DC. Thus peptide vaccines do not pose a particular burden to the patients, neither during preparation nor when they are administered. Compared to other vaccination attempts using whole tumor lysates or RNA, peptides are a more specific approach. Whether this results in better immunoreactivity is still a matter of debate. Side effects, however, are frequently predictable and depend on the putative cross-reactivity of cytotoxic lymphocytes with antigenic targets that are expressed on healthy cells.[19]

As stated earlier, virtually all vaccines, including peptide-based vaccines, do not elicit sufficient immune responses when administered alone. Even more, the injection of peptides alone may induce tolerance to the chosen epitope. To enhance the vaccine's immunogenicity, adjuvants are an important part of the treatment. So far, there is no "standard adjuvant" available, mainly due to the lack of data from clinical trials. However, several adjuvants have been used repeatedly and their role is largely undisputed. One of the most common adjuvants is granulocyte-macrophage colony-stimulating factor (GM-CSF), a growth factor known to recruit antigen-presenting cells that are subsequently put in an active state.[20] Similarly, "Freund's adjuvant," a water-in-oil emulsion, frequently used in combination with mycobacterial components ("complete Freund's adjuvant") has been used for vaccination purposes. Furthermore, the clonal expansion of T cells may be improved with toll-like receptor agonists such as polyinosinic-polycytidylic acid stabilized by lysine and carboxymethylcellulose (poly-ICLC) or imiquimod.[21] Given the importance of adjuvants in the initiation of powerful antitumor immune responses, more research on their optimal composition as well as administration is required.

Current efforts aim at circumventing the potential problems associated with vaccinations restricted to a single peptide. This can be done using several peptides that allow for a therapeutic targeting of multiple tumor antigens. Furthermore, tailored vaccines based on individual tumor assessments may help to define patient-specific vaccination approaches that are more likely to induce powerful immune responses (Fig. 9.1). Several questions associated with the ongoing clinical assessment of vaccines have only been partially addressed so far including the crucial point whether peptide vaccination shall be administered ideally alone or rather be combined with conventional

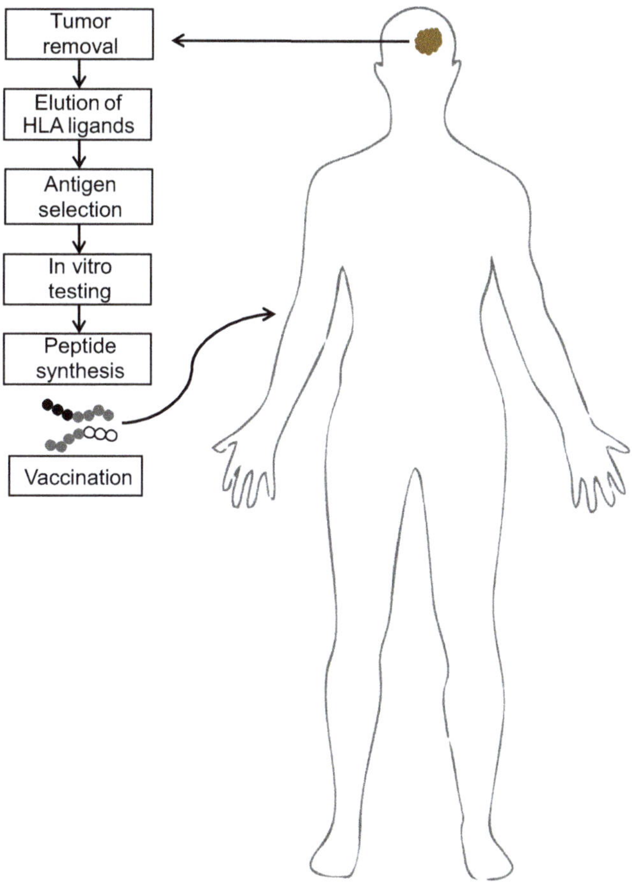

FIGURE 9.1 **Patient-tailored vaccination approach.** Removal of the tumor and subsequent characterization of the human leukocyte antigen ligandome. Selection and in vitro testing of several peptides followed by generation of the individual peptide vaccine.

treatment modalities, particularly radiation therapy and alkylating chemotherapy.[22] Similarly, the selection of optimal candidates and epitopes for peptide vaccination is still an unsolved problem and no standard algorithm has been agreed on. Finally, most trials involving vaccination against brain tumors enroll only patients with partial or gross total tumor resection but do not allow inclusion of patients who have only undergone biopsy. The rationale of such a selection process is based on the assumption that a large, remaining tumor mass confers significant immunosuppression through various tumor-derived signals as well as the obstacle of a huge tumor bulk that may preclude a control of the tumor growth by the immune system. These points need to be taken in consideration when the available clinical data on peptide vaccination are analyzed.

CLINICAL TRIAL DATA

Epidermal Growth Factor Receptor Variant III–Based Vaccination

The EGFR links extracellular signals to important intracellular downstream pathways including the mitogen-activated protein kinase and phosphatidylinositol 3-kinase signaling pathways.[10] EGFR binds to various ligands such as EGF, transforming growth factor-α, heparin-binding EGF-like growth factor, amphiregulin, betacellulin, epigen, and epiregulin 3.[23] Amplification of EGFR is found in 40–50% of all GBMs.[5] Furthermore, gene amplification can be associated with a particular mutation called EGFRvIII (also known as ΔEGFR). EGFRvIII reflects an in-frame deletion of exons 2–7 of the gene.[24] Overall, approximately 25–30% of all GBMs express EGFRvIII that represents a constitutive active form of the receptor that does not require ligand binding for signal transduction. Expression of EGFRvIII is associated with several tumor-driving mechanisms.[25,26] Accordingly, targeting the EGFR pathway has been considered a promising therapeutic approach. However, pharmacological EGFR inhibitors have been largely futile in glioma trials.[27] Although the overall prognostic role of EGFRvIII expression in GBM has been discussed controversially, it seems to be clear that the presence of this mutated receptor decreases the likelihood of long-term survival.[28] As stated earlier, EGFRvIII represents a neoantigen that is exclusively expressed in tumor tissue. Not surprisingly, it has gained increasing interest as a promising target for active immunotherapy.[29] Rindopepimut (CDX-110) is a peptide vaccine that is currently in advanced clinical development for patients with EGFR-positive GBM.[30] No other peptide-based vaccine has undergone such a thorough clinical assessment so far. Rindopepimut represents a 14-mer peptide bound to KLH that is injected intradermally in combination with GM-CSF as an adjuvant.

Three phase II studies in patients with newly diagnosed GBM have been completed. The results indicate that the vaccine yields better survival outcomes than could have been expected from historical and matched controls. ACTIVATE, the first of these phase II studies assessed rindopepimut alone in 18 patients with newly diagnosed, EGFRvIII-expressing GBM following gross-total tumor resection and radiation therapy with concomitant temozolomide treatment. All patients had no signs of tumor progression at the end of radiation therapy. The vaccine was administered without maintenance temozolomide until intolerance or tumor progression.[31] ACTIVATE-II (ACT-II) as well as ACT-III added rindopepimut to temozolomide-based radiochemotherapy based on previously published data suggesting that chemotherapy-induced lymphopenia may increase the immune-stimulating activity of a vaccine.[32,33] No major side effects or dose-limiting toxicity was reported. Injection reactions were the most frequent adverse event but mostly did not require particular measures. All three phase II studies exploring rindopepimut in patients with newly diagnosed GBM suggest a survival benefit compared to matched historical controls.[34,35] A large phase III study, assessing the combination of rindopepimut with standard temozolomide-based radiochemotherapy in patients with newly diagnosed, EGFR-positive GBM (ACT IV, NCT01480479) has completed enrollment.

ACT IV was a two-arm, randomized, double-blind, placebo-controlled study which enrolled 745 patients. The primary endpoint was overall survival (OS) in patients who had minimal residual disease (MRD) after concomitant radiochemotherapy. Treatment with rindopepimut or placebo vaccine started 2 weeks after completion of radiotherapy and continued during and beyond adjuvant temozolomide therapy. The trial was discontinued after an interim analysis. In the cohort of patients with MRD, median OS was 20.1 in the rindopepimut arm and 20.0 months in the placebo arm (HR = 0.99). In the non-MRD cohort, median OS was 14.8 vs. 14.1 months (HR = 0.79; p = 0.066). These data indicate that the addition of rindopepimut to the standard of care does not result in an additional therapeutic benefit.

Preliminary results of a randomized phase II study in patients with relapsed EGFRvIII-positive GBM (ReACT, NCT01498328) suggest that rindopepimut is an active treatment in this setting, too. In this double-blind study, patients with recurrent GBM were treated with bevacizumab in combination with rindopepimut or placebo. Progression-free survival (PFS) at 6 months was 28% in the rindopepimut arm versus 16% in the placebo cohort (p = .1163) and OS was significantly longer in the rindopepimut arm (11.6 versus 9.3 months, p = .0386; HR 0.57) in the intention-to-treat population.[36]

WILMS TUMOR PEPTIDE 1

A single peptide vaccine directed against Wilms tumor peptide 1 (WT-1) has been examined by Hashimoto and colleagues within a phase I study. The vaccine comprised a modified 9-mer peptide of WT-1. This approach is based on the rationale that peptide modifications may allow for enhanced binding to MHC class I and II molecules, which shall ultimately result in increased immunogenicity. Treatment of patients with newly diagnosed GBM was performed using intradermal vaccination of the peptide emulsified in Montanide ISA51. Vaccination was done together with temozolomide chemotherapy and was well tolerated.[37] However, whether this approach results in a survival benefit, needs to be explored in larger trials.

VACCINATION TARGETING IDH-1^{R132H}

It has been recognized only few years ago that mutations in the isocitrate dehydrogenase (IDH)-1 gene are present in the majority of World Health Organization (WHO) grade II and III gliomas.[38,39] These mutations are rarely found in GBM and it is believed that GBMs with IDH mutation have developed from lower-grade tumors. Although wild-type IDH catalyzes isocitrate to α-ketoglutarate, mutated IDH produces the oncometabolite 2-hydroxyglutarate.[40] The most common IDH-1 mutation is R132H (IDH-1^{R132H}). IDH-1^{R132H} contains a neo-epitope that can be exploited for mutation-specific vaccination. An MHC class II-restricted peptide has already been examined in a preclinical sarcoma model. Here, vaccination with a peptide representing amino acids 123–142 (p123–142) of IDH-1^{R132H} resulted in CD4 T cell responses as well as increased antibody titers in humanized mice. Depletion of CD4 cells abrogated the vaccine-mediated regression of tumor growth.[41] Based on these findings, peptide vaccination targeting IDH-1 has now entered clinical development. Several clinical trials are currently being set up to examine the usefulness of this approach in patients with IDH-1-mutant gliomas. Ongoing and planned studies include the RESIST (NCT02193347) trial for patients with WHO grade II gliomas as well as the NOA-16 trial (NCT02454634) for patients suffering from anaplastic glioma or GBM.

MULTIPEPTIDE VACCINATION REGIMENS

Multipeptide vaccines may cover several HLA alleles of a patient including MHC class I and II peptides. Thus multipeptide vaccines are ideally composed in a matter allowing the induction of CD4 as well as CD8 T cell responses. This requires the identification of appropriate peptides binding to MHC class I and II molecules.[42] Although CD4 T cells are not supposed to exhibit direct antitumor activity, they can boost adaptive immune responses relying on CD8 T cell activity. Preclinical data suggest that the presence of CD4 T cells amplifies antitumor immune cell responses.[43,44] Similar to vaccines targeting a single antigen, the peptides are ideally derived from mutated tumor antigens thereby allowing specific immune cell attack of a tumor in the absence of toxicity.

Multipeptide vaccines have emerged in clinical neurooncology within the last years. They can be composed of "pure" peptides or in combinations with DC.[45,46] The latter approach was used in a phase I/II trial aiming at determining the activity of a DC-based multipeptide vaccine in patients with recurrent anaplastic glioma or GBM. Peptides were derived from antigens associated with gliomas including epitopes from ephrin type-A receptor 2 (EphA2), interleukin (IL)-13 receptor-α2 (IL-13Rα2), YKL-40, and glycoprotein 100 (gp100), administered with polyinosinic-polycytidylic acid [poly(I:C)] stabilized by lysine and carboxymethylcellulose (poly-ICLC). All patients had to be HLA-A2 positive. The vaccine was administered intranodally and poly-ICLC was applied by intramuscular injection. No significant toxicity was reported. An immune response against at least one of the targeted antigens was noticed in the blood of 58% of the patients. Furthermore, an increased expression of potentially immune-stimulating cytokines such as interferon (IFN)-α and CXCL10 was observed. Nine out of twenty-two patients achieved PFS for 12 months or more.[47]

Another study investigated the use of subcutaneous vaccinations with peptides derived from glioma-associated antigens in HLA-A2–positive children with newly diagnosed brain stem or high-grade glioma. The peptide cocktail consisted of epitopes from EphA2, IL-13Rα2, and survivin. All three target antigens are overexpressed in pediatric gliomas. The peptides were emulsified in Montanide-ISA-51 and administered every 3 weeks together with intramuscular poly-ICLC. The trial enrolled 26 patients and the vaccine was administered together with radiation therapy or combined radiochemotherapy. The treatment regimen was overall well tolerated with no dose-limiting toxicity. Imaging changes, referred to as pseudoprogression, were observed in five patients and symptom control was obtained using steroids. Patients experiencing pseudoprogression had prolonged OS. Immune responses to the vaccine were assessed by enzyme-linked immunosorbent spot (ELISPOT) analysis. Immune reactivity against the targeted TAA was observed in 13 out of 21 patients and mostly directed against IL-13Rα2 and EphA2 suggesting that the peptides chosen from these antigens were more suitable as a vaccine than the epitope from survivin. Except for two patients who had tumor progression early during the vaccination regimen, all other children experienced at least stable disease.[48]

In a very similar approach, patients with low-grade gliomas were treated within a phase I study to determine the safety and immunogenicity of a vaccine consisting of peptides for epitopes from IL-13Rα2, EphA2, WT-1, and survivin. The peptides were emulsified in Montanide-ISA-51 and administered every 3 weeks for eight courses together with intramuscular injections of poly-ICLC. Booster vaccines were applied every 12 weeks. In line with previous trials using similar vaccines, no dose-limiting toxicity was observed. Using ELISPOT, the investigators found IFN-γ responses against at least three epitopes in the majority of cases.[49]

ICT-107

The clinically most advanced multipeptide vaccine is ICT-107 that is composed of six peptides representing the following tumor antigens: absent in melanoma 2 protein, melanoma-associated antigen 1, tyrosinase related protein-2, gp100, human epidermal growth factor receptor 2 (HER2/neu), and IL-13Rα2. Patient-derived DC were pulsed with glioma-associated MHC class I peptides. Within a pilot phase I study, ICT-107 was administered to 21 patients with newly diagnosed or recurrent GBM. The vaccination was well tolerated and elicited antiglioma immune responses.[50] Subsequently, the concept was put forward and assessed in a randomized phase II study in patients with newly diagnosed GBM. Here, mature, activated DC without peptide loading was used as control treatment. The vaccine was administered together with standard temozolomide-based radiochemotherapy. Similar to the previous phase I study, antigen-specific immune responses were observed. No significant difference in adverse events between control and ICT-107 patients occurred. Most promising survival times were seen for HLA-A2–positive patients.[51] Currently, a randomized phase III study is being planned to demonstrate the benefit from ICT-107 in combination with standard of care in patients with newly diagnosed GBM.

IMA950

IMA950 is a multipeptide vaccine comprising 11 tumor-associated peptides from antigens expressed by GBM cells.[52] Nine of these peptides bind to HLA-A2 that is expressed by ~45% of the Caucasian population, whereas the remaining two peptides bind to HLA-DR, an MHC class II molecule that is present in most patients. The rationale for this approach is to increase the probability of inducing an immune response against the tumor following vaccination. It is expected that this peptide mixture allows for the induction of both CD4- and CD8-based immune reactions. Target antigens were selected following isolation of HLA-peptide complexes from GBM tissue specimens and subsequent analysis of the peptide sequences by mass spectrometry. Final selection of peptides was done using a complex algorithm, which included the comparison of expression levels in tumor versus healthy tissue, the in vitro immunogenicity of the peptides as well as the general possibility of their synthetic production.

IMA950 was explored as an addendum to standard temozolomide-based radiochemotherapy in patients with newly diagnosed GBM. Within a pilot study, 45 HLA-A2–positive patients were enrolled. All patients received standard radiochemotherapy and started intradermal vaccination with IMA950 together with GM-CSF as an adjuvant. Similar to other peptide vaccines, little toxicity was observed, mostly restricted to injection site reactions. Grade 3 fatigue or anaphylaxis occurred in two patients. HLA-multimer analysis demonstrated T cell reactivity to at least one peptide in 90% of the patients. Concomitant steroid administration did not impair such T cell responses. PFS was 74% at 6 months and 31% at 9 months.[53] OS data are pending. An ongoing trial is exploring IMA950 together with poly-ICLC in patients with newly diagnosed GBM (NCT01920191). It needs to be awaited whether IMA950 will be put forward into a randomized trial to prove whether its addition to the standard of care improves the survival of GBM patients.

Altogether, the administration of vaccines that involves several peptides targeting TSA or TAA seems to be safe and allows for the induction of measurable immune responses. Whether these vaccination regimens are superior to vaccines against singles antigens need to be awaited. Furthermore, larger clinical trials are required to define whether multipeptide vaccines result in a survival benefit.

CONCLUSION

Peptide-specific vaccines have made considerable progress in the field of neurooncology within the last years. Peptide vaccines have been well tolerated across various trials and are easy to handle and administer. Single peptide vaccines such as rindopepimut but also multipeptide vaccines such as ICT-107 have undergone profound clinical testing and are in late-stage clinical development. Although it has become clear that peptide vaccines can elicit strong antitumor immune responses, there are still a number of important questions that need clarification including the choice of the most appropriate target antigens and the respective epitopes, the ideal adjuvant that is administrated together with the peptide and, finally, the best way to combine radiation therapy and/or alkylating chemotherapy with vaccination to obtain the best possible activation of the immune system. Further challenges that need to be addressed include immune escape mechanisms of the tumor resulting from antigen loss following vaccination, a process called immunoediting.[54] Even more advanced concepts of peptide-based vaccination may include the generation of patient-tailored vaccines based on the peptidome of a patient's tumor.[55] Although such strategies are technically feasible, they are time and cost sensitive and thus require a thorough planning as well as execution within reasonably designed clinical trials.

References

1. Ostrom QT, Gittleman H, Liao P, et al. CBTRUS statistical report: primary brain and central nervous system tumors diagnosed in the United States in 2007–2011. *Neuro Oncol*. 2014;16(suppl 4):iv1–iv63.
2. Weller M, van den Bent M, Hopkins K, et al. EANO guideline for the diagnosis and treatment of anaplastic gliomas and glioblastoma. *Lancet Oncol*. 2014;15(9):e395–403.
3. Callahan MK, Wolchok JD. At the bedside: CTLA-4- and PD-1-blocking antibodies in cancer immunotherapy. *J Leukoc Biol*. 2013;94(1):41–53.
4. Purcell AW, McCluskey J, Rossjohn J. More than one reason to rethink the use of peptides in vaccine design. *Nat Rev Drug Discov*. 2007;6(5):404–414.
5. Brennan CW, Verhaak RG, McKenna A, et al. The somatic genomic landscape of glioblastoma. *Cell*. 2013;155(2):462–477.
6. Sturm D, Bender S, Jones DT, et al. Paediatric and adult glioblastoma: multiform (epi) genomic culprits emerge. *Nat Rev Cancer*. 2014;14(2):92–107.

7. Montano N, Cenci T, Martini M, et al. Expression of EGFRvIII in glioblastoma: prognostic significance revisited. *Neoplasia*. 2011;13(12):1113–1121.
8. Aldape KD, Ballman K, Furth A, et al. Immunohistochemical detection of EGFRvIII in high malignancy grade astrocytomas and evaluation of prognostic significance. *J Neuropathol Exp Neurol*. 2004;63(7):700–707.
9. Shinojima N, Tada K, Shiraishi S, et al. Prognostic value of epidermal growth factor receptor in patients with glioblastoma multiforme. *Cancer Res*. 2003;63(20):6962–6970.
10. Hatanpaa KJ, Burma S, Zhao D, Habib AA. Epidermal growth factor receptor in glioma: signal transduction, neuropathology, imaging, and radioresistance. *Neoplasia*. 2010;12(9):675–684.
11. Di Tomaso T, Mazzoleni S, Wang E, et al. Immunobiological characterization of cancer stem cells isolated from glioblastoma patients. *Clin Cancer Res*. 2010;16(3):800–813.
12. Mehling M, Simon P, Mittelbronn M, et al. WHO grade associated downregulation of MHC class I antigen-processing machinery components in human astrocytomas: does it reflect a potential immune escape mechanism? *Acta Neuropathol*. 2007;114(2):111–119.
13. Facoetti A, Nano R, Zelini P, et al. Human leukocyte antigen and antigen processing machinery component defects in astrocytic tumors. *Clin Cancer Res*. 2005;11(23):8304–8311.
14. Yeung JT, Hamilton RL, Ohnishi K, et al. LOH in the HLA class I region at 6p21 is associated with shorter survival in newly diagnosed adult glioblastoma. *Clin Cancer Res*. 2013;19(7):1816–1826.
15. Restifo NP, Marincola FM, Kawakami Y, Taubenberger J, Yannelli JR, Rosenberg SA. Loss of functional beta 2-microglobulin in metastatic melanomas from five patients receiving immunotherapy. *J Natl Cancer Inst*. 1996;88(2):100–108.
16. Zagzag D, Salnikow K, Chiriboga L, et al. Downregulation of major histocompatibility complex antigens in invading glioma cells: stealth invasion of the brain. *Lab Invest*. 2005;85(3):328–341.
17. Roth P, Eisele G, Weller M. Immunology of brain tumors. *Handb Clin Neurol*. 2012;104:45–51.
18. Lateef SS, Gupta S, Jayathilaka LP, Krishnanchettiar S, Huang JS, Lee BS. An improved protocol for coupling synthetic peptides to carrier proteins for antibody production using DMF to solubilize peptides. *J Biomol Tech*. 2007;18(3):173–176.
19. Mohme M, Neidert MC, Regli L, Weller M, Martin R. Immunological challenges for peptide-based immunotherapy in glioblastoma. *Cancer Treat Rev*. 2014;40(2):248–258.
20. Heufler C, Koch F, Schuler G. Granulocyte/macrophage colony-stimulating factor and interleukin 1 mediate the maturation of murine epidermal Langerhans cells into potent immunostimulatory dendritic cells. *J Exp Med*. 1988;167(2):700–705.
21. Warger T, Schild H, Rechtsteiner G. Initiation of adaptive immune responses by transcutaneous immunization. *Immunol Lett*. 2007;109(1):13–20.
22. Dietrich PY, Dutoit V, Tran Thang NN, Walker PR. T-cell immunotherapy for malignant glioma: toward a combined approach. *Curr Opin Oncol*. 2010;22(6):604–610.
23. Schneider MR, Wolf E. The epidermal growth factor receptor ligands at a glance. *J Cell Physiol*. 2009;218(3):460–466.
24. Vogt N, Lefevre SH, Apiou F, et al. Molecular structure of double-minute chromosomes bearing amplified copies of the epidermal growth factor receptor gene in gliomas. *Proc Natl Acad Sci USA*. 2004;101(31):11368–11373.
25. Nagane M, Levitzki A, Gazit A, Cavenee WK, Huang HJ. Drug resistance of human glioblastoma cells conferred by a tumor-specific mutant epidermal growth factor receptor through modulation of Bcl-XL and caspase-3-like proteases. *Proc Natl Acad Sci USA*. 1998;95(10):5724–5729.
26. Fan QW, Cheng CK, Gustafson WC, et al. EGFR phosphorylates tumor-derived EGFRvIII driving STAT3/5 and progression in glioblastoma. *Cancer Cell*. 2013;24(4):438–449.
27. Roth P, Weller M. Challenges to targeting epidermal growth factor receptor in glioblastoma: escape mechanisms and combinatorial treatment strategies. *Neuro Oncol*. 2014;16(suppl 8):viii14–19.

28. Weller M, Kaulich K, Hentschel B, et al. Assessment and prognostic significance of the epidermal growth factor receptor vIII mutation in glioblastoma patients treated with concurrent and adjuvant temozolomide radiochemotherapy. *Int J Cancer*. 2014;134(10):2437–2447.
29. Congdon KL, Gedeon PC, Suryadevara CM, et al. Epidermal growth factor receptor and variant III targeted immunotherapy. *Neuro Oncol*. 2014;16(suppl 8):viii20–25.
30. Swartz AM, Li QJ, Sampson JH. Rindopepimut: a promising immunotherapeutic for the treatment of glioblastoma multiforme. *Immunotherapy*. 2014;6(6):679–690.
31. Sampson JH, Heimberger AB, Archer GE, et al. Immunologic escape after prolonged progression-free survival with epidermal growth factor receptor variant III peptide vaccination in patients with newly diagnosed glioblastoma. *J Clin Oncol*. 2010;28(31):4722–4729.
32. Asavaroengchai W, Kotera Y, Mule JJ. Tumor lysate-pulsed dendritic cells can elicit an effective antitumor immune response during early lymphoid recovery. *Proc Natl Acad Sci USA*. 2002;99(2):931–936.
33. Sanchez-Perez LA, Choi BD, Archer GE, et al. Myeloablative temozolomide enhances CD8+ T-cell responses to vaccine and is required for efficacy against brain tumors in mice. *PLoS One*. 2013;8(3):e59082.
34. Sampson JH, Aldape KD, Archer GE, et al. Greater chemotherapy-induced lymphopenia enhances tumor-specific immune responses that eliminate EGFRvIII-expressing tumor cells in patients with glioblastoma. *Neuro Oncol*. 2011;13(3):324–333.
35. Schuster J, Lai RK, Recht LD, et al. A phase II, multicenter trial of rindopepimut (CDX-110) in newly diagnosed glioblastoma: the ACT III study. *Neuro Oncol*. 2015;17(6):854–861.
36. Reardon DA, Schuster J, Tran DD, et al. ReACT: overall survival from a randomized phase II study of rindopepimut (CDX-110) plus bevacizumab in relapsed glioblastoma. *J Clin Oncol*. 2015;33 (suppl; abstr 2009).
37. Hashimoto N, Tsuboi A, Kagawa N, et al. Wilms tumor 1 peptide vaccination combined with temozolomide against newly diagnosed glioblastoma: safety and impact on immunological response. *Cancer Immunol Immunother*. 2015;64(6):707–716.
38. Weller M, Pfister SM, Wick W, Hegi ME, Reifenberger G, Stupp R. Molecular neuro-oncology in clinical practice: a new horizon. *Lancet Oncol*. 2013;14(9):e370–379.
39. Yan H, Parsons DW, Jin G, et al. IDH1 and IDH2 mutations in gliomas. *N Engl J Med*. 2009;360(8):765–773.
40. Ward PS, Patel J, Wise DR, et al. The common feature of leukemia-associated IDH1 and IDH2 mutations is a neomorphic enzyme activity converting alpha-ketoglutarate to 2-hydroxyglutarate. *Cancer Cell*. 2010;17(3):225–234.
41. Schumacher T, Bunse L, Pusch S, et al. A vaccine targeting mutant IDH1 induces antitumour immunity. *Nature*. 2014;512(7514):324–327.
42. Accolla RS, Tosi G. Optimal MHC-II-restricted tumor antigen presentation to CD4+ T helper cells: the key issue for development of anti-tumor vaccines. *J Transl Med*. 2012;10:154.
43. Newcomb EW, Demaria S, Lukyanov Y, et al. The combination of ionizing radiation and peripheral vaccination produces long-term survival of mice bearing established invasive GL261 gliomas. *Clin Cancer Res*. 2006;12(15):4730–4737.
44. Ochiai H, Archer GE, Herndon 2nd JE, et al. EGFRvIII-targeted immunotoxin induces antitumor immunity that is inhibited in the absence of CD4+ and CD8+ T cells. *Cancer Immunol Immunother*. 2008;57(1):115–121.
45. Jackson C, Ruzevick J, Brem H, Lim M. Vaccine strategies for glioblastoma: progress and future directions. *Immunotherapy*. 2013;5(2):155–167.
46. Bregy A, Wong TM, Shah AH, Goldberg JM, Komotar RJ. Active immunotherapy using dendritic cells in the treatment of glioblastoma multiforme. *Cancer Treat Rev*. 2013;39(8):891–907.
47. Okada H, Kalinski P, Ueda R, et al. Induction of CD8+ T-cell responses against novel glioma-associated antigen peptides and clinical activity by vaccinations with α-type 1 polarized dendritic cells and polyinosinic-polycytidylic acid stabilized by lysine and carboxymethylcellulose in patients with recurrent malignant glioma. *J Clin Oncol*. 2011;29(3):330–336.

48. Pollack IF, Jakacki RI, Butterfield LH, et al. Antigen-specific immune responses and clinical outcome after vaccination with glioma-associated antigen peptides and poly-inosinic-polycytidylic acid stabilized by lysine and carboxymethylcellulose in children with newly diagnosed malignant brainstem and nonbrainstem gliomas. *J Clin Oncol*. 2014;32(19):2050–2058.
49. Okada H, Butterfield LH, Hamilton RL, et al. Induction of robust type-I CD8$^+$ T-cell responses in WHO grade 2 low-grade glioma patients receiving peptide-based vaccines in combination with poly-ICLC. *Clin Cancer Res*. 2015;21(2):286–294.
50. Phuphanich S, Wheeler CJ, Rudnick JD, et al. Phase I trial of a multi-epitope-pulsed dendritic cell vaccine for patients with newly diagnosed glioblastoma. *Cancer Immunol Immunother*. 2013;62(1):125–135.
51. Wen P, Reardon D, Phuphanich S, et al. A randomized double blind placebo-controlled phase 2 trial of dendritic cell (DC) vaccine ICT-107 following standard treatment in newly diagnosed patients with GBM. *Neuro Oncol*. 2014;16(suppl 5):v22.
52. Schoor O, Hilf N, Dutoit V, et al. A novel multi-peptide cancer vaccine for treatment of glioblastoma. *Cancer Res*. 2010;70:2396.
53. Halford S, Rampling R, James A, et al. Final results from a Cancer Research UK first in man phase I trial of IMA950 (a novel multi peptide vaccine) plus GM-CSF in patients with newly diagnosed glioblastoma. *Ann Oncol*. 2014;25(suppl 4):iv361–iv372.
54. Pellegatta S, Cuppini L, Finocchiaro G. Brain cancer immunoediting: novel examples provided by immunotherapy of malignant gliomas. *Expert Rev Anticancer Ther*. 2011;11(11):1759–1774.
55. Dutoit V, Herold-Mende C, Hilf N, et al. Exploiting the glioblastoma peptidome to discover novel tumour-associated antigens for immunotherapy. *Brain*. 2012;135(Pt 4):1042–1054.

CHAPTER

10

Immunotoxin Therapy for Brain Tumors

V. Chandramohan[1], *J.H. Sampson*[1], *I.H. Pastan*[2], *D.D. Bigner*[1]

[1]Duke University Medical Center, Durham, NC, United States; [2]National Institutes of Health, Bethesda, MD, United States

OUTLINE

Translational Immunotherapy of Brain Tumors
http://dx.doi.org/10.1016/B978-0-12-802420-1.00010-7

INTRODUCTION

Glioblastoma (GBM) is the most common primary malignant brain tumor of the central nervous system (CNS).[1] The median survival for newly diagnosed GBM patients with the current standard of care, including maximum safe surgical resection, radiotherapy, and concomitant chemotherapy with temozolomide, is 14.6 months.[2] Median survival for recurrent GBM patients with currently available salvage therapies is only nine months.[3] Given the poor survival rate observed with the present approved therapies, there is a need for the development of innovative therapeutics to improve GBM patient survival.

Targeted therapy based on the molecular profiling of tumors has generated improved clinical outcomes in some patients with advanced cancers.[4] Molecular characterization of GBM oncogenesis has identified distinct biological features common to glioma but atypical in normal brain tissue.[5,6] Identification of tumor-associated antigens (TAAs) warrants selective targeting of tumor cells, with low toxicity to the surrounding normal tissue.[7]

HISTORY OF IMMUNOTOXINS

Development of Immunotoxins

The seminal article by Moolten and Cooperband in 1970 set the stage for the concept of targeting cancer cells using antibody–toxin conjugates.[8] As a proof of principle, they tested the feasibility of using mumps antibody–diphtheria toxin conjugate against viral antigens overexpressed on the surface of monkey kidney cells. Their study successfully demonstrated the applicability of antibody–toxin fusion proteins to target tumor cells. The advent of monoclonal antibody (mAb) technology and the identification of tumor antigen-specific mAbs facilitated the construction of mAb-toxin chimeric molecules, termed immunotoxins (ITs), to kill tumor cells.[9] In addition, ITs were also constructed by the fusion of toxin to protein ligands (growth factors and cytokines) that specifically bind to antigens overexpressed on tumor cells.[10] First generation ITs were constructed

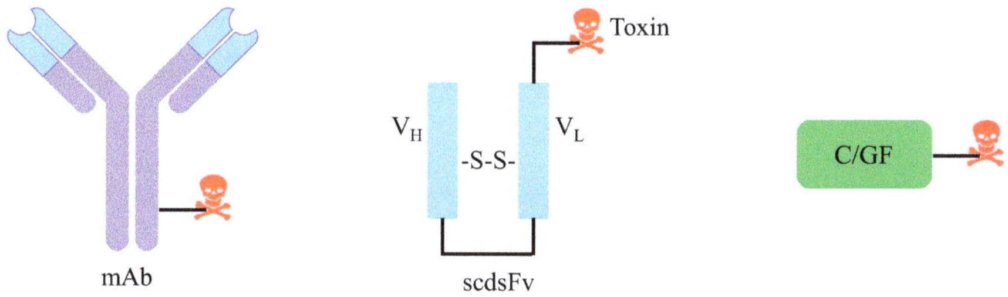

FIGURE 10.1 Different immunotoxin formats used in brain tumor therapy. *C*, cytokine; *GF*, growth factor; *mAb*, monoclonal antibody; *scdsFv*, single chain disulfide variable fragment; *S–S*, disulfide bond; V_H, variable heavy; V_L, variable light.

by chemical coupling of antibody or protein ligand to a toxin[11] (Fig. 10.1). The second generation ITs were created by the direct fusion of protein fragments (single chain variable region Ab fragments, scFv or ligands) to catalytically active toxin domains through recombinant DNA technology[12] (Fig. 10.1). To facilitate multiple cycles of therapy, the third generation ITs, which are smaller in size with reduced immunogenicity, are under development.[13,14]

Structure and Function of Immunotoxins

ITs are composed of two major domains: a targeting domain derived from an antibody or a protein ligand and a cytotoxic domain derived from a toxin molecule. The toxins used in the construction of ITs are natural products of plants and bacteria. Both plant and bacterial toxins share a common mechanism of action, protein synthesis inhibition, and are categorized as ribosome-inactivating proteins (RIPs). RIPs have been classified as type 1, consisting of a single-subunit with enzymatic activity and type 2, consisting of functionally distinct A (enzymatic activity) and B (binding to surface galactose or other surface residues) subunits.[15] The RIPs that are most commonly used in the construction of ITs include the bacterial toxins *Pseudomonas aeruginosa* exotoxin A (PE) and diphtheria toxin (DT), and the plant toxins ricin and saporin (Fig. 10.2). Type 1 RIPs include the plant toxin saporin, whereas type 2 RIPs comprise PE, DT, and ricin (Fig. 10.2). Bacterial toxins PE and DT catalyze the adenosine-5′-diphosphate (ADP)-ribosylation of elongation factor 2 (EF2) at the diphthamide-modified histidine 715, which inhibits translation elongation.[16] Plant toxins ricin and saporin inhibit protein synthesis through RNA N-glycosidase activity, which removes an adenine nucleotide from an exposed loop of 28S ribosomal RNA (rRNA) in the 60S ribosomal subunit.[17–19] This loop participates in EF binding and the cleaved loop no longer supports protein synthesis.[20] A single toxin molecule can inactivate 300 ribosomes in 35 min and in some

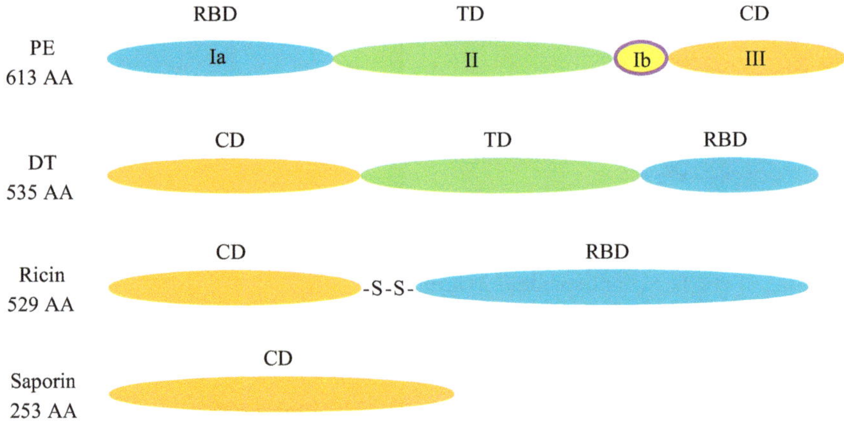

FIGURE 10.2 Graphical representation of bacterial [diphtheria toxin (DT) and *Pseudomonas* exotoxin (PE)] and plant toxins (ricin and saporin), used in immunotoxin construction. *AA*, amino acid; *CD*, catalytic domain; *RBD*, receptor binding domain; *S–S*, disulfide bond; *TD*, translocation domain.

cases a single molecule can kill a cancer cell.[21–24] Recombinant ITs were constructed by genetically conjugating antibody or ligand fragments to enzymatically active subunits of PE, DT, ricin, and saporin.

Toxins Employed in the Construction of Immunotoxins

Pseudomonas Exotoxin A

Pseudomonas exotoxin A is expressed as a 638 amino acid proprotein and upon processing, a 613-amino-acid mature toxin is secreted.[23] The mature toxin is composed of three major functional domains: a receptor binding domain, a translocation domain, and a catalytic domain[25,26] (Fig. 10.2). The N-terminal receptor binding domain consists of domains Ia (1–252 amino acids) and Ib (365–404 amino acids), and it targets the low-density, lipoprotein-receptor–related protein (LRP1) or the closely related LRP1B for cellular uptake of PE by receptor-mediated endocytosis.[27,28] The PE domain II (253–364 amino acids) is involved in toxin translocation and intracellular trafficking. Upon endocytosis, PE is cleaved between Arg279 and Gly280 by furin in endosomes to generate an enzymatically active 37-kDa C-terminal fragment, which is then translocated to the cytosol.[29] Once in the cytosol, the enzymatically active C-terminal domain (405–613 amino acids) catalyzes the ADP-ribosylation and inactivation of EF2, which leads to protein synthesis inhibition and cell death.[30] Initially, the chemical conjugation was used to generate full-length PE-based ITs.[31] Increased understanding of the toxin's structural and functional domains and improved molecular cloning techniques revolutionized the development of recombinant ITs. ITs with higher binding affinity to cancer cells were generated by replacing the receptor binding domain of PE with an antibody, an antibody fragment (scFv), a growth factor, or a cytokine targeting tumors.[10]

Diphtheria Toxin

Diphtheria toxin is secreted as a single polypeptide chain of 535 amino acids by the gram-positive bacterium *Corynebacterium diphtheria*.[32] Diphtheria toxin is composed of an amino terminal catalytic domain A (1–193 amino acids) and carboxyl terminal domain B (194–535 amino acids), which is further divided into receptor binding and translocation domains (Fig. 10.2). The C-terminal receptor-binding (R) domain (385–455 amino acids) is involved in cell surface binding and internalization of toxin by receptor-mediated endocytosis. The human heparin-binding epidermal growth factor-like precursor (HB-EGF) acts as the receptor for DT on the plasma membrane of human cells.[33] Once internalized by receptor-mediated endocytosis, an arginine-rich segment connecting the A and B domains is readily cleaved by trypsin-like enzymes to yield active fragments A and B.[34] The acidic endosomal pH causes a conformational change in the translocation (T) domain (201–384 amino acids), which then inserts into the endosomal membrane and facilitates the transport of the catalytic domain A into the cytoplasm.[35,36] Once it reaches the cytoplasm, the catalytic domain A initiates the ADP ribosylation and inactivation of EF2.[37–39] Modified EF2 can no longer make new proteins, and the cell dies via apoptosis. Recombinant-DT-based ITs are constructed by mutating or replacing the carboxyl terminal cell binding domain of DT with a ligand or the Fv fragment of an antibody to a cancer cell surface receptor.

Ricin

The plant toxin ricin is extracted from the seeds of *Ricinus communis*. Ricin is synthesized as a 576 amino acids precursor protein, consisting of N-terminal signal sequence (35 amino acids), A chain (267 amino acids), linker (12 amino acids), and B chain (262 amino acids)[40,41] (Fig. 10.2). During biosynthesis, the signal sequence and linker are removed to yield the mature protein, which is held together by a single disulfide bond. Ricin A chain carries the rRNA-specific N-glycosidase activity and the B chain, which is a D-galactose- and *N*-acetylgalactosamine-specific lectin that binds to the galactose-containing receptors on the cell surface.[42,43] Upon binding to its cognate receptor, ricin is internalized by endocytosis and undergoes retrograde transport from Golgi to endoplasmic reticulum (ER).[44] Cleavage of ricin A chain from B chain is catalyzed by the enzyme protein disulfide-isomerase in the lumen of ER; ricin A chain then enters the cytoplasm by utilizing the ER associated protein degradation pathway.[44] Once the A chain reaches the cytosol of the target cell, it enzymatically attacks the adenine residue at position 4324 within the 28S rRNA in the 60S ribosomal subunit, leading to the inactivation of ribosomes and protein synthesis inhibition.[18,45,46] ITs are generated by chemical crosslinking of the intact ricin or the recombinant ricin A chain to antibodies directed to antigens overexpressed on the tumor cell surface.

Saporin

Saporin (also known as saporin-S6) is a type I RIP isolated from the seeds of *Saponaria officinalis*. The mature form of saporin is composed of 253 amino acids[19] (Fig. 10.2). Due to the lack of receptor binding domain, saporin by itself enters cells at low efficiency. However, when conjugated to cell surface targeting molecules, saporin actively enters the cells through endocytosis, cleaves 28S rRNA, inhibits protein synthesis, induces apoptosis, and causes cell death.[19] Recombinant ITs were constructed by conjugating saporin to tumor directed antibody, or antibody fragments (Fab, scFv).[19]

IMMUNOTOXINS FOR BRAIN TUMOR THERAPY

The development of targeted cancer therapeutics had its origins in Paul Ehrlich's "magic bullet concept," in which the delivery of drugs selectively targets the disease-causing cells while remaining harmless to normal cells. Ehrlich's vision of magic bullet was realized by the development of mAbs by Köhler and Milstein. Since its inception, the utility of antibodies to specifically diagnose, target, and treat human tumors has been actively pursued. An antigen expressed only by the tumor cell and not by normal cells is hard to discover. However, several cell surface antigens that are expressed in much higher concentrations on glioma cells compared to normal brain tissues have been identified and mAbs targeting these tumor antigens have been developed.[47,48] We discuss the therapeutic utility of several of the glioma-associated antigens below. A summary of the ITs in clinical and preclinical testing for brain tumors is presented in Tables 10.1 and 10.2.

Transferrin Receptor

Transferrin receptor (TR) is a transmembrane glycoprotein expressed by all nucleated cells of the body. Transferrin receptor mediates the cellular uptake of iron from the plasma glycoprotein transferrin.[49] The expression level of TR corresponds to the cellular iron requirements and proliferative status. Overexpression of TR has been observed in several cancer types owing to the increased growth and rapid proliferation capacity of cancer cells.[50] Trowbridge and Domingo investigated the utility of an

TABLE 10.1 List of Immunotoxin Trials for Brain Tumors

Immunotoxin	Target	Payload	Phase	Tumor Treated	References	Status
454A12-rRA	TR	mAb fused to Ricin A chain	Phase I	Eight leptomeningeal neoplasia patients	59	Completed
TransMID (Tf-CRM107)	TR	Tf fused to full length DT	Phase II	Recurrent GBM	61	Completed
NBI-3001 (IL-4(38-37)-PE38KDEL	IL-4R	IL-4 linked to PE38	Phase II	Recurrent GBM	82	Completed
Cintredekin besudotox (IL-13-PE38QQR)	IL-13Rα2	IL-13 linked to PE38	Phase III	Recurrent GBM	102	Completed
IVAX (TP-38)	EGFR	TGF-α linked to PE38	Phase I	GBM, gliosarcoma metastatic spindle cell sarcoma, anaplastic oligodendroglioma	129	Completed
MR1-1 (MR1-1-PE38KDEL)	EGFRvIII	scFv fused to PE38	Phase I	Malignant brain tumors	141	Terminated
D2C7-IT (D2C7-(scdsFv)-PE38KDEL)	EGFR and EGFRvIII	scFv fused to PE38	Phase I	Recurrent GBM	145	Initiated

DT, diphtheria toxin; *EGFR*, epidermal growth factor receptor; *EGFRvIII*, EGFR variant III; *GBM*, glioblastoma; *IL-4R*, interleukin-4 receptor; *IL-13R α2*, interleukin-13 receptor alpha two; *mAb*, monoclonal antibody; *PE38*, *Pseudomonas* exotoxin A; *rRA*, recombinant ricin A chain; *scFv*, single chain variable region antibody fragment; *TGF-α*, transforming growth factor alpha; *TR*, transferrin receptor.

TABLE 10.2 Immunotoxins in Preclinical Testing for Brain Tumors

Immunotoxin	Target	Payload	Tumor Treated	References
F6V-PE38	GPNMB	scFv fused to PE38	GBM and medulloblastoma	149
9.2.27/Mel-14-PE38KDEL	HMW-MAA	scFv fused to PE38	GBM	156
NZ-1-(scdsFv)-PE38KDEL	Podoplanin	scFv fused to PE38	GBM and medulloblastoma	158
EphrinA1-PE38QQR	Ephrin receptor	EphrinA1 linked to PE38	GBM	167
DTAT	uPAR	uPA linked to DT	GBM	169
DTEGF13	EGFR and IL-13Rα2	EGF and IL-13 fused to DT	GBM	170
EGFATFKDEL 7mut	EGFR and uPAR	EGF and uPA fused to PE38	GBM	171
DTAT13	uPAR and IL-13Rα2	uPA and IL-13 fused to DT	GBM	172

ATF, amino terminal fragment; *DT*, diphtheria toxin; *EGF*, epidermal growth factor; *EGFR*, EGF receptor; *GBM*, glioblastoma; *GPNMB*, glycoprotein (Transmembrane) nmb; *HMW-MAA*, high molecular weight-melanoma associated antigen; *IL-13R α2*, interleukin-13 receptor alpha 2; *PE38*, *Pseudomonas* exotoxin A; *scFv*, single chain variable region antibody fragment; *uPA*, urokinase-type plasminogen activator; *uPAR*, uPA receptor.

anti-TR mAb chemically coupled to the subunits of DT or ricin for tumor cell-directed targeted therapy.[51] Since then, there have been several in vitro and in vivo studies investigating TR-based ITs for the treatment of malignant brain tumors.[52,53] The two major TR-targeting ITs utilized in the efficacy studies against TR-expressing brain tumors include (1) Tf-CRM107, a conjugate of human transferrin and full-length DT with two-point mutations in the B chain (CRM107) which inhibits nonspecific binding of toxin to normal cells[54] and (2) 454A12-rRA, a conjugate of human TR-targeted mAb (454A12) and recombinant ricin A chain (rRA). In both in vitro and in vivo testing, Tf-CRM107 exhibited significant potency against human GBM cell lines compared to 454A12.[55–57]

The prevalence of leptomeningeal neoplasia is 5–20%. Leptomeningeal patients have a poor clinical outcome with a median survival of only a few months.[58] A Phase I study to identify the pharmacokinetics, toxicity, and antitumor activity of a single-dose of 454A12-rRA IT delivered by intraventricular injection was conducted in eight patients with leptomeningeal neoplasia (six breast carcinoma, one melanoma, and one leukemia).[59] In ventricular cerebrospinal fluid (CSF), the early and late phase half-life of 454A12-rRA was 44±21 min and 237±86 min, respectively. No drug-related toxicity was observed at dose levels ≤38 μg of 454A12-rRA. Patients presented signs of acute toxicity at ≥120 μg of 454A12-rRA. Subsequent to the intraventricular delivery of 454A12-rRA, a more than 50% drop in tumor cell counts was detected in the lumbar CSF in four of the eight treated patients. However, seven of the eight patients treated with 454A12-rRA exhibited no evidence of CSF tumor clearance and demonstrated tumor progression.[59]

The toxicity, safety, and efficacy of Tf-CRM107 were evaluated in a Phase I, dose escalation, single-arm trial in malignant brain tumor patients refractory to conventional therapy.[60] To bypass blood–brain barrier, reduce systemic toxicity, and achieve high concentration of IT in the tumor locality, Tf-CRM107 was delivered directly into the tumor by high-flow interstitial microinfusion. Eighteen patients (age: 24–61 years) were enrolled in the trial, in which five received a single infusion and 13 received multiple cycles of infusion with Tf-CRM107. Fifteen patients were evaluated for radiographic evidence of tumor regression. Decrease in tumor volume (≥50%) was observed in 9/15 patients (60%) evaluated.[60] The treated tumors showed complete responses (CRs) in an anaplastic astrocytoma (AA) and a recurrent GBM patient. The recurrent GBM patient remained tumor free for 23 months after a single infusion of Tf-CRM107.[60] Phase I results demonstrated that direct tumor delivery of Tf-CRM107 via high-flow interstitial microinfusion can elicit antitumor activity in the absence of severe neurologic or systemic toxicity in malignant brain tumor patients.

A multicenter Phase II trial of Tf-CRM107 by intratumoral convection-enhanced delivery (CED) in patients with recurrent GBM or AA was initiated. The primary objectives of the study were to evaluate the safety and efficacy of Tf-CRM107 in recurrent GBM and AA patients. Participants (age: 15–75 years) enrolled for the study received Tf-CRM107 infusions (0.67 μg/mL, up to 0.40 mL/h) for 4–5 days until a volume of 40 mL was delivered.[61] A second cycle of Tf-CRM107 treatment was delivered to patients 4–10 weeks after initial infusion. Radiographic evidence of a minimum of 50% decrease in tumor volume within 1 year after the second Tf-CRM107 treatment was the primary endpoint of the study. Tf-CRM107 treatment efficacy response was categorized as CR, partial response (PR), stable disease, or progressive disease. Forty-four patients were enrolled for the Phase II study and 31/44 participants completed two treatment cycles with Tf-CRM107. In the 34 evaluable patients, Tf-CRM107 treatment resulted in CRs and PRs in a total of five and seven patients (35%, $P<.0001$), respectively. The median survival time was 37 weeks, 13/44 patients (30%) survived more than one year after initial infusion, and one patient survived 3.1 years. Tf-CRM107 infusions caused symptomatic progressive cerebral edema and seizures in 8/44 and 3/44 patients, respectively, though they responded to medical management.[61] Since the recurrent GBM and AA patients had already failed standard therapy, an objective response rate of at least 5% was chosen for the Phase II study. Tumor response was demonstrated in a significant percentage (35%) of recurrent or refractory GBMa or AA patients, and the tumor response observed through magnetic resonance imaging (MRI) correlated positively with survival. A Phase III clinical trial, involving two intratumoral infusions of Tf-CRM107 (TransMID), to improve overall survival (OS) of patients with nonresectable, recurrent primary GBM was initiated in 2004 by Celtic Pharma. However, the study was discontinued in 2007 because of a failure to demonstrate a positive therapeutic response.

In addition to the density of TRs on the tumor cell surface, the efficacy of Tf-toxin conjugate is influenced both by the Tf-iron saturation state and by the presence of free diferric Tf in the tumor microenvironment. The affinity of Tf-Saporin (Tf-SO6) to TR was greatly diminished when Fe-desaturation happened and resulted in the loss of protein synthesis inhibition.[62] Similarly, when there was excess diferric-Tf in the assay media, the ability of Tf-SO6 to inhibit protein synthesis was completely abolished.[62] The Tf-iron desaturation drawback was overcome by using a genetically engineered Tf with a decreased iron release rate and an increased cellular association and cytotoxicity.[63] However, it is difficult to control the levels of diferric Tf in the tumor microenvironment. The inability of Tf-CRM107 to generate a favorable therapeutic response in the Phase 3 study could be attributed to Tf desaturation and diferric Tf competition. ITs derived from antibodies targeting TR are not susceptible to Tf-iron saturation or the presence of diferric Tf.[62] For these reasons, it would be advantageous to use antibody-derived toxins for TR targeting in future clinical trials.

Interleukin-4 Receptor

Interleukin-4 (IL-4) is a pleiotropic cytokine produced in response to receptor activation by TH2-type $CD4^+$ T cells, basophils, and mast cells. Major functions of IL-4 include regulation of antigen-stimulated naive T cell differentiation and the specificity of immunoglobulin class switching by human B cells to express IgE and IgG4.[64] IL-4 functions by signaling through its receptor IL-4R, which is expressed in several cell types, including hematopoietic, endothelial, muscular, and neuronal cells, and are generally expressed at levels ranging from 100 to 5000 receptors per cell.[64] Two types of IL-4 receptors, Type I and Type II, have been identified so far. Type I receptors consist of a 140-kDa protein, IL-4Rα (also known as IL-4Rβ), and the γ common (γ_c) chain.[65,66] Type II receptors are composed of IL-4Rα and IL-13Rα1 (also known as IL-13Rα′) chains,[67,68] and are expressed mainly in nonhematopoietic cells. Ligand binding induces heterodimerization of IL-4R, leading to the activation of cytoplasmic tyrosine kinases (Janus kinase family, Jak), which in turn induces the phosphorylation of cellular substrates and initiation of intracellular signaling cascade, leading to the induction of gene expression and cellular proliferation.[64]

Several human tumors, including malignant melanoma, breast carcinoma, ovarian carcinoma, renal cell carcinoma, GBM, and neuroblastoma, are known to express IL-4R.[69–71] A chimeric toxin, human interleukin 4 (hIL4)-PE4E, targeting IL-4R, was constructed by fusing human IL-4 and a full-length PE molecule, with mutation in the cell-binding domain of PE. The recombinant hIL-4-PE4E was cytotoxic to a wide range of human cancer cell lines with the half maximal inhibitory concentration (IC_{50}) (50% inhibition of protein synthesis) in the 12–120 pM range.[72,73] The IC_{50} of IL-4-PE4E against neurological tumors was in the range of 85–2000 pM.[71] In comparison to the native hIL-4, there was a 15- to 20-fold decrease in the affinity of the hIL-4-PE4E toxin.[72] This drop in hIL-4-PE4E toxin affinity could be due to the attachment of the large PE molecule to the carboxyl terminus of hIL-4.[72] To increase the affinity of hIL-4-based toxin to its receptor, a circularly permuted recombinant IL-4 was developed by connecting the IL-4 amino acids 38–129 to 1–37 via a five amino acid linker GGNGG. The circularly permuted recombinant IL-4 [IL-4 (38–37)] was then fused to a redesigned PE containing amino acids 253–364 and 381–608, followed by a new ER retention sequence, KDEL.[74,75] This new toxin, IL-4(38–37)-PE38KDEL, exhibited improved affinity, cytotoxicity (IC_{50}: 6–17 pM), and in vivo antitumor activity against IL-4R-expressing cancers.[74,75] Additionally, malignant brain tumor samples expressing the IL-4 receptor were highly sensitive to IL-4(38–37)-PE38KDEL treatment (IC_{50}: 9–180 pM).[76,77] Intratumoral administration of IL-4(38–37)-PE38KDEL into U251 GBM flanks tumors in nude mice induced complete regression of tumors in all of the animals.[78] Toxicity studies in monkeys revealed an absence of IL-4(38–37)-PE38KDEL induced CNS or other systemic abnormalities at 2- and 6-μg/kg toxin concentration.[77]

Based on the preclinical studies, a Phase I open-label, dose-escalation trial (0.2–6 μg/mL) was initiated to determine the safety and efficacy of direct intratumoral infusion of IL-4(38–37)-PE38KDEL (NBI-3001) in nine recurrent malignant glioma patients.[79] No apparent systemic toxicity or neurotoxicity to normal brain was identified in any patient. Six of nine patients treated with IL-4(38–37)-PE38KDEL showed extensive dose-dependent glioma necrosis, as evidenced by decreased gadolinium enhancement on their MRIs. Of the six treated patients, one patient remained disease free for >18 months after a single infusion of IL-4(38–37)-PE38KDEL (6 μg/mL).[79]

A multicenter Phase I open-label, dose-escalation trial was launched to determine the maximum tolerated dose (MTD), volume, and safety of IL-4(38–37)-PE38KDEL delivered intratumorally by CED.[80,81] Twenty-five GBM and six AA patients were treated with IL-4(38–37)-PE38KDEL at one of four dose levels: 6 μg/mL, 9 μg/mL, or 15 μg/mL in

40 mL, and 9 μg/mL in 100 mL.[80,81] No drug-related systemic toxicity was evident in the 31 patients treated with IL-4(38–37)-PE38KDEL. Treatment-related adverse events occurred mainly in CNS. Drug-related CNS toxicity (grade 3 or 4) was observed in 12/31 patients (39%) at all of the four dose levels tested, and in 2/9 patients (22%) at the MTD dose of 6 μg/mL in 40 mL. An increase in serum IgG levels (≥4-fold) was noted in 11/28 patients and this correlated with the corresponding increase in the generation of anti-PE antibodies (3.77–28.48 fold) in patients, pre- and post-IL-4(38-37)-PE38KDEL infusion.[80] The overall median survival for a total of 27 patients (based on data availability) was 8.2 months and the median survival for GBM patients was 5.8 months. Six-month survival was 52% and 48%, for all and GBM patients, respectively. One GBM patient receiving 6 μg/mL in 40 mL was alive three years after IL-4(38–37)-PE38KDEL treatment. Following IL-4(38–37)-PE38KDEL treatment MRI of the brain showed areas of decreased signal intensity consistent with tumor necrosis.[80,81] An increase in MRI contrast enhancement at four weeks post-IL-4(38–37)-PE38KDEL infusion followed by a decrease was noted. IL-4(38–37)-PE38KDEL-induced tumor necrosis and inflammation were hypothesized to be the cause for the observed increase in contrast enhancement.

A multicenter, randomized, open-label Phase II study, to evaluate the safety, tolerability, and the optimal dose of IL-4(38–37)-PE38KDEL administered by CED followed by surgical resection of the tumor (3 weeks post infusion) was initiated in recurrent GBM patients.[82] A total of 30 adult patients (age: ≥18 years) with unilateral, unifocal tumor with a volume of 5–80 mL and a Karnofsky Performance Score (KPS) of ≥70 were enrolled.[83] The results of the study have not been published. Currently, there are no Phase III protocols involving IL-4(38–37)-PE38KDEL.

Interleukin-13 Receptor

Interleukin-13 (IL-13) is a 12-kDa cytokine that is expressed by CD4 T helper 2 (Th2) cells, basophils, eosinophils, mast cells, invariant natural killer T cells, and type 2 innate lymphoid cells.[84] There is a 30% amino acid sequence homology between IL-13 and IL-4 proteins. Hence, these cytokines share a variety of tasks.[85] Functions of IL-13 include major histocompatibility complex class II and CD23 upregulation on monocytes, induction of IgG class switching, and synthesis of IgG and IgM in B cells, and inhibition of inflammatory cytokine production.[86] There are two receptors for IL-13; IL-13Rα1 and IL-13Rα2. IL-13Rα1 is a low-affinity IL-13 receptor. IL-13 upon interaction with IL-13Rα1 binds to IL-4Rα with high affinity and activates intracellular signaling through Janus-activated kinase/signal transducers and activators of transcription 6 pathway in tumor cells, endothelial cells, fibroblast, and immune cells.[87,88] IL-13Rα2 is a membrane-bound protein with a high affinity for IL-13; the IL-13Rα2-IL-13 complex is internalized upon ligand binding.[88] IL-13Rα2 binds IL-13 in an IL-4–independent manner.[89]

The IL-13Rα2 is expressed in many human tumors, including GBM, pediatric brain tumors (astrocytoma, ependymoma, medulloblastoma, and brain stem glioma), squamous cell carcinoma of the head and neck, pancreatic cancer, colorectal cancer, and acquired immunodeficiency syndrome–associated Kaposi sarcoma.[88–96] Human glioma cells overexpressing IL-13R were highly sensitive (IC_{50} at <20 pM) to a chimeric toxin, IL-13-PE38QQR, composed of hIL-13 and a mutated version of PE (PE38QQR). To regulate the chemical coupling of antibody to toxin molecules at specific amino acid residues, the PE38QQR toxin was generated by replacing lysines 590 and 606 of PE with glutamines and lysine 613 with arginine.[97] A second fusion toxin, DT_{390}IL13 (DTIL13), composed of hIL-13 and the first 389 amino acids of DT, exhibited significant cytotoxicity against GBM cell lines. Further, DTIL13 potentiated tumor regression and prolonged survival of subcutaneous and intracranial GBM tumors in nude mice.[98,99] A recombinant IT, anti-IL-13Rα2(scFv)-PE38, consisting of human scFvs against IL-13Rα2, isolated from a phage library, and PE38 was also constructed. The in vitro and in vivo efficacy of anti-IL-13Rα2(scFv)-PE38 against GBM cell line was significantly lower than that of IL-13-PE38.[100]

On the basis of promising preclinical data, several Phase I/II clinical trials were initiated in recurrent malignant glioma patients with IL13-PE38QQR [also known as Cintredekin Besudotox (CB; NeoPharm, Lake Bluff, IL)]. The tolerability, tissue distribution, and methods for optimal intracerebral CED of IL13-PE38QQR were evaluated in three Phase I clinical trials (IL13PEI-002, IL13PEI-103, and IL13PEI-105).[101] A total of 53 patients were enrolled in the United States and Israel and 51 patients [46 GBMs, three AA, one anaplastic oligoastrocytoma, one anaplastic oligodendroglioma (AO)] were treated with IL13-PE38QQR by CED. Following tumor resection, one to three catheters were inserted in intraparenchymal areas that were at risk from residual infiltrating tumors. IL13-PE38QQR was then administered by CED at concentrations of 0.25 μg/mL (21 patients), 0.5 μg/mL (27 patients), and 1.0 μg/mL (3 patients) for 96 h (45 patients), 120 h (3 patients), and 144 h (3 patients). To assess IL13-PE38QQR distribution, intratumoral infusion prior to tumor resection was completed in 18 of 51 patients. Dose-limiting toxicities (DLT) were not observed at IL13-PE38QQR concentrations of 0.25 μg/mL and 0.5 μg/mL and were well tolerated with infusion durations of 5–6 days. However, DLT, including necrotic and inflammatory processes, were observed in patients (2/3) treated with 1 μg/mL of IL13-PE38QQR. The 1 μg/mL of IL13-PE38QQR was defined to be the dose-limiting intraparenchymal concentration whereas the maximum tolerated infusate concentration was established to be 0.5 μg/mL. IL13-PE38QQR and

CED procedure-related adverse events were primarily limited to the CNS and were observed in all patients. There was no correlation between IL13-PE38QQR concentration and the severity of the observed adverse event. An overall median survival of 45.9 weeks for the entire group ($N=51$) and a median survival of 42.7 weeks for GBM patients ($N=46$) were observed. Postoperative catheter placement was important for optimal drug distribution and improved patient survival. Optimal catheter positioning (≥2 optimally positioned catheters) ($N=24$) resulted in improved median survival of 55.6 weeks as opposed to 37.4 weeks (<2 optimally positioned catheters) ($N=19$) for GBM patients. Sixteen recurrent GBM patients ($N=46$) had a prolonged progression-free survival beyond 1 and 2 years, with patient follow-up extending beyond 5 years.[101]

A Phase III randomized evaluation of IL13-PE38QQR by CED compared to Gliadel wafer [GW; Rhone-Poulenc Rorer, Inc. (Paris, France), and Guilford Pharmaceuticals, Inc. (Baltimore, MD)] was conducted in adult recurrent GBM patients (PRECISE Trial).[102] A total of 296 patients were enrolled at 52 centers in the United States, Canada, Europe, and Israel, with OS as the primary endpoint. The primary objective of the study was to determine whether IL13-PE38QQR CED therapy compared to GW treatment is efficacious in improving the OS of recurrent GBM patients. Safety and toxicity assessment were secondary objectives and health-related quality-of-life parameters were the tertiary objective. Patients were randomized to receive IL13-PE38QQR ($N=192$) and GW ($N=104$). Of these 296 patients, a total of 276 patients [intent-to-treat (ITT) population] had histopathologic confirmation of GBM ($N=183$ in IL13-PE38QQR and $N=93$ in GW arms). IL13-PE38QQR was administered at a concentration of 0.5 μg/mL at a total flow rate of 0.75 mL/h over 96 h via 2–4 intraparenchymal catheters placed 2–7 days after gross total tumor resection. Patients randomized for GW group received a maximum of eight wafers, 3.85%/7.7 mg carmustine per wafer, immediately after tumor resection.[102] The total number of patients available for safety and efficacy analysis following IL13-PE38QQR or GW treatment was 269 and 188, respectively. The ITT population had a median survival of 36.4 weeks (9.1 months) and 35.3 weeks (8.8 months) for the IL13-PE38QQR and GW treatment groups ($P=.476$), respectively. The efficacy evaluable population had a median survival of 45.3 weeks (11.3 months) and 39.8 weeks (10 months) for IL13-PE38QQR and GW groups ($P=.310$), respectively.[102] Both treatment groups had similar safety profiles. However, the incidence of vascular disorders (grade ≥3 adverse events) was significantly higher in the IL13-PE38QQR treatment group ($P<.001$). The Phase III PRECISE trial was the largest study using CED with an active comparator in recurrent GBM patients. Despite being well tolerated, IL13-PE38QQR administered via CED had no survival advantage compared with GW.

The clinical success of therapeutic drugs delivered via CED for brain tumor patients depends on the specificity of the drug for its target antigen, optimal catheter positioning, and real-time imaging techniques to validate sufficient drug distribution to the target region. Since earlier studies described high levels of IL-13Rα2 expression in a significant proportion of GBM samples, patients included in the current trial were not selected on the basis of IL-13Rα2 expression. However, a later study revealed IL-13Rα2 to be expressed in only 44–47% of GBM specimens, and the distribution of IL-13Rα2 was highly heterogeneous within different sites of individual tumors.[103] This study underscored the importance of prescreening patients for target antigen expression prior to enrollment into targeted therapies. Catheter configuration, infusion rate and volume, and catheter positioning are some of the parameters that affect the volume of drug distribution by CED. Despite rigorous training of neurosurgeons in catheter placement, only 68% of catheters were placed according to protocol guidelines. Finally, a critical factor in evaluating CED is to ascertain if the delivered drug reached the residual tumor locality to have a therapeutic effect. The current trial did not include real-time imaging to ascertain if the requisite concentration of the therapeutic agent was delivered to the tumor site. Hence, optimization of several important factors, including initial screening of patients for target antigen (IL-13Rα2) expression, optimal catheter positioning, and real-time drug delivery imaging, should increase the success rate of future clinical trials.

To overcome the complications associated with the delivery of drugs via CED for brain tumor therapy, toxin resistant human neural stem cells (hNSCs) engineered to secrete IL13-PE were developed.[104] In in vitro studies, hNSCs-derived IL13-PE exhibited cytotoxic activity against both established and primary GBM cell lines expressing IL-13Rα2; a direct correlation between cytotoxicity and IL-13Rα2 levels on the target cells was noted. Importantly, hNSCs secreting IL13-PE encapsulated in a synthetic extracellular matrix demonstrated statistically significant efficacy compared to direct delivery of IL13-PE ($P = .0093$) in an intracranial GBM resection model.[104]

Epidermal Growth Factor Receptor

The epidermal growth factor receptor (EGFR) is a 170-kDa transmembrane glycoprotein. EGFR is composed of three functional domains—an extracellular ligand-binding domain (ECD), a transmembrane domain, and an intracellular tyrosine kinase catalytic domain. EGFR is activated by binding of its ligands [transforming growth factor (TGF)-α or EGF] to its ECD.[105,106] EGFR ligands are small mitogenic proteins composed of 53 (EGF) and 50 (TGF-α) amino acids. TGF-α and EGF share about 30% amino acid sequence identity, including the conservation of six cysteine residues involved in the formation of three intramolecular disulfide bonds that are essential for their biological activity.[107,108] EGF and TGF-α compete for the same sites for binding on the membrane-bound receptor.[105] Ligand binding induces receptor dimerization and activation of EGFR tyrosine kinase activity[109] that in turn controls epithelial-cell growth and proliferation. Ultimately, the EGFR signal is terminated upon internalization of the receptor–ligand complex.

EGFR overexpression is frequently observed in a wide variety of human cancers, including breast,[110] lung,[111] head and neck,[112] prostate,[113] bladder,[114] colorectal,[115] ovarian carcinoma,[116] and brain tumors.[117,118] EGFR expression in glioma has been demonstrated to be as high as 300-fold compared to its levels in normal control brain specimens.[119] *EGFR* is the most frequently amplified gene in GBM.[120] Correlating with the gene amplification, the protein is overexpressed in about 60–90% of GBM cases. Additionally, in the absence of gene amplification, protein overexpression, potentially due to aberrant translational and posttranslational mechanisms, has been observed in 12–38% of GBM patients.[121] In addition to protecting cells from apoptosis, EGFR signaling induces tumorigenic processes, such as proliferation, angiogenesis, and metastasis.[122]

Several EGFR ligand and PE-based chimeric toxins, for example, TGF-α-PE38 and TGF-α-PE40, were constructed. In in vitro assays, TGF-α-PE38 and TGF-α-PE40 were highly cytotoxic (IC_{50} < 11 pM) against GBM cell lines.[123,124] The activity of the toxins correlated to the number of EGFR on the tumor cell surface. Further, in vivo studies demonstrated the ability of TGF-α-PE38 and TGF-α-PE40 to retard tumor growth and increase the survival of nude mice that were implanted with subcutaneous or intracranial GBM xenografts.[123,124] An anti-EGF receptor antibody (425.3), chemically conjugated to the whole PE molecule, was also investigated for the treatment of human gliomas. Treatment of intracerebral U87MG tumors in nude rats with 4 μg of 425.3-PE produced an increase in survival from 23 to 40 days, with 2/9 rats in the 425.3-PE treatment group surviving >90 days.[125] An EGFR ligand DT-based protein, DAB_{389}EGF, was constructed by fusing the catalytic and translocation domains of DT (amino acid 1–387) to EGF (amino acids 1–53) via a His–Ala linker.[126] DAB_{389}EGF exhibited potent cytotoxicity (IC_{50} = 0.4–50 pM) against EGFR-expressing GBM cell lines. Significant correlation (P < .001) between DAB_{389}EGF cytotoxicity and levels of EGFR were observed in GBM cell lines.[126] Intratumoral DAB_{389}EGF treatment generated tumor regression in a subcutaneous GBM model established with U87MG cells.[127]

The toxicity of TGF-α-PE38 (TP-38, IVAX) was assessed in athymic rats and Rhesus macaques. Athymic rats were treated with 0.222, 0.666, 1, 2, 4, and 8 μg of TP-38. There were no histopathologic abnormalities observed in animals treated with 0.222 μg or 0.666 μg of TP-38. Demyelination and necrosis were detected in animals treated with higher doses of TP-38. On the basis of this study 0.666 μg was determined to be a safe intracerebral dose in athymic rats.[128] Rhesus macaques received 0.2, 2, or 6 μg of TP-38. Necrosis and hemorrhage were observed in one macaque that received 6 μg of TP-38. Therefore, 2 μg of TP-38 was established as the MTD in rhesus macaques.[128] TP-38 efficacy studies were performed against EGFR overexpressing A431 epidermoid carcinoma cells in athymic mice. A431 cells were mixed with phosphate-buffered saline (PBS) alone or 0.03, 0.1, and 0.3 μg of TP-38 toxin and implanted into the caudate nuclei of athymic mice. Mice treated with 0.03 and 0.1 μg of TP-38 survived with no evidence of tumor while those treated with 0.3 μg or higher died due to drug toxicity. Hence, 0.1 μg of TP-38 was concluded as the MTD in athymic mice. On the basis of animal studies, a total dose of 200 μg of TP-38 was extrapolated to be the MTD for human studies. Therefore, a starting total dose of <1.0 μg (at a concentration of 25 ng/mL) was predicted to be a safe range in humans.[128] A Phase I study was initiated in recurrent malignant glioma patients to define the MTD, DLT, and efficacy of TP-38 delivered by CED.[129] Twenty adult patients (17 GBM, 1 gliosarcoma, 1 metastatic spindle cell sarcoma, and 1 AO) were enrolled. On the basis of the presence of radiographic evidence of residual disease, patients were divided into two groups for TP-38 therapy. Patients without residual disease received 25 ng/mL ($N=3$) and 50 ng/mL ($N=2$) of TP-38. Patients with residual disease received 25 ng/mL ($N=3$), 50 ng/mL ($N=6$), and 100 ng/mL ($N=6$) of TP-38. Two catheters were placed into each patient with stereotactic guidance. TP-38 was infused over 50 h at a flow rate of 0.4 mL/h from each catheter for a total volume of 40 mL. In the last eight patients, coinfusion of ^{123}I-labeled human serum albumin (HSA) was performed to monitor TP-38 distribution within the brain.[129] Median survival for all patients after TP-38 treatment was 28 weeks, for patients with residual disease 20.1 weeks, and for those without radiographic evidence of residual disease 33 weeks. Two of the 15 patients in the residual disease group demonstrated radiographic responses with TP-38 treatment. One GBM patient receiving 50 ng/mL of TP-38 was alive and without progression >211 weeks after therapy, and a second GBM patient receiving 25 ng/mL of TP-38 had a nearly CR and was alive >260 weeks from therapy. Coinfusion of ^{123}I-labeled HSA demonstrated that homogeneous distribution of high concentrations of the infusate within human brain could be achieved by CED. However, only 3 of 16 (19%) catheters produced intraparenchymal infusate distribution, whereas the majority demonstrated significant leaks of infusate into the subarachnoid or intraventricular cerebrospinal fluid spaces. Owing to insufficient TP-38 delivery in a majority of the patients, the study was terminated at a dose of 100 ng/mL before arriving at an MTD. Two DLTs, one each at 50 and 100 ng/mL dose levels, were seen and both were neurologic.[129] Intracerebral CED of TP-38 at doses ≤100 ng/mL was well tolerated and generated certain durable radiographic responses. Future studies should concentrate on optimization of CED delivery technique and drug infusion parameters.

Mutant *Epidermal Growth Factor Receptor* Variant III

EGFRvIII is the most common *EGFR* deletion mutant that is present in 20–50% of GBMs with *EGFR* amplification.[130] The mutant *EGFRvIII* contains an in-frame deletion of 801 base pairs of the coding region that corresponds to exons 2–7 of the *EGFR* gene.[131] This deletion creates a novel glycine residue at the fusion junction, generating a tumor-specific protein sequence. EGFRvIII is a constitutively active receptor tyrosine kinase,[132] which is widely expressed in malignant gliomas[133] and carcinomas, including head and neck[134] and breast cancers.[135] EGFRvIII overexpression induces GBM resistance to commonly used chemotherapeutic agents[136] rendering EGFRvIII a desirable target for therapeutic intervention.

The therapeutic efficacy of EGFRvIII-specific mAbs, L8A4, Y10, and H10, chemically coupled to PE35KDEL, was evaluated against a mutant EGFRvIII-expressing cell line. All three ITs were highly efficacious in in vitro cytotoxicity assays, with IC_{50} in the range of 15–50 pM.[137] A scFv specific for EGFRvIII, MR1 (mutant receptor), was isolated from an immunized phage library and was fused to PE38KDEL. The cytotoxic activity (IC_{50}) of MR1-PE38KDEL against EGFRvIII-transfected malignant glioma cells was in the range of 110–160 pM.[138] A dose escalation toxicity study (three doses of 1–20 μg of MR1-PE38KDEL) compared the survival of the MR1-PE38KDEL treatment group with the saline group in nontumor bearing athymic rats. On the basis of survival data, 3 μg of MR1-PE38KDEL was established to be the nontoxic dose.[139] The therapeutic efficacy of three doses of MR1-PE38KDEL (1, 2, and 3 μg) was compared with saline and control IT (3 μg) in an EGFRvIII-expressing (U87MG.ΔEGFR) neoplastic meningitis model in athymic rats.[139] There were 75% (1-μg group) and 57% (2 or 3 μg group) long-term survivors in the MR1-PE38KDEL treatment group. The MR1-PE38KDEL treatment group failed to reach median survival by the termination of the study at 53 days. Hence, the median survival for the MR1-PE38KDEL group was estimated to be > 53 days. All saline or control IT-treated animals died, with median survival of 7 and 10 days, respectively. Histological examination of the 3 μg MR1-PE38KDEL treatment group presented evidence of demyelination. Regional therapy with three doses of 2 μg of the MR1-PE38KDEL was concluded to be the effective dose for the treatment of EGFRvIII-expressing neoplastic meningitis tumors.[139]

An affinity-matured variant of MR1, termed MR1-1, with increased affinity to EGFRvIII, was generated and fused to PE38, to generate MR1-1-PE38 for targeted glioma therapy.[140] MR1-1-PE38 in comparison to the parental MR1-PE38 exhibited improved cytotoxicity against the EGFRvIII-expressing NR6M cell line.[140] A Phase I study to determine the MTD and DLT of MR1-1-PE38KDEL delivered intracerebrally by CED in patients with supratentorial malignant brain tumors expressing EGFRvIII was initiated.[141] The study design included infusion of MR1-1-PE38KDEL by CED using two intracerebral catheters. ^{124}I-labeled HSA was coinfused with gadolinium–diethylene triamine pentaacetic acid (Gd-DTPA) to monitor drug distribution and leakage into the CSF space after infusion. A starting total drug dose of 0.5 μg (1/20th of the MTD in rats) at a fixed flow rate of 0.5 mL/h was infused from each of the two catheters. The distribution of MR1-1-PE38KDEL by CED at a concentration of 25 ng/mL was monitored by coinfusion with the low molecular weight tracer gadolinium and ^{124}I-labeled HSA in a malignant glioma patient.[142] This study established the utility of Gd-DTPA coinfusion to precisely demonstrate MR1-1-PE38KDEL distribution at the tumor site (Fig. 10.3). Monitoring of MR1-1PE38KDEL distribution will result in improved delivery of the drug at the tumor site and enhanced therapeutic efficacy. The MR1-1-PE38KDEL Phase I study was terminated due to low patient accrual.

D2C7 is a unique mAb that reacts with both EGFRwt and EGFRvIII proteins.[143] A novel recombinant IT, D2C7-(scdsFv)-PE38KDEL (D2C7-IT) was constructed by fusing the disulfide stabilized D2C7 scFv (scdsFv) with PE38KDEL.[144] In preclinical studies, the dual-specific IT D2C7-IT demonstrated a strong antitumor response against intracranial GBM

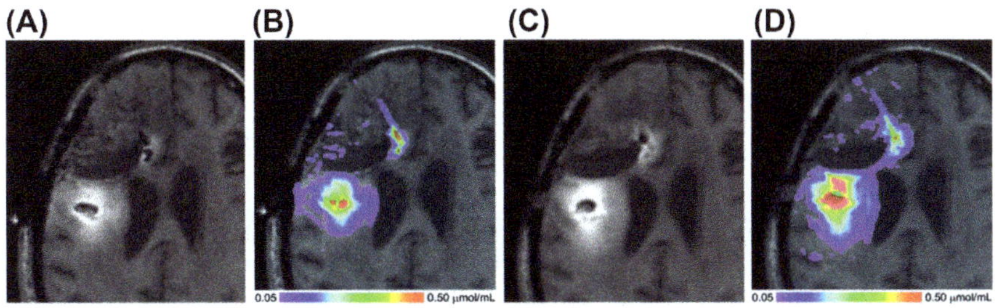

FIGURE 10.3 T1-weighted signal of MR1-1-PE38KDEL (A and C), compared with measured concentration profile of gadolinium–diethylene triamine pentaacetic acid (Gd-DTPA) (B and D). (A) T1-weighted signal of MR1-1-PE38KDEL at 24 h. (B) Concentration of Gd-DTPA at 24 h. (C) T1-weighted signal of MR1-1-PE38KDEL at 72 h. (D) Concentration of Gd-DTPA at 72 h. *Gd-DTPA*, gadolinium-diethylene triamine pentaacetic acid. *Reprinted from Sampson JH, Brady M, Raghavan R, et al. Colocalization of gadolinium-diethylene triamine pentaacetic acid with high-molecular-weight molecules after intracerebral convection-enhanced delivery in humans.* Neurosurgery. *2011;69(3):668–676, with permission from Wolters Kluwer Health.*

xenografts expressing either EGFRwt or both EGFRwt and EGFRvIII.[144] In a preclinical toxicity study in Sprague–Dawley rats, the MTD of D2C7-IT was determined to be between 0.10 and 0.35 μg total doses, and the no observed adverse effect level of D2C7-IT was the 0.05 μg total dose (manuscript in preparation). On the basis of the preclinical toxicity study, a Phase I study has been initiated to determine the MTD of D2C7-IT delivered intratumorally by CED in recurrent malignant glioma patients.[145]

Glycoprotein Nonmetastatic Melanoma Protein B

Glycoprotein nonmetastatic melanoma protein B (GPNMB), a potential GBM marker gene, is a type I transmembrane glycoprotein.[146] Normal brain samples express little or no *GPNMB*. However, *GPNMB* transcript [70% (35/50)] and protein [66% (52/79)] have been detected in a significant proportion of GBM patient samples.[147] A significantly higher risk of death was observed in patients with >3-fold *GPNMB* transcript levels over normal brain and positive immunohistochemistry.[147] Thus, GPNMB is an ideal antigen for targeted therapy of malignant glioma. The GPNMB protein is composed of a long ECD (465 amino acids), a single transmembrane region (21 amino acids), and a short cytoplasmic domain (65 amino acids).[148] A GPNMB-specific scFv clone, G49, was isolated from a human synthetic phage-display library.[149] The G49 scFv was subjected to affinity maturation and a mutant scFv, F6V, with increased affinity (0.3 nM) for GPNMB was obtained. The affinity-matured F6V scFv clone was fused to PE38 to generate F6V-PE38. The F6V-PE38 exhibited potent cytotoxicity against GPNMB-expressing glioma and medulloblastoma cells ($IC_{50} = 8$ pM). Further, F6V-PE38 demonstrated significant antitumor activity ($P < .001$) against GPNMB-expressing malignant glioma (D212MG) and medulloblastoma (D487MED) tumor models.[149]

High Molecular Weight Melanoma-Associated Antigen

The high molecular weight melanoma-associated antigen (HMW-MAA) is a single-pass type I transmembrane protein consisting of a long ECD (2195 amino acids), a hydrophobic transmembrane region (21 amino acids), and a short cytoplasmic tail (77 amino acids).[150] HMW-MAA is expressed as a 250-kDa N-linked glycoprotein that then undergoes extensive glycosylation to generate the >450-kDa proteoglycan.[151] The HMW-MAA expression has been demonstrated in melanomas, breast carcinomas, squamous cell carcinoma, and malignant gliomas and is recognized by the mAb 9.2.27.[152,153] The 9.2.27-PE IT was constructed by chemical conjugation of 9.2.27 mAb to the whole PE molecule. Intratumoral treatment of established human glioma xenografts with 9.2.27-PE extended the survival time of tumor-bearing animals from 30% to 90%.[154] The HMW-MAA is also recognized by the mAb Mel-14.[155] In an immunohistochemical analysis, 57% (28/49) of GBM patient specimens exhibited Mel-14 reactivity.[47] Both 9.2.27 and Mel-14 scFvs were fused to PE38KDEL to generate 9.2.27-PE38KDEL and Mel-14-PE38KDEL for targeted GBM therapy.[156]

Podoplanin

Podoplanin is a 162-amino acid type-I transmembrane sialomucin-like glycoprotein consisting of a serine- and threonine-rich extracellular domain (109 amino acids), a single transmembrane domain (21 amino acids), and a short cytoplasmic domain (10 amino acids).[157] Podoplanin expression has been described in several tumors, including squamous cell carcinoma, skin cancer, chondrosarcoma, germ cell tumors, lung cancer, mesothelioma, CNS tumors, and specific subtypes of vascular tumors.[158,159] Podoplanin is involved in epithelial–mesenchymal transition, metastasis, and invasion.[159,160] Among CNS tumors, podoplanin expression is observed in 83% of GBM and 27% medulloblastoma cases, making it an attractive target for brain tumor therapy.[161] NZ-1 is a mAb that is highly reactive with podoplanin.[162] A recombinant IT was constructed by fusing NZ-1 scFv to PE38KDEL to generate NZ-1-(scdsFv)-PE38KDEL (NZ-1-IT) for targeted brain tumor therapy.[158] In in vitro studies, the NZ-1-IT was highly cytotoxic, with an IC_{50} in the range of 25–446 pM against GBM and medulloblastoma xenograft cells. In in vivo studies with D2159MG GBM and D283MED medulloblastoma tumors, NZ-1-IT demonstrated growth delay, averaging 24 days ($P < .001$) and 21 days ($P < .001$), respectively. Crucially, NZ-1-IT caused a 41% increase in survival ($P \leq .001$) in the D425MED intracranial tumor model.[158]

Ephrin Receptor A2

The Ephrin receptors (Eph) are the largest family of tyrosine kinase receptors. Ephrin receptors, upon activation by their ligand ephrins, participate in cell positioning, tissue and organ patterning, control of normal and tumor cell survival, vascular development, axon guidance, and synaptic plasticity.[163,164] Based on ligand specificity, Eph are divided into nine Eph A- and five Eph B-receptors. The ligand Ephrins are comprised of five glycosylphosphatidylinositol-linked cell surface bound ephrin-A subclass (binds A-type receptor), and three transmembrane ephrin-B subclass (binds B-type receptor).[163–165] The positioning of Eph and ephrins necessitates cell–cell interactions for their binding and activation. Ephrin receptor A2 is overexpressed in a number of solid tumors, including those of the breast, prostate, ovary, melanoma, lung, cervical, colon, esophageal, gastric, bladder, and renal cell.[164] Overexpression of EphA2 was noted in about 90% of GBM specimens and cell lines; however, low levels of its ligand, ephrinA1, were observed.[166] Ephrin receptor A2-targeted therapy for GBM was developed by chemical conjugation of ephrinA1 to PE38QQR. The ephrinA1-PE38QQR cytotoxin exhibited significant killing of GBM cells overexpressing the EphA2, with an IC_{50} of ~10–100 pM.[167]

Urokinase-Type Plasminogen Activator Receptor

The urokinase-type plasminogen activator receptor (uPAR) regulates the proteolysis of extracellular matrix (ECM) by binding to the amino-terminal growth factor domain of its ligand, serine protease urokinase-type plasminogen activator (uPA).[168] Surface bound uPA cleaves its substrate plasminogen and generates the active plasmin. Plasmin in turn cleaves and activates matrix metalloproteases (MMPs). Together, plasmin and MMPs degrade ECM components and aid in growth factor mobilization. Further, uPAR binds to the protein vitronectin and activates intracellular signaling through its coreceptor integrins. Thus, uPAR plays a critical role in cell proliferation, differentiation, migration, and survival.[168] uPAR belongs to the lymphocyte antigen 6 (Ly-6) superfamily of proteins, which are characterized by the Ly-6 and uPAR (Lu) domain. uPAR contains three Lu domains, designated D1, D2, and D3, which are connected by short linkers.[168] uPAR is expressed in many human cancers, including bladder cancer, breast cancer, prostate cancer, colorectal cancer, gastric cancer, lung cancer, pancreatic cancer, GBM, leukemia, and lymphoma.[168] Significantly, uPAR expression in tumors can be detected both on tumor cells and tumor-associated stromal cells, including fibroblasts and macrophages.[168] Tumor expression of uPAR is an indicator of poor prognosis and in certain cases is a predictor of invasion and metastasis. A recombinant fusion protein, diphtheria toxin/urokinase fusion protein (DTAT), was constructed by fusing the catalytic portion of DT to the amino-terminal (AT) fragment (ATF) of uPA. The potency of DTAT was tested against uPAR-positive human GBM cells U118MG and U87MG.[169] DTAT exhibited significant in vitro killing of U118MG ($IC_{50} = 10\,nM$) and U87MG ($IC_{50} < 1\,nM$). Intratumoral treatment of subcutaneous U118MG tumors in athymic nude mice with DTAT (five injections of 20 μg DTAT) caused a statistically significant ($P = .05$) tumor regression in the DTAT group compared to the control treatment groups (PBS and DTIL13).[169]

Bispecific Cytotoxins

Bispecific cytotoxins targeting two different TAAs are being developed to improve the efficacy of anti-GBM therapeutics and to reduce the cost associated with combination therapies. A bispecific cytotoxin targeting

EGFR and IL-13Rα2, DTEGF13, was constructed by fusing IL-13, EGF, and the catalytically active truncated DT in a single polypeptide chain.[170] In an in vitro cytotoxicity assay, DTEGF13 demonstrated significant cytotoxic activity against U87MG (IC_{50} = 0.015 nM) and U118 (IC_{50} = 0.02 nM) GBM cells. The cytotoxicity of DTEGF13 against U87MG and U118MG cells showed an increase in activity compared to the monospecific cytotoxins DTEGF (U87MG IC_{50} = 1.2 nM and U118MG IC_{50} = 3 nM) and DTIL-13 (U87MG IC_{50} = 10 nM and U118MG IC_{50} = 2 nM).[170] Intratumoral injection of DTEGF13 (eight injections of 2.5 μg) generated significant inhibition in the growth of established U87MG tumors in nude mice ($P < .04$).[170] In contrast, U87MG tumors treated with the monospecific DTEGF or DTIL13 grew at the same rate as a nonspecific control toxin treated tumors with no demonstration of tumor growth inhibition.[170]

A second bispecific cytotoxin, EGFATFKDEL 7mut, was created by splicing EGF and ATF of uPA to a version of PE38KDEL in which the seven major immunodominant epitopes in PE38 were mutated to decrease toxin immunogenicity without a loss of catalytic activity.[171] EGFATFKDEL 7mut showed increased cytotoxic activity compared to the monospecific toxins (EGFKDEL and ATFKDEL), against U87MG cells with an IC_{50} of 0.001 nM.[171] Intracranial treatment of human U87MG xenograft, in athymic nude rats with EGFATFKDEL 7mut via CED, led to significant tumor growth delay in two independent experiments ($P < .01$).[171] Two out of five rats in the EGFATFKDEL 7mut treatment group remained tumor free over 130 days posttumor inoculation. A significant drop in antitoxin antibody response was observed between the EGFATFKDEL 7mut ($P < .05$) and EGFATFKDEL toxin immunized animals.[171]

A bispecific cytotoxin DTAT13 was also synthesized by fusing IL-13, ATF, and DT for simultaneous targeting of uPAR and IL-13Rα2-expressing GBM cells.[172] DTAT13 exhibited similar or decreased cytotoxicity compared to the corresponding monospecific toxins against U87MG and U373MG cells. The IC_{50} of DTAT13 and DTAT were 0.23 and 0.24 nM for U87MG cells, respectively.[172] The IC_{50} of DTAT13 and DTIL13 were 0.0007 and 0.00015 nM for U373MG cells, respectively.[172] In a subcutaneous model of GBM established with U373MG or U87MG cells, both DTAT13 and DTIL13 exhibited significant tumor growth inhibition compared to untreated controls ($P < .05$).[172] However, in in vivo toxicity studies, DTAT13 displayed less toxicity than DTAT or DTIL13.[172]

FUTURE DIRECTIONS

Successful delivery of the IT to tumor cells, selective killing of tumor cells in the absence of nonspecific toxicity, and prevention of anti-IT antibody generation are critical for the therapeutic success of a tumor-targeting agent. Several of the molecular targets for ITs while overexpressed on the tumor cells are also expressed at lower levels on normal tissues. The toxic effect of an IT on normal cells is circumvented by direct intratumoral delivery into brain by CED. Although intratumoral drug delivery and the best volume of distribution of a locally administered drug is achieved through CED, specific regions of the brain are naturally difficult to saturate completely with the infusate. Complete tumor coverage with the therapeutics in such cases will therefore depend on the proper placement of catheters. Advances in imaging will help predict therapeutic agent distribution by CED and improved computational models consequently can be used to optimize catheter positioning.[173] Finally, a major disadvantage for IT retreatment is the possibility of developing neutralizing antibodies against toxins. To avoid the production of neutralizing antibodies against PE and to allow for multiple treatment cycles, less immunogenic forms of PE have been generated by identifying and eliminating B- and T cell epitopes on PE.[13,174–176] The new PE molecule is significantly smaller and less immunogenic but still retains full cytotoxic and antitumor activities.

Brain tumors are highly heterogeneous, with multiple genetic alterations, and are therefore defined by a combination of tumor cell antigens. Owing to the overexpression of different tumor antigens, brain tumors tend to recur with single-agent therapy. Hence, to be effective, a therapeutic regimen should incorporate a cocktail of drugs aimed at different cellular targets critical for brain tumor development, growth, and invasion. Bispecific antibodies with different antigenic specificities are being developed to accomplish multitarget therapy. As discussed earlier, the clinical utility and efficacy of bispecific antibody-based therapy is still in its infancy, with the majority of supporting evidence emerging from preclinical studies. The therapeutic success of a bispecific antibody in the clinical setting depends on the identification of the appropriate antigenic target combinations.

Combination therapies that aim to achieve improved tumor cell killing by ITs have been explored as early as 1985.[177] Initial combination therapies aimed to increase the delivery of IT to cytosol to attain total tumor cell killing. More recently, IT-based combination therapies targeting antiapoptotic pathways that are hyperactive in tumor cells are being tested. IT and prosurvival inhibitor combination therapies have demonstrated efficacy in different tumor types, including colon cancer, melanoma, and small cell lung cancer.[178–180]

Furthermore, preclinical studies have underscored the involvement of $CD4^+$ and $CD8^+$ T cells in secondary immune responses in tumor models with MR1-1-(dsFv)-PE38KDEL and IL-13-PE38 ITs, respectively.[181,182] In addition to specifically targeting and destroying tumor cells, MR1-1-(dsFv)-PE38KDEL generated long-lasting immune responses that recognized and eliminated tumor antigen-expressing as well as tumor antigen-free cancer cells.[182] However, the highly lethal end-stage manifesting GBM tumors maintain an immunosuppressive tumor microenvironment. Immune suppression mediated by the GBM microenvironment is a major obstacle to an effective and durable IT-mediated antitumor immune response.[183,184] IT in combination with immune checkpoint inhibitors that will overcome glioma-microenvironment-associated immunosuppression will induce a durable and long-lasting antitumor response.[185,186] Combination therapies involving multiple agents targeting the heterogeneous brain tumor cell population and tumor microenvironment will be extremely powerful and are going to be the future of cancer therapy.

CONCLUSION

Construction of ITs with reduced toxicity, immunogenicity, along with improved delivery, and drug distribution monitoring techniques have revived our interest in ITs as potential antineoplastic agents. ITs target and kill tumor cells based on their unique molecular features or protein expression. However, the high level of intratumoral and intertumoral heterogeneity observed in brain tumors presents an enormous hurdle for ITs to be effective clinically as single agents. Targeting tumor heterogeneity using a multiagent approach might be beneficial for brain tumor patients. Identifying agents that work best in combination with ITs is currently under investigation.

Acknowledgment

This work was supported by the following NIH grants: P01 – 5P01CA154291, Outstanding Investigator Award – 1R35CA197264. This work was also supported in part by the Intramural Research Program of the NIH, National Cancer Institute, Center for Cancer Research.

References

1. Louis DN, Gusella JF. A tiger behind many doors: multiple genetic pathways to malignant glioma. *Trends Genet.* 1995;11(10):412–415.
2. Stupp R, Hegi ME, Mason WP, et al. Effects of radiotherapy with concomitant and adjuvant temozolomide versus radiotherapy alone on survival in glioblastoma in a randomised phase III study: 5-year analysis of the EORTC-NCIC trial. *Lancet Oncol.* 2009;10(5):459–466.
3. Friedman HS, Prados MD, Wen PY, et al. Bevacizumab alone and in combination with irinotecan in recurrent glioblastoma. *J Clin Oncol.* 2009;27(28):4733–4740.
4. Tsimberidou AM, Wen S, Hong DS, et al. Personalized medicine for patients with advanced cancer in the phase I program at MD Anderson: validation and landmark analyses. *Clin Cancer Res.* 2014;20(18):4827–4836.

5. Furnari FB, Fenton T, Bachoo RM, et al. Malignant astrocytic glioma: genetics, biology, and paths to treatment. *Genes Dev*. 2007;21(21):2683–2710.
6. Sathornsumetee S, Reardon DA, Desjardins A, Quinn JA, Vredenburgh JJ, Rich JN. Molecularly targeted therapy for malignant glioma. *Cancer*. 2007;110(1):13–24.
7. Herlyn M, Menrad A, Koprowski H. Structure, function, and clinical significance of human tumor antigens. *J Natl Cancer Inst*. 1990;82(24):1883–1889.
8. Moolten FL, Cooperband SR. Selective destruction of target cells by diphtheria toxin conjugated to antibody directed against antigens on the cells. *Science*. 1970;169(940):68–70.
9. Blythman HE, Casellas P, Gros O, et al. Immunotoxins: hybrid molecules of monoclonal antibodies and a toxin subunit specifically kill tumour cells. *Nature*. 1981;290(5802):145–146.
10. Pastan I, Hassan R, FitzGerald DJ, Kreitman RJ. Immunotoxin treatment of cancer. *Annu Rev Med*. 2007;58:221–237.
11. Vitetta ES, Krolick KA, Miyama-Inaba M, Cushley W, Uhr JW. Immunotoxins: a new approach to cancer therapy. *Science*. 1983;219(4585):644–650.
12. Pastan I, Hassan R, Fitzgerald DJ, Kreitman RJ. Immunotoxin therapy of cancer. *Nat Rev Cancer*. 2006;6(7):559–565.
13. Mazor R, Eberle JA, Hu X, et al. Recombinant immunotoxin for cancer treatment with low immunogenicity by identification and silencing of human T-cell epitopes. *Proc Natl Acad Sci USA*. 2014;111(23):8571–8576.
14. Pastan I, Onda M, Weldon J, Fitzgerald D, Kreitman R. Immunotoxins with decreased immunogenicity and improved activity. *Leuk Lymphoma*. 2011;52(suppl 2):87–90.
15. Stirpe F. Ribosome-inactivating proteins: from toxins to useful proteins. *Toxicon*. 2013;67:12–16.
16. Wayne AS, Fitzgerald DJ, Kreitman RJ, Pastan I. Immunotoxins for leukemia. *Blood*. 2014;123(16):2470–2477.
17. Lyu MA, Cao YJ, Mohamedali KA, Rosenblum MG. Cell-targeting fusion constructs containing recombinant gelonin. *Methods Enzymol*. 2012;502:167–214.
18. Olsnes S, Refsnes K, Pihl A. Mechanism of action of the toxic lectins abrin and ricin. *Nature*. 1974;249(458):627–631.
19. Polito L, Bortolotti M, Mercatelli D, Battelli MG, Bolognesi A. Saporin-S6: a useful tool in cancer therapy. *Toxins*. 2013;5(10):1698–1722.
20. Endo Y, Gluck A, Wool IG. Ribosomal RNA identity elements for ricin A-chain recognition and catalysis. *J Mol Biol*. 1991;221(1):193–207.
21. Carrasco L, Fernandez-Puentes C, Vazquez D. Effects of ricin on the ribosomal sites involved in the interaction of the elongation factors. *Eur J Biochem*. 1975;54(2):499–503.
22. Eiklid K, Olsnes S, Pihl A. Entry of lethal doses of abrin, ricin and modeccin into the cytosol of HeLa cells. *Exp Cell Res*. 1980;126(2):321–326.
23. Weldon JE, Pastan I. A guide to taming a toxin – recombinant immunotoxins constructed from Pseudomonas exotoxin A for the treatment of cancer. *FEBS J*. 2011;278(23):4683–4700.
24. Yamaizumi M, Mekada E, Uchida T, Okada Y. One molecule of diphtheria toxin fragment A introduced into a cell can kill the cell. *Cell*. 1978;15(1):245–250.
25. Allured VS, Collier RJ, Carroll SF, McKay DB. Structure of exotoxin A of *Pseudomonas aeruginosa* at 3.0-Angstrom resolution. *Proc Natl Acad Sci USA*. 1986;83(5):1320–1324.
26. Hwang J, Fitzgerald DJ, Adhya S, Pastan I. Functional domains of *Pseudomonas* exotoxin identified by deletion analysis of the gene expressed in *E. coli*. *Cell*. 1987;48(1):129–136.
27. Kounnas MZ, Morris RE, Thompson MR, FitzGerald DJ, Strickland DK, Saelinger CB. The alpha 2-macroglobulin receptor/low density lipoprotein receptor-related protein binds and internalizes *Pseudomonas* exotoxin A. *J Biol Chem*. 1992;267(18):12420–12423.
28. Pastrana DV, Hanson AJ, Knisely J, Bu G, Fitzgerald DJ. LRP 1 B functions as a receptor for *Pseudomonas* exotoxin. *Biochim Biophys Acta*. 2005;1741(3):234–239.
29. Chiron MF, Fryling CM, FitzGerald DJ. Cleavage of pseudomonas exotoxin and diphtheria toxin by a furin-like enzyme prepared from beef liver. *J Biol Chem*. 1994;269(27):18167–18176.

30. Siegall CB, Chaudhary VK, FitzGerald DJ, Pastan I. Functional analysis of domains II, Ib, and III of *Pseudomonas* exotoxin. *J Biol Chem.* 1989;264(24):14256–14261.
31. FitzGerald DJ, Trowbridge IS, Pastan I, Willingham MC. Enhancement of toxicity of antitransferrin receptor antibody-Pseudomonas exotoxin conjugates by adenovirus. *Proc Natl Acad Sci USA.* 1983;80(13):4134–4138.
32. Greenfield L, Bjorn MJ, Horn G, et al. Nucleotide sequence of the structural gene for diphtheria toxin carried by corynebacteriophage beta. *Proc Natl Acad Sci USA.* 1983;80(22):6853–6857.
33. Naglich JG, Metherall JE, Russell DW, Eidels L. Expression cloning of a diphtheria toxin receptor: identity with a heparin-binding EGF-like growth factor precursor. *Cell.* 1992;69(6):1051–1061.
34. Tsuneoka M, Nakayama K, Hatsuzawa K, Komada M, Kitamura N, Mekada E. Evidence for involvement of furin in cleavage and activation of diphtheria toxin. *J Biol Chem.* 1993;268(35):26461–26465.
35. Bell CE, Eisenberg D. Crystal structure of diphtheria toxin bound to nicotinamide adenine dinucleotide. *Biochemistry.* 1996;35(4):1137–1149.
36. Sandvig K, Olsnes S. Diphtheria toxin entry into cells is facilitated by low pH. *J Cell Biol.* 1980;87(3 Pt 1):828–832.
37. Collier RJ, Pappenheimer Jr AM. Studies on the mode of action of diphtheria toxin. II. Effect of toxin on amino acid incorporation in cell-free systems. *J Exp Med.* 1964;120:1019–1039.
38. Honjo T, Nishizuka Y, Hayaishi O. Diphtheria toxin-dependent adenosine diphosphate ribosylation of aminoacyl transferase II and inhibition of protein synthesis. *J Biol Chem.* 1968;243(12):3553–3555.
39. Van Ness BG, Howard JB, Bodley JW. ADP-ribosylation of elongation factor 2 by diphtheria toxin. Isolation and properties of the novel ribosyl-amino acid and its hydrolysis products. *J Biol Chem.* 1980;255(22):10717–10720.
40. P02879 Ricin precursor – *Ricinus communis* (Castor bean). UniProtKB. UniProt Consortium.
41. Lamb FI, Roberts LM, Lord JM. Nucleotide sequence of cloned cDNA coding for preproricin. *Eur J Biochem.* 1985;148(2):265–270.
42. Newton DL, Wales R, Richardson PT, et al. Cell surface and intracellular functions for ricin galactose binding. *J Biol Chem.* 1992;267(17):11917–11922.
43. Olsnes S, Pihl A. Different biological properties of the two constituent peptide chains of ricin, a toxic protein inhibiting protein synthesis. *Biochemistry.* 1973;12(16):3121–3126.
44. Spooner RA, Watson PD, Marsden CJ, et al. Protein disulphide-isomerase reduces ricin to its A and B chains in the endoplasmic reticulum. *Biochem J.* 2004;383(Pt 2):285–293.
45. Endo Y, Mitsui K, Motizuki M, Tsurugi K. The mechanism of action of ricin and related toxic lectins on eukaryotic ribosomes. The site and the characteristics of the modification in 28 S ribosomal RNA caused by the toxins. *J Biol Chem.* 1987;262(12):5908–5912.
46. Sperti S, Montanaro L, Mattioli A, Stirpe F. Inhibition by ricin of protein synthesis in vitro: 60 S ribosomal subunit as the target of the toxin. *Biochem J.* 1973;136(3):813–815.
47. Kurpad SN, Zhao XG, Wikstrand CJ, Batra SK, McLendon RE, Bigner DD. Tumor antigens in astrocytic gliomas. *Glia.* 1995;15(3):244–256.
48. Bullard DE, Bigner DD. Applications of monoclonal antibodies in the diagnosis and treatment of primary brain tumors. *J Neurosurg.* 1985;63(1):2–16.
49. Trowbridge IS, Newman RA, Domingo DL, Sauvage C. Transferrin receptors: structure and function. *Biochem Pharmacol.* 1984;33(6):925–932.
50. Gatter KC, Brown G, Trowbridge IS, Woolston RE, Mason DY. Transferrin receptors in human tissues: their distribution and possible clinical relevance. *J Clin Pathol.* 1983;36(5):539–545.
51. Trowbridge IS, Domingo DL. Anti-transferrin receptor monoclonal antibody and toxin-antibody conjugates affect growth of human tumour cells. *Nature.* 1981;294(5837):171–173.

52. Recht LD, Griffin TW, Raso V, Salimi AR. Potent cytotoxicity of an antihuman transferrin receptor-ricin A-chain immunotoxin on human glioma cells in vitro. *Cancer Res.* 1990;50(20):6696–6700.
53. Zovickian J, Johnson VG, Youle RJ. Potent and specific killing of human malignant brain tumor cells by an anti-transferrin receptor antibody-ricin immunotoxin. *J Neurosurg.* 1987;66(6):850–861.
54. Greenfield L, Johnson VG, Youle RJ. Mutations in diphtheria toxin separate binding from entry and amplify immunotoxin selectivity. *Science.* 1987;238(4826):536–539.
55. Laske DW, Ilercil O, Akbasak A, Youle RJ, Oldfield EH. Efficacy of direct intratumoral therapy with targeted protein toxins for solid human gliomas in nude mice. *J Neurosurg.* 1994;80(3):520–526.
56. Martell LA, Agrawal A, Ross DA, Muraszko KM. Efficacy of transferrin receptor-targeted immunotoxins in brain tumor cell lines and pediatric brain tumors. *Cancer Res.* 1993;53(6):1348–1353.
57. Johnson VG, Wrobel C, Wilson D, et al. Improved tumor-specific immunotoxins in the treatment of CNS and leptomeningeal neoplasia. *J Neurosurg.* 1989;70(2):240–248.
58. Giannone L, Greco FA, Hainsworth JD. Combination intraventricular chemotherapy for meningeal neoplasia. *J Clin Oncol.* 1986;4(1):68–73.
59. Laske DW, Muraszko KM, Oldfield EH, et al. Intraventricular immunotoxin therapy for leptomeningeal neoplasia. *Neurosurgery.* 1997;41(5):1039–1049. [discussion 1049–1051].
60. Laske DW, Youle RJ, Oldfield EH. Tumor regression with regional distribution of the targeted toxin TF-CRM107 in patients with malignant brain tumors. *Nat Med.* 1997;3(12):1362–1368.
61. Weaver M, Laske DW. Transferrin receptor ligand-targeted toxin conjugate (Tf-CRM107) for therapy of malignant gliomas. *J Neurooncol.* 2003;65(1):3–13.
62. Gosselaar PH, van-Dijk AJ, de-Gast GC, et al. Transferrin toxin but not transferrin receptor immunotoxin is influenced by free transferrin and iron saturation. *Eur J Clin Invest.* 2002;32(suppl 1):61–69.
63. Yoon DJ, Chu DS, Ng CW, et al. Genetically engineering transferrin to improve its in vitro ability to deliver cytotoxins. *J Control Release.* 2009;133(3):178–184.
64. Nelms K, Keegan AD, Zamorano J, Ryan JJ, Paul WE. The IL-4 receptor: signaling mechanisms and biologic functions. *Annu Rev Immunol.* 1999;17:701–738.
65. Kondo M, Takeshita T, Ishii N, et al. Sharing of the interleukin-2 (IL-2) receptor gamma chain between receptors for IL-2 and IL-4. *Science.* 1993;262(5141):1874–1877.
66. Russell SM, Keegan AD, Harada N, et al. Interleukin-2 receptor gamma chain: a functional component of the interleukin-4 receptor. *Science.* 1993;262(5141):1880–1883.
67. Obiri NI, Debinski W, Leonard WJ, Puri RK. Receptor for interleukin 13. Interaction with interleukin 4 by a mechanism that does not involve the common gamma chain shared by receptors for interleukins 2, 4, 7, 9, and 15. *J Biol Chem.* 1995;270(15):8797–8804.
68. Obiri NI, Leland P, Murata T, Debinski W, Puri RK. The IL-13 receptor structure differs on various cell types and may share more than one component with IL-4 receptor. *J Immunol.* 1997;158(2):756–764.
69. Obiri NI, Hillman GG, Haas GP, Sud S, Puri RK. Expression of high affinity interleukin-4 receptors on human renal cell carcinoma cells and inhibition of tumor cell growth in vitro by interleukin-4. *J Clin Invest.* 1993;91(1):88–93.
70. Obiri NI, Siegel JP, Varricchio F, Puri RK. Expression of high-affinity IL-4 receptors on human melanoma, ovarian and breast carcinoma cells. *Clin Exp Immunol.* 1994;95(1):148–155.
71. Puri RK, Leland P, Kreitman RJ, Pastan I. Human neurological cancer cells express interleukin-4 (IL-4) receptors which are targets for the toxic effects of IL4-Pseudomonas exotoxin chimeric protein. *Int J Cancer.* 1994;58(4):574–581.

72. Debinski W, Puri RK, Kreitman RJ, Pastan I. A wide range of human cancers express interleukin 4 (IL4) receptors that can be targeted with chimeric toxin composed of IL4 and *Pseudomonas* exotoxin. *J Biol Chem*. 1993;268(19):14065–14070.
73. Puri RK, Debinski W, Obiri N, Kreitman R, Pastan I. Human renal cell carcinoma cells are sensitive to the cytotoxic effect of a chimeric protein composed of human interleukin-4 and Pseudomonas exotoxin. *Cell Immunol*. 1994;154(1):369–379.
74. Kreitman RJ, Puri RK, Pastan I. A circularly permuted recombinant interleukin 4 toxin with increased activity. *Proc Natl Acad Sci USA*. 1994;91(15):6889–6893.
75. Kreitman RJ, Puri RK, Pastan I. Increased antitumor activity of a circularly permuted interleukin 4-toxin in mice with interleukin 4 receptor-bearing human carcinoma. *Cancer Res*. 1995;55(15):3357–3363.
76. Joshi BH, Leland P, Asher A, Prayson RA, Varricchio F, Puri RK. In situ expression of interleukin-4 (IL-4) receptors in human brain tumors and cytotoxicity of a recombinant IL-4 cytotoxin in primary glioblastoma cell cultures. *Cancer Res*. 2001;61(22):8058–8061.
77. Puri RK, Hoon DS, Leland P, et al. Preclinical development of a recombinant toxin containing circularly permuted interleukin 4 and truncated *Pseudomonas* exotoxin for therapy of malignant astrocytoma. *Cancer Res*. 1996;56(24):5631–5637.
78. Husain SR, Behari N, Kreitman RJ, Pastan I, Puri RK. Complete regression of established human glioblastoma tumor xenograft by interleukin-4 toxin therapy. *Cancer Res*. 1998;58(16):3649–3653.
79. Rand RW, Kreitman RJ, Patronas N, Varricchio F, Pastan I, Puri RK. Intratumoral administration of recombinant circularly permuted interleukin-4-Pseudomonas exotoxin in patients with high-grade glioma. *Clin Cancer Res*. 2000;6(6):2157–2165.
80. Weber F, Asher A, Bucholz R, et al. Safety, tolerability, and tumor response of IL4-*Pseudomonas* exotoxin (NBI-3001) in patients with recurrent malignant glioma. *J Neurooncol*. 2003;64(1–2):125–137.
81. Weber FW, Floeth F, Asher A, et al. Local convection enhanced delivery of IL4-Pseudomonas exotoxin (NBI-3001) for treatment of patients with recurrent malignant glioma. *Acta Neurochir Suppl*. 2003;88:93–103.
82. Rainov NG, Gorbatyuk K, Heidecke V. Clinical trials with intracerebral convection-enhanced delivery of targeted toxins in malignant glioma. *Rev Recent Clin Trials*. 2008;3(1):2–9.
83. ClinicalTrial.gov identifier: NCT00014677.
84. Bao K, Reinhardt RL. The differential expression of IL-4 and IL-13 and its impact on type-2 immunity. *Cytokine*. 2015;75(1):25–37.
85. Zurawski SM, Vega Jr F, Huyghe B, Zurawski G. Receptors for interleukin-13 and interleukin-4 are complex and share a novel component that functions in signal transduction. *EMBO J*. 1993;12(7):2663–2670.
86. Suzuki A, Leland P, Joshi BH, Puri RK. Targeting of IL-4 and IL-13 receptors for cancer therapy. *Cytokine*. 2015;75(1):79–88.
87. Shimamura T, Husain SR, Puri RK. The IL-4 and IL-13 pseudomonas exotoxins: new hope for brain tumor therapy. *Neurosurg Focus*. 2006;20(4):E11.
88. Shimamura T, Fujisawa T, Husain SR, Joshi B, Puri RK. Interleukin 13 mediates signal transduction through interleukin 13 receptor alpha2 in pancreatic ductal adenocarcinoma: role of IL-13 Pseudomonas exotoxin in pancreatic cancer therapy. *Clin Cancer Res*. 2010;16(2):577–586.
89. Mintz A, Gibo DM, Slagle-Webb B, Christensen ND, Debinski W. IL-13Ralpha2 is a glioma-restricted receptor for interleukin-13. *Neoplasia*. 2002;4(5):388–399.
90. Kawakami M, Kawakami K, Takahashi S, Abe M, Puri RK. Analysis of interleukin-13 receptor alpha2 expression in human pediatric brain tumors. *Cancer*. 2004;101(5):1036–1042.

91. Husain SR, Obiri NI, Gill P, et al. Receptor for interleukin 13 on AIDS-associated Kaposi's sarcoma cells serves as a new target for a potent *Pseudomonas* exotoxin-based chimeric toxin protein. *Clin Cancer Res*. 1997;3(2):151–156.
92. Joshi BH, Leland P, Puri RK. Identification and characterization of interleukin-13 receptor in human medulloblastoma and targeting these receptors with interleukin-13-pseudomonas exotoxin fusion protein. *Croat Med J*. 2003;44(4):455–462.
93. Kawakami K, Kawakami M, Joshi BH, Puri RK. Interleukin-13 receptor-targeted cancer therapy in an immunodeficient animal model of human head and neck cancer. *Cancer Res*. 2001;61(16):6194–6200.
94. Debinski W, Gibo DM, Hulet SW, Connor JR, Gillespie GY. Receptor for interleukin 13 is a marker and therapeutic target for human high-grade gliomas. *Clin Cancer Res*. 1999;5(5):985–990.
95. Barderas R, Bartolome RA, Fernandez-Acenero MJ, Torres S, Casal JI. High expression of IL-13 receptor alpha2 in colorectal cancer is associated with invasion, liver metastasis, and poor prognosis. *Cancer Res*. 2012;72(11):2780–2790.
96. Joshi BH, Puri RA, Leland P, et al. Identification of interleukin-13 receptor alpha2 chain overexpression in situ in high-grade diffusely infiltrative pediatric brainstem glioma. *Neuro Oncol*. 2008;10(3):265–274.
97. Debinski W, Obiri NI, Powers SK, Pastan I, Puri RK. Human glioma cells overexpress receptors for interleukin 13 and are extremely sensitive to a novel chimeric protein composed of interleukin 13 and pseudomonas exotoxin. *Clin Cancer Res*. 1995;1(11):1253–1258.
98. Li C, Hall WA, Jin N, Todhunter DA, Panoskaltsis-Mortari A, Vallera DA. Targeting glioblastoma multiforme with an IL-13/diphtheria toxin fusion protein in vitro and in vivo in nude mice. *Protein Eng*. 2002;15(5):419–427.
99. Rustamzadeh E, Hall WA, Todhunter DA, et al. Intracranial therapy of glioblastoma with the fusion protein DTIL13 in immunodeficient mice. *Int J Cancer*. 2006;118(10):2594–2601.
100. Kioi M, Seetharam S, Puri RK. Targeting IL-13Ralpha2-positive cancer with a novel recombinant immunotoxin composed of a single-chain antibody and mutated *Pseudomonas* exotoxin. *Mol Cancer Ther*. 2008;7(6):1579–1587.
101. Kunwar S, Prados MD, Chang SM, et al. Direct intracerebral delivery of cintredekin besudotox (IL13-PE38QQR) in recurrent malignant glioma: a report by the Cintredekin Besudotox Intraparenchymal Study Group. *J Clin Oncol*. 2007;25(7):837–844.
102. Kunwar S, Chang S, Westphal M, et al. Phase III randomized trial of CED of IL13-PE38QQR vs Gliadel wafers for recurrent glioblastoma. *Neuro Oncol*. 2010;12(8):871–881.
103. Jarboe JS, Johnson KR, Choi Y, Lonser RR, Park JK. Expression of interleukin-13 receptor alpha2 in glioblastoma multiforme: implications for targeted therapies. *Cancer Res*. 2007;67(17):7983–7986.
104. Stuckey DW, Hingtgen SD, Karakas N, Rich BE, Shah K. Engineering toxin-resistant therapeutic stem cells to treat brain tumors. *Stem Cells*. 2015;33(2):589–600.
105. Massague J. Epidermal growth factor-like transforming growth factor. II. Interaction with epidermal growth factor receptors in human placenta membranes and A431 cells. *J Biol Chem*. 1983;258(22):13614–13620.
106. Pike LJ, Marquardt H, Todaro GJ, et al. Transforming growth factor and epidermal growth factor stimulate the phosphorylation of a synthetic, tyrosine-containing peptide in a similar manner. *J Biol Chem*. 1982;257(24):14628–14631.
107. Derynck R, Roberts AB, Winkler ME, Chen EY, Goeddel DV. Human transforming growth factor-alpha: precursor structure and expression in *E. coli*. *Cell*. 1984;38(1):287–297.
108. Carpenter G, Cohen S. Epidermal growth factor. *Annu Rev Biochem*. 1979;48:193–216.
109. Ushiro H, Cohen S. Identification of phosphotyrosine as a product of epidermal growth factor-activated protein kinase in A-431 cell membranes. *J Biol Chem*. 1980;255(18):8363–8365.

110. Klijn JG, Berns PM, Schmitz PI, Foekens JA. The clinical significance of epidermal growth factor receptor (EGF-R) in human breast cancer: a review on 5232 patients. *Endocr Rev*. 1992;13(1):3–17.
111. Pavelic K, Banjac Z, Pavelic J, Spaventi S. Evidence for a role of EGF receptor in the progression of human lung carcinoma. *Anticancer Res*. 1993;13(4):1133–1137.
112. Rubin Grandis J, Melhem MF, Barnes EL, Tweardy DJ. Quantitative immunohistochemical analysis of transforming growth factor-alpha and epidermal growth factor receptor in patients with squamous cell carcinoma of the head and neck. *Cancer*. 1996;78(6):1284–1292.
113. Fox SB, Persad RA, Coleman N, Day CA, Silcocks PB, Collins CC. Prognostic value of c-erbB-2 and epidermal growth factor receptor in stage A1 (T1a) prostatic adenocarcinoma. *Br J Urol*. 1994;74(2):214–220.
114. Chow NH, Chan SH, Tzai TS, Ho CL, Liu HS. Expression profiles of ErbB family receptors and prognosis in primary transitional cell carcinoma of the urinary bladder. *Clin Cancer Res*. 2001;7(7):1957–1962.
115. Yasui W, Sumiyoshi H, Hata J, et al. Expression of epidermal growth factor receptor in human gastric and colonic carcinomas. *Cancer Res*. 1988;48(1):137–141.
116. Bartlett JM, Langdon SP, Simpson BJ, et al. The prognostic value of epidermal growth factor receptor mRNA expression in primary ovarian cancer. *Br J Cancer*. 1996;73(3):301–306.
117. Arita N, Hayakawa T, Izumoto S, et al. Epidermal growth factor receptor in human glioma. *J Neurosurg*. 1989;70(6):916–919.
118. Libermann TA, Razon N, Bartal AD, Yarden Y, Schlessinger J, Soreq H. Expression of epidermal growth factor receptors in human brain tumors. *Cancer Res*. 1984;44(2):753–760.
119. Libermann TA, Nusbaum HR, Razon N, et al. Amplification, enhanced expression and possible rearrangement of EGF receptor gene in primary human brain tumours of glial origin. *Nature*. 1985;313(5998):144–147.
120. Fuller GN, Bigner SH. Amplified cellular oncogenes in neoplasms of the human central nervous system. *Mutat Res*. 1992;276(3):299–306.
121. Chaffanet M, Chauvin C, Laine M, et al. EGF receptor amplification and expression in human brain tumours. *Eur J Cancer*. 1992;28(1):11–17.
122. Huang SM, Harari PM. Epidermal growth factor receptor inhibition in cancer therapy: biology, rationale and preliminary clinical results. *Invest New Drugs*. 1999;17(3):259–269.
123. Kunwar S, Pai LH, Pastan I. Cytotoxicity and antitumor effects of growth factor-toxin fusion proteins on human glioblastoma multiforme cells. *J Neurosurg*. 1993;79(4):569–576.
124. Phillips PC, Levow C, Catterall M, Colvin OM, Pastan I, Brem H. Transforming growth factor-alpha-Pseudomonas exotoxin fusion protein (TGF-alpha-PE38) treatment of subcutaneous and intracranial human glioma and medulloblastoma xenografts in athymic mice. *Cancer Res*. 1994;54(4):1008–1015.
125. Engebraaten O, Hjortland GO, Juell S, Hirschberg H, Fodstad O. Intratumoral immunotoxin treatment of human malignant brain tumors in immunodeficient animals. *Int J Cancer*. 2002;97(6):846–852.
126. Liu TF, Cohen KA, Ramage JG, Willingham MC, Thorburn AM, Frankel AE. A diphtheria toxin-epidermal growth factor fusion protein is cytotoxic to human glioblastoma multiforme cells. *Cancer Res*. 2003;63(8):1834–1837.
127. Liu TF, Hall PD, Cohen KA, et al. Interstitial diphtheria toxin-epidermal growth factor fusion protein therapy produces regressions of subcutaneous human glioblastoma multiforme tumors in athymic nude mice. *Clin Cancer Res*. 2005;11(1):329–334.
128. Sampson JH, Akabani G, Archer GE, et al. Progress report of a phase I study of the intracerebral microinfusion of a recombinant chimeric protein composed of transforming growth factor (TGF)-alpha and a mutated form of the *Pseudomonas* exotoxin termed PE-38 (TP-38) for the treatment of malignant brain tumors. *J Neurooncol*. 2003;65(1):27–35.

129. Sampson JH, Akabani G, Archer GE, et al. Intracerebral infusion of an EGFR-targeted toxin in recurrent malignant brain tumors. *Neuro Oncol*. 2008;10(3):320–329.
130. Wikstrand CJ, Reist CJ, Archer GE, Zalutsky MR, Bigner DD. The class III variant of the epidermal growth factor receptor (EGFRvIII): characterization and utilization as an immunotherapeutic target. *J Neurovirol*. 1998;4(2):148–158.
131. Sugawa N, Ekstrand AJ, James CD, Collins VP. Identical splicing of aberrant epidermal growth factor receptor transcripts from amplified rearranged genes in human glioblastomas. *Proc Natl Acad Sci USA*. 1990;87(21):8602–8606.
132. Batra SK, Castelino-Prabhu S, Wikstrand CJ, et al. Epidermal growth factor ligand-independent, unregulated, cell-transforming potential of a naturally occurring human mutant EGFRvIII gene. *Cell Growth Differ*. 1995;6(10):1251–1259.
133. Humphrey PA, Wong AJ, Vogelstein B, et al. Anti-synthetic peptide antibody reacting at the fusion junction of deletion-mutant epidermal growth factor receptors in human glioblastoma. *Proc Natl Acad Sci USA*. 1990;87(11):4207–4211.
134. Sok JC, Coppelli FM, Thomas SM, et al. Mutant epidermal growth factor receptor (EGFRvIII) contributes to head and neck cancer growth and resistance to EGFR targeting. *Clin Cancer Res*. 2006;12(17):5064–5073.
135. Wikstrand CJ, Hale LP, Batra SK, et al. Monoclonal antibodies against EGFRvIII are tumor specific and react with breast and lung carcinomas and malignant gliomas. *Cancer Res*. 1995;55(14):3140–3148.
136. Nagane M, Levitzki A, Gazit A, Cavenee WK, Huang HJ. Drug resistance of human glioblastoma cells conferred by a tumor-specific mutant epidermal growth factor receptor through modulation of Bcl-XL and caspase-3-like proteases. *Proc Natl Acad Sci USA*. 1998;95(10):5724–5729.
137. Lorimer IA, Wikstrand CJ, Batra SK, Bigner DD, Pastan I. Immunotoxins that target an oncogenic mutant epidermal growth factor receptor expressed in human tumors. *Clin Cancer Res*. 1995;1(8):859–864.
138. Lorimer IA, Keppler-Hafkemeyer A, Beers RA, Pegram CN, Bigner DD, Pastan I. Recombinant immunotoxins specific for a mutant epidermal growth factor receptor: targeting with a single chain antibody variable domain isolated by phage display. *Proc Natl Acad Sci USA*. 1996;93(25):14815–14820.
139. Archer GE, Sampson JH, Lorimer IA, et al. Regional treatment of epidermal growth factor receptor vIII-expressing neoplastic meningitis with a single-chain immunotoxin, MR-1. *Clin Cancer Res*. 1999;5(9):2646–2652.
140. Beers R, Chowdhury P, Bigner D, Pastan I. Immunotoxins with increased activity against epidermal growth factor receptor vIII-expressing cells produced by antibody phage display. *Clin Cancer Res*. 2000;6(7):2835–2843.
141. ClinicalTrials.gov Identifier: NCT01009866.
142. Sampson JH, Brady M, Raghavan R, et al. Colocalization of gadolinium-diethylene triamine pentaacetic acid with high-molecular-weight molecules after intracerebral convection-enhanced delivery in humans. *Neurosurgery*. 2011;69(3):668–676.
143. Zalutsky MR, Boskovitz A, Kuan CT, et al. Radioimmunotargeting of malignant glioma by monoclonal antibody D2C7 reactive against both wild-type and variant III mutant epidermal growth factor receptors. *Nucl Med Biol*. 2011;39(1):23–34.
144. Chandramohan V, Bao X, Keir ST, et al. Construction of an immunotoxin, D2C7-(scdsFv)-PE38KDEL, targeting EGFRwt and EGFRvIII for brain tumor therapy. *Clin Cancer Res*. 2013;19(17):4717–4727.
145. ClinicalTrials.gov Identifier: NCT02303678.
146. Loging WT, Lal A, Siu IM, et al. Identifying potential tumor markers and antigens by database mining and rapid expression screening. *Genome Res*. 2000;10(9):1393–1402.
147. Kuan CT, Wakiya K, Dowell JM, et al. Glycoprotein nonmetastatic melanoma protein B, a potential molecular therapeutic target in patients with glioblastoma multiforme. *Clin Cancer Res*. 2006;12(7 Pt 1):1970–1982.

148. http://www.uniprot.org/uniprot/Q14956.
149. Kuan CT, Wakiya K, Keir ST, et al. Affinity-matured anti-glycoprotein NMB recombinant immunotoxins targeting malignant gliomas and melanomas. *Int J Cancer*. 2011;129(1):111–121.
150. http://www.uniprot.org/uniprot/Q6UVK1.
151. Pluschke G, Vanek M, Evans A, et al. Molecular cloning of a human melanoma-associated chondroitin sulfate proteoglycan. *Proc Natl Acad Sci USA*. 1996;93(18):9710–9715.
152. Price MA, Colvin Wanshura LE, Yang J, et al. CSPG4, a potential therapeutic target, facilitates malignant progression of melanoma. *Pigm Cell Melanoma Res*. 2011;24(6):1148–1157.
153. Schrappe M, Klier FG, Spiro RC, Waltz TA, Reisfeld RA, Gladson CL. Correlation of chondroitin sulfate proteoglycan expression on proliferating brain capillary endothelial cells with the malignant phenotype of astroglial cells. *Cancer Res*. 1991;51(18):4986–4993.
154. Hjortland GO, Garman-Vik SS, Juell S, et al. Immunotoxin treatment targeted to the high-molecular-weight melanoma-associated antigen prolonging the survival of immunodeficient rats with invasive intracranial human glioblastoma multiforme. *J Neurosurg*. 2004;100(2):320–327.
155. Carrel S, Accolla RS, Carmagnola AL, Mach JP. Common human melanoma-associated antigen(s) detected by monoclonal antibodies. *Cancer Res*. 1980;40(7):2523–2528.
156. Ayriss JRR, Kuan CT, Keir S, Pastan I, Bigner DD. Mel-14 and 9.2.27 immunotoxins: promising therapeutics for pediatric glioma. In: *Pediatric Neuro-Oncology Basic and Translational Research Conference*. ; 2011. New Orleans, Louisiana.
157. http://www.uniprot.org/uniprot/Q86YL7.
158. Chandramohan V, Bao X, Kato Kaneko M, et al. Recombinant anti-podoplanin (NZ-1) immunotoxin for the treatment of malignant brain tumors. *Int J Cancer*. 2013;132(10):2339–2348.
159. Raica M, Cimpean AM, Ribatti D. The role of podoplanin in tumor progression and metastasis. *Anticancer Res*. 2008;28(5B):2997–3006.
160. Dang Q, Liu J, Li J, Sun Y. Podoplanin: a novel regulator of tumor invasion and metastasis. *Med Oncol*. 2014;31(9):24.
161. Shibahara J, Kashima T, Kikuchi Y, Kunita A, Fukayama M. Podoplanin is expressed in subsets of tumors of the central nervous system. *Virchows Arch*. 2006;448(4):493–499.
162. Ogasawara S, Kaneko MK, Price JE, Kato Y. Characterization of anti-podoplanin monoclonal antibodies: critical epitopes for neutralizing the interaction between podoplanin and CLEC-2. *Hybridoma (Larchmt)*. 2008;27(4):259–267.
163. Kullander K, Klein R. Mechanisms and functions of Eph and ephrin signalling. *Nat Rev Mol Cell Biol*. 2002;3(7):475–486.
164. Boyd AW, Bartlett PF, Lackmann M. Therapeutic targeting of EPH receptors and their ligands. *Nat Rev Drug Discov*. 2014;13(1):39–62.
165. Unified nomenclature for Eph family receptors and their ligands, the ephrins. Eph Nomenclature Committee. *Cell*. 1997;90(3):403–404.
166. Wykosky J, Gibo DM, Stanton C, Debinski W. EphA2 as a novel molecular marker and target in glioblastoma multiforme. *Mol Cancer Res*. 2005;3(10):541–551.
167. Wykosky J, Gibo DM, Debinski W. A novel, potent, and specific ephrinA1-based cytotoxin against EphA2 receptor expressing tumor cells. *Mol Cancer Ther*. 2007;6(12 Pt 1):3208–3218.
168. Smith HW, Marshall CJ. Regulation of cell signalling by uPAR. *Nat Rev Mol Cell Biol*. 2010;11(1):23–36.
169. Vallera DA, Li C, Jin N, Panoskaltsis-Mortari A, Hall WA. Targeting urokinase-type plasminogen activator receptor on human glioblastoma tumors with diphtheria toxin fusion protein DTAT. *J Natl Cancer Inst*. 2002;94(8):597–606.
170. Stish BJ, Oh S, Vallera DA. Anti-glioblastoma effect of a recombinant bispecific cytotoxin cotargeting human IL-13 and EGF receptors in a mouse xenograft model. *J Neurooncol*. 2008;87(1):51–61.

171. Oh S, Tsai AK, Ohlfest JR, Panoskaltsis-Mortari A, Vallera DA. Evaluation of a bispecific biological drug designed to simultaneously target glioblastoma and its neovasculature in the brain. *J Neurosurg*. 2011;114(6):1662–1671.
172. Todhunter DA, Hall WA, Rustamzadeh E, Shu Y, Doumbia SO, Vallera DA. A bispecific immunotoxin (DTAT13) targeting human IL-13 receptor (IL-13R) and urokinase-type plasminogen activator receptor (uPAR) in a mouse xenograft model. *Protein Eng Des Sel*. 2004;17(2):157–164.
173. Vandergrift WA, Patel SJ, Nicholas JS, Varma AK. Convection-enhanced delivery of immunotoxins and radioisotopes for treatment of malignant gliomas. *Neurosurg Focus*. 2006;20(4):E13.
174. Onda M, Beers R, Xiang L, Nagata S, Wang QC, Pastan I. An immunotoxin with greatly reduced immunogenicity by identification and removal of B cell epitopes. *Proc Natl Acad Sci USA*. 2008;105(32):11311–11316.
175. Liu W, Onda M, Lee B, et al. Recombinant immunotoxin engineered for low immunogenicity and antigenicity by identifying and silencing human B-cell epitopes. *Proc Natl Acad Sci USA*. 2012;109(29):11782–11787.
176. Mazor R, Vassall AN, Eberle JA, et al. Identification and elimination of an immunodominant T-cell epitope in recombinant immunotoxins based on Pseudomonas exotoxin A. *Proc Natl Acad Sci USA*. 2012;109(51):E3597–E3603.
177. Antignani A, Fitzgerald D. Immunotoxins: the role of the toxin. *Toxins*. 2013;5(8):1486–1502.
178. Mattoo AR, FitzGerald DJ. Combination treatments with ABT-263 and an immunotoxin produce synergistic killing of ABT-263-resistant small cell lung cancer cell lines. *Int J Cancer*. 2013;132(4):978–987.
179. Risberg K, Fodstad O, Andersson Y. Synergistic anticancer effects of the 9.2.27PE immunotoxin and ABT-737 in melanoma. *PLoS One*. 2011;6(9):e24012.
180. Traini R, Ben-Josef G, Pastrana DV, et al. ABT-737 overcomes resistance to immunotoxin-mediated apoptosis and enhances the delivery of pseudomonas exotoxin-based proteins to the cell cytosol. *Mol Cancer Ther*. 2010;9(7):2007–2015.
181. Kawakami K, Terabe M, Kioi M, Berzofsky JA, Puri RK. Intratumoral therapy with IL13-PE38 results in effective CTL-mediated suppression of IL-13Ralpha2-expressing contralateral tumors. *Clin Cancer Res*. 2006;12(15):4678–4686.
182. Ochiai H, Archer GE, Herndon 2nd JE, et al. EGFRvIII-targeted immunotoxin induces antitumor immunity that is inhibited in the absence of CD4+ and CD8+ T cells. *Cancer Immunol Immunother*. 2008;57(1):115–121.
183. Adams S, Braidy N, Bessede A, et al. The kynurenine pathway in brain tumor pathogenesis. *Cancer Res*. 2012;72(22):5649–5657.
184. Humphries W, Wei J, Sampson JH, Heimberger AB. The role of tregs in glioma-mediated immunosuppression: potential target for intervention. *Neurosurg Clin North Am*. 2010;21(1):125–137.
185. Pyonteck SM, Akkari L, Schuhmacher AJ, et al. CSF-1R inhibition alters macrophage polarization and blocks glioma progression. *Nat Med*. 2013;19(10):1264–1272.
186. Wainwright DA, Chang AL, Dey M, et al. Durable therapeutic efficacy utilizing combinatorial blockade against IDO, CTLA-4, and PD-L1 in mice with brain tumors. *Clin Cancer Res*. 2014;20(20):5290–5301.

CHAPTER

11

Checkpoint Blockade Immunotherapy for Glioblastoma: Progress and Challenges

K.A. Riccione, P. Gedeon, L. Sanchez-Perez, J.H. Sampson

Duke University Medical Center, Durham, NC, United States

OUTLINE

Translational Immunotherapy of Brain Tumors
http://dx.doi.org/10.1016/B978-0-12-802420-1.00011-9

THE TUMOR IMMUNE RESPONSE AND IMMUNE CHECKPOINT BIOLOGY

The tumor immune response occurs through a series of sequential events that form a complex conversation between cancer cells and immune compartments. Tumor antigens are shed from dying cells in the tumor microenvironment (TME), taken up by professional antigen-presenting cells (APCs), and presented to clonal antigen-specific naïve T cells in the tumor draining lymph node (DLN) (Fig. 11.1). These antigen-primed activated T cells then traffic to and infiltrate the tumor exerting tumor-specific cytotoxicity upon tumor antigen recognition. Each of these steps is subject to an intricate balance of activating and inhibitory signals, or checkpoints, that tell T cells to proceed in their effector state or stop in their tracks, thereby fine-tuning the duration and intensity of the tumor immune response.

Checkpoint signaling can occur during the initial priming phase of the immune response or at the effector stage, inhibiting naïve T cell activation or antigen-specific tumoricidal activity, respectively. The priming phase of the immune response occurs in secondary lymphoid tissues when naïve T cells are presented with tumor antigens by resident APCs. Such antigen-induced T cell activation is subject to inhibitory checkpoints provided by receptor–ligand interactions that take place between naïve T cells and resident APCs [e.g., cytotoxic T lymphocyte associated protein 4 (CTLA-4)/B7]. Activated T cells that proceed past the first checkpoint migrate to peripheral tissues and undergo an upregulation in their inhibitory receptors [e.g., programmed-death receptor 1 (PD-1), T cell immunoglobulin and mucin-3 (TIM-3)], which are engaged by checkpoint ligands [e.g., PD-L1, galectin-9 (GAL-9)] expressed by the tumor or tumor stroma (Fig. 11.2).

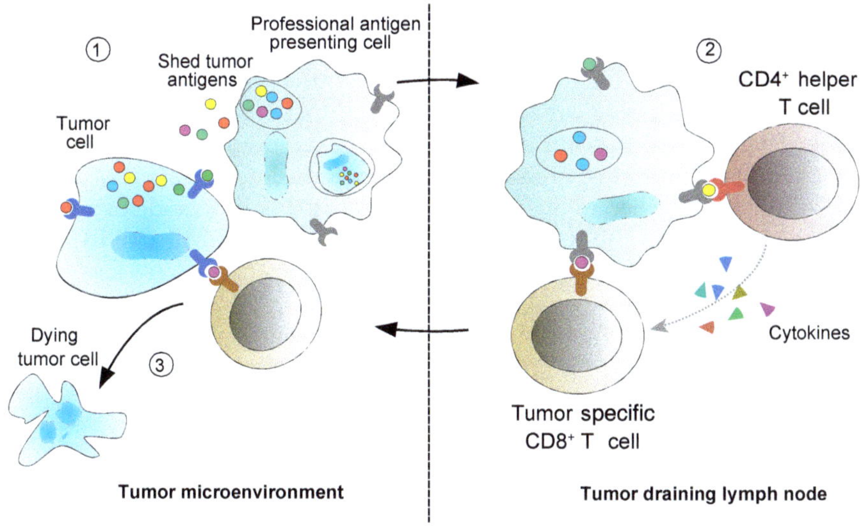

FIGURE 11.1 Presentation of tumor antigens in the tumor microenvironment and tumor draining lymph node. (1) Antigens are shed by dead/dying tumor cells and taken up by professional antigen presenting cells (APCs); (2) activated APCs then proceed to the tumor draining lymph node to present peptide antigens on major histocompability complex (MHC) molecules to naïve CD4+ helper T cells and CD8+ T cells; (3) activated effector CD8+ T cells migrate to the tumor where they recognize tumor antigens presented by MHC on the tumor surface, inducing their cytotoxic activity to kill the tumor.

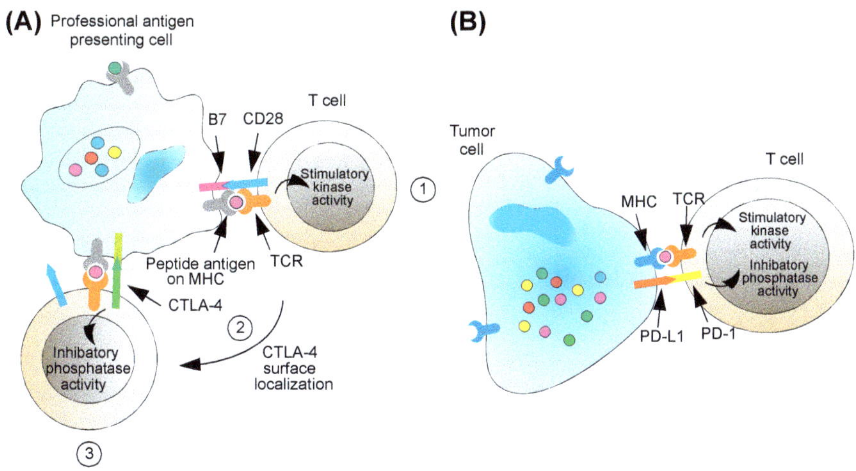

FIGURE 11.2 Immune-modulatory signaling to stimulate or inhibit the T cell response. (A) Regulation of T cell activation occurs by an interplay of inhibitory and stimulatory signals during antigen presentation to naïve T cells: (1) Upon their first encounter with peptide antigens on major histocompatibility complex molecules, naïve T cells require a costimulatory B7/CD28 signal to become activated; (2) cytotoxic T lymphocyte associated protein 4 (CTLA-4) is expressed on the surface of activated (or dysregulated) T cells; (3) CTLA-4 competes for binding with B7 to prevent its interaction with CD28, thus inhibiting T cell activation in the presence of antigen. (B) Activated T cells express surface programmed-death receptor 1 (PD-1), which is bound by PD-L1 expressed on the tumor surface to prevent effector T cell function in the tumor microenvironment.

Under normal physiological conditions, immune checkpoints are crucial in preventing pathological autoimmunity induced by the recognition of self-antigens as well as immune-mediated tissue damage caused by sustained inflammatory responses during pathogenic infection. Many tumors, however, have learned to highjack this otherwise healthy mode of immune regulation to evade and resist tumor-specific cytotoxic T cell activity.

Given the high degree of dysregulated immune checkpoints associated with many malignancies, the concept of eliminating inhibitory signals to unleash a dormant antitumor immune response has emerged as a promising avenue for tumor immunotherapy. The most straightforward strategy for reducing the engagement of checkpoint pathways is to sever ligand/receptor interactions through binding with an obstructive third-party molecule, specific for either the ligand or receptor. The exquisite epitope-binding specificity of monoclonal antibodies (mAbs) and the potential to engineer such mAbs to have high affinity for putatively any desired biological target provide an ideal platform for precisely targeting immune checkpoint molecules. Indeed, pivotal approvals by the United States Food and Drug Administration (FDA) for the immunomodulatory antibodies ipilimumab (anti-CTLA-4, Bristol Myers Squibb) and nivolumab (anti-PD1, Bristol Myers Squibb), which demonstrate significant survival benefits in patients with metastatic melanoma[1] and non–small-cell lung cancer (NSCLC),[2] have validated antibody-based immune checkpoint blockade as a viable treatment modality for cancer. In this chapter, we will provide an overview of immune checkpoint biology and its relevance to malignant glioma, the feasibility of antibody-based immunomodulation as a therapeutic strategy for central nervous system (CNS) malignancies, preclinical and clinical experiences with checkpoint antibodies, and the ongoing progress of checkpoint blockade for glioma immunotherapy (Tables 11.1 and 11.2).

CTLA-4: A Checkpoint for T cell Activation

CTLA-4 regulates the priming of naïve T cells in the initial stages of antigen stimulation by interfering with one of the two signals required for T cell activation. During the priming phase of the immune response, T cell activation requires (1) peptide antigen presented on surface-expressed major histocompatibility complex (MHC) proteins, which engage the T cell receptor and (2) a costimulatory signal from ligands expressed on the surface of APCs, which bind to surface receptors on T cells (e.g., B7 with CD28). The CD28 costimulatory molecule shares its B7 ligand with receptor CTLA-4,[3,4] which rapidly localizes to the surface of T cells upon initial antigen recognition. The overall affinity for CTLA-4 with B7 is higher than that of CD28 with B7, which leads to CTLA-4 outcompeting and blocking costimulatory CD28/B7 interactions[5,6] (Fig. 11.2A). In addition to its direct

TABLE 11.1 Immunomodulatory Antibodies in Clinical Development

Antibody	Target	Isotype	Clinical Status
Ipilimumab	CTLA-4	IgG1	Approved for NSCLC and melanoma
Tremelimumab	CTLA-4	IgG2	Approved for mesothelioma
Pidilizumab	PD-1	IgG1	Phase II for large B cell lymphoma; Phase I/II for DIPG and MM
Pembrolizumab	PD-1	IgG4	Approved for advanced melanoma (BRAF mutant), ipilimumab-refractory; and for platinum-resistant PD-L1^{+} NSCLC with EGFR or ALK abnormality
Nivolumab	PD-1	IgG4	Approved for platinum-resistant NSCLC with EGFR or ALK abnormality; in combination with ipilimumab for advanced melanoma; metastatic treatment-refractory RCC
Durvalumab	PD-L1	Fc optimized for no ADCC/CDC	Phase III for head and neck cancer, bladder cancer, and NSCLC Multiple other Phase I/II studies for solid tumors
Atezolizumab	PD-L1	Fc optimized for no ADCC/CDC	Approved for bladder cancer
Avelumab	PD-L1	IgG1	Phase III for gastric cancer, NSCLC, advanced RCC, urothelial cancer, and refractory ovarian cancer Phase II for Merkel cell carcinoma and Phase I for locally advanced solid tumors
BMS986016	LAG-3		Phase II for NSCLC and gastric cancer Phase I for solid tumors, refractory hematologic neoplasms, recurrent GBM
PDR001	PD-1	IgG4	Phase III for advanced melanoma Phase II for nasopharyngeal carcinoma, thoracic, gastric, pancreatic cancers, NSCLC, and other solid tumors Phase I/II for advanced malignancies
LAG525	LAG-3	Humanized	Phase I for advanced malignancies
MBG453	TIM-3	Humanized	Phase I for advanced malignancies
Enoblituzumab	B7-H3	IgG1	Phase I for SCCHN, NSCLC, and refractory solid tumors
Lirilumab	NK KIR	IgG4	Phase II for AML, CLL, and MDS Phase I for HL, NHL, MM, MDS, and advanced solid tumors

Continued

TABLE 11.1 Immunomodulatory Antibodies in Clinical Development—cont'd

Antibody	Target	Isotype	Clinical Status
Indoximod	IDO1	N/A	Phase II for metastatic pancreatic, metastatic breast, metastatic prostate, NSCLC Phase I/II for AML, melanoma, and malignant brain tumors Phase I for metastatic melanoma, refractory solid tumors
CP-870,893	CD40	IgG2	Phase I for melanoma
MEDI6469	OX40	Murine IgG1	Phase I for advanced SCCHN, metastatic prostate, metastatic colorectal, and metastatic breast
Urelumab	4-1BB	IgG4	Phase II for melanoma, B cell NHL, refractory CLL, bladder cancer, and other solid tumors Phase I for GBM, multiple myeloma, colorectal, metastatic head/neck cancer and other malignancies
Utomilumab	4-1BB	IgG2	Phase III for diffuse large B cell lymphoma Phase II for advanced solid tumors Phase I B cell lymphomas, $CD20^{+}$ NHL, RCC, SCCHN, and advanced solid tumors
Varililumab	CD27	IgG1	Phase II for GBM, RCC, and melanoma Phase I/II for clear cell RCC, urologic/kidney neoplasms, breast, colorectal, SCCHN, NSCLC, low-grade glioma, advanced melanoma Phase I $CD27^{+}$ B cell malignancies, advanced ovarian cancer, prostate cancer, and other solid tumors

Clinical Status current as of December 2016. *ADCC*, antibody-dependent cell-mediated cytotoxicity; *ALK*, anaplastic lymphoma kinase; *AML*, acute myeloid leukemia; *BRAF*, human gene that encodes for serine/threonine-protein kinase B-Raf; *CDC*, complement-dependent cytotoxicity; *CTLA*, cytotoxic T lymphocyte associated protein 4; *CLL*, chronic lymphocytic leukemia; *DIPG*, diffuse intrinsic pontine gliomas; *EGFR*, epidermal growth factor receptor; *GBM*, glioblastoma; *HL*, Hodgkin's lymphoma; *IDO*, indoleamine 2,3 dioxygenase 1; *IgG*, immunoglobulin G; *KIR*, killer inhibitory receptors; *LAG*, lymphocyte activation gene 3; *MDS*, myelodysplastic syndromes; *MM*, multiple myeloma; *N/A*, not applicable; *NHL*, non-Hodgkin's lymphoma; *NK*, natural killer; *NSCLC*, non–small-cell lung cancer; *PD-1*, programmed-death receptor 1; *RCC*, renal cell carcinoma.

blocking of T cell costimulation, ligand-bound CTLA-4 delivers inhibitory signals to the T cell[7,8] by inducing phosphatase activity to counteract kinase signals required for T cell activation.[5] Thus, CTLA-4 signaling during naïve T cell priming inhibits T cell activation and leads to antigenic tolerance. Indeed, as a testament to its immune-regulatory function,

TABLE 11.2 Clinical Trials of Checkpoint Blockade Antibodies for the Treatment of CNS Malignancies

Drug/Disease	Cohorts	Phase/Status	Primary Outcomes	Secondary Outcomes	NCT#
Combined ipilimumab and nivolumab for newly diagnosed GBM with TMZ SOC	**1.** TMZ + ipilimumab **2.** TMZ + nivolumab **3.** TMZ + Ipilimumab + nivolumab	Phase I recruiting	irDLTs	Biomarker analysis (IHC), NCICTCAE, 1-year and 2-year survival rate	02311920
Combined ipilimumab and nivolumab vs. bevacizumab for recurrent GBM	**1.** Nivolumab **2.** Ipilimumab + nivolumab **3.** Bevacizumab	Phase III recruiting	Safety (i and ii) OS (i vs. iii)	PFS and RR (i vs. iii)	02017717
Ipilimumab and nivolumab versus fotemustine for MBM	**1.** Fotemustine **2.** Fotemustine + ipilimumab **3.** Ipilimumab + nivolumab	Phase III recruiting	OS (i vs. ii; i vs. iii)	Safety, irDCR, irPFS, ORR, DoR, irDoR, TRR, irTRR	02460068
Ipilimumab with RT for MBM	**1.** Ipilimumab + RT	Phase II recruiting	1-year survival	PFS, OS, ORR, safety	02460068
Ipilimumab with corticosteroids for MBM	**1.** Ipilimumab **2.** Ipilimumab + corticosteroids	Phase II Completed	Disease control rate	irRC, DoR, ORR, PFS, survival rate, irAEs, OS	00623766
Ipilimumab with RT for MBM	**1.** Ipilimumab + WBRT **2.** Ipilimumab + SRS	Phase I Active	MTD	OS, PFS, AEs/SAEs	01703507
Ipilimumab and nivolumab for MBM	**1.** Ipilimumab + nivolumab	Phase II recruiting	Clinical benefit rate	OS, safety	02320058
Ipilimumab with SRS for MBM	**1.** Ipilimumab + SRS	Phase II Completed	OS	AEs, ORR, PFS	02662725
Ipilimumab before or after SRS for MBM	**1.** SRS followed by ipilimumab **2.** Ipilimumab followed by SRS	Phase II recruiting	Local control rate	Toxicity, OS	02097732
Nivolumab with CMV pp65 DCs for recurrent grade III/IV brain tumors	**1.** Nivolumab prior to resection and nivolumab + DC after resection **2.** Nivolumab + DC prior to resection and nivolumab after	Phase I recruiting	Safety	OS, PFS	02529072

Continued

TABLE 11.2 Clinical Trials of Checkpoint Blockade Antibodies for Treatment of CNS Malignancies—cont'd

Drug/Disease	Cohorts	Phase/Status	Primary Outcomes	Secondary Outcomes	NCT#
Nivolumab with anti-CSF1R mAb for GBM	**1.** Anti-CSF1R **2.** Anti-CSF1R + niovlumab	Phase I recruiting	AEs/SAEs, RD, ORR	OS, PK parameters of anti-CSF1R, TIL and TIM levels, anti-nivolumab and anti-antiCSF1R antibody titers	02526017
Galunisertib with nivolumab for GBM	**1.** Nivolumab + galunisertib	Phase II recruiting	MTD	PK parameters of galunisertib, anti-nivolumab titers, PFS, ORR, DoR, TTR, OS	02423343
Nivolumab with SOC for newly diagnosed GBM	**1.** Nivolumab + TMZ + RT **2.** Placebo + TMZ + RT	Phase III not yet recruiting	OS	PFS	02617589
Neoadjuvant nivolumab	Nivolumab	Phase II recruiting	Tumor PD-L1 expression	ORR, toxicity	02550249
Pembrolizumab and bevacizumab for recurrent HGG	Pembrolizumab + bevacizumab + HFSRT	Phase I recruiting	MTD	ORR	02313272
Pembrolizumab with MLA for recurrent MG	**1.** Pembrolizumab + MLA **2.** Pembrolizumab	Phase I	MTD, DLT, PFS	Toxicity, OS, intratumoral PD-L1 expression, glioma immune response	02311582
Pembrolizumab for recurrent hypermutated MG	Pembrolizumab	Pilot recruiting	ORR	None	02658279
Pembrolizumab with SOC for newly diagnosed GBM	Pembrolizumab + TMZ + RT	Phase I/II recruiting	DLT, PFS	AEs, OS, ORR	02530502

Pembrolizumab ± bevacizumab for recurrent GBM	**1.** Pembrolizumab + bevacizumab **2.** Pembrolizumab	Phase II Active	PFS, MTD	AEs, OS radiographic response	02337491
Pembrolizumab vs. anti-PI3K/Akt for GBM/gliosarcoma	**1.** Pembrolizumab **2.** Anti-PI3K/Akt	Phase I/II Active	PFS	CTCAE	02430363
Pembrolizumab for recurrent GBM	Pembrolizumab	Phase II recruiting	PFS, Effector:Treg ratio	None	02337686
Durvalumab for recurrent, bevacizumab-naïve GBM	**1.** Durvalumab (newly diagnosed) **2.** Bevacizumab + durvalumab	Phase II recruiting	OS, PFS	AEs, PK profile, ORR	02336165
BMS-986016 or urelumab for recurrent GBM	**1.** BMS-986016 (anti-LAG-3) **2.** Urelumab (anti-4-1BB) **3.** BMS-986016 + nivolumab (anti-PD1) **4.** Urelumab + nivolumab	Phase I not yet recruiting	MTD	OS, PFS, ORR	02658981
Indoximod + SOC for primary MG	**1.** Indoximod (anti-IDO1) + TMZ **2.** Indoximod + TMZ (bevacizumab naïve) **3.** Indoximod + TMZ + bevacizumab **4.** Indoximod + TMZ + SRT	Phase I/II	Dosing, PFS	AEs, PK profile, ORR	02052648

AEs, adverse events; *CMV*, cytomegalovirus; *CTCAE*, Common Terminology Criteria for Adverse Events; *DLT*, Dose-limiting toxicities; *GBM*, glioblastoma; *HGG*, high grade glioma; *MG*, malignant glioma; *irDCR*, immune-related disease control rate; *irDLT*, immune-related dose-limiting toxicities; *irDoR*, immune-related duration of response; *irTRR*, immune-related time to response; *MLA*, MRI-guided laser ablation; *MTD*, maximum tolerated dose; *NCT*, National Clinical Trial; *ORR*, objective response rates; *OS*, overall survival; *PFS*, progression free survival; *PK*, protein kinase; *PI3K*, phosphoinositide 3-kinase; *SAE*, significant adverse effects; *SOC*, standard of care; *SRT*, stereotactic radiotherapy; *SRS*, stereotactic radiosurgery; *RR*, response rate; *TMZ*, temozolomide; *WBRT*, whole brain radiotherapy.

Ctla-4 knockout mice are subject to lethal systemic immune hyperactivation caused by the recognition and response to self-antigens.[9,10]

Though CTLA-4 engagement primarily occurs in secondary lymphoid organs, checkpoint signaling through this molecule can take place in peripheral tissues through its action on regulatory T cells (Tregs), a subset of T cells that exert immunosuppressive function at effector sites. The Treg phenotype is characterized by expression of the forkhead box P3 (FoxP3) transcription factor, which exerts direct transcriptional control over the *Ctla-4* gene, such that CTLA-4 is constitutively expressed on Tregs.[11,12] CTLA-4 is integral to the immunosuppressive function of Tregs,[13] as it competes with binding for B7 molecules and leads to their downregulation on APCs.[14] As Tregs are concentrated in peripheral tissues under pathological conditions (including within tumors), the CTLA-4 checkpoint pathway is likely a major player in abrogating T cell effector activity. Indeed, CTLA-4 blockade has been shown to give rise to antitumor immune responses under conditions in which T cell egress from lymph nodes is artificially inhibited, indicating that this checkpoint exerts at least some effects directly at the effector site.[15]

The CTLA-4 checkpoint is central in suppressing antitumor immune responses. Naïve T cells that receive a CTLA-4 signal in lieu of costimulatory CD28 during tumor-antigen presentation fail to differentiate into an effector phenotype. Such CTLA-4-mediated inhibition decreases the repertoire of activated tumor-specific T cells and prevents vaccine-induced T cell activation, thus curtailing the potential immune-stimulatory effects of adjuvant tumor therapies and limiting the potential for active immunotherapy strategies.

PD-1: A Checkpoint for T cell Effector Function

In contrast to CTLA-4, which exerts its inhibitory activity largely during naïve T cell priming, PD-1 checkpoint signaling occurs during the effector stage of the immune response. Surface expression of PD-1 is upregulated in activated T cells,[16] and prolonged antigen exposure, such as that which occurs during chronic viral infection and cancer, leads to high levels of persistent PD-1 expression and a state of T cell exhaustion.[17,18] PD-1 signaling in T cells leads to decreased proliferation, reduced cytokine production, and abrogated cytolytic function.[19] In addition to its role in T cell inhibition, PD-1 is expressed on B cells and natural killer (NK) cells,[20,21] limiting their antibody production and lytic activity, respectively. Furthermore, PD-1 is highly expressed on Tregs to enhance their proliferation in the presence of ligand.[22]

The two ligands for PD-1, PD-L1, and PD-L2, share 37% sequence homology but are subject to divergent regulatory mechanisms and exert different effects upon PD-1 receptor engagement.[23] PD-L1 is expressed on peripheral epithelial cells, activated hematopoietic cells, and tumors,[24,25] and can be further upregulated by the inflammatory cytokine interferon gamma (IFN-γ) produced by activated T cells.[26,27] PD-L2 has a more selective expression pattern, where it is upregulated by interleukin (IL)-4 signaling in activated dendritic cells (DCs) and a subset of macrophages.[28] PD-1/PD-L1 interactions inhibit T cell activation through a downstream phosphatase, in a similar fashion to CTLA-4,[29] whereas the role of PD-L2 signaling in T cells remains unclear. Interestingly, PD-L1 can function as a receptor to deliver inhibitory signals upon interactions with CD80 on activated T cells (and possibly APCs),[30,31] adding another layer of complexity to the immune-regulation imparted by this pathway.

The PD-1 axis provides a multifaceted mechanism of tumor immune evasion. PD-1 is highly expressed on tumor-infiltrating lymphocytes (TILs)[32,33] wherein signaling occurs through the engagement of PD-L1 expressed in the TME (Fig. 11.2B),[34] either on tumor cells themselves or on myeloid cells in the tumor stroma. PD-L1 expression by the tumor can be innate to the tumor phenotype or an adaptive response to avoid immune recognition. Innate PD-L1 expression is imparted by genetic alterations and/or activation of certain signaling pathways common to many cancers [such as protein kinase B (Akt) and signal transducer and activator of transcriptions (STATs)].[35,36] Adaptive expression refers to the upregulation of PD-L1 in response to inflammatory cytokines (e.g., IFN-γ) produced during the tumor immune response, wherein this adaptation renders the tumors newly resistant to immune-mediated cytotoxicity. Indeed, T cell infiltration and IFN-γ levels in the TME positively correlate with PD-L1 expression in human tumors[27,37] including glioblastoma (GBM).[38] The relative importance of PD-1 expression on TILs and innate versus adaptive expression of PD-L1 in conferring tumor immune resistance/evasion remains to be determined.

IMMUNE CHECKPOINTS IN MALIGNANT GLIOMA

Dysregulated immune checkpoint signaling has emerged as an important mechanism underlying the potent immunosuppression and immune evasion associated with GBM. This immunosuppression takes the form of ineffective tumor antigen presentation by APCs,[39] recruitment and induction of intratumoral Tregs,[40] and an exhausted phenotype of effector T cells.[41,42] Indeed, increased expression of checkpoint molecules and immunosuppressive pathways in patients with GBM is associated with decreased overall survival (OS) and progression-free survival (PFS).[43,44]

The PD-1/PD-L1 axis in particular has been implicated as a significant mediator of glioma-associated immunosuppression. Peripheral T cells from patients with grade II through IV astrocytomas have elevated PD-1 expression relative to healthy controls, and PD-1 expression increases with increasing grade.[45] Importantly, PD-1 expression on peripheral T cells decreases upon surgical resection, further implicating a causal role of the tumor in PD-1 upregulation, though the underlying mechanism remains unclear.

Even more pronounced than elevated PD-1 levels is the increased surface expression of PD-L1 on glioma cells, tumor-infiltrating macrophages, and circulating monocytes from GBM patients.[46] In an analysis of 135 GBM specimens, PD-L1 expression was present in over 70% of samples, with a higher proportion in newly diagnosed cases,[47] and was found to be specific to tumor cells and absent from surrounding healthy CNS tissue. Interestingly, this high frequency of expression is in contrast to a rate of ~30% for other tumor types,[48,49] making it a distinctive characteristic of GBM.[47] In this same study, PD-L1 expression was further stratified across cellular subtypes, with the mesenchymal subtype in particular having a higher level of expression, suggesting a potential link between PD-L1 expression and the highly infiltrative nature of GBM.[50]

In GBM tumors positive for PD-L1, its expression is believed to be an innate phenotype, caused by a loss of expression of the tumor suppressor, phosphatase and tensin homology, and resultant activation of the oncogenic phosphatidylinositol-3-OH (PI(3)K) kinase pathway.[35] PD-L1 expression in gliomas is further increased upon inflammatory cytokine exposure,[38] suggesting the possibility for adaptive immune resistance. Tumoral PD-L1 expression has been shown to predict a worse prognosis in several tumor types,[51–53] and a recent report suggests a poorer prognosis in patients with PD-L1 expression on glioma cells.[54] Interestingly, in this same study, PD-L1 expression by surrounding neurons was shown to correlate with a better prognosis, and these neurons were shown to have a tumoricidal effect on glioma cells, though the significance of this finding is unclear. Overall, the relationship between PD-L1 expression and GBM prognosis remains to be determined, but a detrimental effect on the tumor immune response is almost certain.

In a similar vein to PD-1, binding of CTLA-4 on naïve T cells results in an abrogated antiglioma immune response. Polymorphisms in CTLA-4 that alter gene expression and increase CTLA-4-mediated downregulation of T cell activity are correlated with increased susceptibility to gliomas,[55] highlighting the role of this inhibitory pathway in glioma immune resistance. Because CTLA-4 signaling plays a role largely during antigen priming, the inhibitory effects of this checkpoint mechanism are particularly detrimental for strategies designed to enhance or induce tumor immune responses. For instance, in the context of tumor immunotherapy, decreased expression of CTLA-4 on peripheral blood T cells after DC vaccination correlates with longer survival in GBM patients.[56]

Though CTLA-4 signaling exerts inhibitory effects on naïve T cells during antigen priming, the glioma-mediated immunosuppressive action of CTLA-4 is believed to occur largely through the engagement of this pathway in Tregs, which are increased both systemically[40] and within the TME[40,57,58] in patients with GBM. This increased Treg pool corresponds to decreased T cell effector activity and a shift from proinflammatory type 1 cytokines (e.g., IFN-γ and IL-2) to antiinflammatory type 2 cytokines (e.g., IL-4 and IL-5).[40] CTLA-4 on Tregs enhances their immunosuppressive function and prevents T cell responses to tumor antigens.[59]

IMMUNE CHECKPOINT BLOCKADE

Given the prevalence of aberrant immune checkpoint signaling in cancer, checkpoint blockade has emerged as a promising treatment modality for many tumor types. Though the evaluation of this immunotherapy strategy for the treatment of GBM is still in its infancy, the high degree of checkpoint dysregulation in GBM patients coupled with the success of checkpoint blockade in the setting of other malignancies has made this an attractive treatment modality for GBM. Checkpoint blockade is based on the underlying premise that GBM patients possess endogenous T cells specific for glioma antigens, but that aberrant inhibitory interactions make the activity of these cells suboptimal. Indeed, glioma lesions are characteristically heavily infiltrated with T cells,[60,61] and substantial evidence suggests that, under appropriate immune-stimulatory conditions and/or in the absence of overwhelming inhibitory signals, cytotoxic T lymphocytes have the ability to eradicate large, well-established tumors.[62] In this section, we will provide an overview of the clinical successes of antibody-based checkpoint blockade in other tumor types and discuss the feasibility of this strategy for the treatment of GBM.

CTLA-4 Blockade

CTLA-4 was the first checkpoint molecule to garner clinical attention for tumor immunotherapy. Initially, the lack of tumor-specific expression of checkpoint molecules fueled skepticism surrounding the general strategy of immune checkpoint blockade and made this field slow to progress. However, with the first demonstration that CTLA-4 blockade could elicit therapeutic antitumor immunity in several mouse models of malignancy,[63] immune checkpoint modulation rapidly emerged as a highly attractive strategy for tumor immunotherapy. Such preclinical findings encouraged the development and clinical evaluation of single-agent CTLA-4 blocking antibodies first in the setting of advanced melanomas, and currently across a variety of other tumor types.

Two fully humanized anti-CTLA-4 antibodies, ipilimumab, a fully human IgG1 mAb, and tremelimumab, a fully human IgG2 (Pfizer), entered clinical testing as the first checkpoint-blocking drugs for cancer therapy and produced objective clinical responses in ~10% of patients with metastatic melanoma.[64–66] In a Phase III randomized three-arm clinical trial of patients with advanced melanoma that received either a peptide vaccine derived from melanoma-specific gp100, the gp100 vaccine combined with ipilimumab, or ipilimumab alone, there was a 3.5-month increase in survival in both arms receiving ipilimumab compared to the group receiving vaccine alone.[1] Later stage trials of tremelimumab, however, failed to show a survival benefit relative to melanoma standard of care. Such therapeutic discrepancies between ipilimumab and tremelimumab are attributed to the unique properties endowed by their specific isotypes. Ipilimumab depletes CTLA-4^{high} Tregs through antibody-dependent cell-mediated cytotoxicity (ADCC),[67] which requires the specific engagement of Fc receptors bound by the IgG1 isotype, whereas IgG2/Fc-receptor signaling does not engage ADCC pathways. Importantly, such findings corroborate a role for Tregs in the immunosuppressive function of CTLA-4, making intratumoral Tregs a potential predictor of ipilimumab efficacy (and thus the potential contribution of ipilimumab to GBM therapy).

As the first therapy to provide survival benefit in the setting of metastatic melanoma, ipilumumab was approved in 2010 by the US FDA for the treatment of advanced melanoma. More impressive than the 3.5-month survival benefit was the effect of ipilimumab on long-term survival: with just four doses administered in the course of 3 months, 18% of patients survived beyond 2 years (approximately double that which is measured by standard oncological criteria), with durable survival extending to 10 years in some.[68] This effect on long-term survival generated by a relatively short treatment course supports the concept that blocking T cell inhibitory pathways to induce tumor-specific activation instills a long-lasting immunological memory, keeping tumors in check even after the immunomodulatory drug has left the scene.

The success of ipilimumab in metastatic melanoma has led to its evaluation in the setting of a number of other malignancies. In a Phase II study of monotherapy ipilimumab for metastatic renal cell carcinoma (RCC), a partial response rate of 10% ($n=61$) was observed. In Phase III trials for lung cancer ($n=204$ for NSCLC; $n=130$ for extensive disease SCLC), ipilimumab combined with standard of care chemotherapy extended PFS by 1 month, compared to chemotherapy alone,[69,70] with no impact on OS. The greater clinical benefit of ipilimumab observed in patients with advanced melanoma compared to other tumor types is likely due to the higher degree of inherent immunogenicity imparted by melanoma tumors and their physical availability to cellular immune responses. Indeed, early preclinical studies showed a survival benefit in mice bearing partially immunogenic tumors,[63] whereas monotherapy anti-CTLA-4 is less efficacious against poorly immunogenic tumors. Such findings indicate that CTLA-4 blockade likely works through enhancing a basal antitumor immune response. Therefore, though promising, such modest survival benefits in the setting of "less immunogenic" or "less immunologically accessible" malignancies highlight the requirement of combined therapy to induce and/or enhance an antitumor immune response for CTLA-4 blockade to be optimized as a treatment modality. As will later be discussed, this requirement could pose challenges in the setting of CNS malignancies.

PD-1/PD-L1 Blockade

The profound success of ipilimumab for the treatment of advanced melanoma set the stage for immune checkpoint blockade as an up-and-coming tumor therapy, propelling the development and evaluation of drugs targeting the PD-1 pathway. Unlike CTLA-4, whose natural B7 ligand is required for T cell activation, the PD-1 axis offers both its ligand and receptor as a means of checkpoint blockade. Although clinical experience with anti-PD1 and anti-PD-L1 antibodies is less extensive than those targeting CTLA-4, initial results appear promising. Several different anti-PD-1/anti-PD-L1 antibodies are currently in clinical testing, and antitumor activity has been observed with all of them, emphasizing the PD-1 pathway as a dominant tumor immunosuppressive pathway and key target for tumor immunotherapy.

The results of the first anti-PD1 mAb to reach clinical testing, CT-011 (Pidilizumab, CureTech), however, were discouraging. CT-011, a humanized IgG1 anti-PD1 mAb, was evaluated in a dose-escalating study in 17 patients with hematological malignancies and was found to yield clinical benefit in 33% of those treated, with one patient experiencing full remission.[71] CT-011 proceeded to several Phase II trials, including those for relapsed follicular lymphoma ($n=29$) and advanced melanoma ($n=103$). Though 66% objective response rates (ORRs) and 86% tumor regression were observed in the setting of follicular lymphoma,[72] just 6% of advanced melanoma patients experienced ORRs,[48] leading to the notion that PD-1 blockade may not be a viable treatment strategy for solid tumors, or that tumor expression of PD-1 may be necessary for the efficacy of PD-1 blockade. A closer look, however, reveals that the lack of efficacy of CT-011 may be due to the physiology of the IgG1 isotype and its propensity to induce ADCC. In this case, CT-011 would be efficacious against PD-1$^+$ tumors, leading to their elimination by ADCC, although CT-011 treatment in the setting of tumors lacking PD-1 expression may lead to ADCC-mediated depletion of PD-1$^+$ tumor-specific T cells, abrogating the potentially immunomodulatory benefits of disrupting this checkpoint pathway.

Indeed, this phenomenon may prove relevant for glioma immunotherapy, where increased PD-1 expression on TILs is observed.

Despite this apparent setback for anti-PD-1 treatment of solid tumors, several other anti-PD-1 mAbs of differing isotypes and affinities remained on the clinical radar. In a Phase II dose-escalating study of pembrolizumab (human IgG4, Merck) for the treatment of advanced melanoma, 38% ORRs were observed ($n=135$).[73] Similarly, in a Phase I study of pembrolizumab for NSCLC ($n=282$), ORRs were observed in 23% of patients, with 47% ORRs in treatment-naïve patients.[74] As a direct comparison between pembrolizumab and ipilimumab for advanced melanoma, a Phase III study ($n=834$) showed ORRs of ~33%, a 1-year survival rate of ~74%, and decreased toxicity in patients receiving pembrolizumab compared to an ORR of 12% in patients receiving ipilimumab.[75] In a race to hit the market, a third anti-PD-1 antibody, nivolumab (human IgG4, Bristol Myers Squibb), surfaced in clinical trials alongside pembrozilumab. Although pembrozilumab took on the approach of directly challenging ipilimumab, nivolumab entered into clinical evaluation as a combination therapy with ipilimumab for advanced melanoma and showed an unprecedented 2-year survival rate of 79% and ORR of 43%.[76] Similarly, a Phase III study comparing nivolumab with docetaxel chemotherapy for NSCLC ($n=272$) showed an increased OS of 9.2 months versus 6 months and ORR of 20% versus 9%.[2]

Such profound clinical findings led to the recent FDA approval of both drugs for the treatment of advanced melanoma and NSCLC. Both antibodies are approved as a single-agent therapy in patients with unresectable or metastatic melanoma following disease progression after ipilimumab and/or human gene that encodes for serine/threonine-protein kinase B-Raf (BRAF) inhibitor treatment for BRAF mutant tumors. Both antibodies are also approved for NSCLC patients who undergo progression on or after platinum-based chemotherapy. Additionally, nivolumab is approved as a first-line therapy in combination with ipilimumab for the treatment of BRAF wild type tumors and for advanced RCC patients who have received prior antiangiogenic therapy. The added indications of nivolumab (including its use as a first-line treatment for some malignancies) and its seemingly increased prevalence in the market relative to pembrolizumab are likely a result of straightforward comparisons in well-controlled clinical studies, rather than inherent differences in efficacy between the two antibodies, underscoring the importance of trial design for advancing checkpoint blockade strategies into the clinic.

Given the overwhelming success of PD-1 blockade strategies, antibody-based PD-L1 blockade has also emerged as a promising treatment modality. In a similar vein to PD-1 blockade, anti-PD-L1 antibodies would prevent signaling through this checkpoint pathway. However, as PD-L1 is primarily expressed in the TME, several unique biological aspects of anti-PD-L1 antibodies would need to be considered to make this strategy viable. Whereas PD-1 antibodies bind to tumor-specific T cells and could potentially remain there upon migration to the tumor to prevent T cell exhaustion in the TME,[77] anti-PD-L1 antibodies would need to directly penetrate the tumor, without an opportunity to "piggy-back" on infiltrating immune cells. Additionally, PD-L1 antibodies may benefit from a different isotype that induces ADCC, such that PD-L1 can both prevent T cell exhaustion within the TME while also engaging innate immune responses to aid in tumor eradication. Furthermore, PD-L1 blockade would only prevent PD-1/PD-L1 interactions, continuing to allow for PD-1/PD-L2 signaling, such that the impact of PD-L1 blockade in the setting of PD-L2$^+$ malignancies remains unclear.

Several PD-L1 antibodies are currently in early stages of clinical evaluation. A Phase I/II study of durvalumab (MEDI4736, AstraZeneca) for NSCLC has shown antitumor activity,[78] and a Phase III study for squamous cell carcinoma of the head and neck (SCCHN) was recently initiated (NCT02369874). A recent Phase I study of atezolizumab (Roche) for metastatic RCC ($n=63$) showed an improved OS of 28.9 months with ORRs of 15%,[79] and a dose-escalating Phase I study of avelumab (MSB0010718C, Pfizer) has demonstrated safety for the treatment of refractory solid tumors (NCT01772004). Importantly, all three of these antibodies are of the human IgG1 isotype and are expected to provide tumoricidal activity by the induction of ADCC, in addition to inhibition of the PD-1 pathway.

Interestingly, in many PD-1/PD-L1 blockade studies, patients with PD-L1$^+$ tumors experienced higher ORRs compared to tumors with no detectable PD-L1 staining,[80,81] suggesting that blockade of tumor-mediated engagement of the PD-1 pathway may underlie the efficacy of this strategy. However, responses have also been observed in patients with PD-L1-negative tumors, making the role of PD-L1 as a biomarker for anti-PD-L1 efficacy unclear. Importantly, blockade of intratumoral PD-L1 signaling would require that such anti-PD-L1 antibodies reach the TME, which may pose challenges in the setting of CNS malignancies and their availability to antibody penetrance.

FEASIBILITY OF CHECKPOINT BLOCKADE FOR CENTRAL NERVOUS SYSTEM MALIGNANCIES

The Blood–Brain Barrier and Antibody Access

The ability for tumor immunotherapy to confer clinical benefit in the setting of "immunoprivileged" CNS malignancies has historically been called to question. The traditional assumption was once that CNS immune responses were limited due to the blood–brain barrier (BBB), a supposed lack of CNS lymphatic drainage, and low levels of antigen presentation and infiltrative immune cells in the brain.[82] Though this view has been challenged by the finding that immune cells readily migrate to disease sites within the CNS (including gliomas), and the recent discovery of CNS lymphatics,[83] the feasibility of antibody-based therapies for CNS malignancies remains unclear. The skepticism surrounding this approach is largely due to uncertainties in the ability for such large biomolecules to have activity across the BBB, as only low levels of antibodies are known to be present in the CNS in physiologic states.[84] Glioma tumor cells, however, induce compositional changes in the basal lamina and astrocytic components of the neurovascular unit that disrupt the integrity of the BBB, allowing for enhanced penetrance of large soluble molecules to render CNS lesions potentially available to antibody-based therapy. Indeed, several studies have demonstrated that intravenously administered antibodies gain access to intracranial tumors and exert significant therapeutic benefit in the setting of GBM.[85–87] The therapeutic success of a given antibody-based therapy, however, will likely prove to be dependent upon the specific immunomodulatory target in question.

Conventional wisdom holds that CTLA-4 pathways are engaged in secondary lymphoid tissues (between B7 molecules on myeloid-derived APCs and naïve T cells), whereas PD-1 engagement occurs in the TME (between PD-L1/2 on tumor cells/stroma and PD-1 on effector T cells). Therefore, because the mechanism of CTLA-4 blockade in ameliorating antitumor immune inhibition likely occurs in DLNs, CNS penetrance of anti-CTLA-4 mAbs may not be necessary to enhance antiglioma immune responses. On the other hand, because PD-1/PD-L1/2 interactions are believed to occur predominantly at the effector site, blocking antibodies for PD-1 and its ligands would need to cross the BBB and infiltrate the tumor. This may also hold true for CTLA-4 blockade if the underlying mechanism proves to be through the suppression of Treg function within the TME. Though the feasibility of checkpoint blockade for the treatment of CNS malignancies remains unclear, preclinical and clinical experience with anti-CTLA-4, anti-PD-1, and anti-PD-L1 antibodies provide some rationale for further exploring these modalities in GBM.

Clinical experience in patients with melanoma brain metastases has demonstrated that checkpoint blockade can have activity in the CNS. Historically, advanced melanoma patients with metastatic brain lesions have been purposefully excluded from immunotherapy trials, due to the inherent poor prognosis associated with this disease stage and the longstanding notion of CNS immunoprivilege. However, in a retrospective study of a small cohort of patients with melanoma brain metastases (MBM) ($n = 38$), researchers uncovered a partial response rate of 16% in those treated with ipilimumab,[88] leading to the further evaluation of checkpoint blockade strategies in MBM patients. More recently, in a Phase II trial of ipilimumab combined with fotemustine chemotherapy, 50% of patients with MBM showed immune-related disease control, with a 54.2% 1-year survival rate.[89] An ongoing Phase III randomized three-arm trial comparing fotemustine versus combined fotemustine and ipilimumab or the combination of ipilimumab and nivolumab in patients with MBM (NCT02460068) will provide further insight into the feasibility of PD-1 blockade for the treatment of CNS malignancies. Additionally, further studies on the BBB penetrance of macromolecules in the setting of malignant gliomas will shed light on the feasibility of this approach.

Tumor Immunogenicity

The BBB is not the only obstacle that calls into question the viability of checkpoint blockade for GBM therapy. A major unanswered question underlying this approach remains in regards to the dominant antigenic targets that are recognized by newly liberated T cells. Circumstantial evidence suggests that neo-antigens generated from tumor somatic mutations serve as "foreign" targets for the unleashed tumor immune response. Indeed, melanomas with higher mutational loads were shown to be more responsive to CTLA-4 blockade,[90] and anti PD-1/PDL-1 therapy has had more clinical success in carcinogen-induced tumor types (i.e., those with higher median mutational loads).[91] Additionally, for tumor antigens to be recognized by effector T cells, they must be presented on MHC molecules within the TME. Because gliomas, including GBM, have been shown to have relatively low mutation rates[92] and decreased MHC expression in the TME[93–95] (and therefore poor immunogenicity), there remains concern over the feasibility for checkpoint blockade alone to elicit substantial immune-mediated tumor regression.

Safety

While breakthrough T cell activation with immune checkpoint blockade has been shown to translate into antitumor responses, it can also manifest as toxicity in the form of unrestrained T cell responses to self-antigens and resultant autoimmune or immune-related adverse events (irAEs). Reported systemic toxicities include dermatologic and gastrointestinal adverse effects, hepatitis, endocrinopathies, pneumonitis, and renal insufficiency,[96] most being a result of infiltration of highly-activated $CD4^+$ and $CD8^+$ T cells and the increased production of inflammatory cytokines in normal tissues. Local inflammation in the glioma microenvironment and/or collateral T cell responses to self-antigens within in the brain could lead to particularly dangerous morbidities. Though early clinical trials suggest that CNS toxicities associated with brain tumor immunotherapy are rare,[97] neurologic side effects have been reported.[98]

Significant irAEs have been observed for nearly all checkpoint blockade antibodies in clinical development, but in general, the magnitude of irAEs varies depending on the target in question as well as the tumor type, with CTLA-4 blockade being less tolerated in patients than anti-PD-1/PD-L1 therapy. Interestingly, such varying toxicities have proven to be retrospectively predicted by knockout mouse models; whereas, *Ctla-4* knockout mice are subject to rapidly lethal systemic immune hyperactivation, *Pd-1, Pd-L1,* and *Pd-L2* knockout mice have more mild phenotypes. Indeed, ipilimumab has been reported to give rise to a host of inflammatory side effects in up to 60% of patients, with severe toxicities (grade 3/4) in 7% of patients,[1] whereas PD-1/PD-L1 blockade has proven to be better tolerated, with an irAE rate of ~35% and grade 3/4 toxicities occurring in <5% of patients for most tumor types.[99] The comparably reduced toxicity observed with PD-1 blockade may be attributable to the fact that PD-1/PD-L1 interactions take place largely at the effector site (within the TME), whereas the CTLA-4/B7 interaction occurs in systemic lymphoid organs, allowing for access to a more diverse array of tissue-specific antigens expressed outside of the TME.[99]

Common antigens expressed by the tumor and surrounding stroma are believed to be the cause of several grade 3/4 tissue-specific irAEs associated with checkpoint blockade therapy. Specifically, in the setting of RCC, 21% of ipilimumab-treated patients experienced grade 3/4 colitis ($n = 198$),[66] and low-grade dermatologic toxicities, such as pruritus, erythematous rash, or vitiligo, are observed in ~15% of ipilimumab-treated patients with advanced melanoma.[1] Extrapolating this phenomenon to glioma immunotherapy, autoimmune T cell responses in the brain could give rise to debilitating/fatal encephalitis or meningitis. CNS irAEs can potentially be countered with standard of care corticosteroids (e.g., dexamethasone) to reduce brain edema or other irAEs.[100,101] However, given the lesser degree of irAEs observed with PD-1/PD-L1 blockade compared to anti-CTLA-4 strategies, the former may prove to be a more clinically feasible checkpoint blockade strategy for the treatment of CNS malignancies.

Preclinical Evidence

Preclinical studies in murine models of GBM have helped to partially assuage concerns regarding limited antibody penetrance, poor tumor immunogenicity, and the safety of checkpoint blockade. In a mouse model of poorly immunogenic orthotopic low grade glioma, monotherapy CTLA-4 mAb blockade resulted in 80% long-term survival.[102] The authors showed that CTLA-4 blockade increases conventional $CD4^+$ T cell proliferation, in the presence or absence of Tregs, indicating that anti-CTLA-4 activity in this setting likely occurs in secondary lymphoid tissues, rather than by reducing Treg-mediated immunosuppression in the TME. This study suggests that CTLA-4 antibodies may not need to cross the BBB to have a therapeutic effect against CNS malignancies, and that checkpoint blockade remains feasible even in the setting of a weakly immunogenic tumor. A recent study by Rearden et al. has served to further corroborate this notion in murine glioma. The authors showed that single-agent CTLA-4, PD-1, or anti-PD-L1 therapy was well-tolerated and resulted in long-term tumor-free survival in 50%, 20%, and 15% of treated animals, respectively.[103] The differences in the degree of efficacy between these three modalities are attributed to whether the target checkpoint molecule is expressed primarily in lymphoid tissues or in the TME. Importantly, however, the efficacy of anti-PD-L1 blockade against $PD\text{-}L1^+$ intracranial tumors suggests that checkpoint blockade antibodies can cross the BBB to have direct local antitumor effects.

CLINICAL PROGRESS OF CHECKPOINT BLOCKADE FOR GLIOBLASTOMA

Though the evaluation of antibody-based checkpoint blockade for the treatment of GBM is still in its infancy, many ongoing trials will soon shed light on its capacity for clinical translation. In this section, we will provide an overview of ongoing checkpoint blockade trials for the treatment of GBM. Specifically, we will discuss the relevance of GBM standard of care therapies to tumor immunotherapy and how standard of care is being leveraged in many clinical studies. We will also highlight the promise of combination checkpoint blockade strategies and concomitant immunotherapies for further potentiating the antitumor immune response.

Glioblastoma Standard of Care

Current standard of care for GBM is maximal surgical resection followed by radiotherapy and adjuvant temozolomide (TMZ) chemotherapy. Despite this three-pronged approach to treatment, there remains a 5-year survival rate of <5%.[104] Immunotherapies offer a new wave of treatment for GBM to address the shortcomings of standard of care. Initially, radiation and chemotherapy standard of care were believed to be independent or antagonistic to the aims of immunotherapy, due to their preferential cytotoxic effects on proliferating cells and resultant T cell lymphopenia[105–108] as well as their induction of non-inflammatory apoptotic pathways in tumor cells believed to give rise to immune tolerance.[109] Evidence garnered over the last 5–10 years, however, has called these beliefs into question.

A 2008 case study showed that successful vaccine-induced immune activation can be achieved with concomitant TMZ, so long as TMZ is administered outside of the effector window.[110] At this point, TMZ may actually be desirable to the vaccine response, as it can reduce the immunosuppressive Treg compartment that normally increases upon vaccination[110–112] and allow for homeostatic proliferation of vaccine-induced T cells.[113] Additionally, a recent preclinical study showed that TMZ-induced lymphopenia prior to adoptive lymphocyte transfer and/or vaccination gives rise to enhanced antigen-specific T cell numbers via homeostatic proliferation, a phenomenon referred to as "host-conditioning."[114] Radiotherapy has also been shown to augment the antitumor immune response through several mechanisms: (1) the triggering of "danger" signals that activate innate immune cell compartments[115,116]; (2) the phagocytosis of dead/dying tumor cells by APCs, subsequently presented as tumor antigens to naïve T cells[117,118]; and (3) increased expression of MHC molecules by the tumor.[119,120] Such immune-stimulatory effects imparted

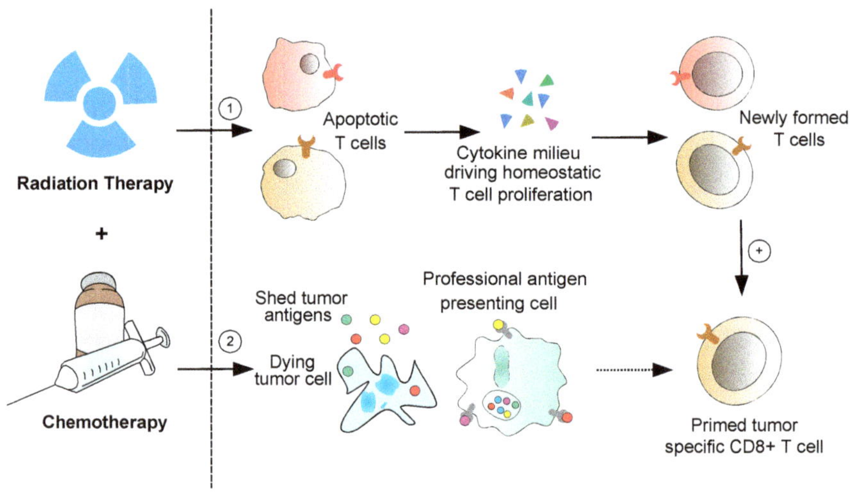

FIGURE 11.3 Synergy between glioblastoma standard of care and immunotherapy. Radiation and chemotherapy promote the death of tumor cells and their subsequent shedding of antigens to be taken up by antigen presenting cells. Additionally, standard of care therapy is known give rise to T cell lymphopenia. During the priming phase of the tumor immune response, the remaining cytokine milieu induces the preferential proliferation of primed tumor-specific T cells, resulting in a larger proportion of effector cells available for tumor killing.

by GBM standard of care have the potential to synergize with checkpoint blockade strategies, by allowing for an increase in the quantity and quality of tumor antigens presented, as well as the enhanced proliferation and effector function of antigen-specific T cells (Fig. 11.3).

Indeed, several preclinical studies have served to validate the notion that GBM standard of care can act as a complement to immune checkpoint blockade strategies. In a murine model of PD-L1$^+$ glioma, Zeng et al. showed that PD-1 mAb blockade combined with focal radiation led to a near doubling in median survival and a ~30% frequency of long-term survival, while neither therapy alone conferred a significant survival benefit.[121] Additionally, when used in conjunction with anti-CTLA-4 antibodies, radiation has been shown to induce immune-mediated tumoricidal activity outside of the irradiated field, a phenomenon known as the abscopal effect.[122,123] Mechanistically, such synergistic effects are likely due to the release of multiple tumor-associated antigens in a pro-inflammatory environment, which serves as a vaccine to induce T cell activation and immunologic memory. The effect of concomitant TMZ on checkpoint blockade, however, is less clear. In a recent preclinical study evaluating the efficacy of triple-therapy checkpoint blockade [anti-CTLA-4, anti-PD-L1, and anti-indoleamine 2,3 dioxygenase 1 (anti-IDO)], TMZ slightly enhanced the efficacy of anti-IDO while decreasing the efficacy of combined anti-CTLA-4/anti-PD-L1 treatments.[124] Such studies suggest that the underlying checkpoint mechanism and treatment regimen are likely crucial for successfully leveraging the tumoricidal and indirect immune-stimulating effects of concomitant TMZ chemotherapy.

Clinical experience in a variety of extracranial cancers has shown that fractionated radiotherapy and chemotherapy synergizes with CTLA-4 and/or PD-1 blockade to produce tumor regression and long-term survival.[122,125] A number of ongoing clinical trials for GBM will help to elucidate the relationship between GBM standard of care and antibody-based checkpoint blockade, including a Phase I trial for evaluating the safety of combined ipilimumab and nivolumab in the setting of TMZ (NCT02311920), a Phase I/II trial for pembrolizumab safety/efficacy combined with TMZ/RT (02530502), and a Phase III trial for nivolumab efficacy in the setting of TMZ/RT standard of care (02617589). Additionally, a Phase II study to evaluate the efficacy and safety of an anti-PD-L1 antibody (durvalumab) with and without standard of care RT will assess the durvalumab pharmacokinetic profile (NCT02336165), providing insight into the relationship between antibody penetrance across the BBB and the feasibility of this strategy against PD-L1$^+$ CNS tumors.

In addition to the direct immune-potentiating effects instilled by GBM standard of care, the disruption to the BBB that surgical resection and concomitant RT/TMZ affords may also be beneficial to checkpoint blockade strategies. To further address concerns over antibody access to the CNS, a Phase I/II trial was recently initiated to evaluate pembrolizumab (anti-PD-1) with combined MRI-guided laser ablation, which is known to disrupt the peritumoral BBB and allow for increased access of new tumor antigens to DLNs and infiltration of immune effector cells to the tumor. Additionally, a Phase II trial of pembrolizumab administered to patients who undergo resection surgery during the course of antibody treatment (NCT02550249) may help to reveal a synergistic relationship between surgical disturbances to the BBB and the efficacy of checkpoint blockade.

Bevacizumab

Bevacizumab (Genentech) is a humanized mAb that targets human vascular endothelial growth factor A (VEGF-A) to prevent tumor angiogenesis. Bevacizumab comprises standard of care for several tumor types, including GBM in some countries, typically for patients with disease progression. Though the interaction between bevacizumab and glioma tumor immunity remains unclear, studies in other tumor types suggest that bevacizumab may have immune-stimulating effects,[126] and could therefore be a synergistic therapy with checkpoint inhibitors. Specifically, one report shows that treatment with bevacizumab increases DC numbers and function in solid tumors and reduces peripheral Treg counts in patients with colorectal cancer.[127] An ongoing Phase I trial of bevacizumab combined with an anti-PD-L1 antibody will help to elucidate the potentially synergistic relationship between these two therapies for the treatment of GBM (NCT02336165).

Combinatorial Therapy

Owing to the clinical successes of CTLA-4 and PD-1 antibodies for NSCLC and metastatic melanoma, several Phase I trials have recently been initiated for CNS malignancies. Specifically, the profound clinical benefit observed upon combined ipilimumab and nivolumab for the treatment of advanced melanoma has opened the doors for safety/tolerability trials of nivolumab alone or in combination with ipilimumab for many other cancer types, including recurrent GBM (NCT02017717). The Phase I trial for GBM comprises two patient cohorts receiving four administrations of varying doses of nivolumab and ipilimumab, followed by nivolumab every 2 weeks thereafter until disease progression or unacceptable toxicity. Additionally, a more direct Phase I study designed to formally compare the efficacy of single-agent ipilimumab or nivolumab with the combination of the two is currently recruiting patients (NCT02311920). This study consists of three arms: (1) nivolumab, (2) ipilimumab, and

(3) nivolumab + ipilimumab, with the primary outcome measures being immune-related dose-limiting toxicities (irDLTs) for all three arms. Though lacking in statistical power, secondary outcomes include 1- and 2-year survival rates, as well as biomarker analysis of tumor immune cell infiltrates, which will shed light on the effect that such treatment has on the TME. Given the purported mechanism of these two antibodies, the combination thereof may allow for checkpoint blockade to occur at both the tumor site and within the tumor DLN to maximize the endogenous antitumor immune response.

Concomitant Immunotherapy

As there remains skepticism regarding the ability for checkpoint blockade antibodies to have activity in the brain and induce endogenous antitumor immune responses against weakly immunogenic glioma tumors, several trials are underway to evaluate the effect of combining checkpoint antibodies with other immune-potentiating agents and/or tumor vaccines. Indeed, checkpoint blockade combined with other immune-stimulatory therapies has been shown to be well-tolerated and provide enhanced efficacy in various tumor types. A Phase I trial of nivolumab combined with a multipeptide vaccine targeting several melanoma-associated antigens showed an increase in antigen-specific $CD8^+$ T cell numbers and was well tolerated in the patient cohort ($n = 33$).[128] Additionally, a Phase II trial of ipilimumab combined with granulocyte-macrophage colony-stimulating factor (G-CSF) (Sagramostim, GenZyme) to promote APC function showed increased OS and lower toxicity compared to ipilimumab alone for the treatment of metastatic melanoma ($n = 122$ per cohort).[129] Numerous other trials are currently underway to evaluate the safety and efficacy of combined ipilimumab or nivolumab with vaccines for the treatment of pancreatic cancer (NCT01896869; NCT02243371), cervical cancer (NCT02426892), prostate cancer (NCT02506114), and other metastatic malignancies (NCT02070406).

A Phase I trial for recurrent high-grade gliomas was recently announced, designed to evaluate the safety and efficacy of nivolumab given alongside DC vaccines (NCT02529072). Additionally, a Phase I/II study of nivolumab combined with an inhibitor of the immunosuppressive transforming growth factor (TGF)-β pathway is currently recruiting patients (NCT02423343). Indeed, several preclinical studies have shown that checkpoint blockade works synergistically alongside combined immune-potentiating agents to enhance survival. For instance, in the setting of murine glioma, CTLA-4 blockade following tumor-lysate vaccination[130] or alongside intratumoral IL-12[131] resulted in 40–50% long-term survival and an enhanced effector T cell response.

EMERGING TARGETS FOR CHECKPOINT BLOCKADE

Although strategies targeting the CTLA-4 and PD-1 pathways have dominated the clinical arena to date, a number of additional T cell inhibitory molecules have emerged as potential targets for checkpoint blockade. Because many checkpoint molecules are coexpressed during increasing states of T cell exhaustion, it is likely that combinatorial blockade will further the clinical success of this approach. In this section, we will review "next-generation" checkpoint pathways, their relevance to CNS malignancies, and their clinical progress to date.

Lymphocyte Activation Gene 3

Lymphocyte activation gene 3 (LAG-3 or CD223) is a CD4-like coreceptor upregulated on activated T cells[132,133] and Tregs,[134] NK cells,[133] B cells,[135] and a subset of DCs.[136] Class II MHC molecules are the only known ligands for LAG-3[132]; signaling through which results in enhanced Treg function[134] and abrogated $CD8^+$ T cell effector function.[137] LAG-3 is commonly coexpressed with PD-1 in states of chronic infection[138] and self-antigen recognition,[139] as T cells progress through increasing stages of exhaustion and/or tolerance. The synergistic relationship between PD-1 and LAG-3 has been shown to have particular relevance to antitumor immunity; whereas single *Pd1* or *Lag3* knockout mice experience delayed tumor growth, double *Pd1/Lag3* knockout mice have been shown to reject tumors altogether.[140]

Though the role of LAG-3 in GBM patients has yet to be explored, the coexpression of LAG-3 and PD-1 has been found to be increased on $CD8^+$ T cells from patients with a variety of other cancer types,[141,142] suggesting that this checkpoint pathway could also be relevant to brain tumors. Given the putative immunosuppressive relationship between PD-1 and LAG-3, an anti-LAG-3 blocking mAb (BMS-986016) has recently entered clinical testing as a monotherapy (NCT02061761) or in combination with nivolumab (NCT01968109) for the treatment of hematological malignancies and advanced/metastatic solid tumors, including GBM (NCT02658981). Indeed, several preclinical studies have shown that combined antibody blockade of LAG-3 and PD-1 results in increased tumor efficacy compared to either agent alone.[140,143,144]

T Cell Immunoglobulin and Mucin-3

T cell immunoglobulin and mucin-3 (TIM-3) is expressed on activated T cells, NK cells, and monocytes. The ligand for TIM-3 is GAL-9 expressed on a variety of cell types in many tissues,[145] through which binding has been shown to cause death of Th1 cells.[146] Similar to LAG-3, TILs from several murine cancer models show the near-universal coexpression of TIM-3 and PD-1,[147] indicating a more exhausted phenotype defined by their reduced proliferation and production of inflammatory cytokines. In this same study, combined blockade of both TIM-3 and PD-1 was more efficacious than blocking either checkpoint alone.

TIM-3 expression levels are increased in the setting of a variety of human malignancies.[148–150] A recent study has reported elevated TIM-3 expression on peripheral $CD4^+$ and $CD8^+$ T cells from glioma patients.[151] Patients with a higher tumor grade showed increased $CD8^+$ TIM-3 expression compared to those with lower grade tumors, and TIM-3 expression levels in both T cell compartments were shown to be correlated with decreased plasma levels of the inflammatory cytokine tumor necrosis factor alpha (TNF-α). An anti-TIM-3 mAb (MBG453, Novartis) has recently entered clinical testing in a Phase I trial for advanced malignancies, as a monotherapy or in combination with an anti-PD-1 mAb (NCT02608268).

B7-H3

B7-H3 (CD276) is a type I transmembrane protein with high sequence and structural homology to other B7 family molecules. Expression of B7-H3 on immune cells is believed to serve a costimulatory role,[152,153] but the function of B7-H3 on tumor cells is less clear. While some reports indicate that B7-H3 expression on tumor cell lines increases their immunogenicity and facilitates tumor rejection,[154] in situ B7-H3 expression in many human tumors has been associated with poor outcome.[155,156] In the setting of human brain tumors, B7-H3 expression levels in gliomas was found to be correlated with tumor grade and was associated with increased immunoreactivity of diffuse intrinsic pontine gliomas.[157]

Based on the findings of tumor-specific B7-H3 overexpression, enoblituzumab (MGA271, MacroGenics), an anti-B7-H3 mAb has recently entered clinical testing as a monotherapy (NCT01391143) or in combination with ipilimumab (NCT02381314) or pembrolizumab (NCT02475213) for refractory solid tumors. Rather than being employed as a checkpoint blocking antibody, however, the mechanism of MGA271 action is presumed to be through ADDC, similar in concept to anti-HER2 antibodies for $HER2^+$ breast cancers. Interestingly, a radiolabeled anti-B7-H3 antibody (^{131}I 8H9) was shown to suppress tumor cell growth via the delivery of therapeutic radiation doses to solid tumors in established intracranial xenografts,[158] and has been reported to have antitumor activity in pediatric patients with recurrent metastatic CNS neuroblastoma.[159] Such studies suggest the feasibility of mAb strategies where direct tumor cell targeting underlies therapeutic activity.

Killer Inhibitory Receptors as an NK Cell Checkpoint

NK cells are innate lymphoid cells that recognize and kill malignant and virus-infected cells through the secretion of preformed perforin/granzyme granules. Unlike T cells, NK cells do not recognize tumor peptide antigens presented on MHC molecules, but instead can become activated in response to the absence of MHC expression on cells. NK cell function is controlled by a complex interplay of activating receptors and killer inhibitory receptors (KIRs) and their ligands.[160] Binding of KIRs to MHC class I molecules transmits a cascade of inhibitory signals that result in NK cell tolerance to self-cells, such that KIRs take on a role of checkpoint molecules for NK cells.

NK cells comprise a substantial proportion of immune cell infiltrates in glioma patients and have been shown to play a role in antitumor immunity.[161] Several murine tumors have been shown to be susceptible to CTL-independent NK-mediated eradication.[162,163] Specifically, in the setting of orthotopic GBM xenograft models, NK cells transplanted intratumorally induced glioma cell apoptosis and gave rise to significant therapeutic effects.[164] Such findings suggest that blocking KIRs on NK cells to enhance their activation could make for a promising treatment modality for tumor immunotherapy. To that end, a fully-human anti-KIR mAb (lirilumab, Bristol Myers Squibb) has entered clinical testing in the setting of hematologic malignancies (NCT02399917, NCT02481297, NCT01687387) and has shown to be efficacious against transgenic humanized murine lymphoma.[163] Additionally, several trials are designed to evaluate the combination of ipilimumab and/or nivolumab with lirilumab in the setting of hematologic malignancies and solid tumors (NCT01592370, NCT02599649, NCT01714739, NCT01750580), which offers the potential for additive antitumor efficacy through simultaneous activation of both innate and adaptive immune compartments. Given the putative role of NK cells in anti-glioma immunity and the preliminary success in directing NK cell tumoricidal activity, future clinical studies should reveal the potential of anti-KIR blockade for glioma immunotherapy.

Indoleamine Dioxygenase as a "Checkpoint Enzyme"

Indolemaine dioxygenase (IDO) is a rate limiting enzyme that catabolizes the breakdown of tryptophan, an amino acid required for T cell survival and effector function. As such, IDO-mediated tryptophan deficiency results in T cell dysfunction/tolerance and subsequent differentiation of $CD4^+$ T cells into Tregs.[165] Because IDO does not exert its effects through ligand/receptor interactions on T cells, this pathway is not regarded as a classical immune checkpoint. However, IDO plays an important role in downregulating antitumor immunity across many tumor types,[166–169] particularly malignant gliomas,[170] making this enzyme an attractive target for immunomodulatory blockade for the treatment of GBM.

Under normal physiological conditions, IDO is found at only low levels in the brain, whereas IDO is expressed in 96% of glioma patients[44] and is correlated with a worse prognosis.[44,171] In a murine model of glioma, single agent blockade of IDO with 1-methyl-D-tryptophan (indoximod) leads to a relatively small increase in OS, while triple therapy with combined PD-L1 and CTLA-4 blockade gives rise to an 80% durable OS benefit.[124] Paradoxically, dual PD-L1/CTLA-4 blockade resulted in a similar survival benefit when compared to triple therapy, but only triple therapy resulted in significantly decreased Treg levels, suggesting that targeting all three pathways likely contributes to alleviating tumor-mediated immunosuppression. Indeed, a Phase I trial for indoximod combined with docetaxel chemotherapy showed partial responses in ~20% of treated patients ($n=22$) with metastatic solid tumors.[172] Phase I/II trials for indoximod in the setting of TMZ chemotherapy and radiation standard of care for the treatment of malignant gliomas (including GBM) are currently recruiting patients (NCT02052648; NCT02502708). Interestingly, increased IDO expression in the TME may also serve as a vaccine target; in a Phase I trial for NSCLC, an IDO-derived peptide conferred an OS of 26 months ($n=15$) compared to 7.7 months in the untreated patient cohort ($n=10$).[173] Thus, in addition to checkpoint blockade enhancing endogenous T cell responses, such data supports the notion of leveraging tumor-associated checkpoint molecules as antigens for inducing antitumor immunity.

ON THE HORIZON: IMMUNOMODULATORY ANTIBODIES FOR T CELL COSTIMULATION

On the flipside to mitigating T cell immunosuppression through checkpoint blockade, antibody-based targeting of T cell costimulatory receptors has the potential to directly enhance their activation upon tumor antigen recognition and their resultant tumor-specific effector function. Costimulation is the "second signal" required for a functional T cell response, which is provided through the engagement of T cell costimulatory receptors with costimulatory ligands upregulated on activated APCs.[174,175] Costimulatory pathways in APCs from cancer-bearing hosts are often abrogated, which can result in T cell apoptosis or anergy to tumor antigens.[176,177] Agonist mAbs for costimulatory receptors on T cells offer a promising strategy for overcoming aberrant APC activity and ensuring delivery of the vital costimulatory signal.

To date, the evaluation of costimulatory mAbs has received far less clinical attention than checkpoint blockade. Such agonist antibodies, however, demonstrate great promise in clinical trials for the treatment of several cancer types and are an attractive strategy for GBM immunotherapy. Targeting costimulatory members of the tumor necrosis factor receptor superfamily specifically upregulated on activated T cells (e.g., OX40, 4-1BB, and CD27), has proven particularly effective in preclinical glioma models. The use of mAbs for OX40[178] and 4-1BB[179] as monotherapies for murine glioma resulted in a 50% cure rate and increased median survival from 31 days to 42 days, respectively. In a separate study, addition of an anti-4-1BB antibody enhanced the antitumor activity of T cells primed with glioma-lysate pulsed DCs.[180] Indeed, given promising preclinical findings in murine glioma and other tumor models,[181–184] there are a number of agonist antibodies for costimulatory receptors currently in clinical development (Table 11.1).

In addition to providing costimulatory signals directly to T cells, promoting activation of APCs through the use of agonist mAbs for CD40 is yet another attractive strategy for enhancing the antitumor immune response. CD40 is constitutively expressed by immature DCs, and its ligation by CD40L results in DC maturation and upregulation of costimulatory ligands, making it crucial for functional antigen presentation.[185,186] Indeed, CD40 agonist mAbs have anti-tumor effects in a variety of murine tumor models.[187–189] Interestingly, CD40 stimulation in gliomas was found to directly inhibit cell proliferation through nuclear factor kappa B signaling and TNF-α production.[190] These studies make mAb targeting of CD40 a promising strategy from both an immunomodulatory perspective and as a direct cancer therapeutic. An anti-CD40 mAb, CP-870,893 (Pfizer), has shown antitumor activity mediated by CD40-expressing macrophages; a clinical trial evaluating the effect of CP-870,893 in combination with anti-CTLA-4 (tremilumumab) in melanoma patients is currently underway (NCT01103635).

CONCLUSION

Checkpoint blockade therapy has emerged as a promising strategy for the treatment of a variety of malignancies. Though this modality has only recently reached the glioma clinical arena, immunomodulatory mAbs have proven safe and effective in the treatment of a variety of other solid tumors, lending credence to the field of tumor immunotherapy as a whole. Combined blockade of CTLA-4 and PD-1 has garnered particular attention recently, given the success of this dual therapy for the treatment of advanced melanoma. Data generated from ongoing clinical studies of such combination therapy for GBM is awaited. Several costimulatory molecules on the surface of immune effector cells are also attractive targets for the reversal of immunosuppression and antibody-based immunomodulation for glioma treatment. Indeed, clinical trials investigating this approach are underway for a variety of solid tumors, and preclinical data in the context of glioma therapy demonstrate encouraging results.

It is evident that given the profound immunosuppression among glioma patients, a multitude of immunosuppressive pathways are amenable to antibody-based blockade or activation. However, antibody-based checkpoint blockade for glioma immunotherapy faces several challenges and points of skepticism, due to particular characteristics unique to CNS malignancies. The success of this approach will likely require combined treatment modalities to improve access across the BBB, enhance the endogenous tumor immune response and/or provide an immunogenic tumor target, and alleviate irAEs in the brain. It is likely that optimal therapy for gliomas will require synergistic standard of care therapy, combination tumor vaccines, and/or multitarget immunomodulatory antibodies. Large-scale clinical trials will be necessary to determine which therapeutic targets provide the safest and most effective antitumor immunotherapy for glioma patients. Additionally, the fine-tuning of response criteria and biomarker correlates of efficacy will further improve the success of glioma immunotherapy clinical trials. Though antibody-based checkpoint blockade may not be the "magic bullet" for cancer therapy, this approach could add another weapon to the arsenal of glioma standard of care to further improve prognosis for this challenging disease.

References

1. Hodi FS, O'Day SJ, McDermott DF, et al. Improved survival with ipilimumab in patients with metastatic melanoma. *N Engl J Med*. 2010;363(8):711–723.
2. Brahmer J, Reckamp KL, Baas P, et al. Nivolumab versus Docetaxel in advanced squamous-cell non-small-cell lung cancer. *N Engl J Med*. 2015;373(2):123–135.
3. Azuma M, Ito D, Yagita H, et al. B70 antigen is a second ligand for CTLA-4 and CD28. *Nature*. 1993;366:76–79.
4. Hathcock KS, Laszlo G, Dickler HB, Bradshaw J, Linsley P, Hodes RJ. Identification of an alternative CTLA-4 ligand costimulatory for T cell activation. *Science*. 1993;262(5135):905–907.
5. Rudd CE, Taylor A, Schneider H. CD28 and CTLA-4 coreceptor expression and signal transduction. *Immunol Rev*. 2009;229(1):12–26.
6. Schwartz RH. Costimulation of T lymphocytes: the role of CD28, CTLA-4, and B7/BB1 in interleukin-2 production and immunotherapy. *Cell*. 1992;71(7):1065–1068.
7. Schneider H, Downey J, Smith A, et al. Reversal of the TCR stop signal by CTLA-4. *Science*. 2006;313(5795):1972–1975.
8. Parry RV, Chemnitz JM, Frauwirth KA, et al. CTLA-4 and PD-1 receptors inhibit T-cell activation by distinct mechanisms. *Mol Cell Biol*. 2005;25(21):9543–9553.
9. Tivol EA, Borriello F, Schweitzer AN, Lynch WP, Bluestone JA, Sharpe AH. Loss of CTLA-4 leads to massive lymphoproliferation and fatal multiorgan tissue destruction, revealing a critical negative regulatory role of CTLA-4. *Immunity*. 1995;3(5):541–547.
10. Waterhouse P, Penninger JM, Timms E, et al. Lymphoproliferative disorders with early lethality in mice deficient in Ctla-4. *Science*. 1995;270(5238):985–988.

11. Hill JA, Feuerer M, Tash K, et al. Foxp3 transcription-factor-dependent and -independent regulation of the regulatory T cell transcriptional signature. *Immunity*. 2007;27(5):786–800.
12. Gavin MA, Rasmussen JP, Fontenot JD, et al. Foxp3-dependent programme of regulatory T-cell differentiation. *Nature*. 2007;445(7129):771–775.
13. Wing K, Onishi Y, Prieto-Martin P, et al. CTLA-4 control over Foxp3 regulatory T cell function. *Science*. 2008;322(5899):271–275.
14. Qureshi OS, Zheng Y, Nakamura K, et al. Trans-endocytosis of CD80 and CD86: a molecular basis for the cell-extrinsic function of CTLA-4. *Science*. 2011;332(6029): 600–603.
15. Spranger S, Koblish HK, Horton B, Scherle PA, Newton R, Gajewski TF. Mechanism of tumor rejection with doublets of CTLA-4, PD-1/PD-L1, or IDO blockade involves restored IL-2 production and proliferation of CD8(+) T cells directly within the tumor microenvironment. *J Immunother Cancer*. 2014;2(3).
16. Ishida Y, Agata Y, Shibahara K, Honjo T. Induced expression of PD-1, a novel member of the immunoglobulin gene superfamily, upon programmed cell death. *EMBO J*. 1992;11:3887–3895.
17. Day CL, Kaufmann DE, Kiepiela P, et al. PD-1 expression on HIV-specific T cells is associated with T-cell exhaustion and disease progression. *Nature*. 2006;443(7109):350–354.
18. Barber DL, Wherry EJ, Masopust D, et al. Restoring function in exhausted CD8 T cells during chronic viral infection. *Nature*. 2006;439(7077):682–687.
19. Riley JL. PD-1 signaling in primary T cells. *Immunol Rev*. 2009;229(1):114–125.
20. Terme M, Ullrich E, Aymeric L, et al. IL-18 induces PD-1-dependent immunosuppression in cancer. *Cancer Res*. 2011;71(16):5393–5399.
21. Fanoni D, Tavecchio S, Recalcati S, et al. New monoclonal antibodies against B-cell antigens: possible new strategies for diagnosis of primary cutaneous B-cell lymphomas. *Immunol Lett*. 2011;134(2):157–160.
22. Francisco LM, Salinas VH, Brown KE, et al. PD-L1 regulates the development, maintenance, and function of induced regulatory T cells. *J Exp Med*. 2009;206(13):3015–3029.
23. Tseng SY, Otsuji M, Gorski K, et al. B7-DC, a new dendritic cell molecule with potent costimulatory properties for T cells. *J Exp Med*. 2001;193(7):839–846.
24. Zou W, Chen L. Inhibitory B7-family molecules in the tumour microenvironment. *Nat Rev Immunol*. 2008;8(6):467–477.
25. Dong H, Strome SE, Salomao DR, et al. Tumor-associated B7-H1 promotes T-cell apoptosis: a potential mechanism of immune evasion. *Nat Med*. 2002;8(8):793–800.
26. Lee SK, Seo SH, Kim BS, et al. IFN-gamma regulates the expression of B7-H1 in dermal fibroblast cells. *J Dermatol Sci*. 2005;40(2):95–103.
27. Taube JM, Anders RA, Young GD, et al. Colocalization of inflammatory response with B7-h1 expression in human melanocytic lesions supports an adaptive resistance mechanism of immune escape. *Sci Transl Med*. 2012;4(127):127ra137.
28. Yamazaki T, Akiba H, Iwai H, et al. Expression of programmed death 1 ligands by murine T cells and APC. *J Immunol*. 2002;169(10):5538–5545.
29. Freeman GJ, Long AJ, Iwai Y, et al. Engagement of the PD-1 immunoinhibitory receptor by a novel B7 family member leads to negative regulation of lymphocyte activation. *J Exp Med*. 2000;192(7):1027–1034.
30. Butte MJ, Keir ME, Phamduy TB, Sharpe AH, Freeman GJ. Programmed death-1 ligand 1 interacts specifically with the B7-1 costimulatory molecule to inhibit T cell responses. *Immunity*. 2007;27(1):111–122.
31. Park JJ, Omiya R, Matsumura Y, et al. B7-H1/CD80 interaction is required for the induction and maintenance of peripheral T-cell tolerance. *Blood*. 2010;116(8):1291–1298.
32. Ahmadzadeh M, Johnson LA, Heemskerk B, et al. Tumor antigen-specific CD8 T cells infiltrating the tumor express high levels of PD-1 and are functionally impaired. *Blood*. 2009;114(8):1537–1544.

33. Sfanos KS, Bruno TC, Meeker AK, De Marzo AM, Isaacs WB, Drake CG. Human prostate-infiltrating CD8+ T lymphocytes are oligoclonal and PD-1+. *Prostate.* 2009;69(15):1694–1703.
34. Patel SP, Kurzrock R. PD-L1 expression as a predictive biomarker in cancer immunotherapy. *Mol Cancer Ther.* 2015;14(4):847–856.
35. Parsa AT, Waldron JS, Panner A, et al. Loss of tumor suppressor PTEN function increases B7-H1 expression and immunoresistance in glioma. *Nat Med.* 2007;13(1):84–88.
36. Marzec M, Gowen BG, Goradia A, et al. Oncogenic kinase NPM/ALK induces through STAT3 expression of immunosuppressive protein CD274 (PD-L1, B7-H1). *Proc Natl Acad Sci USA.* 2008;105:20852–20857.
37. Spranger S, Spaapen RM, Zha Y, et al. Up-regulation of PD-L1, IDO, and T(regs) in the melanoma tumor microenvironment is driven by CD8(+) T cells. *Sci Transl Med.* 2013;5(200):200ra116.
38. Wintterle S, Schreiner B, Mitsdoerffer M, et al. Expression of the B7-related molecule B7-H1 by glioma cells: a potential mechanism of immune paralysis. *Cancer Res.* 2003;63(21):7462–7467.
39. Petersen TR, Dickgreber N, Hermans IF. Tumor antigen presentation by dendritic cells. *Crit Rev™ Immunol.* 2010;30(4):345–386.
40. Fecci PE, Mitchell DA, Whitesides JF, et al. Increased regulatory T-cell fraction amidst a diminished CD4 compartment explains cellular immune defects in patients with malignant glioma. *Cancer Res.* 2006;66(6):3294–3302.
41. Jiang Y, Li Y, Zhu B. T-cell exhaustion in the tumor microenvironment. *Cell Death Dis.* 2015;6:e1792.
42. Mirghorbani M, Van Gool S, Rezaei N. Myeloid-derived suppressor cells in glioma. *Exp Rev Neurother.* 2013;13(12):1395–13406.
43. Yue Q, Zhang X, Ye HX, et al. The prognostic value of Foxp3+ tumor-infiltrating lymphocytes in patients with glioblastoma. *J Neuro Oncol.* 2014;116(2):251–259.
44. Mitsuka K, Kawataki T, Satoh E, Asahara T, Horikoshi T, Kinouchi H. Expression of indoleamine 2,3-dioxygenase and correlation with pathological malignancy in gliomas. *Neurosurgery.* 2013;72(6):1031–1038. discussion 1038–1039.
45. Wei B, Wang L, Zhao XL, Du C, Guo YC, Sun ZG. The upregulation of programmed death 1 on peripheral blood T cells of glioma is correlated with disease progression. *Tumor Biol.* 2014;35(4):2923–2929.
46. Bloch O, Crane CA, Kaur R, Safaee M, Rutkowski MJ, Parsa AT. Gliomas promote immunosuppression through induction of B7-H1 expression in tumor-associated macrophages. *Clin Cancer Res.* 2013;19(12):3165–3175.
47. Berghoff AS, Kiesel B, Widhalm G, et al. Programmed death ligand 1 expression and tumor-infiltrating lymphocytes in glioblastoma. *Neuro Oncol.* 2015;17(8):1064–1075.
48. Weber JS, Kudchadkar RR, Yu B, et al. Safety, efficacy, and biomarkers of nivolumab with vaccine in ipilimumab-refractory or -naive melanoma. *J Clin Oncol.* 2013;31(34):4311–4318.
49. Velcheti V, Schalper KA, Carvajal DE, et al. Programmed death ligand-1 expression in non-small cell lung cancer. *Lab Invest.* 2014;94(1):107–116.
50. Zhang L, Zhang Z, Liu Y, Xue Y, Parney I. Programmed death-ligand 1 (PD-L1) may play a role in malignant glioma infiltration. *Med Hypotheses.* 2015;85:127–129.
51. Thompson RH, Gillett MD, Cheville JC, et al. Costimulatory B7-H1 in renal cell carcinoma patients: indicator of tumor aggressiveness and potential therapeutic target. *Proc Natl Acad Sci USA.* 2004;101(49):17174–17179.
52. Ohigashi Y, Sho M, Yamada Y, et al. Clinical significance of programmed death-1 ligand-1 and programmed death-1 ligand-2 expression in human esophageal cancer. *Clin Cancer Res.* 2005;11(8):2947–2953.
53. Hamanishi J, Mandai M, Iwasaki M, et al. Programmed cell death 1 ligand 1 and tumor-infiltrating CD8+ T lymphocytes are prognostic factors of human ovarian cancer. *Proc Natl Acad Sci USA.* 2007;104(9):3360–3365.

54. Liu Y, Carlsson R, Ambjorn M, et al. PD-L1 expression by neurons nearby tumors indicates better prognosis in glioblastoma patients. *J Neurosci*. 2013;33(35):14231–14245.
55. Wu Q, Zhan X, Dou T, et al. CTLA4 A49G polymorphism shows significant association with glioma risk in a Chinese population. *Biochem Genet*. 2011;49:190–201.
56. Fong B, Jin R, Wang X, et al. Monitoring of regulatory T cell frequencies and expression of CTLA-4 on T cells, before and after DC vaccination, can predict survival in GBM patients. *PLoS One*. 2012;7(4):e32614.
57. El Andaloussi A, Han Y, Lesniak MS. Prolongation of survival following depletion of $CD4^+CD25^+$ regulatory T cells in mice with experimental brain tumors. *J Neurosurg*. 2006;105(3):430–437.
58. Heimberger AB, Abou-Ghazal M, Reina-Ortiz C, et al. Incidence and prognostic impact of $FoxP3^+$ regulatory T cells in human gliomas. *Clin Cancer Res*. 2008;14(16):5166–5172.
59. Walker LS, Sansom DM. The emerging role of CTLA4 as a cell-extrinsic regulator of T cell responses. *Nat Rev Immunol*. 2011;11(12):852–863.
60. Dix AR, Brooks WH, Roszman TL, Morford LA. Immune defects observed in patients with primary malignant brain tumors. *J Neuroimmunol*. 1999;100(1–2):216–232.
61. Rossi ML, Hughes JT, Esiri MM, Coakham HB, Brownell DB. Immunohistological study of mononuclear cell infiltrate in malignant gliomas. *Acta Neuropathol*. 1987;74(3):269–277.
62. Dudley ME, Rosenberg SA. Adoptive-cell-transfer therapy for the treatment of patients with cancer. *Nat Rev Cancer*. 2003;3(9):666–675.
63. Leach DR, Krummel MF, Allison JP. Enhancement of antitumor immunity by CTLA-4 blockade. *Science*. 1996;271(5256):1734–1736.
64. Phan GQ, Yang JC, Sherry RM, et al. Cancer regression and autoimmunity induced by cytotoxic T lymphocyte-associated antigen 4 blockade in patients with metastatic melanoma. *Proc Natl Acad Sci USA*. 2003;100(14):8372–8377.
65. Ribas A, Camacho LH, Lopez-Berestein G, et al. Antitumor activity in melanoma and anti-self responses in a phase I trial with the anti-cytotoxic T lymphocyte-associated antigen 4 monoclonal antibody CP-675,206. *J Clin Oncol*. 2005;23(35):8968–8977.
66. Beck KE, Blansfield JA, Tran KQ, et al. Enterocolitis in patients with cancer after antibody blockade of cytotoxic T-lymphocyte-associated antigen 4. *J Clin Oncol*. 2006;24(15):2283–2289.
67. Selby MJ, Engelhardt JJ, Quigley M, et al. Anti-CTLA-4 antibodies of IgG2a isotype enhance antitumor activity through reduction of intratumoral regulatory T cells. *Cancer Immunol Res*. 2013;1(1):32–42.
68. Schadendorf D, Hodi FS, Robert C, et al. Pooled analysis of long-term survival data from phase II and phase III trials of Ipilimumab in unresectable or metastatic melanoma. *J Clin Oncol*. 2015;33(17):1889–1894.
69. Lynch TJ, Bondarenko I, Luft A, et al. Ipilimumab in combination with paclitaxel and carboplatin as first-line treatment in stage IIIB/IV non-small-cell lung cancer: results from a randomized, double-blind, multicenter phase II study. *J Clin Oncol*. 2012;30(17):2046–2054.
70. Reck M, Bondarenko I, Luft A, et al. Ipilimumab in combination with paclitaxel and carboplatin as first-line therapy in extensive-disease-small-cell lung cancer: results from a randomized, double-blind, multicenter phase 2 trial. *Ann Oncol*. 2013;24(1):75–83.
71. Berger R, Rotem-Yehudar R, Slama G, et al. Phase I safety and pharmacokinetic study of CT-011, a humanized antibody interacting with PD-1, in patients with advanced hematologic malignancies. *Clin Cancer Res*. 2008;14(10):3044–3051.
72. Westin JR, Chu F, Zhang M, et al. Safety and activity of PD1 blockade by pidilizumab in combination with rituximab in patients with relapsed follicular lymphoma: a single group, open-label, phase 2 trial. *Lancet Oncol*. 2014;15(1):69–77.
73. Hamid O, Robert C, Daud A, et al. Safety and tumor responses with lambrolizumab (anti-PD-1) in melanoma. *N Engl J Med*. 2013;369(2):134–144.

74. Garon EB, Gandhi L, Rizvi NA, et al. Antitumor activity of pembrolizumab (PEMBRO; MK-3475) and correlation with programmed death ligand (PD-L1) expression in a pooled analysis of patients (pts) with advanced non small cell lung carcinoma (NSCLC). *Ann Oncol*. 2014;25(suppl 4s).
75. Robert C, Schachter J, Long GV, et al. Pembrolizumab versus Ipilimumab in advanced melanoma. *N Engl J Med*. 2015;372(26):2521–2532.
76. Wolchok JD, Kluger H, Callahan MK, et al. Nivolumab plus ipilimumab in advanced melanoma. *N Engl J Med*. 2013;369(2):122–133.
77. Choi BD, Pastan I, Bigner DD, Sampson JH. A novel bispecific antibody recruits T cells to eradicate tumors in the "immunologically privileged" central nervous system. *Oncoimmunology*. 2013;2(4):e23639.
78. Antonia S, Goldberg SB, Balmanoukian A, et al. Safety and antitumour activity of durvalumab plus tremelimumab in non-small cell lung cancer: a multicentre, phase 1b study. *Lancet Oncol*. 2016;**17**(3) :299–308.
79. McDermott DF, Sosman JA, Sznol M, et al. Atezolizumab, an anti-programmed death-ligand 1 antibody, in metastatic renal cell carcinoma: long-term safety, clinical activity, and immune correlates from a phase Ia study. *J Clin Oncol* 2016:34(8):833–842.
80. Topalian SL, Hodi FS, Brahmer JR, et al. Safety, activity, and immune correlates of anti-PD-1 antibody in cancer. *N Engl J Med*. 2012;366(26):2443–2454.
81. Omuro AM, Faivre S, Raymond E. Lessons learned in the development of targeted therapy for malignant gliomas. *Mol Cancer Ther*. 2007;6(7):1909–1919.
82. Jackson C, Ruzevick J, Phallen J, Belcaid Z, Lim M. Challenges in immunotherapy presented by the glioblastoma multiforme microenvironment. *Clin Dev Immunol*. 2011;2011:732413.
83. Louveau A, Smirnov I, Keyes TJ, et al. Structural and functional features of central nervous system lymphatic vessels. *Nature*. 2015;523(7560):337–341.
84. Male DK. Immunology of brain endothelium and the blood–brain barrier. In: Bradbury MWB, ed. *Physiology and Pharmacology of the Blood–Brain Barrier*. Berlin: Springer; 1992:397–415.
85. Bullard DE, Adams CJ, Coleman RE, Bigner DD. In vivo imaging of intracranial human glioma xenografts comparing specific with nonspecific radiolabeled monoclonal antibodies. *J Neurosurg*. 1986;64(2):257–262.
86. Bourdon MA, Coleman RE, Blasberg RG, Groothuis DR, Bigner DD. Monoclonal antibody localization in subcutaneous and intracranial human glioma xenografts: paired-label and imaging analysis. *Anticancer Res*. 1984;4(3):133–140.
87. Zalutsky MR, Moseley RP, Coakham HB, Coleman RE, Bigner DD. Pharmacokinetics and tumor localization of 131I-labeled anti-tenascin monoclonal antibody 81C6 in patients with gliomas and other intracranial malignancies. *Cancer Res*. 1989;49(10):2807–2813.
88. Konstantinou MP, Dutriaux C, Gaudy-Marqueste C, et al. Ipilimumab in melanoma patients with brain metastasis: a retrospective multicentre evaluation of thirty-eight patients. *Acta Derm-Venereol*. 2014;94(1):45–49.
89. Di Giacomo AM, Ascierto PA, Pilla L, et al. Ipilimumab and fotemustine in patients with advanced melanoma (NIBIT-M1): an open-label, single-arm phase 2 trial. *Lancet Oncol*. 2012;13(9):879–886.
90. Snyder A, Makarov V, Merghoub T, et al. Genetic basis for clinical response to CTLA-4 blockade in melanoma. *N Engl J Med*. 2014;371(23):2189–2199.
91. Topalian SL, Drake CG, Pardoll DM. Immune checkpoint blockade: a common denominator approach to cancer therapy. *Cancer Cell*. 2015;27(4):450–461.
92. Lawrence MS, Stojanov P, Polak P, et al. Mutational heterogeneity in cancer and the search for new cancer-associated genes. *Nature*. 2013;499(7457):214–218.
93. Flugel A, Labeur MS, Grasbon-Frodl EM, Kreutzberg GW, Graeber MB. Microglia only weakly present glioma antigen to cytotoxic T cells. *Int J Dev Neurosci*. 1999;17(5–6):547–556.

94. Tran CT, Wolz P, Egensperger R, et al. Differential expression of MHC class II molecules by microglia and neoplastic astroglia: relevance for the escape of astrocytoma cells from immune surveillance. *Neuropathol Appl Neurobiol*. 1998;24(4):293–301.
95. Zagzag D, Salnikow K, Chiriboga L, et al. Downregulation of major histocompatibility complex antigens in invading glioma cells: stealth invasion of the brain. *Lab Invest*. 2005;85(3):328–341.
96. Villadolid J, Amin A. Immune checkpoint inhibitors in clinical practice: update on management of immune-related toxicities. *Transl Lung Cancer Res*. 2015;4(5):560–575.
97. Xu LW, Chow KK, Lim M, Li G. Current vaccine trials in glioblastoma: a review. *J Immunol Res*. 2014;2014:796856.
98. Plautz GE, Miller DW, Barnett GH, et al. T cell adoptive immunotherapy of newly diagnosed gliomas. *Clin Cancer Res*. 2000;6(6):2209–2218.
99. Gelao L, Criscitiello C, Esposito A, Goldhirsch A, Curigliano G. Immune checkpoint blockade in cancer treatment: a double-edged sword cross-targeting the host as an "innocent bystander". *Toxins*. 2014;6(3):914–933.
100. Weber JS. Practical management of immune-related adverse events from immune checkpoint protein antibodies for the oncologist. *Am Soc Clin Oncol Educ Book*. 2012:174–177.
101. Fecher LA, Agarwala SS, Hodi FS, Weber JS. Ipilimumab and its toxicities: a multidisciplinary approach. *Oncologist*. 2013;18(6):733–743.
102. Fecci PE, Ochiai H, Mitchell DA, et al. Systemic CTLA-4 blockade ameliorates glioma-induced changes to the $CD4^+$ T cell compartment without affecting regulatory T-cell function. *Clin Cancer Res*. 2007;13(7):2158–2167.
103. Reardon DA, Gokhale PC, Klein SR, et al. Glioblastoma eradication following immune checkpoint blockade in an orthotopic, immunocompetent model. *Cancer Immunol Res*. 2016;4(2):124–135.
104. Stupp R. Radiotherapy plus concomitant and adjuvant temozolomide for glioblastoma. *N Engl J Med*. 2005;352:987–996.
105. Rosen EM, Fan S, Rockwell S, Goldberg ID. The molecular and cellular basis of radiosensitivity: implications for understanding how normal tissues and tumors respond to therapeutic radiation. *Cancer Invest*. 1999;17(1):56–72.
106. Gough MJ, Crittenden MR. Combination approaches to immunotherapy: the radiotherapy example. *Immunotherapy*. 2009;1(6):1025–1037.
107. Trinh VA, Patel SP, Hwu WJ. The safety of temozolomide in the treatment of malignancies. *Expert Opin Drug Saf*. 2009;8(4):493–499.
108. Ishikawa E, Yamamoto T, Sakamoto N, et al. Low peripheral lymphocyte count before focal radiotherapy plus concomitant temozolomide predicts severe lymphopenia during malignant glioma treatment. *Neural Med Chir*. 2010;50(8):638–644.
109. Kerr JF, Wyllie AH, Currie AR. Apoptosis: a basic biological phenomenon with wide-ranging implications in tissue kinetics. *Br J Cancer*. 1972;26(4):239–257.
110. Heimberger AB, Sun W, Hussain SF, et al. Immunological responses in a patient with glioblastoma multiforme treated with sequential courses of temozolomide and immunotherapy: case study. *Neuro Oncol*. 2008;10(1):98–103.
111. Banissi C, Ghiringhelli F, Chen L, Carpentier AF. Treg depletion with a low-dose metronomic temozolomide regimen in a rat glioma model. *Cancer Immunol Immunother*. 2009;58(10):1627–1634.
112. Su YB, Sohn S, Krown SE, et al. Selective $CD4^+$ lymphopenia in melanoma patients treated with temozolomide: a toxicity with therapeutic implications. *J Clin Oncol*. 2004;22(4):610–616.
113. Sampson JH, Aldape K, Archer GE, et al. Greater chemotherapy-induced lymphopenia enhances tumor-specific immune responses that eliminate EGFRvIII-expressing tumor cells in patients with glioblastoma. *Neuro Oncol*. 2011;13(3):323–333.

114. Sanchez-Perez LA, Choi BD, Archer GE, et al. Myeloablative temozolomide enhances CD8(+) T-cell responses to vaccine and is required for efficacy against brain tumors in mice. *PLoS One*. 2013;8(3):e59082.
115. McBride WH, Chiang CS, Olson JL, et al. A sense of danger from radiation. *Radiat Res*. 2004;162(1):1–19.
116. Demaria S, Formenti SC. Sensors of ionizing radiation effects on the immunological microenvironment of cancer. *Int J Radiat Biol*. 2007;83(11–12):819–825.
117. Albert ML, Sauter B, Bhardwaj N. Dendritic cells acquire antigen from apoptotic cells and induce class I-restricted CTLs. *Nature*. 1998;392(6671):86–89.
118. Sauter B, Albert ML, Francisco LM, Larsson M, Somersan S, Bhardwaj N. Consequences of cell death: exposure to necrotic tumor cells, but not primary tissue cells or apoptotic cells, induces the maturation of immunostimulatory dendritic cells. *J Exp Med*. 2000;191:423–434.
119. Santin AD, Hermonat PL, Hiserodt JC, et al. Effects of irradiation on the expression of major histocompatibility complex class I antigen and adhesion costimulation molecules ICAM-1 in human cervical cancer. *Int J Radiat Oncol Biol Phys*. 1997;39(3):737–742.
120. Lugade AA, Sorensen EW, Gerber SA, Moran JP, Frelinger JG, Lord EM. Radiation-induced IFN-gamma production within the tumor microenvironment influences antitumor immunity. *J Immunol*. 2008;180(5):3132–3139.
121. Zeng J, See AP, Phallen J, et al. Anti-PD-1 blockade and stereotactic radiation produce long-term survival in mice with intracranial gliomas. *Int Radiat Oncol Biol Phys*. 2013;86(2):343–349.
122. Dewan MZ, Galloway AE, Kawashima N, et al. Fractionated but not single-dose radiotherapy induces an immune-mediated abscopal effect when combined with anti-CTLA-4 antibody. *Clin Cancer Res*. 2009;15(17):5379–5388.
123. Silk AW, Bassetti MF, West BT, Tsien CI, Lao CD. Ipilimumab and radiation therapy for melanoma brain metastases. *Cancer Med*. 2013;2(6):899–906.
124. Wainwright DA, Chang AL, Dey M, et al. Durable therapeutic efficacy utilizing combinatorial blockade against IDO, CTLA-4, and PD-L1 in mice with brain tumors. *Clin Cancer Res*. 2014;20(20):5290–5301.
125. Demaria S, Kawashima N, Yang AM, et al. Immune-mediated inhibition of metastases after treatment with local radiation and CTLA-4 blockade in a mouse model of breast cancer. *Clin Cancer Res*. 2005;11(2 Pt 1):728–734.
126. Mansfield AS, Nevala WK, Lieser EA, Leontovich AA, Markovic SN. The immunomodulatory effects of bevacizumab on systemic immunity in patients with metastatic melanoma. *Oncoimmunology*. 2013;2(5):e24436.
127. Terme M, Colussi O, Marcheteau E, Tanchot C, Tartour E, Taieb J. Modulation of immunity by antiangiogenic molecules in cancer. *Clin Dev Immunol*. 2012;2012:492920.
128. Gibney GT, Kudchadkar RR, DeConti RC, et al. Safety, correlative markers, and clinical results of adjuvant nivolumab in combination with vaccine in resected high-risk metastatic melanoma. *Clin Cancer Res*. 2015;21(4):712–720.
129. Hodi FS, Lee S, McDermott DF, et al. Ipilimumab plus sargramostim vs ipilimumab alone for treatment of metastatic melanoma: a randomized clinical trial. *JAMA*. 2014;312(17):1744–1753.
130. Agarwalla P, Barnard Z, Fecci P, Dranoff G, Curry Jr WT. Sequential immunotherapy by vaccination with GM-CSF-expressing glioma cells and CTLA-4 blockade effectively treats established murine intracranial tumors. *J Immunother*. 2012;35(5):385–389.
131. Vom Berg J, Vrohlings M, Haller S, et al. Intratumoral IL-12 combined with CTLA-4 blockade elicits T cell-mediated glioma rejection. *J Exp Med*. 2013;210(13):2803–2811.
132. Huard B, Gaulard P, Faure F, Hercend T, Triebel F. Cellular expression and tissue distribution of the human LAG-3-encoded protein, an MHC class II ligand. *Immunogenetics*. 1994;39(3):213–217.

133. Triebel F, Jitsukawa S, Baixeras E, et al. LAG-3, a novel lymphocyte activation gene closely related to CD4. *J Exp Med*. 1990;171(5):1393–1405.
134. Huang CT, Workman CJ, Flies D, et al. Role of LAG-3 in regulatory T cells. *Immunity*. 2004;21(4):503–513.
135. Kisielow M, Kisielow J, Capoferri-Sollami G, Karjalainen K. Expression of lymphocyte activation gene 3 (LAG-3) on B cells is induced by T cells. *Eur J Immunol*. 2005;35(7):2081–2088.
136. Workman CJ, Wang Y, El Kasmi KC, et al. LAG-3 regulates plasmacytoid dendritic cell homeostasis. *J Immunol*. 2009;182(4):1885–1891.
137. Grosso JF, Kelleher CC, Harris TJ, et al. LAG-3 regulates $CD8^+$ T cell accumulation and effector function in murine self- and tumor-tolerance systems. *J Clin Invest*. 2007;117(11):3383–3392.
138. Blackburn SD, Shin H, Haining WN, et al. Coregulation of $CD8^+$ T cell exhaustion by multiple inhibitory receptors during chronic viral infection. *Nat Immunol*. 2009;10(1):29–37.
139. Grosso JF, Goldberg MV, Getnet D, et al. Functionally distinct LAG-3 and PD-1 subsets on activated and chronically stimulated CD8 T cells. *J Immunol*. 2009;182(11):6659–6669.
140. Woo SR, Turnis ME, Goldberg MV, et al. Immune inhibitory molecules LAG-3 and PD-1 synergistically regulate T-cell function to promote tumoral immune escape. *Cancer Res*. 2012;72(4):917–927.
141. Matsuzaki J, Gnjatic S, Mhawech-Fauceglia P, et al. Tumor-infiltrating NY-ESO-1-specific $CD8^+$ T cells are negatively regulated by LAG-3 and PD-1 in human ovarian cancer. *Proc Natl Acad Sci USA*. 2010;107(17):7875–7880.
142. Baitsch L, Legat A, Barba L, et al. Extended co-expression of inhibitory receptors by human CD8 T-cells depending on differentiation, antigen-specificity and anatomical localization. *PLoS One*. 2012;7(2):e30852.
143. Huang RY, Eppolito C, Lele S, Shrikant P, Matsuzaki J, Odunsi K. LAG3 and PD1 co-inhibitory molecules collaborate to limit $CD8^+$ T cell signaling and dampen antitumor immunity in a murine ovarian cancer model. *Oncotarget*. 2015;6(29):27359–27377.
144. Jing W, Gershan JA, Weber J, et al. Combined immune checkpoint protein blockade and low dose whole body irradiation as immunotherapy for myeloma. *J Immunother Cancer*. 2015;3(1):2.
145. Wada J, Kanwar YS. Identification and characterization of galectin-9, a novel beta-galactoside-binding mammalian lectin. *J Biol Chem*. 1997;272(9):6078–6086.
146. Zhu C, Anderson AC, Schubart A, et al. The Tim-3 ligand galectin-9 negatively regulates T helper type 1 immunity. *Nature Immunol*. 2005;6(12):1245–1252.
147. Sakuishi K, Apetoh L, Sullivan JM, Blazar BR, Kuchroo VK, Anderson AC. Targeting Tim-3 and PD-1 pathways to reverse T cell exhaustion and restore anti-tumor immunity. *J Exp Med*. 2010;207(10):2187–2194.
148. Yang M, Yu Q, Liu J, et al. T-cell immunoglobulin mucin-3 expression in bladder urothelial carcinoma: clinicopathologic correlations and association with survival. *J Surg Oncol*. 2015;112(4):430–435.
149. Ji P, Chen D, Bian J, et al. Up-regulation of TIM-3 on $CD4^+$ tumor infiltrating lymphocytes predicts poor prognosis in human non-small-cell lung cancer. *Xi Bao Yu Fen Zi Mian Yi Xue Za Zhi*. 2015;31(6):808–811.
150. Xu L, Huang Y, Tan L, et al. Increased Tim-3 expression in peripheral NK cells predicts a poorer prognosis and Tim-3 blockade improves NK cell-mediated cytotoxicity in human lung adenocarcinoma. *Int Immunopharmacol*. 2015;29(2):635–641.
151. Han S, Feng S, Xu L, et al. Tim-3 on peripheral CD4(+) and CD8(+) T cells is involved in the development of glioma. *DNA Cell Biol*. 2014;33(4):245–250.
152. Chapoval AI, Ni J, Lau JS, et al. B7-H3: a costimulatory molecule for T cell activation and IFN-gamma production. *Nat Immunol*. 2001;2(3):269–274.

153. Yi KH, Chen L. Fine tuning the immune response through B7-H3 and B7-H4. *Immunol Rev*. 2009;229:145–151.
154. Luo L, Chapoval AI, Flies DB, et al. B7-H3 enhances tumor immunity in vivo by costimulating rapid clonal expansion of antigen-specific $CD8^+$ cytolytic T cells. *J Immunol*. 2004;173(9):5445–5450.
155. Crispen PL, Sheinin Y, Roth TJ, et al. Tumor cell and tumor vasculature expression of B7-H3 predict survival in clear cell renal cell carcinoma. *Clin Cancer Res*. 2008;14(16):5150–5157.
156. Chavin G, Sheinin Y, Crispen PL, et al. Expression of immunosuppresive B7-H3 ligand by hormone-treated prostate cancer tumors and metastases. *Clin Cancer Res*. 2009;15(6):2174–2180.
157. Zhou Z, Luther N, Ibrahim GM, et al. B7-H3, a potential therapeutic target, is expressed in diffuse intrinsic pontine glioma. *Neuro Oncol*. 2013;11(3):257–264.
158. Modak S, Guo HF, Humm JL, Smith-Jones PM, Larson SM, Cheung NK. Radioimmunotargeting of human rhabdomyosarcoma using monoclonal antibody 8H9. *Cancer Biother Radiopharm*. 2005;20:534–546.
159. Kramer K, Kushner BH, Modak S, et al. Compartmental intrathecal radioimmunotherapy: results for treatment for metastatic CNS neuroblastoma. *J Neuro Oncol*. 2010;97(3):409–418.
160. Lanier LL. Natural killer cell receptor signaling. *Curr Opin Immunol*. 2003;15(3):308–314.
161. Domingues P, Gonzalez-Tablas M, Otero A, et al. Tumor infiltrating immune cells in gliomas and meningiomas. *Brain Behav Immun*. 2015;53, 1–15.
162. Liu RB, Engels B, Arina A, et al. Densely granulated murine NK cells eradicate large solid tumors. *Cancer Res*. 2012;72(8):1964–1974.
163. Kohrt HE, Thielens A, Marabelle A, et al. Anti-KIR antibody enhancement of anti-lymphoma activity of natural killer cells as monotherapy and in combination with anti-CD20 antibodies. *Blood*. 2014;123(5):678–686.
164. Lee SJ, Kang WY, Yoon Y, et al. Natural killer (NK) cells inhibit systemic metastasis of glioblastoma cells and have therapeutic effects against glioblastomas in the brain. *BMC Cancer*. 2015:15.
165. Fallarino F, Grohmann U, You S, et al. The combined effects of tryptophan starvation and tryptophan catabolites down-regulate T cell receptor zeta-chain and induce a regulatory phenotype in naive T cells. *J Immunol*. 2006;176(11):6752–6761.
166. Uyttenhove C, Pilotte L, Theate I, et al. Evidence for a tumoral immune resistance mechanism based on tryptophan degradation by indoleamine 2,3-dioxygenase. *Nat Med*. 2003;9(10):1269–1274.
167. Holmgaard RB, Zamarin D, Li Y, et al. Tumor-expressed IDO recruits and activates MDSCs in a Treg-dependent manner. *Cell Rep*. 2015;13(2):412–424.
168. Yang C, Zhou Y, Zhang L, Jin C, Li M, Ye L. Expression and function analysis of indoleamine 2 and 3-dioxygenase in bladder urothelial carcinoma. *Int J Clin Exp Pathol*. 2015;8(2):1768–1775.
169. Ferns DM, Kema IP, Buist MR, Nijman HW, Kenter GG, Jordanova ES. Indoleamine-2,3-dioxygenase (IDO) metabolic activity is detrimental for cervical cancer patient survival. *Oncoimmunology*. 2015;4(2).
170. Zhai L, Lauing KL, Chang AL, et al. The role of IDO in brain tumor immunotherapy. *J Neuro Oncol*. 2015;123(3):395–403.
171. Wainwright DA, Balyasnikova IV, Chang AL, et al. IDO expression in brain tumors increases the recruitment of regulatory T cells and negatively impacts survival. *Clin Cancer Res*. 2012;18(22):6110–6121.
172. Soliman HH, Jackson E, Neuger T, et al. A first in man phase I trial of the oral immunomodulator, indoximod, combined with docetaxel in patients with metastatic solid tumors. *Oncotarget*. 2014;5(18):8136–8146.

173. Iversen TZ, Engell-Noerregaard L, Ellebaek E, et al. Long-lasting disease stabilization in the absence of toxicity in metastatic lung cancer patients vaccinated with an epitope derived from indoleamine 2,3 dioxygenase. *Clin Cancer Res*. 2014;20(1):221–232.
174. Ranheim EA, Kipps TJ. Activated T cells induce expression of B7/BB1 on normal or leukemic B cells through a CD40-dependent signal. *J Exp Med*. 1993;177(4):925–935.
175. Roy M, Aruffo A, Ledbetter J, Linsley P, Kehry M, Noelle R. Studies on the interdependence of gp39 and B7 expression and function during antigen-specific immune responses. *Eur J Immunol*. 1995;25(2):596–603.
176. Jenkins M, Schwartz R. Antigen presentation by chemically modified splenocytes induces antigen-specific T cell unresponsiveness in vitro and in vivo. *J Exp Med*. 1987;165:302–319.
177. Nagaraj S, Gabrilovich DI. Tumor escape mechanism governed by myeloid-derived suppressor cells. *Cancer Res*. 2008;68(8):2561–2563.
178. Kjaergaard J, Tanaka J, Kim JA, Rothchild K, Weinberg A, Shu S. Therapeutic efficacy of OX-40 receptor antibody depends on tumor immunogenicity and anatomic site of tumor growth. *Cancer Res*. 2000;60(19):5514–5521.
179. Newcomb EW, Lukyanov Y, Kawashima N, et al. Radiotherapy enhances antitumor effect of anti-CD137 therapy in a mouse Glioma model. *Radiat Res*. 2010;173(4):426–432.
180. Kuhnol C, Herbarth M, Foll J, Staege MS, Kramm C. CD137 stimulation and p38 MAPK inhibition improve reactivity in an in vitro model of glioblastoma immunotherapy. *Cancer Immunol Immunother*. 2013;62(12):1797–1809.
181. Weinberg AD, Rivera MM, Prell R, et al. Engagement of the OX-40 receptor in vivo enhances antitumor immunity. *J Immunol*. 2000;164(4):2160–2169.
182. Melero I, Shuford WW, Newby SA, et al. Monoclonal antibodies against the 4-1BB T-cell activation molecule eradicate established tumors. *Nat Med*. 1997;3(6):682–685.
183. Hirschhorn-Cymerman D, Rizzuto GA, Merghoub T, et al. OX40 engagement and chemotherapy combination provides potent antitumor immunity with concomitant regulatory T cell apoptosis. *J Exp Med*. 2009;206(5):1103–1116.
184. Shuford WW, Klussman K, Tritchler DD, et al. 4-1BB costimulatory signals preferentially induce $CD8^+$ T cell proliferation and lead to the amplification in vivo of cytotoxic T cell responses. *J Exp Med*. 1997;186(1):47–55.
185. Schoenberger SP, Toes REM, van der Voort EIH, Offringa R, Melief CJM. T-cell help for cytotoxic T lymphocytes is mediated by CD40-CD40L interactions. *Nature*. 1998;393(6684):480–483.
186. Mackey MF, Gunn JR, Ting PP, et al. Protective immunity induced by tumor vaccines requires interaction between CD40 and its ligand, CD154. *Cancer Res*. 1997;57(13):2569–2574.
187. Diehl L, den Boer AT, Schoenberger SP, et al. CD40 activation in vivo overcomes peptide-induced peripheral cytotoxic T-lymphocyte tolerance and augments anti-tumor vaccine efficacy. *Nature Med*. 1999;5(7):774–779.
188. French RR, Chan HTC, Tutt AL, Glennie MJ. CD40 antibody evokes a cytotoxic T-cell response that eradicates lymphoma and bypasses T-cell help. *Nature Med*. 1999;5(5):548–553.
189. Li FB, Ravetch JV. Inhibitory Fc gamma receptor engagement drives adjuvant and anti-tumor activities of agonistic CD40 antibodies. *Science*. 2011;333(6045):1030–1034.
190. Zhang Y, Huang T, Hu Y, Wang Y. Activation of CD40 by soluble recombinant human CD40 ligand inhibits human glioma cells proliferation via nuclear factor-kappaB signaling pathway. *J Huazhong Univ Sci Technol*. 2012;32(5):691–696.

CHAPTER

12

Dendritic Cell Therapy for Brain Tumors

R.G. Everson, J.P. Antonios, A.M. Tucker, L.M. Liau

David Geffen School of Medicine at UCLA, Los Angeles, CA, United States

OUTLINE

Translational Immunotherapy of Brain Tumors
http://dx.doi.org/10.1016/B978-0-12-802420-1.00012-0

INTRODUCTION

Standard of care treatment for glioblastoma (GBM) does not substantially alter the poor prognosis of the disease.[4] Combating a disease that is characterized by rapid proliferation, resistance to chemo- and radiotherapy, and diffuse spread into the surrounding brain will require new treatments which can overcome these features.[4] Immunotherapy is an attractive approach for the treatment of brain tumors, as immune cells might be able to seek out and selectively destroy invading malignant cells while sparing normal brain tissues. Of the various forms of immunotherapy, dendritic cell (DC)-based therapies are well situated to be used to induce antitumor immunity given the specialized ability of DCs to both prime and/or boost innate and adaptive immune responses.[4] Clinical trials underway have demonstrated the successful application of DC vaccines to induce antitumor response, support memory T cell development, and increase overall survival (OS) in patients.[4] This chapter discusses the promise that DC-based immunotherapy has shown for treating malignant brain tumors.

TUMOR MICROENVIRONMENT

Many of the immunosuppressive transformations that occur within the tumor microenvironment are targeted toward blocking antigen presenting cells (APCs) such as DCs and their downstream functions (reviewed in Ref. 5,6). Low expression of MHC molecules on glioma cells may safeguard them from displaying immunogenic targets for immune recognition.[7] The elaboration of inhibitory cytokines in the microenvironment such as interleukin (IL)-10 and IL-6, from both tumor cells and microglia, alters the course of DCs maturation and the balance of Th1/Th2 ratio effectively reduces the number of potent T cell activators in the local environment.[8,9,10]

Assuming an immune response can be primed, tumors exist in an environment that lacks the necessary costimulatory signals to maintain an immune response and, meanwhile, T cells may be rendered anergic upon recognition through the activation of checkpoint inhibitors, such as programmed cell death-1 (PD-1) and cytotoxic T-lymphocyte-associated antigen 4 (CTLA-4).[11] Inhibitory immune cells, such as Tregs and myeloid-derived suppressor cells (MDSCs), also play a role in preventing an antitumor immune response.[12,13] Signal transducer and activator of transcription 3 (STAT3) signaling in tumor cells blocks proinflammatory cytokine signaling and amplifies Tregs.[14] Tregs and MDSCs not only block the antitumoral immune response but also change the inflammatory tumor microenvironment.[15,16] Finally, assuming all else fails, substances such as Tenascin-C elaborated by tumors into the extracellular matrix can form a physical barrier which prevents efficient immune cell to tumor cell contact.[17]

DENDRITIC CELLS IN IMMUNOLOGY

DCs are a subset of hematopoietic bone marrow–derived white blood cells, which act as specialized APCs to link the innate and adaptive immune systems through initiation of antigen specific T cell responses.[18]

Though DCs are known most commonly as professional APCs, their immature state is characterized by high levels of phagocytic activity and not antigen presentation.[19] Bone marrow–derived progenitor cells initially transform into immature DCs that are characterized by high levels of phagocytic activity rather than antigen presentation or the potential to active T cells.[20]

These cells are localized in most organs and tissues, especially where pathogen entry is likely, and act as local sentinels and primary immune responders. Controversy exists as to whether or not they generally reside in the brain, with some arguing that DCs are capable of trafficking to the brain. Others ascribe many of the functions of DCs in the brain to microglia. Regardless, immature DCs continually sample their surrounding environment and nearby cells for pathogenic material. Engagement of particular pathogen-associated molecular pattern receptors in the environment, such as the toll-like receptors (TLRs), that are typically found in association with pathogens or tissue damage [e.g., high mobility group box 1 (HMGB1) protein, demethylated DNA, double-stranded RNA] act as a second, "danger" signal to begin the process of maturation.[21–23] This second signal may help steer the subsequent T cell response toward either active immunity or unresponsive tolerance.

This transition to maturation is characterized by the ability to increase antigen presentation, engage costimulatory molecules, migrate to lymphoid organs, and secrete polarizing cytokines [e.g., IL-12 and interferon alpha (IFN-α)].[24] DCs upregulate chemotactic factors such as C-C chemokine receptor type 7 (CCR7) that enable the DC to migrate through the circulation to a lymphoid organ such as the spleen or the lymph nodes.[25] There, upregulated MHC class I and class II molecules on DCs prompt antigen presentation to $CD4^+$ helper T cells, $CD8^+$ killer T cells, and antibody producing B cells. This presentation is boosted by upregulated costimulatory molecules such as CD80 (B7.1), CD86 (B7.2), and CD40 on the surface of DCs to activate these effector cells and others to induce homing to the target organ.[26,27]

The capability that distinguishes DCs as professional APCs is the ability to cross-present antigens in the context of both MHC class I and class II.[28–30] Orchestrating antigen-specific responses through multiple effector arms makes DCs pivotal in inducing antitumor immunity, as multiple effector cell types are likely necessary to create an effective immune response. Other APCs may be able to activate certain T lymphocyte populations (mainly $CD8^+$), but it appears that the coactivation of $CD4^+$ and $CD8^+$ provided by DCs is especially important in the formation of immunologic memory.[31,32] DCs are also thought to be able to activate natural killer (NK) and NKT cells as well, thus allowing the killing of target cells with and without the expression of class I MHC.[33,34]

DENDRITIC CELL–BASED VACCINE: CURRENT APPROACHES

Most of the DC vaccines to date have used the strategy of ex vivo generation and antigen loading of DCs[35] (Fig. 12.1). As cellular products, there are innumerable variations in the methods of manufacture. There is also no consensus as to the characteristics of the ideal starting or final product. Numerous subsets of DCs have been identified, including plasmacytoid DCs, epidermal Langerhans cells, classical ($CD11b^+CD8^-$) DCs, $CD8^+$ DCs, and monocyte-derived DCs.[36] Preclinical investigations have highlighted the differing capabilities of these cells for use in cancer immunotherapy. However, most clinical trials to date have used monocyte-derived DCs.[37]

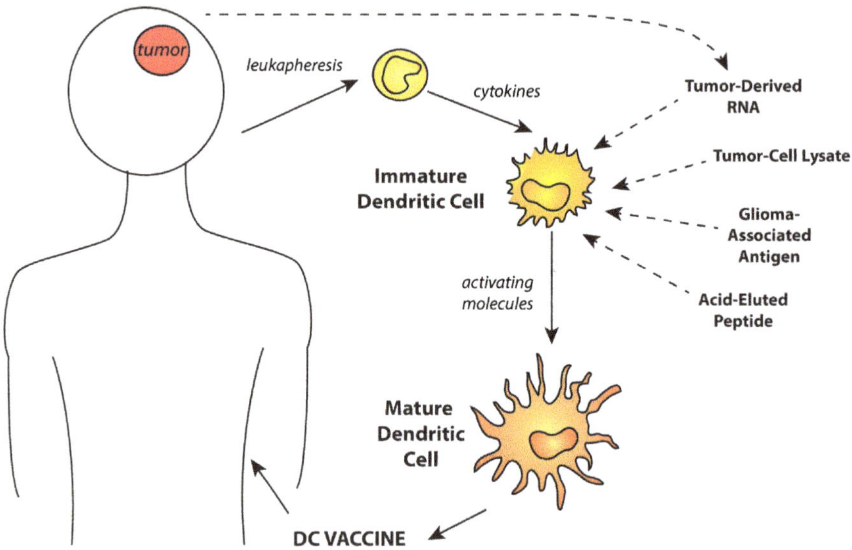

FIGURE 12.1 **Ex vivo generation and antigen loading of dendritic cells (DCs).** Following patient leukapheresis, autologous immature DCs are pulsed with tumor lysate, glioma-associated antigen, acid-eluted peptide, or tumor-derived RNA. To upregulate costimulatory molecules, present antigens, and express chemokine receptors associated with migration, these cells are then matured and activated in culture. Finally, mature DCs are administered intradermally into the patient.

To create a vaccine, these DCs can be generated by culturing patient-derived peripheral blood mononuclear cells obtained from leukapheresis with the cytokines, granulocyte-macrophage colony–stimulating factor and interleukin-4 (IL-4).[38–41] The resulting phenotype of these cells most closely resembles conventional DCs, which are characterized by high levels of CD11 and MHC class II expression and can stimulate T cell proliferation. These immature DCs can then be loaded with antigenic material from a variety of sources such as tumor-specific peptides, tumor lysate, or tumor-derived RNA. These are then matured and activated with upregulation of costimulatory molecules, antigen presentation, and chemokine receptors associated with migration. Maturation of the resulting cells with various TLR ligands or inflammatory cytokines has been reported to overcome these limitations by resulting in a number of different functional phenotypes that have yet to be fully explored in a clinical immunotherapy setting.

ANTIGEN TARGETS

The optimal source of antigens for loading onto DCs remains controversial. Tumor cells give rise to proteins that are not normally expressed in most tissues of the body such as mutations, abnormally or differentially expressed endogenous proteins, differentiation antigens, or in some cases viral proteins. The presence of unique, identifying antigens renders tumor cells susceptible to immune recognition.[42] Selection of tumor-specific antigens expressed is essential in avoiding either down-regulation of the immune response or an autoimmune attack on healthy cells.[43] Identifying these can be difficult given that there is a lack of knowledge of distinct antigens for most cancers.[44] The heterogeneity of tumors presents an additional layer of complexity since within any given tumor cells may display different arrays of antigens.[43] The use of a DC vaccine primed with a single antigen may only eradicate part of the tumor. In killing off all tumor cells with recognized antigens, the immune system naturally selects for a tumor cell population with an immunoevasive antigenic genetic profile.[45] For this reason, it is preferable to ensure the use of antigens that are crucial to cell viability, or which confers metastatic properties to the cells.

Gliomas have been shown to express a variety of tumor-associated antigens (TAAs)—normal/nonmutated proteins that are upregulated in tumors compared to normal tissues. DC vaccines can be generated by loading DCs with peptides or RNA encoding these antigens. In clinical trials, these have included epidermal growth factor receptor (EGFR), interleukin-13 receptor $\alpha 2$ (IL-13R$\alpha 2$), glycoprotein 100 (gp100), tyrosinase-related protein 2

(TRP-2), human epidermal growth factor receptor 2 (HER-2), melanoma-associated antigen 1 (MAGE-1), survivin, absent in melanoma 2 (AIM-2), ephrin type-A receptor 2 (EphA2), chitinase-3-like protein 1 (CHI3L1, also known as YKL-40) among others.[46–50]An immune response against these antigens would hopefully have differential effects against tumors. However, as the TAAs are also expressed in normal tissues, a delicate balance exists between the induction of tolerance and autoimmunity when attempting to vaccinate against these targets. Active vaccination approaches against self-antigens often lead to the generation of weak clones in the case that tolerance is overcome.[51] Mainly these vaccines use short peptides that, when pulsed onto DCs, can bind directly to MHC molecules without further processing. However, this system requires that particular HLA types be present for these antigens to be properly presented. Long peptides or whole proteins could overcome this problem and minimize tumor evasion but may not be specific for particular target epitopes. Although this approach has been used in several clinical trials, no documented cases of autoimmune responses against normal tissues have been reported, which has led some to posit that DC vaccination may not be potent enough to induce such an immune response.[4,52]

Ideally, a truly tumor-specific target should be used to overcome the above-mentioned limitations. However, whether or not such antigens exist are somewhat controversial. EGFRvIII has been used in a DC vaccine and found to be both safe and immunogenic.[53–55] Targeting this antigen is limited to patients that express it (around 35%), and questions remain as to whether it can lead to complete immunologic rejection of tumors given that the target may not be expressed in every cell in a given tumor and tumors recurring after therapy appear to be antigen-loss variants.[56–58] Nonetheless, clinical trials targeting this antigen with both DC and peptide vaccine approaches have yielded dramatic clinical results and represent a significant step forward in glioma immunotherapy. A somewhat similar controversy exists surrounding the targeting of viral antigens in glioma. Several investigators have reported that malignant gliomas appear to contain sequences of and express the viral gene products of cytomegalovirus (CMV).[59,60] Although clinical trials of conventional CMV therapy (valganciclovir) appear to have failed, a recent study using a DC vaccination approach to target the viral antigen pp65 appeared to successfully induce immune responses against the target that were associated with extended survival in treated patients.[61,62]

Because of the lack of a single, universally expressed antigen on human GBM and the marked heterogeneity of malignant gliomas, "total" approaches of antigen loading of DCs have also been used. This strategy involves the use of large numbers of tumor antigens that have not been specifically identified. These methods involve the loading of DCs with materials such as autologous tumor lysate, peptides eluted from MHC from tumor specimens, RNA isolated from tumor, or the fusion of DCs with glioma cells.[50,63–74] Precise identification of the antigen targets is not necessary for this strategy, as it essentially relies on the immune system faithfully sorting out what is self from nonself and directing the response against the tumor while sparing normal tissues. The resulting immune response can target any number of antigens, including TAAs (mentioned earlier), as well as antigens generated from individual-specific mutations. This multitargeted approach would theoretically limit the possibility of antigen-loss variants.[45]

Each approach has benefits and drawbacks. Although not all patient tumors express all target antigens, measuring and monitoring a targeted antitumor immune response is certainly more easily interpretable than vaccines against undefined targets. Certain methods of generating "total" vaccines rely on material from tumors that may be limited and require surgery to obtain.[66] In Phase I and II trials, successes with both strategies have been reported.[75] Information gained from next-generation sequencing is shedding further insight into the identification and categorization of potential immunotherapy targets and is being incorporated into the design of next-generation DC vaccines.

ROUTES OF ADMINISTRATION

The optimal dosage, timing, and route of administration may also be a determining factor in DC vaccine therapy. Investigators have used a number of strategies. The majority of vaccines were administered within a few weeks of surgical resection.[52] One study, however, administered vaccines as far out as 30 weeks after surgery,[69] and most vaccine protocols defined that autologous mature DC loaded with autologous tumor cells (ATCs), ATC lysates, ATC peptides, or ATC homogenates were to be administered in the axilla and/or inguinal region of the patient.[52,75] Others have used ultrasound guided injection directly into lymph nodes or intratumorally via an Ommaya reservoir.[70] Yamanaka et al. compared intradermal injection with the combination of intradermal and intratumoral injection and found that patients in the combination group exhibited longer OS. Conditioning of the injection site appears also to play a role. Several investigators have used TLR agonists at the site in attempts to boost immune response.[48,67,69,76] One study assessing the use of preconditioning of the injection site with tetanus toxoid demonstrated increased migration of injected DCs in lymph nodes as measured by detection of ^{111}In-labeled DCs.[61] The optimal dose is also unknown. Given the rarity of adverse effects, few trials have reached maximum tolerated doses. However, Prins et al. reported that the highest dose was not necessarily correlated with the best clinical outcomes.[69] These findings raise the question of whether these traditional concepts of dose and delivery are even applicable to DC vaccine therapy. DCs and the effector cells they generate have the ability to travel and the magnitude of the immune response may be scalable by the immune system to match the target burden.

SAFETY AND COMPLICATIONS

Although mostly in Phase I and II trials, DC-based immunotherapy appears to be well tolerated and overall safe. At the outset of creation of DC immunotherapy, especially for systems that did not have specified targets (i.e., tumor lysate vaccines), there was concern over the possibility of inducing deleterious autoimmune responses.[77] Unlike in other tissues, immune-mediated damage to normal bystander tissues in the brain can catastrophically affect neurologic function and be life threatening. If even a minor component of a therapy demonstrates off-target effects as unfortunately occurred in some engineered adoptive cell transfer trials, such runaway immune responses in the CNS can be fatal.[78] However, these concerns have not borne out in clinical trials. This contrasts with clinical trials of the newly emerging checkpoint inhibitor therapies (CTLA-4 and PD-1-blocking monoclonal antibodies) in which up to 50% of patients reportedly experienced immune-related toxicity.[79]

Common side effects are usually minor and self-limited. Pain, itching, and swelling at the injection site is common, occurring approximately 20% of the time. More severe dermatologic manifestations, such as the development of papules at the injection site, were rarer. Common systemic side effects include transient elevations in liver function tests (AST/ALT) in less than 10%.[52,75]

Life-threatening complications are also rare and, when reported, appear to be more commonly linked with sequelae from craniotomy and resection of tumor than from the vaccine itself (meningeal irritation, tumor hemorrhage), but the possibility that the vaccine could cause these cannot be completely excluded. Of note, one trial reported an adverse event related to immune monitoring procedures in the clinical trial. Wheeler et al. reported that one patient experienced a cutaneous GBM at the site of irradiated tumor cell inoculation for delayed-type hypersensitivity (DTH) testing.[49]

IMMUNE RESPONSE

There are no universally agreed upon surrogate markers to predict efficacy of DC vaccines. Numerous techniques have been studied, including DTH tests, relative immune phenotypes of circulating lymphocytes, T cell proliferation, and cytotoxic T-lymphocytes (CTL) assays, NK cell assays, IFN-γ production [serum, enzyme-linked immunosorbent spot (ELISPOT), mRNA expression, fluorescence-activated cell sorting], and recent thymic emigrant assay.[37] These and other differences in trial design have made comparing the downstream effects of the vaccines impossible. Nearly every trial has used a unique method for producing and administering DC vaccines. In these trials, immune monitoring data have rarely correlated with long-term survivors. Given the complexities of the immune system, every aspect of an antitumor immune response may not be evaluable by immune monitoring tests. Antibodies generated through humoral immunity and immune effector cells such as monocytes and natural killer cells are often not evaluated in favor of $CD4^+$ and $CD8^+$ T cells assay, when these may be the more important mediators of the immune response for a particular vaccine. However, almost regardless the approach and monitoring tools and assays used, some indication of immunological activity is found in about 50% of patients while autoimmunity has been largely avoided.[4,52,75]

CLINICAL RESPONSE AND SURVIVAL BENEFITS

In a systematic review of all published clinical DC trials, an objective response (achieving either a complete response or partial response by the Response Evaluation Criteria In Solid Tumors criteria) was reported in about 15% of cases, which is a higher frequency than for the other cancers where DCs have been extensively studied (melanoma, renal, prostate).[4] Another systematic review found that vaccination with DCs loaded with ATCs resulted in increased median OS in patients with recurrent GBM (71.6–138.0 weeks) as well as those newly diagnosed (65.0–230.4 weeks) compared to average survival of 58.4 weeks.[52]

A further issue that confounds this especially in GBM is that standard of care protocols call for maximal safe resection of tumors. Some clinical trials have determined that patients who have gone on to have long-term survival had minimal residual and stable disease at the time of vaccination.[66] In the setting of a completely resected tumor, measuring a response by imaging is difficult as the main manifestation of the development of protective immunity would be the absence of recurrence, rather than the shrinkage of a measurable tumor mass.

Additionally, corticosteroids, the mainstay therapy for ameliorating clinical symptoms caused by vasogenic edema from brain tumors, are substantially immunosuppressive, particularly for DCs.[80–82] These drugs are essentially what one would administer to prevent active vaccination from occurring. Clinical immunotherapy trials therefore often stipulate that patients be weaned off steroids completely, which may not be possible in all clinical scenarios and/or lead to selection bias.

Patients that showing signs of progressive disease are often restarted on steroids as well, which would likely negate any positive effects of DC therapy.

Questions remain surrounding the apparent dissociation between objective responses as measured by standard criteria and the long-term survival seen in subsets of patients on these clinical trials, with some authors arguing that immunotherapeutic agents should not be evaluated by the same criteria as standard cytotoxic drugs, where a rapid reduction in tumor size is anticipated. In contrast, patients benefitting from immunotherapy trials appear rather to derive their benefit from long-term stable disease and prevention of progression. Results from currently ongoing Phase III clinical trials will hopefully provide more certainty about the clinical efficacy of DC therapy.

RADIOLOGIC FINDINGS

Imaging findings in immunotherapy can be deceiving as well. On magnetic resonance imaging (MRI) an active immune response in a GBM will be difficult to distinguish from inflammation and edema which could cause increases in tumor size and enhancement. Termed pseudoprogression, this state has been increasingly recognized after a variety of treatments, including temozolomide as well as immunotherapy.[83] It typically occurs 2–3 months following treatment, and may be evident in some form in 15–50% of patients.[84–86] There have been multiple investigations to use novel MRI sequence modalities to distinguish a pseudoprogression from an immune response from true tumor progression, but none has been validated to date. In a study by Vrabec et al., patients with recurrent GBM were treated with dendritic vaccine therapy in conjunction with surgery. Using MR perfusion imaging, the increased regional cerebral blood volume (rCBV) was found to correlate strongly with increased vascularization indicating actual glioma progression. Tumors with lower rCBV values predicted pseudoprogression. Thus, MR perfusion and diffusion imaging techniques may assist in the reliable differentiation of reactive inflammatory processes and recurrent tumor, even in the setting of immunotherapy.[87] Due to the increasing frequency of immunotherapy trials for GBM, the Response Assessment in Neuro-Oncology Working Group (RANO) has developed of a new subset of guidelines, named Immune RANO.[88] These guidelines argue that imaging findings do not always preclude continued benefit from immunotherapies. Instead it weighs clinical symptoms over imaging findings and provides guidance on waiting for three-month confirmatory scans before applying the label of progressive disease.

CLINICAL TRIALS

For the past 20 years, clinical trials have been performed using DC vaccines for the treatment of gliomas.[89] They have proven to be feasible, safe, and elicit immunological responses. Numerous excellent reviews have cataloged the experience with them so far. Recently, several excellent systematic reviews and a meta-analysis have been performed.[35–37,52,75,90–95] The reader is directed to those publications. Table 12.1 presents all active or enrolling clinical trials of DC therapy for brain tumors at the present. Due to the scope of this chapter, experiences learned from clinical trials conducted at our institution are presented with relevant comparisons made to other prominent trials in the field.

In 1999, Liau et al. described the first case report of a GBM patient treated with a DC vaccine, which consisted of three intradermal injections of autologous DCs pulsed with allogeneic acid–eluted peptides.[64] The acid elution technique extracts peptides bound to MHC class I molecules, allowing for the isolation of a the unique tumor specific compliment of antigens to use for the vaccine. Although clinical benefit was difficult to assess in this single case report, an increased CD3 T cell infiltrate into the tumor was noted. In a Phase I dose escalation trial, 12 more malignant glioma patients were treated with DCs loaded with acid-eluted peptide vaccine that was found to be safe and well tolerated, as no cases of either autoimmunity or dose limiting toxicity were reported with up to 10^7 DCs injected per dose.[66] Patients in this study exhibited improved progression-free survival (PFS) and OS compared to historical controls, with several long-term survivors (36 months). For the patients that did recur, specimens demonstrated a $CD3^+$ T cell infiltrate predominantly composed of memory T cells ($CD8^+$/$CD45RO^+$). Data from these initial trials also suggested this vaccine works best in the setting of minimal residual disease after resection. Subsequently, Yu et al. also reported on GBM patients treated with intradermal DCs loaded with acid-eluted tumor peptides with similar findings.[72] However, this approach is limited by the difficulties related to the need to culture large numbers of primary tumor cells elute enough peptide for production of the vaccine.

Subsequent trials have used autologous tumor lysate from surgical specimens pulse onto DCs for generation of the vaccine. In this technique, the patient's tumor is lysed using multiple freeze/thaw cycles to generate cellular fragments that are cocultured with DCs to allow for their phagocytosis and subsequent antigen cross-presentation. Prins et al. reported a Phase I/II clinical trial that involved patients with both newly diagnosed and recurrent GBM receiving autologous tumor lysate-pulsed DCs, as well as

TABLE 12.1 Active or Enrolling Clinical Trials of Dendritic Cell Vaccine Therapy for Brain Tumors

ClinicalTrials. gov Identifier	Phase	Tumors Treated	Sponsor	Antigen Source	Adjuvants and Combinations
NCT01808820	I	HGG adult and peds	University of Miami	ATL	Imiquimod
NCT01635283	II	LGG	Jonsson Comprehensive Cancer Center	ATL	
NCT01792505	I	HGG	Cedars-Sinai Medical Center	ATL	Imiquimod
NCT01204684	II	HGG	Jonsson Comprehensive Cancer Center	ATL	Imiquimod or poly ICLC
NCT02010606	I	GBM	Cedars-Sinai Medical Center	Lysate from allogeneic tumor stem cells	
NCT01902771	I	Pediatric brain tumors	University of Miami	ATL	Imiquimod
NCT01567202	II		Huashan Hospital	Autogeneic A2B5$^+$ stem-like cells	
NCT00045968	III	GBM	Northwest Biotherapeutics	ATL	
NCT02332889	I/II	Pediatric HGG,MB, CNS PNET	University of Louisville	NY-ESO-1, MAGE-A1 and MAGE-A3 peptides	Poly-ICLC, Decitabine
NCT02465268	II	GBM	University of Florida	CMV-pp65 RNA	Td
NCT02649582	I/II	GBM	University Hospital, Antwerp	WT-1 RNA	TMZ
NCT02049489	I	GBM	ImmunoCellular Therapeutics, Ltd.	CD133 peptides	
NCT02366728	II	GBM	Duke University	CMV-pp65-LAMP RNA	Td, Basiliximab

TABLE 12.1 Active or Enrolling Clinical Trials of Dendritic Cell Vaccine Therapy for Brain Tumors—cont'd

ClinicalTrials. gov Identifier	Phase	Tumors Treated	Sponsor	Antigen Source	Adjuvants and Combinations
NCT00639639	I/II	Brain tumors	Duke University	CMV pp65-LAMP mRNA	Td, ALT
NCT02146066	Open label	Progressive GBM	Northwest Biotherapeutics	ATL	
NCT01280552	II	GBM	ImmunoCellular therapeutics, Ltd.	AIM-2, MAGE-1, TRP-2, gp100, HER-2, IL-13Ra2 peptides	
NCT01280552	III	GBM	ImmunoCellular therapeutics, Ltd.	AIM-2, MAGE-1, TRP-2, gp100, HER-2, IL-13Ra2 peptides	
NCT00626483	I	Brain tumors	Duke University	CMV pp65-LAMP mRNA	Basiliximab
NCT01326104	I	MB, PNET	University of Florida	Total tumor RNA	ALT

ALT, autologous lymphocyte transfer; *ATL*, autologous tumor lysate; *GBM*, glioblastoma; *HGG*, high grade glioma; *LAMP*, lysosomal associated membrane protein; *LGG*, low-grade glioma; *MAGE*, melanoma-associated antigen; *MB*, medullobastoma; *NY-ESO-1*, New York esophageal squamous cell carcinoma-1; *PNET*, primitive neuroectodermal tumor; *Poly-ICLC*, polyinosinic-polycytidylic acid stabilized by lysine and carboxymethylcellulose; *Td*, tetanus-diptheria toxoid; *TRP*, tyrosinase-related protein.

adjuvant imiquimod or polyinosinic-polycytidylic acid stabilized by lysine and carboxymethylcellulose (poly-ICLC) during booster vaccinations.[69] Safety of the vaccines and the adjuvants were demonstrated, with no significant adverse events reported. Patients in this trial had substantially better PFS and OS rates compared to historical controls, with a median survival of 31.4months for newly diagnosed GBM patients. Improved outcomes appeared also to stratify by tumor molecular subtype as measured by gene expression profiling, with increased survival shown for patients within the mesenchymal subgroup of GBM. This subclass of tumors are characterized by preexisting immune infiltrate as well as specific gene expression and mutational profiles indicating that these factors may play a critical role in determining the susceptibility of a tumor to immunotherapy.[96]

Interestingly, in one patient from these trials, a robust human cytomegalovirus (HCMV)-specific $CD8^+$ T cell response to the pp65 CMV immunodominant epitope was detected immediately after injection of autologous tumor lysate-pulsed DC in a patient who went on to enjoy long-term survival.[97] This same antigen CMV-pp65 was the target of a vaccine in Mitchell et al.[61] Several other groups including the group in Belgium have also used autologous tumor lysate to create a vaccine and reported in trials with larger groups of patients with similar results.[67,68,76,98] Currently, a Phase III multicenter trial of DC vaccination for newly diagnosed GBM is underway, with approximately 348 patients to be enrolled (NCT00045968).[92]

Comparing vaccines using autologous tumor lysate to glioma-associated peptides was the focus of another clinical trial at our center.[48] In that, HLA-A2.1-specific synthetic peptide antigens for TRP-2, gp100, her-2/neu, and survivin were pulsed onto DCs. The haplotype specificity requirement, however, led to quite low accrual into the trial. Expression of the targets in the tumor was not an eligibility requirement, but subsequently, it was determined that all patients expressed at least one antigen in their tumor. Production of the vaccine was otherwise similar except for the use of a maturation cocktail containing tumor necrosis factor alpha (TNF-α), IL-6, IL-1β, and prostaglandin E2 (PGE2). Ultimately, tetramer-positive $CD8^+$ T cells were observed in glioma-associated antigen DC patients, but no correlations were evident between those responses and OS. This small group did not compare favorably with the lysate group regarding PFS and OS. Several factors could account for the observed differences in this TAA-directed trial and others, including ones targeting several overlapping peptides (AIM-2, MAGE-1, TRP-2, gp100, HER-2, and IL-13Ra) which observed much better results including the exact antigen targets, expression levels in the patients treated, and maturation cocktails used.[47]

FUTURE DIRECTIONS

In the future, potentiating DC vaccines will require utilizing the insights gained from study of DC biology to inform improved methods of generating vaccine products with better immune activating capacity. This will likely be done by manipulating the growth factors, cytokines, and cellular subsets used to create a more powerful vaccine as well as combine DC vaccination with other synergistic therapies.

Although not completely understood, creating DC vaccines that induce a Th1 type response appears to be beneficial. Many studies have also made use of cytokines and prostaglandins (IL-1β, IL-6, TNF_{α}, and PGE-2, the so-called Jonuleit cocktail) to induce further maturation after loading of tumor antigen.[10] Some studies report that DCs produced in such a manner do not ultimately produce IL-12, which is thought to be critical for $CD8^+$ and NK responses and may encourage tolerance to a higher degree.[99,100] Other factors have been tested to replace these. For additional activation of the injected DC, several TLR agonists have been tested clinically of which imiquimod/resiquimod (TLR 7 agonist), poly I:C (TLR 3 agonist), and clinical grade lipopolysaccharide (TLR4 agonist) are the most common. However, none of these adjuvants to date have proven superior.

CD8+ T cells primed by DCs also confront numerous barriers to successful destroying their target including checkpoint mechanisms such as CTLA-4 and PD-1 as well as inhibitory cell populations such as Tregs or MDSCs. Tregs might be targeting with agents such as daclizumab (anti CD-25) and agents targeting MDSCs are in development.[15,101] Checkpoint inhibitor therapies have been developed and have shown tremendous promise in other cancers such as melanoma and lung cancer.[102,103] Preclinical glioma models have corroborated this evidence, with Wainwright et al. demonstrating that blockade of CTLA-4, PD-1, and IDO in established intracranial glioma models resulted in decreased tumor-infiltrating T regulatory cells and increased long-term survival mediated by cytotoxic T lymphocytes.[104] PD-1, an immune regulatory receptor associated with inhibition of T cells with prolonged antigen exposure (a phenomenon noted in chronic disease manifestations, such as viral infections and cancer), plays an important role in regulating the effector phase of the T cell response.[11] Activation of this receptor by PD-L1 and PD-L2, expressed by glioma cells and regulatory cells in the tumor microenvironment, promotes immune suppression. Early studies have shown that inhibition of the PD-1 receptor is associated with a significantly increased immune response to cancer. Zeng et al. showed that PD-1 blockade in the setting of radiation therapy led to long-term survival in mice implanted with a murine glioma cell line.[105] These results suggest promise for the inhibition of these immune-regulatory mechanisms for the potentiation of immune response induced by DC-based vaccination strategies.

CONCLUSION

DCs must function in a microenvironment that favors downregulation of an immune response, especially in the brain (Fig. 12.2). Insights gained from the study of autoimmune disease in the CNS have proven useful in understanding regulation of antitumor immunity in CNS. Antigen loss and immune evasion of tumor targets are the subject of many studies using next generation sequencing modalities to understand the repertoire of unique, immunologically targetable antigens expressed by a tumor. Defining strategies to bypass these obstacles is the object of intense studies to improve the clinical efficacy of vaccination via DCs.

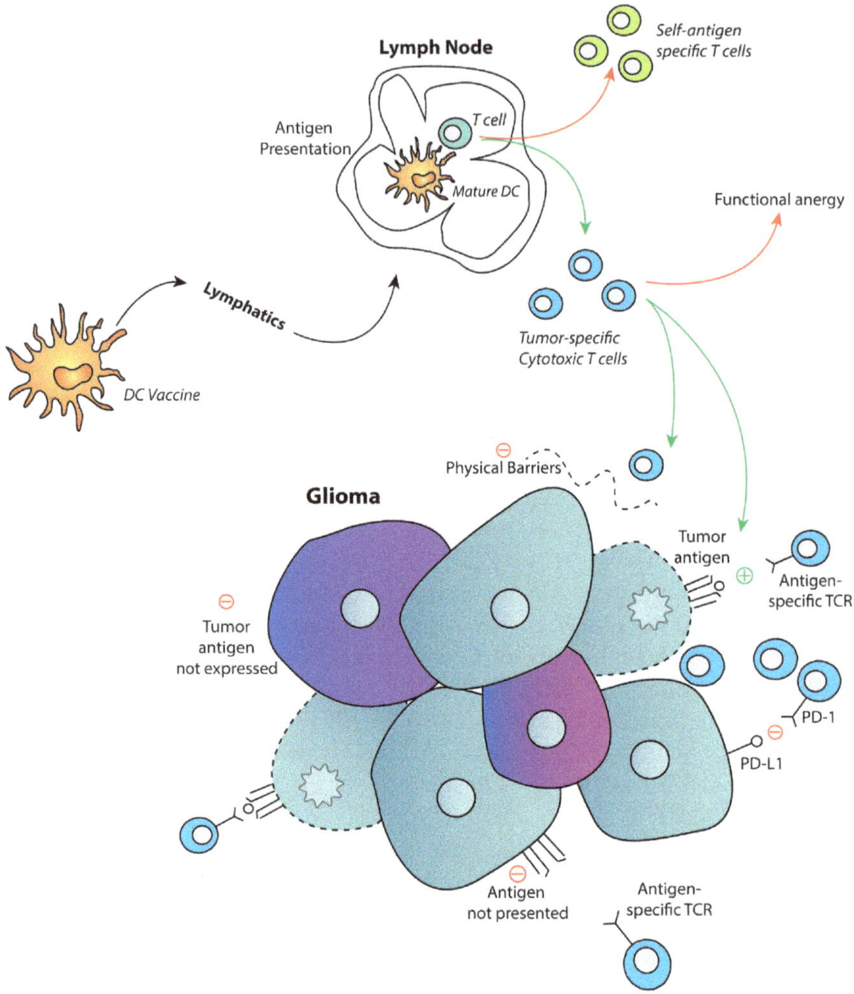

FIGURE 12.2 **Brain tumor microenvironment favors immune response downregulation.** Autologous dendritic cells (DCs) migrate to lymphatics where they seek out antigen-specific T cells to initiate a proliferative burst. Tumor-specific T cells will then traffic to the brain tumor environment to effect a cytotoxic response. Mechanisms in the brain tumor environment, however, limit binding of the T cell antigen-specific T cell receptor with tumor antigen, thus restricting tumor cell cytolysis. Among these include physical barriers, restricted tumor cell antigen expression, and negative costimulatory mechanisms such as programmed death-1/ligand-1.

References

1. Stupp R, Mason WP, van den Bent MJ, et al. Radiotherapy plus concomitant and adjuvant temozolomide for glioblastoma. *N Eng J Med*. 2005;352(10):987–996.
2. Hanahan D, Weinberg RA. Hallmarks of cancer: the next generation. *Cell*. 2011;144(5):646–674.
3. Anguille S, Smits EL, Bryant C, et al. Dendritic cells as pharmacological tools for cancer immunotherapy. *Pharmacol Rev*. 2015;67(4):731–753.
4. Anguille S, Smits EL, Lion E, van Tendeloo VF, Berneman ZN. Clinical use of dendritic cells for cancer therapy. *Lancet Oncol*. 2014;15(7):e257–e267.
5. Varker KA, Kondadasula SV, Go MR, et al. Multiparametric flow cytometric analysis of signal transducer and activator of transcription 5 phosphorylation in immune cell subsets in vitro and following interleukin-2 immunotherapy. *Clin Cancer Res*. 2006;12(19):5850–5858.
6. Ihle JN, Kerr IM. Jaks and Stats in signaling by the cytokine receptor superfamily. *Trends Genet*. 1995;11(2):69–74.
7. Rolle CE, Sengupta S, Lesniak MS. Mechanisms of immune evasion by gliomas. *Adv Exp Med Biol*. 2012;746:53–76.
8. Chomarat P, Banchereau J, Davoust J, Palucka AK. IL-6 switches the differentiation of monocytes from dendritic cells to macrophages. *Nat Immunol*. 2000;1(6):510–514.
9. Palucka K, Banchereau J. Cancer immunotherapy via dendritic cells. *Nat Rev Cancer*. 2012;12(4):265–277.
10. Steinbrink K, Wolfl M, Jonuleit H, Knop J, Enk AH. Induction of tolerance by IL-10-treated dendritic cells. *J Immunol*. 1997;159(10):4772–4780.
11. Callahan MK, Postow MA, Wolchok JD. CTLA-4 and PD-1 pathway blockade: combinations in the clinic. *Front Oncol*. 2014;4:385.
12. Ooi YC, Tran P, Ung N, et al. The role of regulatory T-cells in glioma immunology. *Clin Neurol Neurosurg*. 2014;119:125–132.
13. Mirghorbani M, Van Gool S, Rezaei N. Myeloid-derived suppressor cells in glioma. *Expert Rev Neurother*. 2013;13(12):1395–1406.
14. Assi H, Espinosa J, Suprise S, et al. Assessing the role of STAT3 in DC differentiation and autologous DC immunotherapy in mouse models of GBM. *PLoS One*. 2014;9(5):e96318.
15. Maes W, Rosas GG, Verbinnen B, et al. DC vaccination with anti-CD25 treatment leads to long-term immunity against experimental glioma. *Neuro Oncol*. 2009;11(5):529–542.
16. Maes W, Verschuere T, Van Hoylandt A, Boon L, Van Gool S. Depletion of regulatory T cells in a mouse experimental glioma model through anti-CD25 treatment results in the infiltration of non-immunosuppressive myeloid cells in the brain. *Clin Dev Immunol*. 2013;2013:952469.
17. Jachetti E, Caputo S, Mazzoleni S, et al. Tenascin-C protects cancer STEM-like cells from immune surveillance by arresting T-cell activation. *Cancer Res*. 2015;75(10):2095–2108.
18. Steinman RM, Banchereau J. Taking dendritic cells into medicine. *Nature*. 2007;449(7161):419–426.
19. Everson RG, Graner MW, Gromeier M, et al. Immunotherapy against angiogenesis-associated targets: evidence and implications for the treatment of malignant glioma. *Expert Rev Anticancer Ther*. 2008;8(5):717–732.
20. Steinman RM, Idoyaga J. Features of the dendritic cell lineage. *Immunol Rev*. 2010;234(1):5–17.
21. Guermonprez P, Valladeau J, Zitvogel L, Thery C, Amigorena S. Antigen presentation and T cell stimulation by dendritic cells. *Annu Rev Immunol*. 2002;20:621–667.
22. Barton GM, Medzhitov R. Control of adaptive immune responses by Toll-like receptors. *Curr Opin Immunol*. 2002;14(3):380–383.

23. O'Neill DW, Adams S, Bhardwaj N. Manipulating dendritic cell biology for the active immunotherapy of cancer. *Blood*. 2004;104(8):2235–2246.
24. Banchereau J, Steinman RM. Dendritic cells and the control of immunity. *Nature*. 1998;392(6673):245–252.
25. Verdijk P, Aarntzen EH, Punt CJ, de Vries IJ, Figdor CG. Maximizing dendritic cell migration in cancer immunotherapy. *Expert Opin Biol Ther*. 2008;8(7):865–874.
26. Andrews DM, Andoniou CE, Scalzo AA, et al. Cross-talk between dendritic cells and natural killer cells in viral infection. *Mol Immunol*. 2005;42(4):547–555.
27. Ni K, O'Neill HC. The role of dendritic cells in T cell activation. *Immunol Cell Biol*. 1997;75(3):223–230.
28. Fehres CM, Unger WW, Garcia-Vallejo JJ, van Kooyk Y. Understanding the biology of antigen cross-presentation for the design of vaccines against cancer. *Front Immunol*. 2014;5:149.
29. Rock KL, Clark K. Analysis of the role of MHC class II presentation in the stimulation of cytotoxic T lymphocytes by antigens targeted into the exogenous antigen-MHC class I presentation pathway. *J Immunol*. 1996;156(10):3721–3726.
30. Rock KL, Gamble S, Rothstein L. Presentation of exogenous antigen with class I major histocompatibility complex molecules. *Science*. 1990;249(4971):918–921.
31. Pizzurro GA, Barrio MM. Dendritic cell-based vaccine efficacy: aiming for hot spots. *Front Immunol*. 2015;6:91.
32. Michaud HA, Eliaou JF, Lafont V, Bonnefoy N, Gros L. Tumor antigen-targeting monoclonal antibody-based immunotherapy: Orchestrating combined strategies for the development of long-term antitumor immunity. *Oncoimmunology*. 2014;3(9):e955684.
33. Gottschalk C, Mettke E, Kurts C. The role of invariant natural killer T cells in dendritic cell licensing, cross-priming, and memory $CD8^+$ T cell generation. *Front Immunol*. 2015;6:379.
34. Pampena MB, Levy EM. Natural killer cells as helper cells in dendritic cell cancer vaccines. *Front Immunol*. 2015;6:13.
35. Mineharu Y, Castro MG, Lowenstein PR, Sakai N, Miyamoto S. Dendritic cell-based immunotherapy for glioma: multiple regimens and implications in clinical trials. *Neurol Med-Chir*. 2013;53(11):741–754.
36. Mac Keon S, Ruiz MS, Gazzaniga S, Wainstok R. Dendritic cell-based vaccination in cancer: therapeutic implications emerging from murine models. *Front Immunol*. 2015;6:243.
37. Van Gool SW. Brain tumor immunotherapy: what have we learned so far? *Front Oncol*. 2015;5:98.
38. Markowicz S, Engleman EG. Granulocyte-macrophage colony-stimulating factor promotes differentiation and survival of human peripheral blood dendritic cells in vitro. *J Clin Invest*. 1990;85(3):955–961.
39. Babatz J, Rollig C, Oelschlagel U, et al. Large-scale immunomagnetic selection of CD14+ monocytes to generate dendritic cells for cancer immunotherapy: a phase I study. *J Hematother Stem Cell Res*. 2003;12(5):515–523.
40. Morse MA, Zhou LJ, Tedder TF, Lyerly HK, Smith C. Generation of dendritic cells in vitro from peripheral blood mononuclear cells with granulocyte-macrophage-colony-stimulating factor, interleukin-4, and tumor necrosis factor-alpha for use in cancer immunotherapy. *Ann Surg*. 1997;226(1):6–16.
41. Shurin MR, Salter RD. *Dendritic cells in cancer*. New York, NY: Springer; 2009.
42. Schumacher TN, Schreiber RD. Neoantigens in cancer immunotherapy. *Science*. 2015;348(6230):69–74.
43. Jamal-Hanjani M, Thanopoulou E, Peggs KS, Quezada SA, Swanton C. Tumour heterogeneity and immune-modulation. *Curr Opin Pharmacol*. 2013;13(4):497–503.
44. Overwijk WW, Wang E, Marincola FM, Rammensee HG, Restifo NP. Mining the mutanome: developing highly personalized immunotherapies based on mutational analysis of tumors. *J Immunother Cancer*. 2013;1:11.

45. Dunn GP, Old LJ, Schreiber RD. The three Es of cancer immunoediting. *Annu Rev Immunol*. 2004;22:329–360.
46. Okada H, Kalinski P, Ueda R, et al. Induction of CD8+ T-cell responses against novel glioma-associated antigen peptides and clinical activity by vaccinations with α-type 1 polarized dendritic cells and polyinosinic-polycytidylic acid stabilized by lysine and carboxymethylcellulose in patients with recurrent malignant glioma. *J Clin Oncol*. 2011;29(3):330–336.
47. Phuphanich S, Wheeler CJ, Rudnick JD, et al. Phase I trial of a multi-epitope-pulsed dendritic cell vaccine for patients with newly diagnosed glioblastoma. *Cancer Immunol Immunother*. 2012.
48. Prins RM, Wang X, Soto H, et al. Comparison of glioma-associated antigen peptide-loaded versus autologous tumor lysate-loaded dendritic cell vaccination in malignant glioma patients. *J Immunother*. 2013;36(2):152–157.
49. Wheeler CJ, Black KL, Liu G, et al. Vaccination elicits correlated immune and clinical responses in glioblastoma multiforme patients. *Cancer Res*. 2008;68(14):5955–5964.
50. Yu JS, Wheeler CJ, Zeltzer PM, et al. Vaccination of malignant glioma patients with peptide-pulsed dendritic cells elicits systemic cytotoxicity and intracranial T-cell infiltration. *Cancer Res*. 2001;61(3):842–847.
51. Buhrman JD, Slansky JE. Improving T cell responses to modified peptides in tumor vaccines. *Immunol Res*. 2013;55(1–3):34–47.
52. Bregy A, Wong TM, Shah AH, Goldberg JM, Komotar RJ. Active immunotherapy using dendritic cells in the treatment of glioblastoma multiforme. *Cancer Treat Rev*. 2013;39(8):891–907.
53. Gedeon PC, Choi BD, Sampson JH, Bigner DD. Rindopepimut: anti-EGFRvIII peptide vaccine, oncolytic. *Drugs Future*. 2013;38(3):147–155.
54. Choi BD, Archer GE, Mitchell DA, et al. EGFRvIII-targeted vaccination therapy of malignant glioma. *Brain Pathol*. 2009;19(4):713–723.
55. Sampson JH, Archer GE, Mitchell DA, Heimberger AB, Bigner DD. Tumor-specific immunotherapy targeting the EGFRvIII mutation in patients with malignant glioma. *Semin Immunol*. 2008;20(5):267–275.
56. Padfield E, Ellis HP, Kurian KM. Current therapeutic advances targeting EGFR and EGFRvIII in glioblastoma. *FrontOncol*. 2015;5:5.
57. Gan HK, Kaye AH, Luwor RB. The EGFRvIII variant in glioblastoma multiforme. *J Clin Neurosci*. 2009;16(6):748–754.
58. Sampson JH, Heimberger AB, Archer GE, et al. Immunologic escape after prolonged progression-free survival with epidermal growth factor receptor variant III peptide vaccination in patients with newly diagnosed glioblastoma. *J Clin Oncol*. 2010;28(31):4722–4729.
59. Dziurzynski K, Chang SM, Heimberger AB, et al. Consensus on the role of human cytomegalovirus in glioblastoma. *Neuro Oncol*. 2012;14(3):246–255.
60. Cobbs CS, Harkins L, Samanta M, et al. Human cytomegalovirus infection and expression in human malignant glioma. *Cancer Res*. 2002;62(12):3347–3350.
61. Mitchell DA, Batich KA, Gunn MD, et al. Tetanus toxoid and CCL3 improve dendritic cell vaccines in mice and glioblastoma patients. *Nature*. 2015;519(7543):366–369.
62. Stragliotto G, Rahbar A, Solberg NW, et al. Effects of valganciclovir as an add-on therapy in patients with cytomegalovirus-positive glioblastoma: a randomized, double-blind, hypothesis-generating study. *Int J Cancer*. 2013;133(5):1204–1213.
63. Gilboa E. DC-based cancer vaccines. *J Clin Invest*. 2007;117(5):1195–1203.
64. Liau LM, Black KL, Prins RM, et al. Treatment of intracranial gliomas with bone marrow-derived dendritic cells pulsed with tumor antigens. *J Neurosurg*. 1999;90(6):1115–1124.
65. Liau LM, Black KL, Martin NA, et al. Treatment of a patient by vaccination with autologous dendritic cells pulsed with allogeneic major histocompatibility complex class I-matched tumor peptides. Case Report. *Neurosurg Focus*. 2000;9(6):e8.

66. Liau LM, Prins RM, Kiertscher SM, et al. Dendritic cell vaccination in glioblastoma patients induces systemic and intracranial T-cell responses modulated by the local central nervous system tumor microenvironment. *Clin Cancer Res*. 2005;11(15):5515–5525.
67. Ardon H, Van Gool S, Lopes IS, et al. Integration of autologous dendritic cell-based immunotherapy in the primary treatment for patients with newly diagnosed glioblastoma multiforme: a pilot study. *J Neurooncol*. 2010;99(2):261–272.
68. De Vleeschouwer S, Fieuws S, Rutkowski S, et al. Postoperative adjuvant dendritic cell-based immunotherapy in patients with relapsed glioblastoma multiforme. *Clin Cancer Res*. 2008;14(10):3098–3104.
69. Prins RM, Soto H, Konkankit V, et al. Gene expression profile correlates with T-cell infiltration and relative survival in glioblastoma patients vaccinated with dendritic cell immunotherapy. *Clin Cancer Res*. 2011;17(6):1603–1615.
70. Yamanaka R, Abe T, Yajima N, et al. Vaccination of recurrent glioma patients with tumour lysate-pulsed dendritic cells elicits immune responses: results of a clinical phase I/II trial. *Br J Cancer*. 2003;89(7):1172–1179.
71. Yamanaka R, Homma J, Yajima N, et al. Clinical evaluation of dendritic cell vaccination for patients with recurrent glioma: results of a clinical phase I/II trial. *Clin Cancer Res*. 2005;11(11):4160–4167.
72. Yu JS, Liu G, Ying H, Yong WH, Black KL, Wheeler CJ. Vaccination with tumor lysate-pulsed dendritic cells elicits antigen-specific, cytotoxic T-cells in patients with malignant glioma. *Cancer Res*. 2004;64(14):4973–4979.
73. Caruso DA, Orme LM, Neale AM, et al. Results of a phase 1 study utilizing monocyte-derived dendritic cells pulsed with tumor RNA in children and young adults with brain cancer. *Neuro Oncol*. 2004;6(3):236–246.
74. Kobayashi T, Yamanaka R, Homma J, et al. Tumor mRNA-loaded dendritic cells elicit tumor-specific $CD8^+$ cytotoxic T cells in patients with malignant glioma. *Cancer Immunol Immunother*. 2003;52(10):632–637.
75. Cao JX, Zhang XY, Liu JL, et al. Clinical efficacy of tumor antigen-pulsed DC treatment for high-grade glioma patients: evidence from a meta-analysis. *PLoS One*. 2014;9(9):e107173.
76. Ardon H, Van Gool SW, Verschuere T, et al. Integration of autologous dendritic cell-based immunotherapy in the standard of care treatment for patients with newly diagnosed glioblastoma: results of the HGG-2006 phase I/II trial. *Cancer Immunol Immunother*. 2012.
77. Bigner DD, Pitts OM, Wikstrand CJ. Induction of lethal experimental allergic encephalomyelitis in nonhuman primates and guinea pigs with human glioblastoma multiforme tissue. *J Neurosurg*. 1981;55(1):32–42.
78. Morgan RA, Chinnasamy N, Abate-Daga D, et al. Cancer regression and neurological toxicity following anti-MAGE-A3 TCR gene therapy. *J Immunother*. 2013;36(2):133–151.
79. Hodi FS, O'Day SJ, McDermott DF, et al. Improved survival with ipilimumab in patients with metastatic melanoma. *N Engl J Med*. 2010;363(8):711–723.
80. Pufall MA. Glucocorticoids and cancer. *Adv Exp Med Biol*. 2015;872:315–333.
81. Piemonti L, Monti P, Allavena P, et al. Glucocorticoids affect human dendritic cell differentiation and maturation. *J Immunol*. 1999;162(11):6473–6481.
82. Matasic R, Dietz AB, Vuk-Pavlovic S. Dexamethasone inhibits dendritic cell maturation by redirecting differentiation of a subset of cells. *J Leukoc Biol*. 1999;66(6):909–914.
83. Brandsma D, Stalpers L, Taal W, Sminia P, van den Bent MJ. Clinical features, mechanisms, and management of pseudoprogression in malignant gliomas. *Lancet Oncol*. 2008;9(5):453–461.
84. Knudsen-Baas KM, Moen G, Fluge O, Storstein A. Pseudoprogression in high-grade glioma. *Acta Neurol Scand Suppl*. 2013;196:31–37.
85. Linhares P, Carvalho B, Figueiredo R, Reis RM, Vaz R. Early pseudoprogression following chemoradiotherapy in glioblastoma patients: the value of RANO evaluation. *J Oncol*. 2013;2013:690585.

86. Taal W, Brandsma D, de Bruin HG, et al. Incidence of early pseudo-progression in a cohort of malignant glioma patients treated with chemoirradiation with temozolomide. *Cancer*. 2008;113(2):405–410.
87. Vrabec M, Van Cauter S, Himmelreich U, et al. MR perfusion and diffusion imaging in the follow-up of recurrent glioblastoma treated with dendritic cell immunotherapy: a pilot study. *Neuroradiology*. 2011;53(10):721–731.
88. Okada H, Weller M, Huang R, et al. Immunotherapy response assessment in neuro-oncology: a report of the RANO working group. *Lancet Oncol*. 2015;16(15):e534–542.
89. Constantino J, Gomes C, Falcao A, Cruz MT, Neves BM. Antitumor dendritic cell-based vaccines: lessons from 20 years of clinical trials and future perspectives. *Transl Res*. 2016;168:74–95.
90. Lampson LA. Brain tumor immunotherapy: seeing the brain in the body. *Drug Discov Today*. 2013;18(7–8):399–406.
91. Hdeib A, Sloan AE. Dendritic cell immunotherapy for solid tumors: evaluation of the DCVax® platform in the treatment of glioblastoma multiforme. *CNS Oncol*. 2015;4(2):63–69.
92. Polyzoidis S, Ashkan K. DCVax®-L–developed by Northwest Biotherapeutics. *Human Vaccin Immunother*. 2014;10(11):3139–3145.
93. Batich KA, Swartz AM, Sampson JH. Enhancing dendritic cell-based vaccination for highly aggressive glioblastoma. *Expert Opin Biol Ther*. 2015;15(1):79–94.
94. Wang X, Zhao HY, Zhang FC, Sun Y, Xiong ZY, Jiang XB. Dendritic cell-based vaccine for the treatment of malignant glioma: a systematic review. *Cancer Invest*. 2014;32(9):451–457.
95. Antonios JP, Everson RG, Liau LM. Dendritic cell immunotherapy for brain tumors. *J Neurooncol*. 2015;123(3):425–432.
96. Verhaak RG, Hoadley KA, Purdom E, et al. Integrated genomic analysis identifies clinically relevant subtypes of glioblastoma characterized by abnormalities in *PDGFRA, IDH1, EGFR*, and *NF1*. *Cancer Cell*. 2010;17(1):98–110.
97. Prins RM, Cloughesy TF, Liau LM. Cytomegalovirus immunity after vaccination with autologous glioblastoma lysate. *N Engl J Med*. 2008;359(5):539–541.
98. Ardon H, De Vleeschouwer S, Van Calenbergh F, et al. Adjuvant dendritic cell-based tumour vaccination for children with malignant brain tumours. *Pediatr Blood Cancer*. 2010;54(4):519–525.
99. Zobywalski A, Javorovic M, Frankenberger B, et al. Generation of clinical grade dendritic cells with capacity to produce biologically active IL-12p70. *J Transl Med*. 2007;5:18.
100. Curtsinger JM, Johnson CM, Mescher MF. CD8 T cell clonal expansion and development of effector function require prolonged exposure to antigen, costimulation, and signal 3 cytokine. *J Immunol*. 2003;171(10):5165–5171.
101. Sampson JH, Schmittling RJ, Archer GE, et al. A pilot study of IL-2Rα blockade during lymphopenia depletes regulatory T-cells and correlates with enhanced immunity in patients with glioblastoma. *PLoS One*. 2012;7(2):e31046.
102. Pilotto S, Carbognin L, Karachaliou N, et al. Moving towards a customized approach for drug development: lessons from clinical trials with immune checkpoint inhibitors in lung cancer. *Transl Lung Cancer Res*. 2015;4(6):704–712.
103. Lau PK, Ascierto PA, McArthur G. Melanoma: the intersection of molecular targeted therapy and immune checkpoint inhibition. *Curr Opin Immunol*. 2016;39:30–38.
104. Wainwright DA, Chang AL, Dey M, et al. Durable therapeutic efficacy utilizing combinatorial blockade against IDO, CTLA-4, and PD-L1 in mice with brain tumors. *Clin Cancer Res*. 2014;20(20):5290–5301.
105. Zeng J, See AP, Phallen J, et al. Anti-PD-1 blockade and stereotactic radiation produce long-term survival in mice with intracranial gliomas. *Int J Radiat Oncol Biol Phys*. 2013;86(2):343–349.

CHAPTER

13

Adoptive Immunotherapy Against Brain Tumors

C. Flores, D. Mitchell, P.A. Wells

University of Florida, Gainesville, FL, United States

OUTLINE

INTRODUCTION

The aim of immunotherapy is to establish cancer immunity while enabling its propagation and sustainability. Adoptive immunotherapy through the transfer of autologous tumor reactive T lymphocytes is an immediate establishment of often potent antitumor responses. This approach is a balance between the recognition of nonself antigens and the prevention of autotoxicity.

http://dx.doi.org/10.1016/B978-0-12-802420-1.00013-2

Immunotherapy has proven to be the most promising cancer therapy in both murine and human studies of metastatic melanoma. Adoptive cellular transfer using ex vivo expanded tumor infiltrating lymphocytes (TIL) has proven to have the highest cure rates in patients with metastatic melanoma and renal cancer.[1,2] Good clinical outcomes in response to adoptive cellular therapies are correlated with antitumor function, intratumoral migration, and persistence of adoptively transferred tumor-specific T cells at the tumor site. Despite significant antitumor efficacy against solid tumors, adoptive cellular therapy has had major limitations for development against brain tumors. A significant challenge in developing immunotherapy against brain tumors is that ubiquitously expressed single conserved tumor-specific antigens (TSA) and tumor-associated antigens expressed exclusively by brain tumors, but not normal tissues, are largely uncharacterized, yet to prime an immune response, tumor-specific T cells must have T cell receptors (TCRs) with sufficient affinity for tumor antigens to enable their activation. In addition, brain tumors have extremely limited anatomical accessibility, proving difficult for the application of TIL immunotherapy. Maximal antitumor efficacy of adoptive transfer platforms against solid tumors has also been reliant on myeloablative host conditioning with total body irradiation and hematopoietic stem cell rescue in both murine and human studies. The curative potential of adoptive cell therapy supersedes the burden of the described restrictions.

Crossing the Barrier

The vascular endothelium plays a key role in lymphocyte trafficking. Evidence demonstrates that the tumor vasculature establishes an active barrier that effector T cells must cross to recognize and kill tumor targets.[3,4] Infiltration of T cells into the tumor microenvironment is a critical hurdle that must be overcome in order for effective antitumor responses to occur. Successful transmigration through the tumor endothelial barrier is required for T cells to have maximal antitumor activity. To move through the barrier, T cells must extravasate through the endothelium to the tumor by rolling and adhering to endothelial cells. Although T cells can accumulate in the tumor stroma, they often fail to penetrate deeply into tumors.[3]

Investigators propose that within the tumor microenvironment, angiogenic molecules including vascular endothelial growth factor might prevent the expression of adhesion molecules on endothelial cells.[5] Endothelins and their receptors are expressed on a number of solid tumors including glioma and breast, ovarian, renal, and colon cancer.[6,7] Endothelin-B receptor expressed on tumor cells was demonstrated to suppress T cell homing, endothelial adhesion, and T cell infiltration in tumors.[8] Overexpression of endothelin-B receptor blocked T cell adhesion to the endothelium and was correlated with absence of TILs in solid tumor.[8]

The Encounter

The simple engagement of T cells with peptide–major histocompatibility complexes (MHCs) is insufficient for T cell activation. T cells must encounter antigen in the context of antigen presenting cells (APCs), such as dendritic cells, to provide costimulatory signals B7.1 and B7.2 to become fully activated.[9] Thus, tumors are invisible to naïve T cells until the T cells are activated as a result of cross-priming by dendritic cells that present tumor antigens. Simultaneous recognition of antigen presented on MHC and costimulatory ligands by T cells initiates cytokine production, proliferation, and cytotoxic function of activated T cells. A highly efficacious adoptive immunotherapeutic approach against intracranial glioma incorporates dendritic cell vaccination with transfer of tumor-specific T cells.[10]

Once effector T cells cross the endothelial barrier, they must navigate tumor stroma where they encounter target tumor cells expressing cognate antigen. T cells depend largely on recognition of targets through MHC–TCR interactions. Tumors, including brain tumors, express unique protein products that can be recognized by the immune system ranging from overexpressed self-antigens, including cancer testis and other immune privileged site antigens, to neo-antigens that result from nonsynonymous somatic mutations.

APPROACHES

Tumor Infiltrating Lymphocytes

Adoptive cell therapy using TILs, T cells isolated and expanded from solid tumor, has resulted in high response rates and complete durable overall responses in metastatic melanoma. The response rates to TIL therapy have established the principle that infusion of autologous T lymphocytes can cure advanced cancers in mice and humans.[11–13] However, the antitumor efficacy of this approach has been limited to some peripheral tumors.

Adoptive cell therapy using TILs has been the most effective cell-based immunotherapy to induce complete overall responses in patients with metastatic melanoma.[14] To isolate TILs, solid tumor is first resected and digested into single cell suspensions or divided into tumor fragments. Cells or tumor fragments are individually grown ex vivo in the presence of IL-2 for 2–3 weeks, resulting in tumor-reactive lymphocyte populations. Individual populations can then be further expanded using a rapid expansion protocol where T lymphocytes are grown in the presence of irradiated feeder cells, antibody conjugated beads that target CD3 on the T cells, and IL-2. Cells are then ready for infusion back into the patient after expansion.[12]

Polyclonal TIL populations include T cells that have the specificity for tumor antigens but determining which of these cells are responsible for tumor regression is challenging. Increasing evidence suggests that neoepitopes are the primary drivers of antitumor responses by TILs in melanoma.[11] Melanomas are characterized by high numbers of mutational events and neoepitopes that can be recognized by T cells found in TILs that constitute the dominant antitumor populations in TILs.[15,16] TIL-derived T cell clones that can target mutation antigens are reported to have high avidity to target tumor antigen and lead to robust tumor recognition with rare instances of autoimmune toxicity.[11,17] For melanomas and some other epithelial malignancies, TIL therapy has led to improved clinical outcomes.[15,16,18] Other cancers may be rational targets for TIL therapy depending on their level of expression of highly immunogenic epitopes.

Ex Vivo Activated Tumor-Specific T Cells

The presence of human cytomegalovirus (CMV) in glioblastoma (GBM) has been a controversial issue. CMV antigens were first detected in histologic samples of GBM in 2002.[19] Since then studies using various modalities of detection including immunohistochemistry, in situ hybridization, quantitative real-time PCR with viral DNA sequencing, and western blot analysis have been conducted and confirm that CMV is present in GBM and absent in normal brain.[20] Investigators have leveraged the immunogenicity of CMV antigens, and have developed adoptive cell therapy using autologous peripheral blood mononuclear cells to generate CMV-specific T cells in GBM patients.[21,22] These CMV-reactive T cells demonstrate cytolytic activity with high specificity against autologous tumors.[21,22]

In a clinical study, autologous CMV-specific T cells were successfully expanded from 13 out of 19 patients, and in total, 10 patients received at least three T cell infusions, and 11 received at least one infusion of CMV-specific T cells.[21] Of the 11 patients, the median overall survival was 33.58 months with a range of 133–2428 days.[21] Interestingly, tumor-infiltrating CMV-specific T cells expressed higher levels of inhibitory markers, programmed cell death-1 (PD-1), T cell immunoglobulin and mucin-3 (TIM-3), and cytotoxic T-lymphocyte-associated antigen 4 (CTLA-4) than their counterparts in the peripheral blood, indicating that the immunosuppressive tumor microenvironment reduces functionality even of antigen-specific T cells.[21]

The development of brain tumor-specific T cells is challenging, one major hurdle being that TSA for brain tumors are largely uncharacterized. To bypass this, investigators use dendritic cells pulsed with the total tumor RNA from glioma [TTRNA-pulsed dendritic cells (DCs)] to in vitro expand tumor-specific T cells.[10] The use of RNA-pulsed DCs to generate tumor-specific T cells has been demonstrated in both human and murine systems.[10,22–24] Preclinical studies demonstrate that adoptive transfer of tumor-specific T cells followed by RNA-pulsed DC vaccine is highly efficacious against temodar and irradiation resistant intracranial glioma under myeloablative host conditioning and hematopoietic stem cell rescue.[10] This platform of adoptive cellular transfer of tumor reactive T cells efficiently circumvents the limitations of immunotherapy by expanding a polyclonal population of tumor-specific T cells without the limitations of conserved antigens, and by expanding tumor-specific T cells without the need for TILs. Adoptive transfer of TTRNA-T cells into a highly invasive, chemotherapy and radiation resistant, intracranial astrocytoma bearing host leads to significant prolongation of median survival and long-term cures. This strategy of adoptive cell therapy along with TIL therapy reveals the importance of host conditioning and hematopoietic stem cell rescue to enhance the efficacy of cell-based immunotherapeutic strategies.

Designer T Cells

The combination of molecular biology with synthetic biology and its application to cancer has led to the creation of new biological structures that enhances antitumor functionality. The two forms of engineered T cells in clinical trials have successfully engineered T cells with enhanced affinity and cytotoxic function against autologous tumor. The first approach has affinity-enhanced TCRs. Endogenous TCR repertoire typically has low affinity to self-based cancer antigens.[25] The transfer of antigen-specific TCR genes by retroviral transduction of TCR α- and β-chains into lymphocytes leads to the specific migration and in vivo expansion of T cells in response to cognate antigen.[26]

Cancer/testis (CT) antigens are found to be immunogenic in patients and have tissue-restricted expression to testis and transformed cells.[27,28] These are expressed in several cancers including bladder cancer, non-small-cell lung cancer, prostate, ovary, and melanoma.[27] NYE-SO-1 is a member of the CT family of antigens expressed in many common tumors, making it an attractive target for immunotherapeutic strategies. A clinical trial using NYE-SO-1-specific HLA-A2-restricted TCR has been tested in patients with metastatic synovial cell sarcoma, melanoma, and myeloma.[29,30] Objective clinical responses were observed in 4/6 patients with metastatic synovial cell sarcoma and 5/11 treated patients with melanoma with no evidence of autotoxicity.[29] In another clinical study, investigators engineered autologous T cells to express high affinity receptors to NYE-SO-1 and LAGE-1 (NYE-SO-1-LAGE-1 TCR-engineered T cells).[30] Twenty patients with NYE-SO-1 and LAGE-1 positive multiple myeloma (MM) were treated with autologous NYE-SO-1-LAGE-1 TCR-engineered

T cells following autologous stem cell transplant. The TCR-engineered cells trafficked to bone marrow, expanded, and persisted in vivo. These cells were also capable of specific recognition and cytolytic function against tumor cells. The median progression-free survival in treated patients was 19.1 months, and promising results were observed in 16 out of 20 patients.[30]

Chimeric Antigen Receptors

An increased understanding of T cell recognition and T cell receptor signaling has led to the generation of T cells engineered with synthetic chimeric antigen receptors (CARs) that are able to target tumors with high specificity.[11,31] The basic design of CARs is to link an extracellular ligand recognition domain to an intracellular signaling module that includes CD3ζ. Upon binding to cognate antigen, T cell activation is induced. The modular structure has been developed from first generation CARs with only a CD3ζ signaling domain, to second- and third-generation CARs where costimulation is mimicked by linking CD3ζ signaling to signaling endodomains such as CD28, 4-1BB, and OX40.[32] Full T cell activation occurs after TCR recognition by the MHC molecule on APCs coupled with costimulation. Artificially providing one or more costimulatory signals augments proliferation and cytokine production of CAR-T cells.[31,32] Second- and third-generation CARs can also reprogram the T cells to survive in various immunosuppressive tumor microenvironments.

The first studies using CAR therapy in cancer patients showed limited persistence of the T cells and no efficacy.[31,33,34] However, the use of second-generation CAR T cells in hematologic malignancies has reproducibly demonstrated clinical responses. In a landmark publication, Porter et al.[35] engineered CAR T cells to target CD19 in B-cell neoplasms. Anti-CD19 CAR T cells were endowed with CD137 (4-1BB) signaling domain and were found to have potent effects in patients with B cell malignancies, chronic lymphocytic leukemia (CLL), and acute lymphoblastic leukemia (ALL). CAR T cells using a CD137 signaling domain have been found to persist for more than 3 years to date in patients treated for CLL, and more than 2 years in patients treated for ALL.[31] This long-term persistence of CAR T cells is associated with promising clinical response rates, with more than 50% achieving partial or complete remission.[31] Studies with second generation CD19 CAR-T cells with CD28 or 4-1BB signaling moieties have demonstrated robust antitumor efficacy in both murine and human systems.[13,32,36,37]

It has been observed that patients with high levels of CAR T cell engraftment and proliferation are associated with tumor regression. Studies using CD19 CAR T cells that also encode for CD28 signaling domains do not show long-term persistence of the engineered cells. Thus studies have also explored the use of CAR T cells with CD28 as a bridge after bone marrow transplant during patient recovery from lymphopenia.[31]

Epidermal Growth Factor Receptor Variant III Chimeric Antigen Receptors

A challenge in therapeutic development against brain tumors is that tumor-associated antigens and tumor-specific antigens in brain tumors are largely uncharacterized with the exception of epidermal growth factor receptor variant III (EGFRvIII). This is an important issue in the development of CAR strategies because off-tumor expression of target antigens have been shown to cause profound autotoxicity. EGFRvIII, an in frame deletion within the extracellular domain of EGFR, is a tumor-associated cell surface receptor not expressed in normal tissues.[38] An estimated 30–45% of GBM patients express EGFRvIII making it the most common variant of EGFR that is observed in human tumors, and is associated with poor long-term survival.[39] Investigators at the University of Pennsylvania have designed viral transduced CAR T cells to target both murine and human GBM.[25,40] Their design includes scFv fused to the CD8α hinge and transmembrane domain with intracellular CD3ζ and 4-1BB costimulatory domains.[40] Efficacy and immune reactivity of EGFRvII CAR T cells were extensively studied both in vitro and in xenograft systems. Immunodeficient mice bearing human tumors showed tumor regression with EGFRvIII CAR T cell infusion. Although 80% of injected CAR T cells were $CD4^+$, and only 20% were $CD8^+$, the majority of engineered CAR T cells found at the tumor site were $CD8^+$.[40] EGFRvIII expression has also been associated with glioma progenitor cells, making it a more attractive therapeutic target.[41] EGFRvIII CAR T cells demonstrate promising results; however, this mutation is only expressed in a proportion of GBM patients. In addition, brain tumors are notoriously heterogeneous, potentially allowing for immune escape mechanisms from therapies using single antigen targets.

Erythropoietin-Producing Hepatocellular Carcinoma A2 Chimeric Antigen Receptors

The erythropoietin-producing hepatocellular carcinoma A2 (EphA2) receptor is a member of the Eph family of receptor tyrosine kinases that is overexpressed gliomas.[42] Its expression is associated with poor outcomes and its overexpression has been characterized to enhance tumorigenesis.[43] Investigators have engineered CAR T cells against EphA2 using internal CD3ζ and CD28ζ intracellular domain.[44] EphA2 CAR T cells have cytotoxic activity against EphA2 expressing glioma cells during in vitro cytotoxicity assays and tumor regression was observed in xenograft models of glioma.[44]

Interleukin-13 Receptor Alpha 2 Chimeric Antigen Receptors

The interleukin 13 receptor α2 (IL13Rα2) is a glioma restricted cell-surface epitope that is expressed on >80% of high-grade gliomas including GBM, but not otherwise detected within the central nervous system.[45] IL13Rα2 is expressed by stem-like tumor cells and differentiated malignant cells is most closely associated with expression of mesenchymal signature genes. Investigators have genetically engineered $CD8^+$ cytotoxic T lymphocytes to express membrane tethered IL13 CARs.[46] Here they incorporate an IL13 E13Y mutein amino acid substitution for selective binding to IL13Rα2, termed IL13-zetakine.[47] $IL13\text{-}zetakine^+ CD8^+$ T cells demonstrate IL13Rα2-specific antitumor effector function.[46,48] In orthotopic xenograft models of GBM, adoptive transfer of $IL13\text{-}zetakine^+$ T cells results in the regression of tumor growth.[46,48] These IL13-zetakines signal through an engineered immune synapse composed of membrane-tethered cytokine muteins bound to cell-surface cytokine receptors. This CAR recognizes IL13Rα2 and initiates cytotoxic killing via intracellular CD3-ζ T cell activating domain. The IL13-zetakine T cells secrete Th1 cytokines and proliferate upon engagement of IL13Rα2 expressing targets.[47] In addition, $IL13\text{-}zetakine^+ CD8^+$ T cells have the capacity to target stem-like tumor cells and prevent tumor initiation.[48]

A small clinical study using $IL13\text{-}zetakine^+ CD8^+$ T cells has been conducted in three patients with recurrent GBM (NCT00730613).[47] Peripheral blood mononuclear cells of the leukapheresis product of eligible patients were used to engineer $CD8^+$ T cells to express IL13-zetakine.[47] Cells demonstrated direct killing of target IL13Rα2 expressing target cells in vitro.[47] These cells were administered directly into the patients' resection cavity following recovery from surgery. In one patient who recurred after experimental therapy, investigators were able to compare IL13Rα2 expression in initial resection versus recurrent tumor post-IL13-zetakine T cell therapy. Tumor expression of IL13Rα2 was significantly reduced in treated tumor, 14 weeks after CAR therapy.[47] This study indicated the safety and feasibility of using IL13-zetakine CAR T cells for treatment of malignant GBM.

Human Epidermal Growth Factor 2 Chimeric Antigen Receptors

Human epidermal growth factor 2 (HER2) is a tumor-associated antigen that is expressed by up to 80% of GBMs, but not healthy neurons or glia.[49] Investigators have genetically engineered autologous T lymphocytes using retroviral vector encoding HER2, generating HER2-specific CAR T cells.[50,51] Upon recognition of HER2 on autologous glioma tumor cells, proliferation and IFN-γ release were observed by HER2-CAR T cells. These T cells have potent responses against autologous glioma cells including $CD133^+$ glioma stem-like cells.[51]

In a study using peripheral blood mononuclear cells from 10 newly diagnosed GBM patients, HER2-CAR T cells were generated using second-generation CAR with an extracellular domain derived from the high-affinity HER2 monoclonal antibody FRP5 and signaling domains from costimulatory CD28 and the CD3ζ chain.[50,51] All patients' HER2-CAR T cells killed HER2-expressing target cells with specificity and secreted IFN-γ and IL-2, while nontransduced cells did not demonstrate reactivity.[51] Genetically engineered HER2-CAR T cells were capable of recognition and killing of autologous GBM cells, where 9 out of 10 tumors expressed HER2.[51] In vivo studies using orthotopic GBM xenograft models demonstrated that administration of HER2-CARs led to tumor regression with 50% long-term survivors.[51]

These same investigators also demonstrate the utility of HER2-CAR T cells in adoptive cell therapy against medulloblastoma.[50] Although HER2 is expressed by 40% of medulloblastomas, it is expressed at low levels, rendering immunotherapy using monoclonal antibodies ineffective. HER2-CAR T cells were generated as described earlier. In functionality assays, these cells were then cultured against human medulloblastoma lines Daoy and D283, IFN-γ secretion, and cytotoxicity by HER2-CAR T cells were detected.[50] Adoptive transfer of these Her2-specific CAR T cells also provided survival benefit in an othotopic xenograft model of Daoy medulloblastoma.[50]

Improving Adoptive Immunotherapy

Strategies for improving TIL therapy include lymphodepletive host conditioning regimens that lead to substantial increases in the duration of clinical responses in patients who receive lymphodepleting preparative regimens prior to cell infusion.[3,12] Investigations in humans and murine models of melanoma suggest several mechanisms by which host conditioning contributes to the efficacy of adoptive cell therapy. Lymphodepletive host conditioning depletes negative regulatory cells including regulatory T cells (Tregs) and myeloid-derived suppressor cells (MDSC). Circulating MDSCs in the peripheral blood have been described to suppress T cell activation and proliferation in glioma and melanoma patients.[52] However, studies describe that MDSCs found in peripheral blood, but not tumor, can suppress T cell proliferation in melanoma patients.[53] This suggests a role for circulating rather than tumor-resident myeloid cells in inhibiting T cell responses. The removal of peripheral MDSCs by lymphodepletive host conditioning thus aids in the proliferation of adoptively transferred T lymphocytes.

Regulatory T cells characterized by expression of $CD4^+CD25^+FoxP3^+$ have also been studied extensively in tumor bearing hosts in both human and murine systems. Accumulation of Tregs within the tumor bed has been well described in GBM, and their depletion in patients and tumor bearing mice that are recovering from chemotherapy-induced lymphopenia correlates with enhanced antitumor immunity.[54–56] In the context of TIL therapy, there is a negative correlation between levels of $CD4^+$ Tregs and clinical responses in patients during immune reconstitution treated with autologous TILs.[56] In mice, small numbers of Tregs can abrogate effective $CD8^+$ T cell-mediated adoptive cell therapy.[57]

Conditioning not only reduces competing and suppressive host cells, but also increases homeostatic cytokines IL-7 and IL-15 which cause homeostatic proliferation of adoptively transferred lymphocytes.[58] In mice, dose escalation from nonmyeloablative to myeloablative conditioning increases the efficacy of adoptively transferred antitumor $CD8^+$ T cells.[10] In two adoptive cell therapy platforms, one using TILs and the other using T cells expanded using RNA-pulsed dendritic cells, hematopoietic stem cell-mediated mechanisms play a role in mediating antitumor efficacy.[10,58] Furthermore, in murine studies dose escalation of temozolomide as a conditioning agent led to increased $CD8^+$ T cell responses.[59] In three nonrandomized clinical trials using TIL therapy against melanoma, patients were either pretreated with chemotherapy alone, chemotherapy plus 2Gy total body irradiation, or chemotherapy plus 12Gy total body irradiation.[12,14,60] The overall response rates were 49%, 52%, and 72%, respectively, whereas complete response rates were 12%, 20%, and 40% respectively.[12,14,60] The more intensive preparative regimens may enhance the efficacy of adoptive cell therapy, however, the long-term toxic effects of myeloablative preparations should be taken under consideration.

References

1. Rosenberg SA. Cell transfer immunotherapy for metastatic solid cancer—what clinicians need to know. *Nat Rev Clin Oncol*. 2011;8(10):577–585.
2. Wrzesinski C, Paulos CM, Kaiser A, et al. Increased intensity lymphodepletion enhances tumor treatment efficacy of adoptively transferred tumor-specific T cells. *J Immunother*. 2010;33(1):1–7.
3. Dudley ME, Wunderlich JR, Robbins PF, et al. Cancer regression and autoimmunity in patients after clonal repopulation with antitumor lymphocytes. *Science*. 2002;298(5594):850–854.
4. Lurquin C, Lethe B, De Plaen E, et al. Contrasting frequencies of antitumor and anti-vaccine T cells in metastases of a melanoma patient vaccinated with a MAGE tumor antigen. *J Exp Med*. 2005;201(2):249–257.
5. Min JK, Lee YM, Kim JH, et al. Hepatocyte growth factor suppresses vascular endothelial growth factor-induced expression of endothelial ICAM-1 and VCAM-1 by inhibiting the nuclear factor-kappaB pathway. *Circ Res*. 2005;96(3):300–307.
6. Hsieh WT, Yeh WL, Cheng RY, et al. Exogenous endothelin-1 induces cell migration and matrix metalloproteinase expression in U_{251} human glioblastoma multiforme. *J Neurooncol*. 2014;118(2):257–269.
7. Motz GT, Coukos G. Deciphering and reversing tumor immune suppression. *Immunity*. 2013;39(1):61–73.

8. Buckanovich RJ, Facciabene A, Kim S, et al. Endothelin B receptor mediates the endothelial barrier to T cell homing to tumors and disables immune therapy. *Nat Med.* 2008;14(1):28–36.
9. Greenwald RJ, Freeman GJ, Sharpe AH. The B7 family revisited. *Annu Rev Immunol.* 2005;23:515–548.
10. Flores C, Pham C, Snyder D, et al. Novel role of hematopoietic stem cells in immunologic rejection of malignant gliomas. *Oncoimmunology.* 2015;4(3):e994374.
11. Hinrichs CS, Rosenberg SA. Exploiting the curative potential of adoptive T cell therapy for cancer. *Immunol Rev.* 2014;257(1):56–71.
12. Rosenberg SA. Raising the bar: the curative potential of human cancer immunotherapy. *Sci Transl Med.* 2012;4(127):127–128.
13. Rosenberg SA, Kochenderfer JN. Personalized cell transfer immunotherapy for B-cell malignancies and solid cancers. *Mol Ther.* 2011;19(11):1928–1930.
14. Rosenberg SA, Restifo NP. Adoptive cell transfer as personalized immunotherapy for human cancer. *Science.* 2015;348(6230):62–68.
15. Robbins PF, Lu YC, El-Gamil M, et al. Mining exomic sequencing data to identify mutated antigens recognized by adoptively transferred tumor-reactive T cells. *Nat Med.* 2013;19(6):747–752.
16. Tran E, Turcotte S, Gros A, et al. Cancer immunotherapy based on mutation-specific $CD4^+$ T cells in a patient with epithelial cancer. *Science.* 2014;344(6184):641–645.
17. Kvistborg P, Shu CJ, Heemskerk B, et al. TIL therapy broadens the tumor-reactive $CD8^+$ T cell compartment in melanoma patients. *Oncoimmunology.* 2012;1(4):409–418.
18. Dudley ME, Gross CA, Langhan MM, et al. $CD8^+$ enriched "young" tumor infiltrating lymphocytes can mediate regression of metastatic melanoma. *Clin Cancer Res.* 2010;16(24):6122–6131.
19. Cobbs CS, Harkins L, Samanta M, et al. Human cytomegalovirus infection and expression in human malignant glioma. *Cancer Res.* 2002;62(12):3347–3350.
20. Nair SK, Sampson JH, Mitchell DA. Immunological targeting of cytomegalovirus for glioblastoma therapy. *Oncoimmunology.* 2014;3:e29289.
21. Schuessler A, Smith C, Beagley L, et al. Autologous T cell therapy for cytomegalovirus as a consolidative treatment for recurrent glioblastoma. *Cancer Res.* 2014;74(13): 3466–3476.
22. Nair SK, De Leon G, Boczkowski D, et al. Recognition and killing of autologous, primary glioblastoma tumor cells by human cytomegalovirus pp65-specific cytotoxic T cells. *Clin Cancer Res.* 2014;20(10):2684–2694.
23. Yang S, Archer GE, Flores CE, Mitchell DA, Sampson JH. A cytokine cocktail directly modulates the phenotype of DC-enriched anti-tumor T cells to convey potent anti-tumor activities in a murine model. *Cancer Immunol Immunother.* 2013;62(11):1649–1662.
24. Pham CD, Flores C, Yang C, et al. Differential immune microenvironments and response to immune checkpoint blockade amongst molecular subtypes of murine medulloblastoma. *Clin Cancer Res.* 2016;22(3):582–595.
25. Johnson LA, Morgan RA, Dudley ME, et al. Gene therapy with human and mouse T cell receptors mediates cancer regression and targets normal tissues expressing cognate antigen. *Blood.* 2009;114(3):535–546.
26. Morgan RA, Dudley ME, Yu YY, et al. High efficiency TCR gene transfer into primary human lymphocytes affords avid recognition of melanoma tumor antigen glycoprotein 100 and does not alter the recognition of autologous melanoma antigens. *J Immunol.* 2003;171(6):3287–3295.
27. Scanlan MJ, Simpson AJ, Old LJ. The cancer/testis genes: review, standardization, and commentary. *Cancer Immun.* 2004;4:1.
28. Zhao Y, Zheng Z, Robbins PF, Khong HT, Rosenberg SA, Morgan RA. Primary human lymphocytes transduced with NY-ESO-1 antigen-specific TCR genes recognize and kill diverse human tumor cell lines. *J Immunol.* 2005;174(7):4415–4423.

29. Robbins PF, Morgan RA, Feldman SA, et al. Tumor regression in patients with metastatic synovial cell sarcoma and melanoma using genetically engineered lymphocytes reactive with NY-ESO-1. *J Clin Oncol*. 2011;29(7):917–924.
30. Rapoport AP, Stadtmauer EA, Binder-Scholl GK, et al. NY-ESO-1-specific TCR-engineered T cells mediate sustained antigen-specific antitumor effects in myeloma. *Nat Med*. 2015;21(8):914–921.
31. June CH, Maus MV, Plesa G, et al. Engineered T cells for cancer therapy. *Cancer Immunol Immunother*. 2014;63(9):969–975.
32. Srivastava S, Riddell SR. Engineering CAR-T cells: design concepts. *Trends Immunol*. 2015;36(8):494–502.
33. Kershaw MH, Westwood JA, Parker LL, et al. A phase I study on adoptive immunotherapy using gene-modified T cells for ovarian cancer. *Clin Cancer Res*. 2006;12 (20 Pt 1):6106–6115.
34. Lamers CH, Sleijfer S, Vulto AG, et al. Treatment of metastatic renal cell carcinoma with autologous T-lymphocytes genetically retargeted against carbonic anhydrase IX: first clinical experience. *J Clin Oncol*. 2006;24(13):e20–22.
35. Porter DL, Levine BL, Kalos M, Bagg A, June CH. Chimeric antigen receptor-modified T cells in chronic lymphoid leukemia. *N Engl J Med*. 2011;365(8):725–733.
36. Kochenderfer JN, Rosenberg SA. Chimeric antigen receptor-modified T cells in CLL. *N Engl J Med*. 2011;365(20):1937–1938. author reply 1938.
37. Kochenderfer JN, Wilson WH, Janik JE, et al. Eradication of B-lineage cells and regression of lymphoma in a patient treated with autologous T cells genetically engineered to recognize CD19. *Blood*. 2010;116(20):4099–4102.
38. Wikstrand CJ, McLendon RE, Friedman AH, Bigner DD. Cell surface localization and density of the tumor-associated variant of the epidermal growth factor receptor, EGFRvIII. *Cancer Res*. 1997;57(18):4130–4140.
39. Hegi ME, Rajakannu P, Weller M. Epidermal growth factor receptor: a re-emerging target in glioblastoma. *Curr Opin Neurol*. 2012;25(6):774–779.
40. Johnson LA, Scholler J, Ohkuri T, et al. Rational development and characterization of humanized anti-EGFR variant III chimeric antigen receptor T cells for glioblastoma. *Sci Transl Med*. 2015;7(275):222–275.
41. Morgan RA, Johnson LA, Davis JL, et al. Recognition of glioma stem cells by genetically modified T cells targeting EGFRvIII and development of adoptive cell therapy for glioma. *Hum Gene Ther*. 2012;23(10):1043–1053.
42. Wykosky J, Gibo DM, Stanton C, Debinski W. EphA2 as a novel molecular marker and target in glioblastoma multiforme. *Mol Cancer Res*. 2005;3(10):541–551.
43. Zelinski DP, Zantek ND, Stewart JC, Irizarry AR, Kinch MS. EphA2 overexpression causes tumorigenesis of mammary epithelial cells. *Cancer Res*. 2001;61(5):2301–2306.
44. Chow KK, Naik S, Kakarla S, et al. T cells redirected to EphA2 for the immunotherapy of glioblastoma. *Mol Ther*. 2013;21(3):629–637.
45. Mintz A, Gibo DM, Slagle-Webb B, Christensen ND, Debinski W. IL-13Ralpha2 is a glioma-restricted receptor for interleukin-13. *Neoplasia*. 2002;4(5):388–399.
46. Kahlon KS, Brown C, Cooper LJ, Raubitschek A, Forman SJ, Jensen MC. Specific recognition and killing of glioblastoma multiforme by interleukin 13-zetakine redirected cytolytic T cells. *Cancer Res*. 2004;64(24):9160–9166.
47. Brown CE, Badie B, Barish ME, et al. Bioactivity and safety of IL13Ralpha2-redirected chimeric antigen receptor $CD8^+$ T cells in patients with recurrent glioblastoma. *Clin Cancer Res*. 2015;21(18):4062–4072.
48. Brown CE, Starr R, Aguilar B, et al. Stem-like tumor-initiating cells isolated from IL13Ralpha2 expressing gliomas are targeted and killed by IL13-zetakine-redirected T cells. *Clin Cancer Res*. 2012;18(8):2199–2209.
49. Liu G, Ying H, Zeng G, Wheeler CJ, Black KL, Yu JS. HER-2, gp100, and MAGE-1 are expressed in human glioblastoma and recognized by cytotoxic T cells. *Cancer Res*. 2004;64(14):4980–4986.

50. Ahmed N, Ratnayake M, Savoldo B, et al. Regression of experimental medulloblastoma following transfer of HER2-specific T cells. *Cancer Res*. 2007;67(12):5957–5964.
51. Ahmed N, Salsman VS, Kew Y, et al. HER2-specific T cells target primary glioblastoma stem cells and induce regression of autologous experimental tumors. *Clin Cancer Res*. 2010;16(2):474–485.
52. Raychaudhuri B, Rayman P, Ireland J, et al. Myeloid-derived suppressor cell accumulation and function in patients with newly diagnosed glioblastoma. *Neuro-oncology*. 2011;13(6):591–599.
53. Gros A, Turcotte S, Wunderlich JR, Ahmadzadeh M, Dudley ME, Rosenberg SA. Myeloid cells obtained from the blood but not from the tumor can suppress T cell proliferation in patients with melanoma. *Clin Cancer Res*. 2012;18(19):5212–5223.
54. Mitchell DA, Cui X, Schmittling RJ, et al. Monoclonal antibody blockade of IL-2 receptor alpha during lymphopenia selectively depletes regulatory T cells in mice and humans. *Blood*. 2011;118(11):3003–3012.
55. Sampson JH, Schmittling RJ, Archer GE, et al. A pilot study of IL-2Ralpha blockade during lymphopenia depletes regulatory T cells and correlates with enhanced immunity in patients with glioblastoma. *PLoS One*. 2012;7(2):e31046.
56. Yao X, Ahmadzadeh M, Lu YC, et al. Levels of peripheral $CD4^+FoxP3^+$ regulatory T cells are negatively associated with clinical response to adoptive immunotherapy of human cancer. *Blood*. 2012;119(24):5688–5696.
57. Antony PA, Restifo NP. $CD4^+CD25^+$ T regulatory cells, immunotherapy of cancer, and interleukin-2. *J Immunother*. 2005;28(2):120–128.
58. Wrzesinski C, Paulos CM, Gattinoni L, et al. Hematopoietic stem cells promote the expansion and function of adoptively transferred antitumor CD8 T cells. *J Clin Invest*. 2007;117(2):492–501.
59. Sanchez-Perez L, Suryadevara CM, Choi BD, Reap EA, Sampson JH. Leveraging chemotherapy-induced lymphopenia to potentiate cancer immunotherapy. *Oncoimmunology*. 2014;3(7):e944054.
60. Restifo NP, Dudley ME, Rosenberg SA. Adoptive immunotherapy for cancer: harnessing the T cell response. *Nat Rev Immunol*. 2012;12(4):269–281.

CHAPTER

14

Chimeric Antigen Receptor Therapy of Brain Tumors

L. Sanchez-Perez[1], *C.M. Suryadevara*[1], *B.D. Choi*[2], *L.A. Johnson*[3]

[1]Duke University Medical Center, Durham, NC, United States; [2]Massachusetts General Hospital and Harvard Medical School, Boston, MA, United States; [3]University of Pennsylvania, Philadelphia, PA, United States

OUTLINE

Translational Immunotherapy of Brain Tumors
http://dx.doi.org/10.1016/B978-0-12-802420-1.00014-4

CHIMERIC ANTIGEN RECEPTORS THROUGH THE GENERATIONS

First Generation

Chimeric antigen receptors (CARs) were originally conceptualized by Zelig Eshhar and colleagues in 1989.[1] Their findings, published in *Proceedings of the National Academy of Sciences*, described the generation of an "immunoglobulin-T cell receptor chimeric molecule" by splicing the heavy and light chain variable regions of a monoclonal antibody (mAb) into a T lymphocyte cell line, along with the constant region of a T cell receptor (TCR). They later modified their approach by generating a single-chain fragment encoding both heavy and light variable regions joined by a linker sequence (scFv), negating the need for multiple gene transfers to achieve antibody-like receptor specificity.[2] Upon scFv binding to cognate antigen, the CAR signaled through the CD3 zeta (ζ) chain to activate the receptor-bearing T lymphocyte. These first generation constructs, defined by the inclusion of a single intracellular signaling domain, were evaluated in early clinical trials using CARs targeting folate-receptor in patients with ovarian cancer,[3] carbonic anhydrase IX (CAIX) in patients with renal cancer,[4] and CD171/L1-CAM in pediatric patients with neuroblastoma.[5] These trials uniformly failed to demonstrate antitumor efficacy, and notably lacked long-term survival of circulating CARs in patient blood. The CAIX trial did, however, lead to cases of unexpected biliary tract toxicity, illustrating the potential for acute in vivo off tumor, on target effects in patients.

The first indication of positive clinical outcome using CARs came in a report published in 2009 by Malcolm Brenner and colleagues at Baylor College of Medicine in Houston, TX. A first generation CAR targeting disialoganglioside GD2 in pediatric patients with neuroblastoma induced a complete response in 1 out of 11 total patients.[5,6] Immune monitoring revealed that all patients had remarkably low counts of circulating CAR T cells after treatment, with 0.1% detectable in blood by PCR 24h after infusion. These results suggested that CAR T cell therapy had the potential to induce effective antitumor responses, however, an emerging theme was the challenge of promoting long-term CAR T cell engraftment in vivo in patients.

Concurrent studies using transgenic TCR-based immunotherapy also provided critical information during this period that supported a refinement of

CAR design to support in vivo survival and persistence. TCRs recognize intracellular antigens in the form of peptides processed and presented in the context of MHC molecules on the cell surface. Similar to CARs, TCR engineered T cells were able to redirect patient lymphocytes to recognize and destroy cancer, utilizing transgenic TCRα and β chain genes rather than mAb scFv fragments. Studies conducted by Steven Rosenberg's group at the National Cancer Institute (NCI) provided insight regarding the role of receptor affinity for antigen and in vivo survival of engineered T cells in patients.[7–9] These studies showed that T cells modified to express TCR genes targeting gp100 or melanoma antigen recognized by T cells (MART1) were able to elicit antitumor effects in patients with advanced metastatic melanoma with objective tumor responses [according to RECIST (Response Evaluation Criteria in Solid Tumors) criteria] in 17% and 30% of patients, respectively.[9] Notably, these gene-engineered T cells were found in the circulating blood of patients at remarkably high proportions (up to 80% of circulating T cells), even 1 month after treatment. The long-term persistence of TCR-transgenic T cells compared to the relatively short lifespan of first generation CARs also supported a reformulation of CAR design. Despite this discrepancy, the CAR T cell platform remained popular given its capacity to bypass human leukocyte antigen presentation of tumor antigens, which limits the eligible patient population, and is often dysregulated in cancer and used as a mechanism of immune escape.

Second Generation Chimeric Antigen Receptors

Subsequent efforts were focused on improving the function and survival of CAR T cells, based on an understanding of fundamental T cell biology. Normally, T cells are activated through a two-step mechanism, which helps the immune system to differentiate between dangerous pathogens and normal self-tissues. Upon a strong stimulation of the TCR zeta (ζ) chain, the T cell activation pathway is triggered, resulting in production of effector molecules including type 1 cytokines interferon-gamma, tumor necrosis factor-alpha, and lysis of target cells through production of perforin and granzymes, as well as release of CD107a/LAMP1. A second signal is then initiated that promotes proliferation and survival of the T cell to ensure persistent defense against an offending pathogen. This is achieved through T cell interaction with supporting immune cells like macrophages or dendritic cells expressing costimulatory molecules like CD70, CD80, or CD86 that bind to CD27 or CD28, or 4-1BB on the surface of T cells.

When TCR zeta signaling occurs in tandem with these costimulatory signals, the T cell receives prosurvival instructions, continuing long term with eventual transition to effector memory cells. However, in the absence of such costimulation, an antiautoimmune cycle is launched, whereby the activated T cells quickly undergo activation-induced cell death (AICD). This process ensures that T cells will exhibit continued activation and function against a desired target, but will quickly be eliminated if found to be self-reactive.

Speculation that the second signal of costimulation may be what was missing in the CAR triggered T cells, researchers from the groups of Carl June and Michel Sadelain generated second generation CARs, which included intracellular signaling from a costimulatory molecule such as CD28,[10,11] 4-1BB, or OX40.[12–14] Essentially, the second generation CARs enhanced the first generation CARs by adding intracellular costimulatory signaling regions; whereas first generation CARs consisted of the scFv plus CD3ζ, second generation CARs also included one of several moieties such as CD28, 4-1BB, or OX40. Opinions differed on which costimulatory region was superior, however, evidence supported that CD28 conferred increased antitumor function to T cells, wherease 4-1BB provided longer survival, through resistance to AICD. Following these initial discoveries, additional second generation costimulatory molecules were evaluated by different groups, including ICOS,[15] CD2,[16] CD27,[17] KIR,[18] CD40L,[19] and CD80.[20]

Chimeric Antigen Receptors, The Next Generation

As a definition, the third generation CAR includes more than one costimulatory molecule in addition to the scFv and CD3ζ signal, as depicted in Fig. 14.1. Steven Rosenberg's group at the NCI published preclinical work on several third generation CARs encoding both CD28 and 4-1BB intracellular signaling domains in preclinical models. These include targeting vascular endothelial growth factor receptor on tumor vasculature,[21] fibroblast activation protein (FAP) in solid tumor stroma,[22] CD19 in leukemia,[23] and epidermal growth factor receptor variant three (EGFRvIII) mutation on high grade glioma.[24]

In collaboration with the NCI, Laura Johnson's group at Duke University showed in a syngeneic mouse model that third generation EGFRvIII CARs were able to recognize and eliminate cells expressing their respective targets, persist long-term, and induce an endogenous immune response de novo against additional non-EGFRvIII tumor antigens.[25] In a glioma model, EGFRvIII CARs were able to completely eliminate EGFRvIII expressing glioma tumors in VM/Dk mice in a dose-dependent fashion, and these cured mice later showed protection against tumor rechallenge with EGFRvIII-negative matched tumors. This showed for the first time that single antigen-targeted CAR T cells could elicit additional endogenous adaptive immunity.

The first use of long-lived FAP-targeting CARs in vivo was based upon promising data in vitro against cell lines. However, in syngeneic mouse models, these CARs also demonstrated severe autotoxicity caused by cross-reactivity with shared antigen on normal tissues.[22] Specifically, Tran et al. at the NCI showed in both C57Bl/6 and Balb/C mice engrafted with different tumor types, including melanoma, renal, breast, and colorectal cancers, that FAP CARs had limited antitumor efficacy, but caused profound toxicity linked to destruction of FAP-expressing bone marrow stromal cells.[22] Intriguingly, in similar research by Wang et al. at the University of Pennsylvania (UPENN),

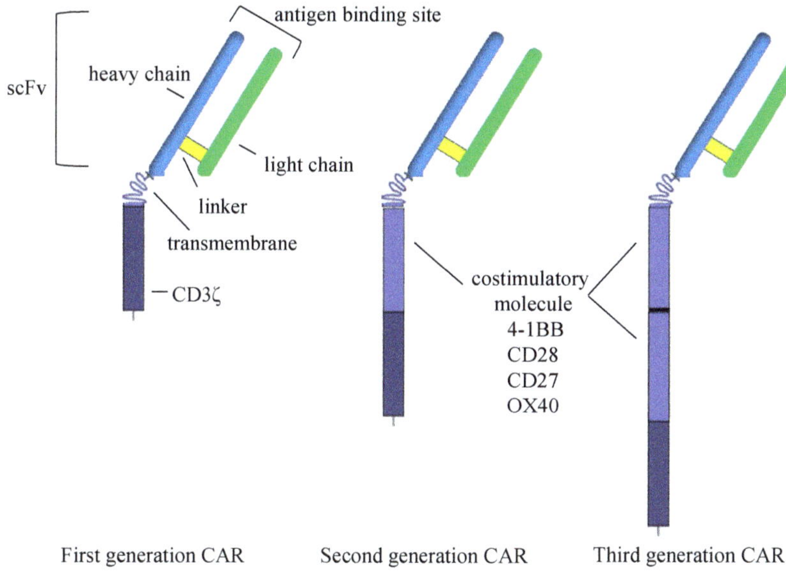

FIGURE 14.1 **Chimeric Antigen Receptor generations.** First generation CARs include an extracellular antigen binding site generally formed by the VH and VL portions of a mAb, joined by a linker sequence to form an scFv. The scFv region is connected to the intracellular signaling portion by a TM domain, followed by a CD3ζ signaling region (first generation), or a CD3ζ preceeded by one (second generation) or more than one (third generation) costimulatory molecules in cis formation. *CAR*, chimeric antigen receptor; *mAb*, monoclonal antibody; *scFv*, single-chain Fragment variable region; *VH*, variable heavy chain; *VL*, variable light chain.

FAP CARs in syngeneic mice demonstrated antitumor effects, but without autotoxicity.[26] Differences in experimental approach between the two groups included use of a murine (m)CD28 signaling region alone at UPENN versus a mCD28 plus m4-1BB at the NCI, variability in cell number, and different anti-FAP mAb clone-derived scFv. Wang et al. also reported that the mCAR T cells did not persist more than a week in vivo. The contrasts between the results in these two groups suggested the potential for a treatment window effect, whereby targeting the same antigen could have differential outcomes depending on differences in (1) receptor affinity or epitope selection, (2) costimulatory signal combination, and/or (3) CAR T cell dosage.

SUCCESS OF TARGETING HEMATOGENOUS MALIGNANCIES: CD19 CHIMERIC ANTIGEN RECEPTORS

After the initial demonstration of unexpected toxicity seen with use of the CAIX-targeting CAR in patients with renal cancer,[4] and skin toxicities observed in patients receiving melanocyte-targeted TCRs,[9] more suitable, "safe" targets were sought out.

Leukemias consist of cancers of the blood, specifically white blood cells such as T and B lymphocytes. Leukemias involving the B lymphocyte subset include chronic lymphoid leukemia (CLL), acute lymphoblastic leukemia (ALL), diffuse large B cell lymphoma, and mantle-cell lymphoma, among others. Although treatment with chemotherapy and radiation can cure some patients with these diseases, those who progress generally have little hope other than a bone marrow transplant (BMT), which itself carries a high risk of mortality of 30–35%. Even those patients for whom BMT is successfully tolerated and initially effective will often relapse, leaving no other treatment options. The harbinger of CAR engineered T cell success came two decades after Zelig Eshhar's first proposal to tie the recognition specificity of an antibody to the destructive power of a T cell in the form of CARs targeting the universal B-cell antigen, CD19.

Although targeting differentiation antigens on B cells (CD19, CD20) were a popular focus for many researchers, three main groups translated the first leukemia-targeting CARs into clinical trials and published their initial results in 2010–2012. At the NCI Surgery Branch, Drs. James Kochenderfer and Steven Rosenberg reported CD19-directed, CD28 costimulated second generation CAR treatment of a patient with advanced follicular lymphoma, who experienced a clinical regression that lasted throughout the 39 weeks of follow-up study, along with elimination of all circulating normal B lymphocytes.[27] From Memorial Sloan-Kettering Cancer Center, Drs. Renier Brentjens and Michel Sadelain reported on a patient with CLL who experienced a severe adverse event within hours of receiving second generation CD19 CAR T cells with CD28 costimulation signaling, resulting in the patient's subsequent death.[28] Results of an additional nine patients treated on this trial were published separately the following year.[29] These patients with CLL or ALL showed no objective antitumor responses, but the CAR T cell engraftment and evidence of T cell cytokine production were deemed promising. Also in 2011, Drs. David Porter and Carl June from the University of Pennsylvania reported on the first patient with CLL treated with CD19-targeted, 4-1BB costimulated CARs in the *New England Journal of Medicine*.[30] This patient experienced a complete remission of tumor, as well as elimination of circulating B lymphocytes that was ongoing at 10 months after treatment. What was particularly remarkable about this case was the relatively small dose of CAR T cells administered (1×10^5/kg) compared with the other two groups (4×10^8 CAR T cells total, and 3×10^7/kg for Rosenberg and Sadelaine, respectively), and the long-term engraftment of CAR T cells at approximately 1% of peripheral blood lymphocyte for the duration of the follow-up, compared with undetectable levels after approximately 2 months in the other groups.

As technologies and clinical experience using CD19 targeted CARs to treat leukemia patients matured, more publications appeared from these groups as well as others, reporting increased response rates and increasing diversity of B-cell malignancies responding to therapy. By 2015, publications reported a total of more than 100 patients treated with CD19 targeting CARs, with response rates of up to 95% in pediatric patients with ALL, and disease indications reporting positive outcomes in leukemias as unexpected as CD19-dim/negative multiple myeloma.[29,31–33] Consistently high response rates including putative cures have engendered substantial interest from commercial sources, with numerous small and large biotechnology and pharmaceutical companies investing heavily in CAR therapy, including Novartis, Juno, Kite, Bluebird, and others. Currently, there are more than 30 CD19-targeting CAR clinical trials taking place in North America, Europe, Asia, and Australia.

SOLID TUMORS AS A CHIMERIC ANTIGEN RECEPTOR TARGET

Although the treatment of CD19 expressing hematogenous cancers has been largely successful, CARs targeting solid, nonhematogenous malignancies face a unique set of challenges in their successful implementation. In contrast to a liquid blood-borne tumor environment, a solid tumor microenvironment presents a hostile milieu and desmoplastic stroma that work together to inhibit the penetration and persistence needed to elicit an effective antitumor immune response. In addition, perhaps the greatest barrier is that although several antigenic targets are currently being explored for solid tumors, with a few exceptions, to date there has been a relative lack of well-described, tumor-associated, or tumor-specific antigens.

CHIMERIC ANTIGEN RECEPTOR THERAPY FOR BRAIN TUMORS

Brain tumors in particular face an added potential issue: crossing the blood–brain barrier (BBB). In the past, the brain has been identified as an immune-privileged site, protected by a tightly woven barrier of pericytes protecting the brain tissues from exposure to proteins, cells, and micro- or macro-organisms transiting through the blood. The ability of the BBB to keep out large molecule drugs, or antibodies is well documented, with brain levels typically showing less than 1/1000th the blood concentration of circulating drug treatments. As mentioned, T cell immunotherapy for brain tumors is currently limited by a paucity of antigens that can be safely targeted.[34,35] CAR technology has been utilized for the development of CAR T cells targeting glioma-associated antigens such as: EGFR,[36–38] the cancer-specific EGFRvIII,[39–42] interleukin-13 receptor alpha-2 (IL-13Rα2),[43–45] receptor tyrosine-protein kinase erbB-2 also known as human epidermal growth factor receptor 2 (HER2/Neu),[46–49] erythropoietin-producing

TABLE 14.1 Chimeric Antigen Receptor Clinical Trials for Patients With Brain Tumors

Study	Phase	Identifier	Sponsor
Pilot study of autologous T cells redirected to EGFRVIII-with a chimeric antigen receptor in patients with EGFRVIII+ glioblastoma (GBM)	Phase I trial	NCT02209376	Abramson Cancer Center of the University of Pennsylvania, Philadelphia, PA, 2 sites: UPENN and UCSF.
Chimeric antigen receptors (CAR) T cell receptor immunotherapy targeting EGFRvIII for patients with malignant gliomas expressing EGFRvIII	Phase I trial	NCT01454596	National Cancer Institute, Bethesda, MD
Genetically modified T cells in treating patients with recurrent or refractory malignant glioma (IL13Rα2)	Phase I trial	NCT02208362	City of Hope Medical Center, Duarte, CA
CAR T cells in treating patients with malignant gliomas overexpressing EGFR	Phase I trial	NCT02331693	RenJi Hospital, China
T cells expressing HER2-specific CARs for patients with GBM	Phase I trial	NCT02442297	Baylor College of Medicine, Houston, TX
Cytomegalovirus-specific cytotoxic T lymphocytes expressing CAR targeting HER2 in patients with GBM	Phase I trial	NCT01109095	Baylor College of Medicine, Houston, TX

human hepatocellular carcinoma type-A receptor 2 (EphA2),[50] and the glioma stem cell (GSC) markers Prominin-1 (CD133)[51,52] and neural cell adhesion molecule L1 (L1-CAM).[53,54] These CARs have been developed and evaluated in preclinical studies, and several have entered phase I clinical studies for the treatment of different brain tumors, and are listed in Table 14.1. Here we review preclinical and clinical CAR T cell studies currently underway for the treatment of malignant brain tumors.

Preclinical Chimeric Antigen Receptor T Cell Implementation

CAR design has been based largely on the availability of existing antibodies, but has recently extended to novel antibodies targeting the above-mentioned antigens. Most of these antibodies were of murine origin, developed by immunizing mice with the human target antigen bound to adjuvants such as keyhole limpet hemocyanin to develop high affinity antibodies with high specificity for the human target. These approaches were designed to leverage molecular differences between mouse and human molecules to produce murine-origin antihuman antibodies, which minimize the potential for cross-reactivity against mouse orthologs. These CARs represent the vast majority used in studies today, typically displaying specificity against human targets expressed on human cell lines, xenografted onto mice. Therefore, preclinical evaluation of these CARs has required use of severely immunocompromised animal models such as nonobese diabetic, severe combined immunodeficient (NOD SCID) mice or NOD SCID IL-2R$\gamma_C^{-/-}$ (NSG) mice.[43,48] The use of these models carries pros and cons. The most obvious advantage is the ability to evaluate the same therapeutic agent that would be used for clinical studies; these models utilize human tumor and human CAR T cells. One major disadvantage, however, is that immunocompromised animal models preclude the ability to adequately evaluate the immunobiology relating to CAR T cell therapy.[55,56] Moreover, accurate assessment over safety and the potential for autoimmune toxicity is severely compromised. Human CAR T cells may also experience limited or altered kinetics in engraftment in immunocompromised models, and failure to persist may prevent the induction of CAR-dependent autoimmune pathology that could otherwise occur in the clinical scenario, and therefore limits the utility of this model in both acute and long-term chronic toxicity studies.

Preclinical Evaluation of Chimeric Antigen Receptors in Brain Tumor Models

EGFR CARs

EGFR is a 170-kDa cell surface receptor tyrosine kinase that recognizes several similar growth factors: epidermal growth factor (EGF), heparin-binding EGF-like growth factor, transforming growth factor α, amphiregulin, epiregulin, and betacellulin.[57] EGFR is naturally expressed in the developing brain and plays a critical role in neurogenesis.[58] In adulthood, EGFR expression in the brain is dramatically limited by comparison, and is detected at low levels in the sub ventricular zone and the subgranular zone within the hippocampus, areas where stem cells reside.[59] However, a large proportion of high-grade primary malignant brain tumors such as glioblastoma (GBM) has been shown to express EGFR and this pattern of gene expression is associated with an aggressive tumor phenotype and poor prognosis.[60] Over 50% of patients with GBM will express EGFR with a relative uniform distribution within the tumor bed.[61,62] Moreover, tumors expressing EGFR typically exhibit gene amplification and express significantly higher levels of expression compared with normal tissue naturally expressing EGFR.[63] Therefore, the limited expression of EGFR in the normal brain and an abundant and uniform expression of this receptor in GBM tumors has made this a potential attractive target for CAR therapy.

Recently, Zhou et al., published an EGFR-specific second generation CAR containing the CD28 intracellular motif, linked to the antihuman EGFR scFv of mAb 108.[36] This CAR construct was evaluated for its efficacy to recognize EGFR expressing cell lines in vitro, mainly lung cancers, and in prolonging survival of immunocompromised NOD SCID mice with experimental lung metastases in vivo. Although it is proposed to be applicable to treat EGFR expressing gliomas, this CAR was not experimentally evaluated in a brain tumor model. At MD Anderson Cancer Center, two EGFR-specific second generation CARs containing the CD28 intracellular motif linked to scFv antibody sequences derived from the clinically approved antihuman EGFR antibodies Cetuximab, and Nimotuzumab have been manufactured and studied preclinically.[37] Although toxicities have been observed with passive antibody treatment of Cetuximab,[64] and the corresponding EGFR-CAR[65] showed skin toxicity in a human skin-grafted mouse model, less toxicity is expected with Nimotuzumab-based CARs. In a series of recent publications, different groups looked at the impact of affinity-tuning scFvs to show differential recognition of tissues with low versus high expression of EGFR. Nimotuzumab-based EGFR-CARs showed selectivity toward tissue with high EGFR expression such as tumor, and capable of sparing low expressing tissue.[37] This selectivity is hypothesized to increase safety and clinical applicability of EGFR-CARs. Whether there is a window of expression whereby high EGFR-expressing tumors can be treated without normal EGFR expressing cell toxicity remains to be seen, and will likely need a clinical trial to determine, due to the scarcity of translatable preclinical models. All three of the EGFR-specific CARs mentioned here, as well as a novel recently developed CAR,[38] recognize the human molecule and do not cross-react with the mouse ortholog, and therefore it remains to be seen whether these CAR T cells can mount efficacy without causing unintended and overt toxicity in peripheral organs naturally expressing EGFR, including the liver, skin, lung, and gut.

EGFRvIII CARs

The type III variant of the EGF receptor, EGFRvIII, is a 145-kDa cell surface protein exclusively expressed in cancer and not normal tissue.[66,67] EGFRvIII is produced by a genetic in-frame deletion of exons 2–7, resulting in the ligation of exons 1 and 8 and the creation of a novel epitope with a glycine residue at the fusion junction.[66] This splice variant exhibits ligand-independent constitutive signaling and can promote ongenetic transformation of nearby cells through paracrine signaling.[68] EGFRvIII is expressed clinically in at approximately 30% of GBMs with variable intratumoral expression. EGFRvIII is also coexpressed with wild-type EGFR in approximately 50% of cases.[61] Despite the relatively low frequency of EGFRvIII expression, it presents a highly attractive therapeutic target given its restricted expression in malignant tissue.

EGFRvIII-CARs have been produced by a number of groups since the antigen was first characterized. First generation EGFRvIII CARs were produced first by Bullain et al.[39] and later by Ohno et al.[40] These CARs contained the CD3ζ chain linked to the mouse antihuman MR1 and 3C10 scFvs, respectively.[39,40] These early studies demonstrated the feasibility and promising efficacy of EGFRvIII CARs against malignant gliomas. Importantly, we gained insight into the potential of CAR T cell therapy by two delivery routes; Bullain et al. coadministered tumor cells and CARs and also studied direct intratumoral T cell infusion, whereas studies published by Ohno et al. described the utilization of intravenous CAR T cell delivery. These studies in sum were performed in NOD SCID mice, and although both CAR constructs were derived from different parent antibodies, they both showed antitumor function against the same human glioma cell line, corroborating the clinical applicability of this approach for GBM.

More recently, EGFRvIII-specific second and third generation CARs have been designed to include some combination of the costimulatory domains ICOS,[42] CD28, or 4-1BB.[24,25,41,65] These CARs demonstrate increased persistence in vivo in immunocompromised mice and have also proven capable of prolonging survival of mice with well-established and highly invasive orthotopic gliomas after systemic infusion.[25,41,65] As these agents have evolved for clinical use, CAR constructs have been redesigned to incorporate humanized or fully human scFvs. Human scFvs offer the potential to overcome the development of human antimouse antibodies, which could theoretically limit the efficacy of CARs due to their immunogenicity.[69] Similar approaches in melanoma patients using murine-derived antigen-specific gp100 TCRs were found to elicit humoral responses directed against murine TCR sequences, and the same could be expected against murine-based B-cell receptor sequences found in the scFv.[9,70] Additionally, using a human-based scFv would reduce the risk of inducing allergic responses in patients, as was recently observed in a clinical trial of an antimesothelin CAR, upon repeated patient infusions of RNA-electroporated murine-based scFv CARs.[71] One recently published EGFRvIII scFv, derived from human mAb clone 139, is fully human with low to nonexistent cross-reactivity to human EGFR and has recently been developed as a third generation CAR[24,41] and is currently undergoing evaluation in a Phase I clinical trial for patients with recurrent GBM at the NCI (NCT01454596), utilizing intravenous treatment following nonmyelodepleting lymphodepletion with cyclophosphamide and fludaribine and supplemented with IL-2. A second clinical trial targeting EGFRvIII in GBM utilizes a humanized version of the 3C10 scFv (now called 2173),[65] and is taking place as a dual-site study at UPENN and UCSF (NCT02209376), delivering cells intravenously to newly diagnosed patients with incomplete resections in the context of standard of care chemotherapy and radiation, or patients with recurrent disease. Two additional Phase I studies are expected to begin in 2016–2017 at Duke University, designed to evaluate the safety of the third generation EGFRvIII CARs developed at the NCI containing clone 139. One study will evaluate the impact of systemic delivery of EGFRvIII CARs for primary GBM in the context of standard of care chemotherapy and radiation. The second study will be conducted in patients with recurrent disease and will evaluate the safety of intracranially delivered EGFRvIII CARs. Table 14.1 includes all CAR-based clinical trials currently underway for patients with GBM.

The preclinical studies mentioned earlier provided proof-of-concept of the utility of adoptive T cell immunotherapy against solid tumors in the brain, demonstrating the capacity of CAR-engineered T cells to migrate to tumors within the CNS and mount efficacy in an organ previously considered void of immunosurveillance. Although elegantly designed, the use of immunocomprised mice and xenogeneic tumor systems previously limited the extent to which we could understand which host elements were involved in antitumor efficacy and which were required for translation into the clinic.[56] Studies evaluating adoptive T cell therapy using TCR transgenic T cells have highlighted the relevance of host preconditioning as a critical determinant of antitumor efficacy and clinical outcome.[72,73] Seminal studies conducted at the NCI and elsewhere have demonstrated that preconditioning hosts with lymphodepletive regimens is requisite to potentiate transferred T cell engraftment and functionality.[74] This is achieved, at least in part, by eliminating endogenous immune cells that act as cellular cytokine sinks and make key homeostatic cytokines, such as IL-7 and IL-15, widely available to enhance the survival and proliferation of infused cells.[72]

Host preconditioning was found to be required in syngeneic mouse models of CD19-specific CARs to treat B-cell cancers. However, the role of host preremained unknown due to a lack of available syngeneic CAR immunotherapy model systems. Previously, a spontaneously arising VM/Dk murine glioma cell line, sma560 expressing the mouse EGFR was transformed to express EGFRvIII mutation, and used in syngeneic immunocompetent mouse studies at Duke University.[75,76] Recently, a fully murine third generation EGFRvIII CAR incorporating mouse CD28, 4-1BB, and CD3ζ was used with the VM/Dk mice and sma560 tumors to evaluate the ability to treat gliomas in a more physiologically relevant model. These studies have shown that CAR T cell therapy in this context requires lymphodepletive preconditioning to mediate complete tumor eradication and elicit long-term cures in mice with well-established GBM tumors implanted in the brain.[25] Moreover, these studies, which were performed using a CAR derived from the 139 scFv, demonstrated that mice previously cured of EGFRvIII-positive tumors were resistant to a secondary tumor challenge with EGFRvIII-negative tumors.[25] This suggests that

CAR-induced epitope spreading generated an endogenous host immune response against new tumor antigens.[77] The ability to induce a localized, tumor-specific, de novo host response via tumor-specific CAR could eliminate the concern of antigen loss variants and may provide a personalized vaccination effect in situ.

These tools have provided a means of intelligently deciphering the mechanisms and factors involved in achieving CAR T cell cures against primary solid brain tumors. As we continue to gain knowledge surrounding the interplay between engineered T cells, the tumor microenvironment, mechanisms of immune suppression, and the induction of epitope spreading, immunocompetent syngeneic systems such as the one described here will continue to be invaluable in helping us translate this therapy into a clinically effective strategy.

IL-13Rα2 CARs

IL-13Rα2 is a 42-kDa high affinity cell-surface receptor for the ligand IL-13, and is exclusively expressed in certain epithelial, lymphoid, and brain tumors, including GBM and medulloblastoma.[78,79] Although IL-13Rα2 expression has been reported tissues other than tumors, the expression levels in tumor cells are order of magnitudes higher, particularly in GBM were it has been exhaustively studied.[80,81] Quantitative surface expression has shown that up to 30,000 copies of the IL-13Rα2 can be expressed per tumor cell.[82] In addition to its high level of expression within each cell, IL-13Rα2 is expressed in ~58% of grade IV gliomas and up to ~80% in GBM patients.[83,84] Like EGFRvIII, IL-13Rα2 is heterogeneously expressed within tumor masses; tumors are typically segregated by areas of dense, low, and no expression.[83,84] However, its relatively confined expression within tumors and broad expression pattern across the GBM patient population make IL-13Rα2 an attractive target for CAR therapy.

Early studies initiated in the 2000s by Michael Jensen's group developed an IL-13Rα2 targeting CAR, which is known today as the zetakine CAR.[43] This initial study employed a novel CAR configuration, in which instead of an scFv, recombinant (r)IL-13 was fused to the intracellular zeta chain. This configuration resulted in a CAR with high affinity for IL-13Rα2, which was shown to confer cytotoxicity against gliomas in vitro, and mediate long-term survival when delivered intratumorally in NOD SCID mice bearing intracranial glioma.[43] These initial studies were performed utilizing a mutant rIL-13 containing a single amino acid substitution (IL13.E13Y) to confer selective specificity to the IL-13Rα2 and prevent recognition by IL-13Rα1 and IL-4Rα, both of which can also bind IL13 and are widely expressed in leukocytes and lymphoid organs.[78] Follow-up studies have introduced dual mutations to the IL-13 (IL13.E13K.R109K) to further enhance specificity,[45] and have also developed second generation CAR configurations containing the CD28 domain, which have been shown to enhance the persistence and antitumor efficacy of the zetakine CARs.[85,86] Based on these studies, the zetakine CAR has entered a Phase I clinical trial for patients with GBM, and is listed in Table 14.1.[87,88] A recent preclinical study by Stephen Gottschalk's group out of Baylor College of Medicine utilized an scFv against the IL-13Rα2 in their CAR design, along with different combinations of CD3z, 4-1BB, CD28, and OX40 signaling, to generate first, second, and third generation CARs with recognition of IL13Rα2, however, cross-recognition against IL13Rα1 was not evaluated, so specificity of the scFv over the rIL13 zetakine CARs was not discussed.[89] These CARs demonstrated in vitro tumor cytotoxicity and efficacy in vivo, the second and third generation CARs comparably performing better than the first Gen CAR.

HER2 CARs

Receptor tyrosine-protein kinase ErbB-2, also commonly known as human epidermal growth factor receptor 2 (HER2/Neu) is a 185-kDa orphan receptor member of the EGFR family, which has constitutive activity.[90] In adulthood, HER2 is naturally expressed in many tissues including the epithelium of the gastro-intestinal, respiratory, reproductive, and urinary tract, as well as placenta, breast, and skin.[91] However, HER2 expression in adult brain is absent in both neuronal and glial tissue and limited to embryonic development.[92] This proto-oncogene is, however, turned on and expressed at the cell surface at high levels in primary malignant high grade brain tumors such as GBM, medulloblastoma, and meningiomas. HER2 expression in GBM and medulloblastoma patients has been documented to range between 20% and 80%,[93] and 40%,[94] respectively. HER2 can be coexpressed at times with EGFR, but in contrast to EGFR, HER2 does not exhibit gene amplification.[95] Intratumoral expression of HER2 is variable and can have a modest degree of heterogeneity, however, given that HER2 expression is completely absent in the CNS and its vast expression in high grade tumors HER2 has become an attractive target for CAR therapy for the treatment of malignant brain tumors.[48,96,97]

As early as the 1990s, first generation CARs were developed by fusing a mouse scFv derived from the N29[46] and FRP5[47] antihuman HER2 hybridoma with the CD3ζ chain. Although the studies performed with the N29-derived HER2 CAR were restricted to in vitro lysis of HER2-expressing tumors, the studies conducted utilizing the FRP5 scFv demonstrated that HER2 CARs could mediate in vivo tumor regressions.[46] However, these studies were limited to subcutaneous tumor models of NIH3T3 cell lines stably expressing the human HER2 molecule and treated systemically with HER2 CARs in BALB/c nude mice.[46] Although this study was the first to demonstrate the principle of HER2 CARs, the results obtained from in vivo mouse experiments were modest at best, mostly due to limited human T cell engraftment in BALB/c nude mice.[46]

The availability of NOD SCID mice, in which human T cells can engraft relatively longer, has allowed evaluation of first and second generation HER2 CARs. Preclinical work from the NCI utilized a CAR based on the same scFv as in the clinically approved Herceptin mAb, which has been administered safely to almost half a million women with breast cancer. Zhao et al. found in a NOD SCID xenogenic mouse model HER2 CARs were found to treat breast cancer without causing animal toxicity.[98] A follow-up clinical trial using this same CAR resulted in rapid multiple organ failure and death of the first colon cancer patient treated.[99] This fatality was attributed to previously unidentified HER2 expression on normal lung tissue.

Other studies using CD28-containing FRP5-based second generation HER2 CARs against established GBM[48] and medulloblastoma[49]; tumors which naturally express the HER2 molecule. These studies conclusively demonstrated that HER2 CARs were capable of recognizing tumors with endogenous HER2 expression levels, and mediating long-term cures.[48,49] In light of the broad expression of HER2 in peripheral tissues outside the CNS, the safety of FRP5 based HER2 CARs has been evaluated in HER2 transgenic mice, which have HER2 expression in the mammary tissue and within the brain in the cerebellum.[100] These studies found no adverse effects despite high levels of HER2 expression.[100] With these results, FRP5-based HER2 CARs are currently being evaluated in two clinical dose-escalation Phase I trials to determine the safety of this approach in $HER2^+$ GBM patients. In addition to the intravenous delivery of FRP5-based HER2 CARs, intratumoral and intracavitary routes are also being explored.

EphA2 CAR

The erythropoietin-producing human hepatocellular carcinoma type-A receptor 2 (EphA2) is a 130-kDa cell surface tyrosine kinase receptor whose ligands belongs to the Ephryn family, including Ephryn A1, A2, and A5. EphA2 expression in adults is restricted to low levels in proliferating epithelium and areas of the brain, bone marrow, lung, thymus, small intestine, urinary bladder, colon, liver, kidney, and spleen.[101,102] Although its expression in normal tissues is low, EphA2 is overexpressed in 60–90% of anaplastic astrocytomas and primary and recurrent GBM.[103,104] EphA2 expression has been associated with poor outcome, due largely to increased invasiveness.[105] Tumor expression of EphA2 is relatively homogenous in GBM patients; up to 75% of cells within GBM express EphA2 with high intensity.[103,104] Therefore, the prevalence of EphA2 in high grade astrocytomas, and its high level and homogeneous expression within tumors has made it an attractive target for CAR therapy.

A second generation CD28 costimulated EphA2 CAR containing the scFv from the mAb 4H5 (a humanized version of the EphA2 mAb EA2) has been developed and evaluated preclinically in immunodeficient SCID mouse models.[50] This study demonstrated in vitro recognition and cytotoxicity of glioma cell lines expressing EphA2. Adoptive transfer of EphA2 CARs into mice with established intracranial tumors resulted in prolonged survival when compared with mice receiving control CARs.[50] This is the only published report to date utilizing EphA2 CARs but more studies are likely to follow.

Glioma Stem Cell Targeted Chimeric Antigen Receptors

Prominin-1, also known as CD133, is a cell surface receptor for cholesterol,[106] with high expression in putative GSCs.[52] CD133 is a five-membrane-spanning glycoprotein with a calculated molecular weight of 97 kDa, but due to splice variants and posttranslational modifications, its weight can range between 80 and 120 kDa.[107] CD133 is expressed widely in epithelial tissue of the gastrointestinal track as well as in liver, gallbladder, pancreas, and testis. However, CD133 expression in the adult normal brain is limited to low levels in neuronal and glial stem cells,[108] but is highly expressed in GSCs[52] and has increased expression in recurrent tumors, making CD133 an attractive target for CAR therapy. A third generation CD133-specific CAR containing the scFv of an antihuman CD133 has been developed and tested in the preclinical setting.[51] In vitro cytotoxicity assay confirmed specificity to $CD133^+$ GSCs and cytotoxicity upon antigen recognition. However, in vivo evaluation of CD133 CARs showed only modest enhancement of survival despite three consecutive intratumoral injections in nude mice with established intracranial $CD133^+$ human xenografts. Immunological assessment of tumor-infiltrating CD133 CAR T cells demonstrated expression of the terminal differentiation marker CD57, although CAR T cell exhaustion was not observed upon a functional assessment.[51] Recently, studies have suggested that the CD133 CAR could lose recognition of glycosylated epitopes in $CD133^+$ progeny cells from GSCs, owed in part to the AC133 mAb that this CAR was constructed from.[107] More studies are needed to properly evaluate this CAR in vivo, both for efficacy and safety, as this CAR may be able to recognize CD133 expressed in normal tissues.

The neural cell adhesion molecule L1 (L1-CAM), or CD171, is a 140-kDa cell surface adhesion molecule, but can be heavily glycosylated to weigh between 200 and 220 kDa.[109] L1-CAM is involved in brain development through the binding of axonin on developing neurons, as well as extracellular matrix proteins such as laminin and neurocan, and therefore regulating neuron to neuron adhesion, survival, migration, and axon guidance and neurite extension.[110] L1-CAM is also highly expressed on kidneys, soft tissue, and has an aberrant expression in solid cancers such as neuroblastoma,[111] ovarian,[112] melanoma,[113,114] pancreatic,[115] and colon cancer where it has been associated with invasiveness and poor prognosis.[116] In gliomas,[117] L1-CAM has been detected at levels higher than neuronal tissue in GSCs, where it plays a role in maintenance of tumor growth and survival of these putative GSCs.[118,119] An L1-CAM CAR has recently been developed for the treatment of primary solid tumors,[53,54] although it has been approached with caution, given the safety concerns over its vast expression in neuronal tissue. A first generation CAR containing the scFv of the human specific CE7 mAb fused to the CD3ζ chain has been shown to be specific for L1-CAM, and interestingly, recognized tumor tissue but failed to recognize normal healthy tissue expressing L1-CAM.[54] It is believed that this tumor-specific recognition may be explained by a previously unknown alternative splice variant with expression restricted to tumors only.[54] However, more studies will need to be conducted to evaluate the recognition of GBM GSC cell lines and potential efficacy. Although L1-CAM CARs implementation in Phase I clinical studies for brain tumors awaits, additional studies are evaluating L1-CAM CAR function against tumors from neuroectodermal origin, such as neuroblastoma.[5]

THE QUESTION OF DELIVERY

The brain has long been considered an immune privileged organ largely due to the presence of the blood–brain barrier (BBB),[120,121] which naturally excludes circulating immune cells and soluble molecules larger than 400 Da[122,123] present in systemic circulation to protect the brain from being exposed to inflammation and potentially fatal autoimmunity and toxins.[124] However, although the original concept of immune privilege precluded the capacity of circulating antigen-specific T cells to survey the brain for the presence of pathogens or malignant cell conversion, recent observations and further understanding of brain anatomy and the immune system have proven this view incorrect.[125–127] Recent studies demonstrate that antigens expressed by glial cells within the brain parenchyma can travel through defined subarachnoid routes and reach cervical lymph nodes via the nasal mucosal path.[128,129] Upon reaching these lymph nodes, these can be picked up and presented by resident dendritic cells and initiate the process of adaptive T cell immunity.[130] Once T cells are activated, they can traffic through the brain despite the presence of an intact BBB, and can be retained for long periods of time when engaging their cognate antigen.[127,131–133] Therefore, it would be expected that CAR T cells, which are activated as part of the procedure to engineer T cells with the CAR, could migrate effectively to brain tumors. Recent studies in both immunocompromised and immunocompetent syngeneic animal models have shown CAR T cells can migrate into tumor deposits infiltrating the brain parenchyma and eradicate experimental brain tumors when infused systemically.[25,40,41,65] There is also clinical evidence that TCR gene-engineered

T cells can eliminate metastatic brain tumors in patients receiving adoptive cell transfer immunotherapy.[9,134] Therefore, the current view of the brain as an immune privileged organ has changed significantly in recent years. It is currently accepted that activated T cells are not limited in their capacity to reach brain tumor deposits, including those areas hiding behind an intact BBB. Although the studies mentioned earlier have provided support for the systemic administration of CAR T cells aimed at targeting brain tumors, many studies today utilize intracavitary or intratumoral CAR T cell injections. However, the reason for these studies is generally more so due to safety concerns than brain tumor trafficking limitations.

POTENTIAL CHALLENGES

The decision (or limitation) in choosing the appropriate animal model for CAR evaluation is dependent on whether the CAR in question is derived from an antibody with reactivity against the human or murine homolog of the target antigen. One strategic method to overcome this potential limitation has been to develop CARs derived from antibodies that have cross-reactivity for both the human and mouse target molecules, which makes it possible to evaluate the antigen-reactive portion of CAR in both xenogeneic and syngeneic mouse systems. Indeed, the same single-chain variable fragment can be linked to either mouse or human T cell transmembrane and signaling moieties in this context. This cross-reactivity has been exploited in the context of EGFRvIII CARs, which recognize the same EGFRvIII tumor-specific mutation in both mice and humans.[25] The added value of studying CAR T cells in syngeneic systems has been instrumental in allowing us to evaluate the short-term and long-term safety and toxicity of these CAR T cells without the concern of GvHD and potential of cross-reactivity against other antigens expressed on normal tissues. Furthermore, immunocompetent animal models have provided critical insights into understanding novel mechanisms by which endogenous immune mechanisms can contribute to tumor rejection and how these mechanisms can be induced by CAR T cell therapy,[77] observations that would have been impossible to make in immunocompromised mice.

Therefore, as newer CARs are developed, an emphasis should be placed on deriving these constructs from antibodies with cross-reactivity against antigens expressed in animals. This will be critical in understanding what barriers remain to achieving durable and long-term cures and how these barriers can be manipulated for therapeutic gain. When cross-reactive clones are unavailable, the translation of CARs into the clinic should be executed with maximal caution, as the potential for antibody cross-reactivity with human molecules on normal tissues still exists and can only be evaluated in patients.

The tumor-specific antigen, EGFRvIII, in many ways serves as an ideal model for immunotherapy antigens, as its expression is restricted to malignant cells and no other site of expression has been detected. The outcomes of the two current clinical trials treating patients with GBM by targeting this antigen (Table 14.1) have not yet been published. The EGFR, Her2, EphA2, and (to a lesser extent) IL-13Rα2 targets are minimally expressed in peripheral sites, but could theoretically result in adverse effects or even death, as was the unfortunate outcome for one patient with colorectal cancer, who received a high dose of Her2 CAR T cells (1×10^{11}), with maximal prior lymphodepletion via chemotherapy and 600 rads total body irradiation.[99] This patient presented with respiratory distress within minutes following CAR T cell infusion, and expired 5 days later. This severe adverse event is believed to be due to the low level of antigen expression of Her2 in lung epithelium, which caused unintended CAR T cell cytotoxicity and irreversible damage to the lungs and eventual death.[99] Therefore, to overcome this terrible toxicity, direct intracavitary, and intratumoral approaches are currently being proposed and evaluated for CARs specific against EGFR,[36–38] Her2,[46–49] EphA2,[50] and IL-13Rα2,[43–45] with the expectation that the isolated nature of the brain with regards to circulation would keep these autoreactive T cells in the brain.

Systemic CAR-T cell infusion is a minimally invasive and viable route for CAR T cell delivery. However, this is limited by safety concerns over the gene expression pattern of targeted antigen. Although intracavitary or intratumoral CAR T cell infusion initially presented itself as an alternative to restrict systemic toxicity, this theory is starting to be questioned. The isolated nature of the brain from circulation was initially hypothesized to be due to the lack of discrete CNS lymphatics, which would preclude brain-infiltrating cells, such as T cells, or solutes present in the cerebrospinal fluid to drain into lymph nodes and reach circulation.[135] Importantly, recent studies have elegantly shown that the brain contains highly organized and structurally defined lymphatics in which T cells and solutes can travel to reach the cervical lymph nodes.[129,136,137] It remains to be seen whether these cells that reach the lymph nodes are eventually deposited back into circulation, and what relevance this will have for CAR T cells delivered loco-regionally in the brain.

As immunotherapies increase in potency, even greater attention will have to be placed on the ability to target tumor antigens that are not shared with normal, healthy tissues. In the brain, for instance, targeting of self-tissues has been shown to incite uncontrolled CNS autoimmunity, which has been well-described in models of experimental autoimmune encephalomyelitis (EAE). Lethal EAE has been well documented in both human and nonhuman primates, as well as in mouse models in the context of immunotherapy for brain tumors, which stems largely from the fact that gliomas express many of the same antigens as the surrounding local tissues. The risk of incurring this adverse effect can be reduced by carefully selecting tumor-specific antigens that are completely absent from all normal tissues. Examples of this in the setting of solid brain tumors include the tumor-specific mutation EGFRvIII[67] and mutations in isocitrate dehydrogenase 1 (IDH1).[138] In addition, conserved mutations in H3.3[139] and BRAF[140,141] have also been described as potential targets for pediatric GBM and pilocytic astrocytomas, respectively, the latter have also been shown to express nonrecurrent, sporadic variants of ATRX/DAXX,[139,142] TP53,[139] NF1[143] and PDGFRα.[143] One drawback to many of these mutations is that they do not display cell surface epitopes, which are also required for CAR targeting. Viral antigens associated with the human cytomegalovirus have also been demonstrated to be expressed in high percentages of GBMs yet absent from surrounding brain[144,145] and may represent an additional nonself CAR target. Almost certainly as innovations in high throughput genetic screening advance, additional antigens will be discovered that may serve as targets immunotherapeutic approaches for brain and other malignancies.

FUTURE PERSPECTIVES

CAR design has evolved dramatically since they were first conceptualized to include multiple signaling domains that favorably influence various aspects of T cell behavior[146,147] (Fig. 14.1). First-generation CARs generally included an antigen-binding extracellular domain linked to an intracellular CD3ζ chain, which is required to initiate T cell cytotoxicity akin to endogenous TCR mechanisms.[148] These CARs were severely limited in vivo as T cells failed to survive in the absence of costimulation. Second- and third-generation CARs have built on this early work to provide stimulation through the addition of CD28,[149] OX-40,[150] ICOS,[42] or 4-1BB[14] within the CAR construct. Incorporation of one- (second generation) or two- (third-generation) signaling domains in any combination has been shown to confer improved T cell proliferation, survival, cytokine secretion, and tumor lysis.[151,152] Second-generation and third-generation CARs against multiple brain tumor targets have been generated and are the focus of the remainder of this review.

Previous studies have identified several MHC-restricted antigens capable of eliciting T cell responses,[104,153] but the effectiveness of this approach can be hampered by immune-evasion through the loss of cell surface MHC expression on brain tumors.[154] Fortuitously, collaborative studies spanning more than two decades of research and development have recently culminated into the production of novel antibody-based technology that can exquisitely target surface antigens without the need for MHC molecules.[155] Improved knowledge over the architecture and precision of humoral immune responses have coincided with the development of molecular techniques by which we can produce mAbs or artificial derivatives within closed laboratory systems against a virtually infinite repertoire of target antigens.[146,147] Importantly, the ability to target antigens independent of MHC restriction has opened the possibility of targeting several surface antigens that are widely expression and overrepresented in brain tumors.[61,81,93,103,118,119,156–158]

References

1. Gross G, Waks T, Eshhar Z. Expression of immunoglobulin-T-cell receptor chimeric molecules as functional receptors with antibody-type specificity. *Proc Natl Acad Sci USA*. 1989;86(24):10024–10028.
2. Eshhar Z, Bach N, Fitzer-Attas CJ, et al. The T-body approach: potential for cancer immunotherapy. *Springer Semin Immunopathol*. 1996;18(2):199–209.
3. Kershaw MH, Westwood JA, Parker LL, et al. A phase I study on adoptive immunotherapy using gene-modified T cells for ovarian cancer. *Clin Cancer Res*. 2006;12(20 Pt 1):6106–6115.
4. Lamers CH, Sleijfer S, Vulto AG, et al. Treatment of metastatic renal cell carcinoma with autologous T-lymphocytes genetically retargeted against carbonic anhydrase IX: first clinical experience. *J Clin Oncol*. 2006;24(13):e20–22.
5. Park JR, Digiusto DL, Slovak M, et al. Adoptive transfer of chimeric antigen receptor re-directed cytolytic T lymphocyte clones in patients with neuroblastoma. *Mol Ther*. 2007;15(4):825–833.
6. Louis CU, Savoldo B, Dotti G, et al. Antitumor activity and long-term fate of chimeric antigen receptor-positive T cells in patients with neuroblastoma. *Blood*. 2011;118(23):6050–6056.
7. Johnson LA, Heemskerk B, Powell Jr DJ, et al. Gene transfer of tumor-reactive TCR confers both high avidity and tumor reactivity to nonreactive peripheral blood mononuclear cells and tumor-infiltrating lymphocytes. *J Immunol*. 2006;177(9):6548–6559.
8. Morgan RA, Dudley ME, Wunderlich JR, et al. Cancer regression in patients after transfer of genetically engineered lymphocytes. *Science*. 2006;314(5796):126–129.
9. Johnson LA, Morgan RA, Dudley ME, et al. Gene therapy with human and mouse T-cell receptors mediates cancer regression and targets normal tissues expressing cognate antigen. *Blood*. 2009;114(3):535–546.
10. Krause A, Guo HF, Latouche JB, Tan C, Cheung NK, Sadelain M. Antigen-dependent CD28 signaling selectively enhances survival and proliferation in genetically modified activated human primary T lymphocytes. *J Exp Med*. 1998;188(4):619–626.
11. Finney HM, Lawson AD, Bebbington CR, Weir AN. Chimeric receptors providing both primary and costimulatory signaling in T cells from a single gene product. *J Immunol*. 1998;161(6):2791–2797.
12. Finney HM, Akbar AN, Lawson AD. Activation of resting human primary T cells with chimeric receptors: costimulation from CD28, inducible costimulator, CD134, and CD137 in series with signals from the TCR zeta chain. *J Immunol*. 2004;172(1):104–113.
13. Imai C, Mihara K, Andreansky M, et al. Chimeric receptors with 4-1BB signaling capacity provoke potent cytotoxicity against acute lymphoblastic leukemia. *Leukemia*. 2004;18(4):676–684.
14. Milone MC, Fish JD, Carpenito C, et al. Chimeric receptors containing CD137 signal transduction domains mediate enhanced survival of T cells and increased antileukemic efficacy in vivo. *Mol Ther*. 2009;17(8):1453–1464.

15. Guedan S, Chen X, Madar A, et al. ICOS-based chimeric antigen receptors program bipolar TH17/TH1 cells. *Blood*. 2014;124(7):1070–1080.
16. Cheadle EJ, Rothwell DG, Bridgeman JS, Sheard VE, Hawkins RE, Gilham DE. Ligation of the CD2 co-stimulatory receptor enhances IL-2 production from first-generation chimeric antigen receptor T cells. *Gene Ther*. 2012;19(11):1114–1120.
17. Song DG, Ye Q, Poussin M, Harms GM, Figini M, Powell Jr DJ. CD27 costimulation augments the survival and antitumor activity of redirected human T cells in vivo. *Blood*. 2012;119(3):696–706.
18. Wang E, Wang LC, Tsai CY, et al. Generation of potent T cell immunotherapy for cancer using DAP12-based, multichain, chimeric immunoreceptors. *Cancer Immunol Res*. 2015;3(7):815–826.
19. Curran KJ, Seinstra BA, Nikhamin Y, et al. Enhancing antitumor efficacy of chimeric antigen receptor T cells through constitutive CD40L expression. *Mol Ther*. 2015;23(4):769–778.
20. Stephan MT, Ponomarev V, Brentjens RJ, et al. T cell-encoded CD80 and 4-1BBL induce auto- and transcostimulation, resulting in potent tumor rejection. *Nat Med*. 2007;13(12):1440–1449.
21. Chinnasamy D, Yu Z, Kerkar SP, et al. Local delivery of interleukin-12 using T cells targeting VEGF receptor-2 eradicates multiple vascularized tumors in mice. *Clin Cancer Res*. 2012;18(6):1672–1683.
22. Tran E, Chinnasamy D, Yu Z, et al. Immune targeting of fibroblast activation protein triggers recognition of multipotent bone marrow stromal cells and cachexia. *J Exp Med*. 2013;210(6):1125–1135.
23. Kochenderfer JN. Genetic engineering of T cells in leukemia and lymphoma. *Clin Adv Hematol Oncol*. 2014;12(3):190–192.
24. Morgan RA, Johnson LA, Davis JL, et al. Recognition of glioma stem cells by genetically modified T cells targeting EGFRvIII and development of adoptive cell therapy for glioma. *Hum Gene Ther*. 2012;23(10):1043–1053.
25. Sampson JH, Choi BD, Sanchez-Perez L, et al. EGFRvIII mCAR-modified T cell therapy cures mice with established intracerebral glioma and generates host immunity against tumor-antigen loss. *Clin Cancer Res*. 2014;20(4):972–984.
26. Wang LC, Lo A, Scholler J, et al. Targeting fibroblast activation protein in tumor stroma with chimeric antigen receptor T cells can inhibit tumor growth and augment host immunity without severe toxicity. *Cancer Immunol Res*. 2014;2(2):154–166.
27. Kochenderfer JN, Yu Z, Frasheri D, Restifo NP, Rosenberg SA. Adoptive transfer of syngeneic T cells transduced with a chimeric antigen receptor that recognizes murine CD19 can eradicate lymphoma and normal B cells. *Blood*. 2010;116(19):3875–3886.
28. Brentjens RJ, Riviere I, Park JH, et al. Safety and persistence of adoptively transferred autologous CD19-targeted T cells in patients with relapsed or chemotherapy refractory B-cell leukemias. *Blood*. 2011;118(18):4817–4828.
29. Brentjens RJ, Davila ML, Riviere I, et al. CD19-targeted T cells rapidly induce molecular remissions in adults with chemotherapy-refractory acute lymphoblastic leukemia. *Sci Transl Med*. 2013;5(177):177ra138.
30. Porter DL, Levine BL, Kalos M, Bagg A, June CH. Chimeric antigen receptor-modified T cells in chronic lymphoid leukemia. *N Engl J Med*. 2011;365(8):725–733.
31. Davila ML, Riviere I, Wang X, et al. Efficacy and toxicity management of 19-28z CAR T cell therapy in B cell acute lymphoblastic leukemia. *Sci Transl Med*. 2014;6(224):224ra225.
32. Garfall AL, Maus MV, Hwang WT, et al. chimeric antigen receptor T cells against CD19 for multiple myeloma. *N Engl J Med*. 2015;373(11):1040–1047.
33. Maude SL, Frey N, Shaw PA, et al. Chimeric antigen receptor T cells for sustained remissions in leukemia. *N Engl J Med*. 2014;371(16):1507–1517.
34. Mitchell DA, Fecci PE, Sampson JH. Adoptive immunotherapy for malignant glioma. *Cancer J*. 2003;9(3):157–166.

35. Reardon DA, Wucherpfennig KW, Freeman G, et al. An update on vaccine therapy and other immunotherapeutic approaches for glioblastoma. *Expert Rev Vaccines*. 2013;12(6):597–615.
36. Zhou X, Li J, Wang Z, et al. Cellular immunotherapy for carcinoma using genetically modified EGFR-specific T lymphocytes. *Neoplasia*. 2013;15(5):544–553.
37. Caruso HG, Hurton LV, Najjar A, et al. Tuning sensitivity of CAR to EGFR density limits recognition of normal tissue while maintaining potent antitumor activity. *Cancer Res*. 2015;75(17):3505–3518.
38. Liu X, Jiang S, Fang C, et al. Affinity-tuned ErbB2 or EGFR chimeric antigen receptor T cells exhibit an increased therapeutic index against tumors in mice. *Cancer Res*. 2015;75(17):3596–3607.
39. Bullain SS, Sahin A, Szentirmai O, et al. Genetically engineered T cells to target EGFRvIII expressing glioblastoma. *J Neurooncol*. 2009;94(3):373–382.
40. Ohno M, Natsume A, Ichiro Iwami K, et al. Retrovirally engineered T cell-based immunotherapy targeting type III variant epidermal growth factor receptor, a glioma-associated antigen. *Cancer Sci*. 2010;101(12):2518–2524.
41. Miao H, Choi BD, Suryadevara CM, et al. EGFRvIII-specific chimeric antigen receptor T cells migrate to and kill tumor deposits infiltrating the brain parenchyma in an invasive xenograft model of glioblastoma. *PLoS One*. 2014;9(4):e94281.
42. Shen CJ, Yang YX, Han EQ, et al. Chimeric antigen receptor containing ICOS signaling domain mediates specific and efficient antitumor effect of T cells against EGFRvIII expressing glioma. *J Hematol Oncol*. 2013;6:33.
43. Kahlon KS, Brown C, Cooper LJ, Raubitschek A, Forman SJ, Jensen MC. Specific recognition and killing of glioblastoma multiforme by interleukin 13-zetakine redirected cytolytic T cells. *Cancer Res*. 2004;64(24):9160–9166.
44. Brown CE, Starr R, Aguilar B, et al. Stem-like tumor-initiating cells isolated from IL13Rα2 expressing gliomas are targeted and killed by IL13-zetakine-redirected T cells. *Clin Cancer Res*. 2012;18(8):2199–2209.
45. Kong S, Sengupta S, Tyler B, et al. Suppression of human glioma xenografts with second-generation IL13R-specific chimeric antigen receptor-modified T cells. *Clin Cancer Res*. 2012;18(21):5949–5960.
46. Stancovski I, Schindler DG, Waks T, Yarden Y, Sela M, Eshhar Z. Targeting of T lymphocytes to Neu/HER2-expressing cells using chimeric single chain Fv receptors. *J Immunol*. 1993;151(11):6577–6582.
47. Altenschmidt U, Klundt E, Groner B. Adoptive transfer of in vitro-targeted, activated T lymphocytes results in total tumor regression. *J Immunol*. 1997;159(11):5509–5515.
48. Ahmed N, Salsman VS, Kew Y, et al. HER2-specific T cells target primary glioblastoma stem cells and induce regression of autologous experimental tumors. *Clin Cancer Res*. 2010;16(2):474–485.
49. Ahmed N, Ratnayake M, Savoldo B, et al. Regression of experimental medulloblastoma following transfer of HER2-specific T cells. *Cancer Res*. 2007;67(12):5957–5964.
50. Chow KK, Naik S, Kakarla S, et al. T cells redirected to EphA2 for the immunotherapy of glioblastoma. *Mol Ther*. 2013;21(3):629–637.
51. Zhu X, Prasad S, Gaedicke S, Hettich M, Firat E, Niedermann G. Patient-derived glioblastoma stem cells are killed by CD133-specific CAR T cells but induce the T cell aging marker CD57. *Oncotarget*. 2015;6(1):171–184.
52. Bao S, Wu Q, McLendon RE, et al. Glioma stem cells promote radioresistance by preferential activation of the DNA damage response. *Nature*. 2006;444(7120):756–760.
53. Gonzalez S, Naranjo A, Serrano LM, Chang WC, Wright CL, Jensen MC. Genetic engineering of cytolytic T lymphocytes for adoptive T cell therapy of neuroblastoma. *J Gene Med*. 2004;6(6):704–711.
54. Hong H, Stastny M, Brown C, et al. Diverse solid tumors expressing a restricted epitope of L1-CAM can be targeted by chimeric antigen receptor redirected T lymphocytes. *J Immunother*. 2014;37(2):93–104.

55. Greenwood JD. Xenogeneic PBL-scid mice: their potential and current limitations. *Lab Anim Sci*. 1993;43(2):151–155.
56. Alcantar-Orozco EM, Gornall H, Baldan V, Hawkins RE, Gilham DE. Potential limitations of the NSG humanized mouse as a model system to optimize engineered human T-cell therapy for cancer. *Hum Gene Ther Methods*. 2013;24(5):310–320.
57. Roskoski Jr R. The ErbB/HER family of protein-tyrosine kinases and cancer. *Pharmacol Res*. 2014;79:34–74.
58. Galvez-Contreras AY, Quinones-Hinojosa A, Gonzalez-Perez O. The role of EGFR and ErbB family related proteins in the oligodendrocyte specification in germinal niches of the adult mammalian brain. *FronT cell Neurosci*. 2013;7:258.
59. Aguirre A, Rubio ME, Gallo V. Notch and EGFR pathway interaction regulates neural stem cell number and self-renewal. *Nature*. 2010;467(7313):323–327.
60. Srividya MR, Thota B, Arivazhagan A, et al. Age-dependent prognostic effects of EGFR/p53 alterations in glioblastoma: study on a prospective cohort of 140 uniformly treated adult patients. *J Clin Pathol*. 2010;63(8):687–691.
61. Wikstrand CJ, McLendon RE, Friedman AH, Bigner DD. Cell surface localization and density of the tumor-associated variant of the epidermal growth factor receptor, EGFRvIII. *Cancer Res*. 1997;57(18):4130–4140.
62. Chandramohan V, Bao X, Keir ST, et al. Construction of an immunotoxin, D2C7-(scdsFv)-PE38KDEL, targeting EGFRwt and EGFRvIII for brain tumor therapy. *Clin Cancer Res*. 2013;19(17):4717–4727.
63. Ekstrand AJ, James CD, Cavenee WK, Seliger B, Pettersson RF, Collins VP. Genes for epidermal growth factor receptor, transforming growth factor alpha, and epidermal growth factor and their expression in human gliomas in vivo. *Cancer Res*. 1991;51(8):2164–2172.
64. Harding J, Burtness B. Cetuximab: an epidermal growth factor receptor chemeric human-murine monoclonal antibody. *Drugs Today (Barcelona)*. 2005;41(2):107–127.
65. Johnson LA, Scholler J, Ohkuri T, et al. Rational development and characterization of humanized anti-EGFR variant III chimeric antigen receptor T cells for glioblastoma. *Sci Transl Med*. 2015;7(275):275ra222.
66. Sugawa N, Ekstrand AJ, James CD, Collins VP. Identical splicing of aberrant epidermal growth factor receptor transcripts from amplified rearranged genes in human glioblastomas. *Proc Natl Acad Sci USA*. 1990;87(21):8602–8606.
67. Choi BD, Archer GE, Mitchell DA, et al. EGFRvIII-targeted vaccination therapy of malignant glioma. *Brain Pathol*. 2009;19(4):713–723.
68. Batra SK, Castelino-Prabhu S, Wikstrand CJ, et al. Epidermal growth factor ligand-independent, unregulated, cell-transforming potential of a naturally occurring human mutant EGFRvIII gene. *Cell Growth Differ*. 1995;6(10):1251–1259.
69. Lamers CH, Willemsen R, van Elzakker P, et al. Immune responses to transgene and retroviral vector in patients treated with ex vivo-engineered T cells. *Blood*. 2011;117(1):72–82.
70. Davis JL, Theoret MR, Zheng Z, Lamers CH, Rosenberg SA, Morgan RA. Development of human anti-murine T cell receptor antibodies in both responding and nonresponding patients enrolled in TCR gene therapy trials. *Clin Cancer Res*. 2010;16(23):5852–5861.
71. Maus MV, Haas AR, Beatty GL, et al. T cells expressing chimeric antigen receptors can cause anaphylaxis in humans. *Cancer Immunol Res*. 2013;1(1):26–31.
72. Gattinoni L, Finkelstein SE, Klebanoff CA, et al. Removal of homeostatic cytokine sinks by lymphodepletion enhances the efficacy of adoptively transferred tumor-specific $CD8^+$ T cells. *J Exp Med*. 2005;202(7):907–912.
73. Klebanoff CA, Khong HT, Antony PA, Palmer DC, Restifo NP. Sinks, suppressors and antigen presenters: how lymphodepletion enhances T cell-mediated tumor immunotherapy. *Trends Immunol*. 2005;26(2):111–117.

74. Dudley ME, Wunderlich JR, Robbins PF, et al. Cancer regression and autoimmunity in patients after clonal repopulation with antitumor lymphocytes. *Science.* 2002;298(5594):850–854.
75. Heimberger AB, Crotty LE, Archer GE, et al. Bone marrow-derived dendritic cells pulsed with tumor homogenate induce immunity against syngeneic intracerebral glioma. *J Neuroimmunol.* 2000;103(1):16–25.
76. Serano RD, Pegram CN, Bigner DD. Tumorigenic cell culture lines from a spontaneous VM/Dk murine astrocytoma (SMA). *Acta Neuropathol.* 1980;51(1):53–64.
77. Johnson LA, Sanchez-Perez L, Suryadevara CM, Sampson JH. Chimeric antigen receptor engineered T-cells can eliminate brain tumors and initiate long-term protection against recurrence. *Oncoimmunology.* 2014;3(7):e944059.
78. Thaci B, Brown CE, Binello E, Werbaneth K, Sampath P, Sengupta S. Significance of interleukin-13 receptor alpha 2-targeted glioblastoma therapy. *Neuro Oncol.* 2014;16(10):1304–1312.
79. Sengupta S, Thaci B, Crawford AC, Sampath P. Interleukin-13 receptor alpha 2-targeted glioblastoma immunotherapy. *Biomed Res Int.* 2014;2014:952128.
80. Debinski W, Gibo DM, Hulet SW, Connor JR, Gillespie GY. Receptor for interleukin 13 is a marker and therapeutic target for human high-grade gliomas. *Clin Cancer Res.* 1999;5(5):985–990.
81. Mintz A, Gibo DM, Slagle-Webb B, Christensen ND, Debinski W. IL-13Rα2 is a glioma-restricted receptor for interleukin-13. *Neoplasia.* 2002;4(5):388–399.
82. Debinski W, Obiri NI, Powers SK, Pastan I, Puri RK. Human glioma cells overexpress receptors for interleukin 13 and are extremely sensitive to a novel chimeric protein composed of interleukin 13 and pseudomonas exotoxin. *Clin Cancer Res.* 1995;1(11):1253–1258.
83. Jarboe JS, Johnson KR, Choi Y, Lonser RR, Park JK. Expression of interleukin-13 receptor alpha2 in glioblastoma multiforme: implications for targeted therapies. *Cancer Res.* 2007;67(17):7983–7986.
84. Brown CE, Warden CD, Starr R, et al. Glioma IL13Rα2 is associated with mesenchymal signature gene expression and poor patient prognosis. *PLoS One.* 2013;8(10):e77769.
85. Hegde M, Corder A, Chow KK, et al. Combinational targeting offsets antigen escape and enhances effector functions of adoptively transferred T cells in glioblastoma. *Mol Ther.* 2013;21(11):2087–2101.
86. Krebs S, Chow KK, Yi Z, et al. T cells redirected to interleukin-13Rα2 with interleukin-13 mutein–chimeric antigen receptors have anti-glioma activity but also recognize interleukin-13Rα1. *Cytotherapy.* 2014;16(8):1121–1131.
87. Yaghoubi SS, Jensen MC, Satyamurthy N, et al. Noninvasive detection of therapeutic cytolytic T cells with ^{18}F-FHBG PET in a patient with glioma. *Nat Clin Pract Oncol.* 2009;6(1):53–58.
88. Brown CE, Badie B, Barish ME, et al. Bioactivity and safety of IL13Rα2-redirected chimeric antigen receptor CD8^{+} T cells in patients with recurrent glioblastoma. *Clin Cancer Res.* 2015;21(18):4062–4072.
89. Krenciute G, Krebs S, Torres D, et al. Characterization and functional analysis of scFv-based CARs to redirect T Cells to IL13Rα2-positive glioma. *Mol Ther.* 2015.
90. Citri A, Yarden Y. EGF-ERBB signalling: towards the systems level. *Nat Rev Mol Cell Biol.* 2006;7(7):505–516.
91. Hynes NE, Lane HA. ERBB receptors and cancer: the complexity of targeted inhibitors. *Nat Rev Cancer.* 2005;5(5):341–354.
92. Kornblum HI, Yanni DS, Easterday MC, Seroogy KB. Expression of the EGF receptor family members ErbB2, ErbB3, and ErbB4 in germinal zones of the developing brain and in neurosphere cultures containing CNS stem cells. *Dev Neurosci.* 2000;22(1–2):16–24.

93. Liu G, Ying H, Zeng G, Wheeler CJ, Black KL, Yu JS. HER-2, gp100, and MAGE-1 are expressed in human glioblastoma and recognized by cytotoxic T cells. *Cancer Res.* 2004;64(14):4980–4986.
94. Gajjar A, Hernan R, Kocak M, et al. Clinical, histopathologic, and molecular markers of prognosis: toward a new disease risk stratification system for medulloblastoma. *J Clin Oncol.* 2004;22(6):984–993.
95. Schlegel J, Stumm G, Brandle K, et al. Amplification and differential expression of members of the erbB-gene family in human glioblastoma. *J Neurooncol.* 1994;22(3):201–207.
96. Rainusso N, Brawley VS, Ghazi A, et al. Immunotherapy targeting HER2 with genetically modified T cells eliminates tumor-initiating cells in osteosarcoma. *Cancer Gene Ther.* 2012;19(3):212–217.
97. Ahmed N, Brawley VS, Hegde M, et al. Human epidermal growth factor receptor 2 (HER2)-specific chimeric antigen receptor-modified T cells for the immunotherapy of HER2-positive sarcoma. *J Clin Oncol.* 2015;33(15):1688–1696.
98. Zhao Y, Wang QJ, Yang S, et al. A herceptin-based chimeric antigen receptor with modified signaling domains leads to enhanced survival of transduced T lymphocytes and antitumor activity. *J Immunol.* 2009;183(9):5563–5574.
99. Morgan RA, Yang JC, Kitano M, Dudley ME, Laurencot CM, Rosenberg SA. Case report of a serious adverse event following the administration of T cells transduced with a chimeric antigen receptor recognizing ERBB2. *Mol Ther.* 2010;18(4):843–851.
100. Wang LX, Westwood JA, Moeller M, et al. Tumor ablation by gene-modified T cells in the absence of autoimmunity. *Cancer Res.* 2010;70(23):9591–9598.
101. Walker-Daniels J, Hess AR, Hendrix MJ, Kinch MS. Differential regulation of EphA2 in normal and malignant cells. *Am J Pathol.* 2003;162(4):1037–1042.
102. Wykosky J, Debinski W. The EphA2 receptor and ephrinA1 ligand in solid tumors: function and therapeutic targeting. *Mol Cancer Res.* 2008;6(12):1795–1806.
103. Wykosky J, Gibo DM, Stanton C, Debinski W. Interleukin-13 receptor alpha 2, EphA2, and Fos-related antigen 1 as molecular denominators of high-grade astrocytomas and specific targets for combinatorial therapy. *Clin Cancer Res.* 2008;14(1):199–208.
104. Zhang JG, Eguchi J, Kruse CA, et al. Antigenic profiling of glioma cells to generate allogeneic vaccines or dendritic cell-based therapeutics. *Clin Cancer Res.* 2007;13(2 Pt 1):566–575.
105. Wang LF, Fokas E, Bieker M, et al. Increased expression of EphA2 correlates with adverse outcome in primary and recurrent glioblastoma multiforme patients. *Oncol Rep.* 2008;19(1):151–156.
106. Corbeil D, Roper K, Fargeas CA, Joester A, Huttner WB. Prominin: a story of cholesterol, plasma membrane protrusions and human pathology. *Traffic.* 2001;2(2):82–91.
107. Miraglia S, Godfrey W, Yin AH, et al. A novel five-transmembrane hematopoietic stem cell antigen: isolation, characterization, and molecular cloning. *Blood.* 1997;90(12):5013–5021.
108. Codega P, Silva-Vargas V, Paul A, et al. Prospective identification and purification of quiescent adult neural stem cells from their in vivo niche. *Neuron.* 2014;82(3):545–559.
109. Patel K, Kiely F, Phimister E, Melino G, Rathjen F, Kemshead JT. The 200/220 kDa antigen recognized by monoclonal antibody (MAb) UJ127.11 on neural tissues and tumors is the human L1 adhesion molecule. *Hybridoma.* 1991;10(4):481–491.
110. Kamiguchi H, Lemmon V. Neural cell adhesion molecule L1: signaling pathways and growth cone motility. *J Neurosci Res.* 1997;49(1):1–8.
111. Figarella-Branger DF, Durbec PL, Rougon GN. Differential spectrum of expression of neural cell adhesion molecule isoforms and L1 adhesion molecules on human neuroectodermal tumors. *Cancer Res.* 1990;50(19):6364–6370.
112. Fogel M, Gutwein P, Mechtersheimer S, et al. L1 expression as a predictor of progression and survival in patients with uterine and ovarian carcinomas. *Lancet.* 2003;362(9387):869–875.

113. Thies A, Schachner M, Moll I, et al. Overexpression of the cell adhesion molecule L1 is associated with metastasis in cutaneous malignant melanoma. *Eur J Cancer*. 2002;38(13):1708–1716.
114. Fogel M, Mechtersheimer S, Huszar M, et al. L1 adhesion molecule (CD 171) in development and progression of human malignant melanoma. *Cancer Lett*. 2003;189(2):237–247.
115. Geismann C, Morscheck M, Koch D, et al. Up-regulation of L1CAM in pancreatic duct cells is transforming growth factor β1- and slug-dependent: role in malignant transformation of pancreatic cancer. *Cancer Res*. 2009;69(10):4517–4526.
116. Fang QX, Lu LZ, Yang B, Zhao ZS, Wu Y, Zheng XC. L1, β-catenin, and E-cadherin expression in patients with colorectal cancer: correlation with clinicopathologic features and its prognostic significance. *J Surg Oncol*. 2010;102(5):433–442.
117. Izumoto S, Ohnishi T, Arita N, Hiraga S, Taki T, Hayakawa T. Gene expression of neural cell adhesion molecule L1 in malignant gliomas and biological significance of L1 in glioma invasion. *Cancer Res*. 1996;56(6):1440–1444.
118. Bao S, Wu Q, Li Z, et al. Targeting cancer stem cells through L1CAM suppresses glioma growth. *Cancer Res*. 2008;68(15):6043–6048.
119. Cheng L, Wu Q, Huang Z, et al. L1CAM regulates DNA damage checkpoint response of glioblastoma stem cells through NBS1. *EMBO J*. 2011;30(5):800–813.
120. Ransohoff RM, Engelhardt B. The anatomical and cellular basis of immune surveillance in the central nervous system. *Nat Rev Immunol*. 2012;12(9):623–635.
121. Carson MJ, Doose JM, Melchior B, Schmid CD, Ploix CC. CNS immune privilege: hiding in plain sight. *Immunol Rev*. 2006;213:48–65.
122. Mayhan WG, Heistad DD. Permeability of blood–brain barrier to various sized molecules. *Am J Physiol*. 1985;248(5 Pt 2):H712–H718.
123. Ziylan YZ, Robinson PJ, Rapoport SI. Blood–brain barrier permeability to sucrose and dextran after osmotic opening. *Am J Physiol*. 1984;247(4 Pt 2):R634–R638.
124. Neuwelt EA, Bauer B, Fahlke C, et al. Engaging neuroscience to advance translational research in brain barrier biology. *Nat Rev Neurosci*. 2011;12(3):169–182.
125. Fecci PE, Heimberger AB, Sampson JH. Immunotherapy for primary brain tumors: no longer a matter of privilege. *Clin Cancer Res*. 2014;20(22):5620–5629.
126. Engelhardt B, Ransohoff RM. Capture, crawl, cross: the T cell code to breach the blood–brain barriers. *Trends Immunol*. 2012;33(12):579–589.
127. Engelhardt B, Ransohoff RM. The ins and outs of T-lymphocyte trafficking to the CNS: anatomical sites and molecular mechanisms. *Trends Immunol*. 2005;26(9):485–495.
128. Harris MG, Hulseberg P, Ling C, et al. Immune privilege of the CNS is not the consequence of limited antigen sampling. *Sci Rep*. 2014;4:4422.
129. Louveau A, Smirnov I, Keyes TJ, et al. Structural and functional features of central nervous system lymphatic vessels. *Nature*. 2015;523(7560):337–341.
130. Na SY, Cao Y, Toben C, et al. Naive CD8 T cells initiate spontaneous autoimmunity to a sequestered model antigen of the central nervous system. *Brain*. 2008;131(Pt 9):2353–2365.
131. Callahan MK, Williams KA, Kivisakk P, Pearce D, Stins MF, Ransohoff RM. CXCR3 marks $CD4^+$ memory T lymphocytes that are competent to migrate across a human brain microvascular endothelial cell layer. *J Neuroimmunol*. 2004;153(1–2):150–157.
132. Kivisakk P, Mahad DJ, Callahan MK, et al. Human cerebrospinal fluid central memory $CD4^+$ T cells: evidence for trafficking through choroid plexus and meninges via P-selectin. *Proc Natl Acad Sci USA*. 2003;100(14):8389–8394.
133. Kivisakk P, Trebst C, Liu Z, et al. T cells in the cerebrospinal fluid express a similar repertoire of inflammatory chemokine receptors in the absence or presence of CNS inflammation: implications for CNS trafficking. *Clin Exp Immunol*. 2002;129(3):510–518.
134. Hong JJ, Rosenberg SA, Dudley ME, et al. Successful treatment of melanoma brain metastases with adoptive cell therapy. *Clin Cancer Res*. 2010;16(19):4892–4898.

135. Laman JD, Weller RO. Drainage of cells and soluble antigen from the CNS to regional lymph nodes. *J Neuroimmune Pharmacol*. 2013;8(4):840–856.
136. Aspelund A, Antila S, Proulx ST, et al. A dural lymphatic vascular system that drains brain interstitial fluid and macromolecules. *J Exp Med*. 2015;212(7):991–999.
137. Mezey E, Palkovits M. Neuroanatomy: Forgotten findings of brain lymphatics. *Nature*. 2015;524(7566):415.
138. Yan H, Parsons DW, Jin G, et al. IDH1 and IDH2 mutations in gliomas. *N Engl J Med*. 2009;360(8):765–773.
139. Schwartzentruber J, Korshunov A, Liu XY, et al. Driver mutations in histone H3.3 and chromatin remodelling genes in paediatric glioblastoma. *Nature*. 2012;482(7384):226–231.
140. Jones DT, Kocialkowski S, Liu L, et al. Tandem duplication producing a novel oncogenic *BRAF* fusion gene defines the majority of pilocytic astrocytomas. *Cancer Res*. 2008;68(21):8673–8677.
141. Schindler G, Capper D, Meyer J, et al. Analysis of *BRAF* V600E mutation in 1,320 nervous system tumors reveals high mutation frequencies in pleomorphic xanthoastrocytoma, ganglioglioma and extra-cerebellar pilocytic astrocytoma. *Acta Neuropathol*. 2011;121(3):397–405.
142. Heaphy CM, de Wilde RF, Jiao Y, et al. Altered telomeres in tumors with ATRX and DAXX mutations. *Science*. 2011;333(6041):425.
143. Verhaak RG, Hoadley KA, Purdom E, et al. Integrated genomic analysis identifies clinically relevant subtypes of glioblastoma characterized by abnormalities in PDGFRA, IDH1, EGFR, and NF1. *Cancer Cell*. 2010;17(1):98–110.
144. Mitchell DA, Xie W, Schmittling R, et al. Sensitive detection of human cytomegalovirus in tumors and peripheral blood of patients diagnosed with glioblastoma. *Neuro Oncol*. 2008;10(1):10–18.
145. Cobbs CS, Harkins L, Samanta M, et al. Human cytomegalovirus infection and expression in human malignant glioma. *Cancer Res*. 2002;62(12):3347–3350.
146. Bridgeman JS, Hawkins RE, Hombach AA, Abken H, Gilham DE. Building better chimeric antigen receptors for adoptive T cell therapy. *Curr Gene Ther*. 2010;10(2):77–90.
147. Kershaw MH, Westwood JA, Darcy PK. Gene-engineered T cells for cancer therapy. *Nat Rev Cancer*. 2013;13(8):525–541.
148. Eshhar Z, Waks T, Gross G, Schindler DG. Specific activation and targeting of cytotoxic lymphocytes through chimeric single chains consisting of antibody-binding domains and the gamma or zeta subunits of the immunoglobulin and T cell receptors. *Proc Natl Acad Sci USA*. 1993;90(2):720–724.
149. Moeller M, Haynes NM, Trapani JA, et al. A functional role for CD28 costimulation in tumor recognition by single-chain receptor-modified T cells. *Cancer Gene Ther*. 2004;11(5):371–379.
150. Hombach AA, Heiders J, Foppe M, Chmielewski M, Abken H. OX40 costimulation by a chimeric antigen receptor abrogates CD28 and IL-2 induced IL-10 secretion by redirected $CD4^+$ T cells. *Oncoimmunology*. 2012;1(4):458–466.
151. Carpenito C, Milone MC, Hassan R, et al. Control of large, established tumor xenografts with genetically retargeted human T cells containing CD28 and CD137 domains. *Proc Natl Acad Sci USA*. 2009;106(9):3360–3365.
152. Zhong XS, Matsushita M, Plotkin J, Riviere I, Sadelain M. Chimeric antigen receptors combining 4-1BB and CD28 signaling domains augment PI3kinase/AKT/Bcl-XL activation and $CD8^+$ T cell-mediated tumor eradication. *Mol Ther*. 2010;18(2):413–420.
153. Okada H, Kalinski P, Ueda R, et al. Induction of $CD8^+$ T cell responses against novel glioma-associated antigen peptides and clinical activity by vaccinations with α-type 1 polarized dendritic cells and polyinosinic-polycytidylic acid stabilized by lysine and carboxymethylcellulose in patients with recurrent malignant glioma. *J Clin Oncol*. 2011;29(3):330–336.

154. Facoetti A, Nano R, Zelini P, et al. Human leukocyte antigen and antigen processing machinery component defects in astrocytic tumors. *Clin Cancer Res*. 2005;11(23):8304–8311.
155. Lienert F, Lohmueller JJ, Garg A, Silver PA. Synthetic biology in mammalian cells: next generation research tools and therapeutics. *Nat Rev Mol Cell Biol*. 2014;15(2):95–107.
156. Torp SH, Helseth E, Dalen A, Unsgaard G. Epidermal growth factor receptor expression in human gliomas. *Cancer Immunol Immunother*. 1991;33(1):61–64.
157. Huang Z, Cheng L, Guryanova OA, Wu Q, Bao S. Cancer stem cells in glioblastoma – molecular signaling and therapeutic targeting. *Protein Cell*. 2010;1(7):638–655.
158. Autelitano F, Loyaux D, Roudieres S, et al. Identification of novel tumor-associated cell surface sialoglycoproteins in human glioblastoma tumors using quantitative proteomics. *PLoS One*. 2014;9(10):e110316.

Index

'*Note*: Page numbers followed by "f" indicate figures and "t" indicate tables.'

I

T

U

V

W

X

www.ingramcontent.com/pod-product-compliance
Ingram Content Group UK Ltd.
Pitfield, Milton Keynes, MK11 3LW, UK
UKHW022206051026
15624UKWH00001B/18

9 780128 024201